COMPUTER SIMULATION USING PARTICLES

Computer Simulation Using Particles

R W Hockney
Emeritus Professor, University of Reading

and

J W Eastwood
Culham Laboratory, Abingdon

Adam Hilger, Bristol and New York

British Library Cataloguing in Publication Data

Hockney, R. W.
Computer simulation using particles.
1. Digital computer simulation 2. System analysis 3. Physics——Mathematical models
4. Physics——Data processing
I. Title II. Eastwood, James W.
530′.0724 QA76.9.C65

ISBN 0-85274-392-0

Library of Congress Catalog Card Number: 87-28779

First printed 1988
Reprinted 1989

Published under the Adam Hilger imprint by IOP Publishing Ltd
Techno House, Redcliffe Way, Bristol BS1 6NX
335 East 45th Street, New York, NY 10017-3483, USA

Printed in Great Britain by J W Arrowsmith Ltd, Bristol

To Oscar
Founder of the Subject

CONTENTS

FOREWORD

The ability to follow the motion of many particles in their own and applied fields has long been crucial to both theoretical and experimental studies. Despite competition from fluid calculations, simulations that rely on tracing the motion of tens to thousands of particles are likely to remain central to such studies for some time. Hybrid models (particles plus fluid) are becoming increasingly attractive, especially in plasmas, where the electron and ion timescales are disparate because the mass ratio, m_e/m_i, is so small. The long history of numerical studies of many-particle motion in theories and in applications appears in many separate publications, which deserve to be collected and published together, as the present authors have now done.

Many studies traced particle motion in vacuum electron devices, such as the early work on magnetrons by Hartree and Buneman and on the traveling-wave tube by Nordsieck in the 1940s, followed by the improved models of Poulter, Tien, and Rowe in the 1950s. These authors followed a wavelength or cycle's worth of electrons through their devices, with runs of hours on computers that were primitive by today's standards, yet they developed methods for particle advancing and the solution of the corresponding fields that aided later developments on much faster computers. Their works first explained the non-linear, large-amplitude behavior of microwave amplifiers and oscillators. Tien also showed the frequency dependence of smoothing of fluctuations from a space-charge-limited cathode due to potential-minimum (Debye cloud) dynamics. Such device studies are now done routinely, sometimes by students learning to program.

About 1960 Buneman and Dawson showed the way to doing many-particle plasma simulations in one dimension. In particular, Dawson performed various plasma computer experiments based on classical kinetic theory that caught the fancy of plasma theorists. This step forward presaged applications of simulation to the solution of the many collisionless plasma problems essential to the realization of controlled thermonuclear fusion—except that one dimension was not enough and their direct Coulomb-force calculations were not practicable in two and three dimensions.

In 1963 Hockney, working with Buneman at Stanford University, came up with a fast direct solver for Poisson's equation using a spatial grid in two dimensions, which immediately made two-dimensional electrostatic simulations attractive and economical. The next several years were filled with rapid advances in plasma simulations in two and three dimensions, including theories for the effects of spatial grids and temporal differencing and in applications to instabilities and transport in fusion plasmas, at Stanford, Berkeley, and Princeton and at the national laboratories in America, Britain, Germany, and Japan.

The early art progressed rapidly to a science through theoretical analyses and a great variety of inventions (efficient and accurate algorithms, flexible architecture) which were perfected in applications. The contributions of Hockney and Eastwood in this period and in the 1970s are many and essential to anyone concerned with particle simulations.

The authors have assembled their work in an attractive and useful form, dealing first in detail with the finite spatial and temporal meshes, as well as overall program design. They next present a range of interesting physical models and corresponding simulation models, with computer experiments, drawn from their own extensive research. The plasma work simulates the long-range Coulomb forces. The P^3M work adds the short-range forces and allows more applications. The semiconductor device simulation adds a variety of collisional effects. The astrophysics extends the $1/r^2$ force work to gravitating bodies, with million-year time steps, in contrast to the picosecond steps of plasma simulations. Last, they present their work on phase changes in solids and liquids.

In many present-day applications, such as very large and costly potential fusion reactors, theory and design are pushed to the limit, but both are greatly aided in producing trustworthy results through computer simulation, much of it many-particle studies. Texts to provide both teaching and reference bases are badly needed. The authors have chosen well in offering selected results they believe to be important. The models and methods are certain to be useful for many problems yet to come in current and future areas of interest.

Charles K. Birdsall
Berkeley, April 1980

PREFACE TO THE PAPERBACK EDITION

During the period since *Computer Simulation Using Particles* first appeared, investigation of physical phenomena by means of computer simulation has grown continuously. The growth is apparent in the birth of new computational physics and computational engineering journals, and in the increasing proportion of computational papers presented at theory conferences. It is further evidenced by the development both in the USA and in Europe of 'supercomputer' centres devoted to large-scale numerical modelling of physical phenomena.

A significant fraction of this development in computational science relies on particle modelling: the theory, algorithms and implementation techniques discussed in this book are as relevant to the topics now being studied as they are to the examples described in this text. Indeed, many of the current topics of study —in astrophysics, space and plasma physics, electronics and condensed matter studies—are either direct applications or extensions of techniques described herein. In addition, the material presented provides the foundation for applications that we have not covered in detail, such as implicit electromagnetic codes, new fluid particle methods and RF device modelling.

We have received numerous requests for our book since the hardback edition went out of print, and we hope that this new lower-cost paperback edition will enable our readers to get at the book without having to photocopy the library copy or translate the Russian edition! We believe that the material we cover provides a valuable reference text for the professional practitioner, and a useful guide for the student through the problems of the design, implementation, analysis and interpretation of results of multidimensional particle simulations.

Computer Simulation Using Particles was originally written as a textbook for a final-year undergraduate course in scientific computing at Reading University. It is equally suitable for MSc courses in computational science and simulation, which are becoming increasingly necessary to support the emphasis on large-scale scientific simulation at the new 'supercomputer' centres. Indeed, the material is of wider interest now than when our book was first conceived, and we are confident that the availability of this special student edition will help both teachers and students of computational science.

July 1987

R. W. Hockney
J. W. Eastwood

PREFACE

The investigation of phenomena by computer simulation is now an established and rapidly growing practice in science and engineering. The most obvious examples are in the physical sciences where computer models are used to study the evolution of spiral structure in galaxies, the stability of hot gas plasmas in fusion machines, and turbulence in fluids. Applications to chemistry include the study of molten salts, aqueous solutions, and phase changes. The application of the same simulation techniques to the dynamics of large biological molecules, such as enzymes, is likely to have an important impact on biology in the near future. In engineering the ability to model the flow of electrons in a semiconductor device, such as a field effect transistor, is potentially an important design tool in the hands of the electronic engineer, with cost and time savings over traditional laboratory techniques.

Central to studies by means of such computer experiments is the representation of the system in terms of an appropriate mathematical model. The mathematical model, when discretized, provides both the apparatus and the material under investigation in the computer experiment. Methods of discretization may be broadly classified under three headings: finite difference approximations, finite element approximations, and particle models. There are numerous texts on finite difference and finite element methods, but very few on particle models.

We concern ourselves in this book with particle models, although some of the material could equally well be placed in the other two classes. Particle models are generally, but not exclusively, used for evolutionary calculations, i.e., for the temporal development of an initial value–boundary value problem. They provide a suitable discrete form wherever the system is by nature corpuscular or is succinctly described by a lagrangian formulation. We identify three types of particle model: particle-particle (purely lagrangian), particle-mesh (Euler–Lagrange), and the hybrid particle-particle–particle-mesh model. Each is appropriate in different situations. When properly used, particle models can show distinct advantages over other methods, although the converse is true when they are not properly used.

Our intentions in this book are to show where the various particle methods are appropriate and how to use such methods effectively. We present the material to reflect the steps that have to be taken to obtain meaningful numerical results; namely, the steps from physical phenomena to mathematical model to numerical algorithm to simulation program to computer experiment. We have tried to show and, hopefully, have succeeded in showing how quantitative criteria can be developed and applied in obtaining the best compromise at each step between costs and quality, rather than just cataloguing the various possibilities. Likewise, the examples of applications of particle methods we give are chosen to illustrate the diversity of applications of particle models, rather than to give an exhaustive survey.

We envisage two possible routes through the book for the reader. The first, or introductory, route omits the more detailed mathematical considerations presented in Chapters 4 to 8. This route gives a discussion of the purpose of computer experiments, a general introduction to particle models, a top-downwards methodology for the development and testing of large simulation programs, a straightforward example to illustrate the process of performing the steps listed above, and illustrations of the use of particle models. The mathematical content of Chapters 4–8 is designed to give the more advanced student the material to enable him to analyze, develop, and efficiently use multidimensional particle models.

We foresee that the introductory route—with perhaps some of the more advanced material depending on the background of the student—is likely to be suitable as an undergraduate text for a course in the scientific applications of computers, to complement courses on commercial data processing. The advanced material, although not really difficult, requires a fair mathematical facility with the Fourier transform and is likely to be more at home in an M.Sc. course in "Numerical Analysis," "Computational Science," or "Computational Physics." The course of about forty lectures might then be called "Particle-Mesh Simulation Techniques" and form a natural complement to traditional courses in, e.g., "Numerical Solution of Partial Differential Equations," or "Computational Fluid Dynamics."

We believe that the text is a reasonably comprehensive description of the current state of the art in particle-mesh techniques. It includes a full discussion of the NGP and CIC methods that have been in use extensively since the beginning of the subject in the 1960s, together with the significant improvements on these made about 1973 (the P^3M algorithm) that have, as yet, not been widely used. For this reason we feel that the text will find a place in the library of professional computational scientists working in research establishments and universities.

It would not be possible to name all those who have helped directly and indirectly with the preparation of this book. However we must express our thanks particularly to John Dawson and his group, George Efstathiou, James Hamilton, David Brownrigg, Bob Berman, Mohammad Amini, and Rob Warriner, who have generously allowed us to use their results, in some cases prior to publication. In addition to the above, many others have read, corrected, and suggested improvements to the manuscript, among whom we would like to mention Sverre Aarseth, Stephen Beard, C. K. Birdsall, Oscar Buneman, Canute Moglestue, Keith Roberts, Peter Robson, John Vokes, Bryan Wilson, and Oliver Wintringham. A very special note of thanks is due to Jill Dickinson, without whose dedicated, accurate, and indeed beautiful typing this manuscript could not have been finished even remotely close to the deadline. All those whose work we may have unintentionally misquoted, misrepresented, or inadequately referenced we ask not to judge us too harshly and to let us know so that, in the happy event of a second edition, these errors can be corrected. For the same reason we would like to be notified of any errors or misprints. Our colleagues at Reading University we thank for bearing with understanding our continued absences during the long and difficult preparation of the manuscript.

The inspiration behind our interest in simulation using particles is undoubtedly that of Oscar Buneman, to whom we gratefully dedicate this book. His ideas run like a thread through all aspects of the subject. The work of the authors has developed over the last eighteen years, principally at Stanford and Reading Universities, NASA Langley Research Center, IBM Research Yorktown Heights, and the SRC Rutherford/Atlas Laboratory near Oxford; and we would like to thank our colleagues in these institutions, including particularly the staff of the computer centers, for their stimulating interaction and patient help. In addition to the encouragement of these institutions and the generous allocations of computer time on the fastest computers, specific work has been funded by many agencies, including the U.S. Atomic Energy Commission, U.S. Air Force, and the U.S. National Aeronautics and Space Administration. In the U.K. the support of the Science Research Council through several of its committees and the Atomic Energy Authority Fusion Laboratory at Culham has extended over nearly ten years and is particularly appreciated.

Close contact with laboratory experiment is essential for successful work in computational science and we wish to record our appreciation to personnel at the Royal Signals and Radar Establishments (Malvern and Baldock) and at the Plessey Allen Clark Research Centre (Caswell) who have provided laboratory data and motivation for much of the work on semiconductor devices. Both the Ministry of Defence and the Plessey Co. Ltd have also provided financial support for staff and students.

Reading University
August 1979

R. W. Hockney
J. W. Eastwood

CHAPTER
ONE

COMPUTER EXPERIMENTS USING PARTICLE MODELS

1-1 INTRODUCTION

In this book, we shall be concerned with an area of scientific investigation where the computer plays a central role. The topics we shall consider are usually classified under the heading "computational physics," although such a title is a little misleading since the material covered overlaps many established subject areas: physics, chemistry, engineering, numerical analysis, and computer science all play a part in computational physics. For this reason, we prefer the more general term "*computational science.*"

The starting point of the computational science approach to scientific investigation is a mathematical model of the physical phenomenon of interest. The equations of the mathematical model are cast into a discrete algebraic form which is amenable to numerical solution. The discrete algebraic equations describe the *simulation model* which, when expressed as a sequence of computer instructions, provides the computer *simulation program.* The computer plus program then allow the evolution of the model physical system to be investigated in *computer experiments.*

Computer simulation may be regarded as the theoretical exercise of numerically solving an initial-value–boundary-value problem. At time $t = 0$, the initial state of the system is specified in some finite region of space (*the computational box*) on the surface of which prescribed boundary conditions hold. The simulation consists of following the temporal evolution of the configuration. The main part of the calculation is the *timestep cycle* in which the state of the physical system is incremented forwards in time by a small timestep, DT. The experimental aspect, and thence the name computer experiment, arises when we consider the problems of measurements. Even the simplest simulation calculation generates large amounts of data which require an experimental approach to obtain results in a digestible form.

Although the amount of data which can be handled by computers is large, it is nevertheless finite. Much of the ingenuity of computational scientists is devoted to obtaining good simulation models of the physical systems within the constraints of the available finite computer resources. Methods of discretization used in obtaining simulation models include finite-difference methods (Richtmyer and Morton, 1967), finite-element methods (Strang and Fix, 1973), and the methods to be dealt with in this book, *particle methods*.

"*Particle models*" is a generic term for the class of simulation models where the discrete representation of physical phenomena involves the use of interacting particles. The name "particle" arose because in most applications the particles may be identified directly with physical objects. Each particle has a set of attributes, such as mass, charge, vorticity, position, momentum. The state of the physical system is defined by the attributes of a finite ensemble of particles and the evolution of the system is determined by the laws of interactions of the particles. A feature which makes particle models computationally attractive is that a number of the particle attributes are conserved quantities and so need no updating as the computer simulation evolves in time.

The relationship between the particles of the simulation model and particles in physical systems is determined largely by the interplay of finite computer resources and the length and timescales of importance in the physical systems. Examples of physical systems which have been successfully simulated using particle models are discussed on this basis in Sec. 1-3. The unifying aspects of the diversity of phenomena listed are similarities of the mathematical models of the physical systems (Sec. 1-4) and similarities of the numerical schemes used (Sec. 1-5).

1-2 THE COMPUTER EXPERIMENT

The simulation approach to scientific investigation is made possible by high-speed digital computers. However, the fact that investigation by simulation is feasible does not always mean that it is the best, or even a desirable approach. As with theoretical and experimental investigation, we must ask what are the limitations and benefits of the computer experiment. Only in those circumstances where a computer experiment is deemed to be the sensible approach need we involve ourselves with more detailed questions concerning the design of the simulation model and the execution of the computer experiment.

1-2-1 The Role of the Computer Experiment

In order to get a picture of the value of simulations, let us look at the traditional pattern of evolution of science and at how that evolution may be accelerated by the use of computer experiments.

Experimental (and observational) work is primarily concerned with the accumulation of factual information. Theoretical work is mainly directed towards ordering such information into logically coherent patterns to provide predictive laws for the behavior of matter. Both theory and experiment rely on the checks and stimuli of each other to direct them. We may identify a threefold direct influence of theory on experiment. Theory leads to experiments,

1. to determine facts which mathematical manipulation of a set of principles suggests are relevant to the description of particular phenomena,
2. to test theoretical predictions and assumptions, and
3. to determine empirical quantities which theory requires.

Experimental work has a similar guiding influence on theory: theory is manipulated to confront experiment, it is used to predict factual information of intrinsic value, and it is extended to encompass an ever-increasing body of data into a unified structure. The normal practice of science is to develop, in increasingly complex situations, the confrontation of theory with experience. Anomalies and uncertainties become centres of attention for both theory and experiment, their resolution leading either to an extension of current theories, or to the recasting of principles in order to provide new and more powerful theories to describe adequately the new phenomena.

The demands of modern technology have provided further stimuli for both theory and experiment to push beyond regimes where the mathematics is manageable and experiments are straightforward. Computers are extensively used to overcome the difficulties—for instance, in the numerical evaluation of mathematical functions and in automated data collection and processing. These activities, which form part of traditional scientific investigation, are not our concern here. We shall be looking at the use of the computer as a new medium in which experiments—computer experiments—are performed to bridge the gaps between theory and experiment.

Computer experiments use mathematical models for simulating the behavior of complex systems. The computer plus program become both the apparatus and material under investigation. The properties of the material are, within the limitations of the finite and discrete representation on the computer, at the disposal of the "computer experimenter." The apparatus of the computer experiment provides a perfectly controlled environment, where even the measurement of physical quantities causes no disturbance. Consequences of primary physical laws in complex situations can be easily investigated, particularly since the subset of laws pertaining to such situations may be freely changed. Nonlinearity, a large (but finite) number of degrees of freedom, and lack of symmetry are not obstacles to investigation by simulation.

The value of the computer experiment is greatest in those areas where the gap between theoretically and experimentally realizable objectives is large. The mathematical development of theories usually relies on linearity or power series expansions, a high degree of symmetry, a small number of parameters, and simplifying approximations. Experiments in the laboratory are faced with the full complexity of nature: conditions can be difficult to control, measurements can be difficult to make, and consequently results are often difficult to interpret unambiguously. The computer experiment, by virtue of its regime of applicability, becomes the link between theory and experiment.

Computer experiments, like their counterparts in the laboratory, may be divided roughly into three categories. First are those simulations designed to predict accurately the workings of complex devices in order to allow many variations to be evaluated before expensive technology is employed in constructing the optimal choice. A second class of computer experiment comprises those experiments to obtain information not readily accessible to laboratory experiment—such is the case in the simulation of galaxies or of submicron electronic devices when very large or very small length and timescales are involved. Finally, there is the class of computer experiment, the "theoretical experiment," which provides a valuable guide for theoretical development. Complex situations, as for example are found in plasma studies, may be dissected into simpler components to help unravel the wealth of phenomena and obtain a clear theoretical picture. Moreover, the clear picture need no longer be just in the theoretician's mind, for animated film output from computer experiments gives a precise picture of time evolution for all to see.

The combination of computer experiment, physical experiment, and theory proves much more effective in obtaining physically useful results than any one approach or pair of approaches. To obtain results, theory uses mathematical analysis and numerical evaluation, physical experiment uses apparatus and data analysis, and the computer experiment uses computer plus simulation program. Weaknesses of each method of investigation are complemented by strengths of the others. The role of the computer experiment is determined by its strengths: complementing theoretical investigation where nonlinearity, many degrees of freedom, or lack of symmetry are of importance and complementing experimental studies where devices are expensive, data is inaccessible, or phenomena are very complex.

1-2-2 Setting Up Computer Experiments

The starting point for all computer experiments is some physical phenomenon. The objective is to obtain physically useful results from the computer experiments. Between these two points we may identify a number of distinct design steps. These steps are summarized in Fig. 1-1.

Each step introduces constraints. Invariably the mathematical formulation is only an approximate description of the physical phenomenon. The computational scientist cannot just take equations at face value, despite the fact that the development of the mathematic description of physical phenomena lies in the realm of the theoretician. He must know what simplifying assumptions are made in order to identify the regime of validity of the equations and thence of his simulation model.

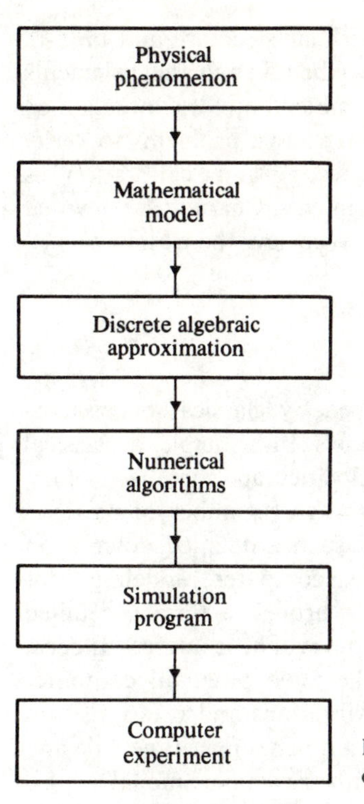

Figure 1-1 The principal steps in setting up a computer experiment.

More severe restrictions arise in the step to the discrete algebraic approximation, where the continuous differential or integral equations of the mathematical model are replaced by algebraic approximations in order to allow numerical solution on computers. This step introduces questions concerning the consequences of finite timestep, discrete spatial meshes and, for particle models, the limited number of particles. (These matters are discussed in general in Chapters 4 to 8 and for particular applications in Chapters 9 to 12.) The onus is always on the numerical experimenter to demonstrate that the results from the simulation model are physically meaningful—we shall see in later chapters that numerical errors can have disastrous effects!

Impinging on the choice of discretization is the question of the numerical algorithm. Discretization replaces continuous variables of the mathematical model by arrays of values and differential equations by algebraic equations. Unless the tens (and sometimes hundreds) of thousands of algebraic equations arising can be rapidly solved on the computer, the proposed simulation becomes impracticable.

The steps leading to the numerical algorithm involve a bootstrapping process in which the best compromise between the quality of the representation and the computational cost is sought. Only when this has been found can the apparatus—the simulation program—be built. As with real apparatus, a great deal of care is needed in "building" a program in order to ensure that it behaves in the desired manner. A well-engineered program should be easy to read, easy to use, and, if necessary, easy to modify. It should have a clear modular structure and built-in diagnostics. The analogy with laboratory practice may be taken further; elements of the program should be tested as they are assembled into the preplanned structure. The complete program should be tested on known problems to verify both the algorithms and the code. Only when the testing and "calibration" is complete is the apparatus ready for performing experiments. Chapter 3 is devoted to a description of programming techniques designed to achieve these aims.

1-3 LENGTH AND TIME SCALES

Physical phenomena which are adequately described by classical or pseudo-classical theory can be simulated using particle models. In principle, a classical system may be described in terms of the positions, velocities, and force laws of the particles of which it is composed. Unfortunately, the vast number of particles involved in quite simple situations generally makes such a detailed description useless both for understanding the system and for computer modeling. For example, if we were investigating the flow of water through a nozzle, detailed information of the positions and velocities of every atom is of little interest. Indeed, if we tried to perform such a calculation using the most powerful computer available, we would end up seeing only molecular vibrations and rotations in a microscopic element of the water—the flow being on a much longer timescale and greater length scale than we could hope to simulate with the detailed model.

The secret of success in computer experiments is to devise a model which is sufficiently detailed to reproduce faithfully the important physical effects and yet not so detailed as to make the calculation impracticable. The best choice of model depends on the relevant physical length and timescales, examples of which are listed in Table 1-1 for a great variety of physical systems which have been successfully modeled using particles. In the case of water flowing through a nozzle, a model which gives bulk flow velocities may be quite adequate, for example in the simulation of fluids using vortex elements. At the other extreme, the "molecular dynamics" experiments are performed to study the structure of liquids, in which short length and timescales of molecular motions are important.

There is a clear one-to-one correspondence between the physical and computer model particles in the molecular dynamics simulation: Each atom corresponds to a particle and the attributes of the particle are those of the atom (charge, mass, position, velocity). Similar remarks hold for simulations of stellar clusters and clustering of galaxies, where stars and galaxies, respectively, correspond to particles. The relevant length scale is the mean interparticle separation and the timescale is determined by oscillation or orbital periods, as appropriate. Clustering of stars and of galaxies and molecular dynamics calculations are described in detail in Chapters 11 and 12, respectively.

At the other extreme, the identity of the atomic building blocks in the vortex fluid simulation model is completely lost. The mathematical description treats the fluid as continuous, incompressible, and inviscid. The discretization of the

Table 1-1 Examples of physical systems represented by particle models

Scale lengths L are in meters and times T in seconds. N_p is the number of particles in L^3.

Example	Computer particles	Particle attributes	Physical N_p	Physical L	Physical T	Computer model N_p	Computer model L	Computer model T
1. *Correlated systems*								
Covalent liquids	Atoms or molecules	Strength constants related to quantum-mechanical dipole and quadruple interactions, mass, force, position, velocity	10^5	10^{-8}	10^{-12}	$10^3 - 10^4$	$10^{-8} - 10^{-9}$	10^{-12}
Ionic liquids	Positive ions, Negative ions	Charge, mass, force, position, velocity, radius	10^5	10^{-8}	10^{-12}	$10^3 - 10^4$	$10^{-8} - 10^{-9}$	10^{-12}
Stellar clusters	Stars	Mass, position, velocity, force, radius	$10^2 - 10^3$	10^{17}	10^{15}	$10^2 - 10^3$	10^{17}	10^{15}
Galaxy clusters	Galaxies	Mass, position, velocity, force, radius	$10^4 - 10^5$	10^{23}	10^{17}	$10^4 - 10^5$	10^{23}	10^{17}
2. *Collisionless systems*								
Collisionless plasma	"Superparticle" $\simeq 10^8$ electrons or 10^8 ions	Charge, mass, position, velocity, radius	$10^9 - 10^{12}$	$10^{-5} - 10^{-2}$	$10^{-9} - 10^{-12}$	$< 10^5$	$10^{-5} - 10^{-2}$	$10^{-9} - 10^{-12}$
Galaxies—spiral structures	"Superparticle" $\simeq 10^6$ stars	Mass, position, velocity, radius	$10^{10} - 10^{11}$	10^{21}	10^{13}	$< 10^5$	10^{21}	10^{13}
3. *Collisional systems*								
Submicron Semiconductor devices (microscopic Monte-Carlo model)	"Superparticle" $= 10^4$ electron wavepackets	Charge, mass, position, wavenumber, radius	10^8	10^{-7}	10^{-10}	$< 10^5$	10^{-7}	10^{-10}
4. *Collision-dominated systems*								
Semiconductor devices (diffusion model)	"Superparticle" $= 10^4$ electrons or holes	Charge, position	10^9	10^{-6}	10^{-9}	$< 10^5$	10^{-6}	10^{-9}
Inviscid, incompressible fluids (vortex)	Vortex element	Vorticity, position	continuum	$10^{-3} - 10^6$	$10^{-3} - 10^5$	$< 10^5$	$10^{-3} - 10^6$	$10^{-3} - 10^5$

mathematical model reintroduces particles (fluid elements), which now have the attributes of position and vorticity. Length and timescales are determined by the structure and velocity of the fluid flow. These scale lengths must be much greater than atomic scale lengths for the model to be meaningful.

A third type of particle model lies between the two extremes exemplified by the molecular dynamics and vortex models. In these models, the computer particles retain much of the identity of the atomic constituents of the material being simulated, yet do not correspond one-to-one as do atoms to particles in the molecular dynamics calculation. Dilute plasmas, galaxy, and microscopic semiconductor device simulations fall into this category.

The "electrons" and "ions" in a collisionless plasma simulation correspond to millions of physical electrons or ions. The mathematical description of the collisionless plasma (Vlasov's equation) replaces the ions and electrons by a continuous-phase fluid where all information on the graininess of the plasma is smoothed out. Provided that the physical phenomena have wavelengths which are long compared with the average electrons and ion spacings and timescales which are short compared with the time for the graininess to have significant effect (the collision time), then this description gives an accurate representation. Discretization of the Vlasov fluid leads to the concept of computational "superparticles." Graininess introduced by the paucity of superparticles is minimized by smoothing the short-range forces. One way of regarding the superparticles is as finite-sized clouds of electrons or ions, the position of the superparticles being the centre of mass of the clouds and their velocities being the mean velocities of the clouds. Another interpretation, one which illuminates the link with fluid particle models, is to regard the superparticles as blobs of incompressible phase fluid moving in (position-velocity) phase space.

The length and time scales in a collisionless plasma model are determined by the frequencies and wavelengths of natural oscillations in a collisionless plasma; these are the plasma frequency and Debye length, respectively. The model is valid until the graininess of the superparticle system causes deviations from the behavior of a collisionless plasma—that is to say, for times less than the collision time. Fortunately, the smoothing of the short-range interaction considerably increases the collision time in the model (see Sec. 9-2-1). In Chapter 9 we give the properties of a collisionless electrostatic model of a plasma and describe its use in the study of the diffusion of plasma across a magnetic field. We also include a model in which the particles represent small current elements and describe its use to study the interaction of the solar plasma wind with the earth's magnetic field and the consequential formation of the geomagnetic tail.

The modeling of galaxies of stars is a further example of a collisionless simulation. Superparticles in a galaxy model represent millions of real stars. Length and time scales are determined by the dimensions and motion of the spiral arm structures. Collisional effects are again made small by smoothing the force of interaction at short range, and by having a large number of superparticles within the range of the smoothing. The principal application of the collisionless galaxy model, the study of spiral structure, is described in Chapter 11.

Semiconductor simulation models differ from the collisionless plasma and galaxy models in that important short-timescale effects of the crystal lattice (scattering) are taken into account by altering the equations of motion. The diffusion model of a semiconductor device is appropriate if the dimensions of the device are large compared with the electron mean free path. Here the attributes of the particle are charge and position—the velocities being given by the product of the electric field and mobility. In submicron semiconductor devices, the time for electrons to reach the steady-state velocity corresponding to the mobility becomes comparable to the lifetime of the electron in the device. Consequently, the mobility approximation must be replaced by a microscopic description in which statistical "Monte-Carlo" methods are used to simulate the scattering of the electrons by the crystal lattice and the electron inertia must be retained. The attributes of the particle are then charge, mass, position, and wavenumber. The microscopic Monte-Carlo model of a gallium arsenide field effect transistor is taken as a case study and described in detail in Chapter 10.

1-4 PHYSICAL SYSTEMS

The hierarchy of physical systems and particle models listed in Table 1-1, obtained by consideration of length and time scales, can be given a formal basis by employing the mathematical analysis of classical statistical mechanics; this statistical link between the various models is outlined below in Secs. 1-4-1 to 1-4-4. Although the statistical mechanics concepts used there are valuable for understanding the relationship between the various particle models, a detailed understanding of the analysis is not assumed throughout the book.

From a computer modeling point of view, the unified framework provided by statistical mechanics appears as the similarity of mathematical models and thence as the similarity of numerical algorithms and of computer programs for a wide range of physical phenomena. All of the mathematical models to which particle simulation methods are applied can be formulated as a set of particles interacting through fields. Particles have a number of conserved attributes (e.g., mass, charge) and variable attributes (e.g., position, velocity). The variable attributes evolve according to the equations of motion whose driving terms are given by the field equations.

The most important of the field equations is Poisson's equation, Eq. (1-7), which pertains to all the examples but the first listed in Table 1-1. The viability of particle simulation relies on methods to rapidly solve it or a similar elliptic partial differential equation and it is for this reason that we devote a large part of Chapter 6 to methods for its solution.

Poisson's equation gives a scalar (potential) field. A second field equation relates a vector field to this scalar field: For correlated, collisionless, and collisional systems, the vector field is a force (or acceleration) field, whilst for collision dominated systems the vector field is the velocity field. When the vector field is a force field, Newton's law of motion $d\mathbf{v}/dt \propto \mathbf{F}$ is integrated to determine the velocity of the particles. In all cases, the position is obtained from the velocity by integrating the equation $d\mathbf{x}/dt = \mathbf{v}$. Suitable numerical methods for these time integrations are given in Chapter 4.

A natural consequence of the similarity of the mathematical models is the similarity of the error analysis and numerical schemes. One benefit of the unified approach to a wide diversity of physical phenomena is that advances in numerical methods in one area of application more rapidly find their way into others. The methods of error analysis and algorithm design described in Chapters 4 to 8 are applicable to a wide range of computer simulation models in addition to those we consider in Chapters 9 to 12.

The principal benefit of recognizing computational science in general and particle models in particular as entities appears when we consider the computer program. Programs developed to study particular phenomena may, with minor modifications, be used to study something completely different. For instance, the same computer program may be used to study the clustering of galaxies (Chapter 11) and ionic liquids (Chapter 12). Other examples include the simulation of the flow of electrons in a strong magnetic field and the motion of an incompressible, inviscid fluid (Sec. 1-4-4); and of collisionless plasmas (Chapter 9) and spiral structures in galaxies (Chapter 11). It is by recognizing the similarities of mathematical, numerical, and computation problems, and by embodying the similarities in standardized structured programs (cf. Chapter 3) that large economies in man power and computer resources can be made.

1-4-1 Correlated Systems

The first three cases listed in Table 1-1, in which there exist one-to-one correspondences between physical and computer particles, are examples of correlated systems. Computer experiments with these systems do statistical mechanics the hard way, following the path of the system in sN-dimensional phase (Γ) space, where s is the number of degrees of freedom of each of the N particles comprising the system.

The computer experiments themselves are deterministic, rather than statistical. An initial state is defined by the set of positions and velocities $\{\mathbf{x}_i, \mathbf{v}_i; i = 1, N\}$ of the particles at time $t = 0$. The force, $\mathbf{F}_i$, on the ith particle is the sum of the forces due to the remaining $N-1$ particles, plus any external forces which may be present.

$$\mathbf{F}_i = \sum_{j \neq i} \mathbf{F}_{ij} + \mathbf{F}_i^{\text{ext}} \tag{1-1}$$

$\mathbf{F}_{ij}$ is the force of particle j on particle i. A familiar example is Coulomb's law, giving the electrostatic force between two charged particles as

$$\mathbf{F}_{ij} = \frac{q_i q_j}{4\pi\epsilon_0} \frac{(\mathbf{x}_i - \mathbf{x}_j)}{|\mathbf{x}_i - \mathbf{x}_j|^3} \tag{1-2}$$

where $\mathbf{x}_i, \mathbf{x}_j$ and q_i, q_j are positions and charges of particles i and j, respectively.

The potential energy of particle i, Ψ_i, may similarly be described in terms of pairwise additive potentials ψ_{ij}:

$$\Psi_i = \sum_{j \neq i} \psi_{ij} + \Psi_i^{\text{ext}} \tag{1-3}$$

where, for point charges

$$\psi_{ij} = \frac{q_i q_j}{4\pi\epsilon_0} \frac{1}{|\mathbf{x}_i - \mathbf{x}_j|} \tag{1-4}$$

Forces and potentials may equally be regarded as fields—that is, quantities which, although arising from particle attributes, are not particle entities. Force and potential energy fields pervade all space, where the force field is minus the gradient of the potential energy field

$$\mathbf{F} = -\nabla\Psi \tag{1-5}$$

The force on and potential energy of particle i are given respectively by

$$\mathbf{F}_i = \mathbf{F}(\mathbf{x})|_{\mathbf{x}=\mathbf{x}_i} \qquad \Psi_i = \Psi(\mathbf{x})|_{\mathbf{x}=\mathbf{x}_i} \tag{1-6}$$

The potential field is related to the source distribution by the *field equation*, e.g., the field equation corresponding to Eq. (1-4) is Poisson's equation

$$\nabla^2\phi = -\rho/\epsilon_0 \tag{1-7}$$

where $\Psi_i = q_i\phi|_{\mathbf{x}=\mathbf{x}_i}$, ϕ is the electrostatic potential and ρ is charge density.

The *action at a distance*, Eqs. (1-1) to (1-4), and *force at a point* formulations, Eqs. (1-5) to (1-7), are equivalent. For a given particle distribution they give the same forces. Specified external forces and potentials in the action-at-a-distance expressions are replaced by specified boundary conditions in the field equation.

The description of the physical systems is completed by boundary conditions and equations of motion. Boundary conditions define the external forces and the volume of space (*the computational box*) in which the particles move. The equations of motion, for example Newton's laws of motion,

$$\frac{d\mathbf{x}_i}{dt} = \mathbf{v}_i \qquad \frac{d}{dt}(m_i\mathbf{v}_i) = \mathbf{F}_i \tag{1-8}$$

give the rules for the dynamics of particles.

The statistical mechanics description of correlated systems on which particle computer experiments may be performed is formally given by Liouville's conservation equation for probability density ρ in Γ space:

$$\frac{\partial\rho}{\partial t} + \sum_{i=1}^{N} \mathbf{v}_i \cdot \frac{\partial\rho}{\partial\mathbf{x}_i} + \sum_{i=1}^{N} \frac{\mathbf{F}_i}{m_i} \cdot \frac{\partial\rho}{\partial\mathbf{v}_i} = 0 \tag{1-9}$$

where $\rho(\mathbf{x}_1 \cdots \mathbf{x}_N, \mathbf{v}_1 \cdots \mathbf{v}_N, t)\, d\mathbf{x}_1 \cdots d\mathbf{x}_N\, d\mathbf{v}_1 \cdots d\mathbf{v}_N$ is the probability that the system is in the volume $[(\mathbf{x}_1, \mathbf{x}_1 + d\mathbf{x}_1) \cdots]$ of Γ space at time t.

A single computer experiment corresponds to ρ being a δ-function in Γ space, the location of which moves along some path in Γ space as time advances.

Theoretically, a series of experiments should be performed to provide an ensemble to define ρ. In practice, time averaging generally replaces ensemble averaging in computing expectation values for near equilibrium configuration, although in nonequilibrium calculations the importance of initial conditions cannot be overlooked.

Analytically, Eq. (1-9) is no more manageable than the complete set of particle orbits. To make progress, more information must be discarded. This is done by defining a hierarchy of distribution functions and obtaining a hierarchy of kinetic equations (the BBGKY chain) by integrating Eq. (1-9) over successively fewer particle coordinates. We restrict ourselves to a brief outline and refer readers elsewhere for further details.†

The state of an N-particle ensemble at time t can be specified by the exact one-particle distribution function:

$$F(\mathbf{x}, \mathbf{v}, t) = \sum_{i=1}^{N} \delta(\mathbf{x}-\mathbf{x}_i(t))\delta(\mathbf{v}-\mathbf{v}_i(t)) \tag{1-10}$$

where $(\mathbf{x}, \mathbf{v})$ now defines a point in the six-dimensional (μ) phase space of a single particle. A reduced statistical description is obtained by ensemble averaging. The ensemble averaged one-particle distribution function is

$$\begin{aligned} f_1(\mathbf{x}, \mathbf{v}, t) &= \langle F(\mathbf{x}, \mathbf{v}, t)\rangle = \int \rho F d\mathbf{x}_1 \cdots d\mathbf{x}_N\, d\mathbf{v}_1 \cdots d\mathbf{v}_N \\ &= N \int \rho(\mathbf{x}, \mathbf{x}_2 \cdots \mathbf{x}_N, \mathbf{v}, \mathbf{v}_2 \cdots \mathbf{v}_N) d\mathbf{x}_2 \cdots d\mathbf{x}_N\, d\mathbf{v}_2 \cdots d\mathbf{v}_N \end{aligned} \tag{1-11}$$

The second integral in Eq. (1-11) is obtained from the first by integrating over the δ functions and relabeling. $f_1(\mathbf{x}, \mathbf{v}, t)d\mathbf{x}\, d\mathbf{v}$ gives the mean number of particles in volume $d\mathbf{x}\, d\mathbf{v}$ at position $(\mathbf{x}, \mathbf{v})$ in μ space at time t. In a similar manner, the ensemble averaged two-particle distribution function is given by

$$\begin{aligned} f_2(\mathbf{x}, \mathbf{v}, \mathbf{x}', \mathbf{v}', t) = N(N-1) \int \rho(\mathbf{x}, \mathbf{x}', \mathbf{x}_3 \cdots \mathbf{x}_N, \mathbf{v}, \mathbf{v}', \mathbf{v}_3 \cdots \mathbf{v}_N) \\ \times d\mathbf{x}_3 \cdots d\mathbf{x}_N\, d\mathbf{v}_3 \cdots d\mathbf{v}_N \end{aligned} \tag{1-12}$$

and $f_2\, d\mathbf{x}\, d\mathbf{v}\, d\mathbf{x}'\, d\mathbf{v}'$ is interpreted as the mean product of numbers of particles in $d\mathbf{x}\, d\mathbf{v}$ and $d\mathbf{x}'\, d\mathbf{v}'$. Further distribution functions f_3, f_4, etc., may be likewise defined, each successive one retaining more information.

A differential equation describing the evolution of distribution function f_s is obtained from Liouville's equation, Eq. (1-9), by integrating over variables associated with $N-s$ particles. The equation for f_1 is

$$\frac{\partial f_1}{\partial t} + \mathbf{v}_1 \cdot \frac{\partial f_1}{\partial \mathbf{x}_1} + \frac{\mathbf{F}_1^{\text{ext}}}{m_1} \cdot \frac{\partial f_1}{\partial \mathbf{v}_1} + \int \frac{\mathbf{F}_{12}}{m_1} \cdot \frac{\partial f_2}{\partial \mathbf{v}_1} d\mathbf{x}_2\, d\mathbf{v}_2 = 0 \tag{1-13}$$

The equation for f_2 involves a term in f_3, and that in f_s involves a term in f_{s+1}. Some approximations are required to close the chain of equations and it is for this reason that much computational effort has been expended investigating f_2 and f_3.

If the two-particle distribution function is integrated over all velocities and

† For example, Part I, Montgomery and Tidman, 1964, Chapter 6; Isihara, 1971, Chapter 12; Clemmow and Dougherty, 1969, Chapter 12; Hansen and McDonald, 1976.

expressed as a function of the scalar separation r between a pair of particles, one obtains the pair or radial distribution function. This function is used extensively to describe the local structure of solids and liquids and is the normal output from computer simulations of such correlated systems. The variation of the radial distribution function during the melting of an electron film (Sec. 12-2-5 and Fig. 12-4), and in melting and glass formation in a KCl microcrystal (Secs. 12-3-5 and 12-3-6 and Fig. 12-9) is given in Chapter 12. In the study of the clustering of galaxies in the universe, however, it is necessary also to measure the three-particle and four-particle correlation functions (cf. Chapter 11, Section 4).

1-4-2 Uncorrelated (Collisionless) Systems

The lowest-order closure of the hierarchy of kinetic equations gives the Vlasov equation. In this approximation, particles are assumed to be uncorrelated, allowing the two-particle distribution function to be written as the product of one-particle distribution functions:

$$f_2(\mathbf{x}, \mathbf{v}, \mathbf{x}', \mathbf{v}', t) = f(\mathbf{x}, \mathbf{v}, t) f_1(\mathbf{x}', \mathbf{v}', t) \tag{1-14}$$

Conditions under which the approximation Eq. (1-14) is appropriate can be demonstrated by simple physical arguments. $f_2\, d\tau\, d\tau'$ can be interpreted as the probability of simultaneously finding a particle in volume $d\tau = d\mathbf{x}\, d\mathbf{v}$ at $(\mathbf{x}, \mathbf{v})$ and volume $d\tau' = d\mathbf{x}'\, d\mathbf{v}'$ at $(\mathbf{x}', \mathbf{v}')$ in μ space. The probability of occupation of $d\tau'$ will be affected significantly by a particle in $d\tau$ only if the potential energy of particle interactions is large compared with their kinetic energy. We define a range a for the interactions by equating potential energy of interaction to typical kinetic energies. For plasmas and galaxies, a is given respectively by

$$a_p = \frac{e^2}{4\pi\epsilon_0 k_B T} \qquad a_g = \frac{Gm^2}{\frac{1}{2}mv^2} \tag{1-15}$$

where e is charge, k_B is Boltzmann's constant, ϵ_0 is the permittivity of free space, G is the gravitational constant, m is mass, and v is speed. Clearly, if a is small compared with average interparticle spacings the statistical independence of occupation of different volumes of μ space implied by Eq. (1-14) will be a good approximation. If we let n be the density of particles (electrons) in a plasma and σ be the surface density of stars in a galactic disk, then measures of the quality of the collisionless approximate (Eq. 1-14) are given by

$$\epsilon_p = a_p n^{1/3} \qquad \epsilon_g = a_g \sigma^{1/2} \tag{1-16}$$

Typical speeds v involved in rotating gravitating disks can be estimated by equating centrifugal and gravitational accelerations at the edge of a uniform gravitating disk of radius R, giving $a_g \sim (\sigma R)^{-1}$. Substituting for a_p and a_g in Eq. (1-16) gives

$$\epsilon_p \sim (n\lambda_D^3)^{-2/3} \qquad \epsilon_g \sim (\sigma R^2)^{-1/2} \tag{1-17}$$

where $\lambda_D = (\epsilon_0 k_B T/ne^2)^{1/2}$ is the Debye length. Thus, for Eq. (1-14) to be valid, we require the number of electrons in a Debye cube $n\lambda_D^3$ or the number of stars in the galactic disk σR^2 to be large: typically, $n\lambda_D^3 \sim 10^9$ and $\sigma R^2 \sim 10^{10}$.

Substituting Eq. (1-14) into Eq. (1-13) and dropping the subscript "1"s yields the Vlasov equation (see, for example, Clemmow and Dougherty 1969, for further details).

$$\frac{df}{dt} \equiv \frac{\partial f}{\partial t} + \mathbf{v} \cdot \frac{\partial f}{\partial \mathbf{x}} + \left\{ \frac{\mathbf{F}^{\text{ext}}}{m} + \int \frac{\mathbf{F}}{m} f(\mathbf{x}', \mathbf{v}', t)\, d\mathbf{x}'\, d\mathbf{v}' \right\} \cdot \frac{\partial f}{\partial \mathbf{v}} = 0 \tag{1-18}$$

The integral term in Eq. (1-18) describes the self-consistent accelerations of the electrons in a plasma or stars in a galactic disk. For example, in the case of a plasma, using Eq. (1-2) for **F** we obtain

$$\int \frac{q^2}{4\pi\epsilon_0 m} \frac{(\mathbf{x} - \mathbf{x}')}{|\mathbf{x} - \mathbf{x}'|^3} f(\mathbf{x}', \mathbf{v}', t)\, d\mathbf{x}'\, d\mathbf{v}' = \frac{q}{m} \mathbf{E} = -\frac{q}{m} \nabla \phi \tag{1-19}$$

or equivalently

$$\nabla^2 \phi = -\frac{q}{\epsilon_0} \int f(\mathbf{x}, \mathbf{v}, t)\, d\mathbf{v} \tag{1-20}$$

where **E** is the electric field and ϕ is the electrostatic potential arising from the ensemble-averaged charge density distribution $\rho = q \int f(\mathbf{x}, \mathbf{v}, t)\, d\mathbf{v}$ (where a sum over electron and ion species is implied). Equation (1-18) is collisionless in the sense that the interaction of individual particles has disappeared; it is as if the points in μ space representing the physical particles have been smeared out to create a continuous "phase" fluid in μ space.

The collisionless plasma and galaxy computer models reintroduce particles in discretizing the continuous-phase fluid, although, as noted in the previous section, there are many fewer particles in the computer model than in the physical situation. Ascribing a finite size W to the computer particles greatly reduces the force of interaction between particles separated by distances less than W. In this case, if the range parameter a is much less then W then the computer model will be collisionless to a good approximation since the number of superparticles per superparticle (nW^3 or σW^2) is reasonably large (see Sec. 9-2-1).

1-4-3 Collisional Systems

Phenomena whose wavelengths are long compared with the mean separation of physical particles and yet are not adequately described by the collisionless approximation are also amenable to simulation using particle models. The appropriate mathematical description for such phenomena is obtained by adding a small correction term, the correlation function g, to the approximation given in Eq. (1-14):

$$f_2(\mathbf{x}, \mathbf{v}, \mathbf{x}', \mathbf{v}', t) = f_1(\mathbf{x}, \mathbf{v}, t) f_2(\mathbf{x}', \mathbf{v}', t) + g(\mathbf{x}, \mathbf{v}, \mathbf{x}', \mathbf{v}', t) \tag{1-21}$$

in which case Eq. (1-18) is replaced by

$$\frac{df}{dt} = -\int \frac{\mathbf{F}}{m} \cdot \frac{\partial}{\partial \mathbf{v}} g \, d\mathbf{x}' \, d\mathbf{v}' = \left(\frac{\partial f}{\partial t}\right)_c \tag{1-22}$$

Equation (1-22) is known as the "Boltzmann equation." The left-hand side consists of the same expression as in the Vlasov equation, whilst the right-hand side, expressed formally as the rate of change of f due to collisions, describes the effects of the graininess of the medium.

Several collisional plasma models have been devised which use computer superparticles to represent f. The collision term in Eq. (1-22) has been represented by exchanging momentum and energy (Oliphant and Nielson, 1970) or by exchanging numbers of electrons per superparticle and energy (Gula and Chu, 1973) between neighboring superparticles. The microscopic semiconductor device model listed in Table 1-1 is a further example of a collisional plasma model, where now the plasma is the gas of conduction electrons moving in the semiconductor and collisions include quantum-mechanical interactions.

On the scale lengths involved, the classical Boltzmann equation, Eq. (1-22), gives a good description of the conduction electrons in semiconductors, provided that the consequences of the energy bands is accounted for by introducing an effective mass $m^* = \hbar^2/(\partial^2 \mathscr{E}/\partial k^2)$, where $\mathscr{E}(\mathbf{k})$ is the energy band structure characteristic of the material (see, e.g., Dekker 1958). If a parabolic approximation is used for the energy bands (Sec. 10-2-2) then the quantum-mechanical wave vector $\mathbf{k}$ and energy $\mathscr{E}$ needed to describe the scattering processes are related to particle velocities by

$$m^*\mathbf{v} = \hbar\mathbf{k} \qquad \mathscr{E} = \frac{\hbar^2 k^2}{2m^*} \tag{1-23}$$

1-4-4 Collision-dominated Systems

Simulation models based on the Boltzmann equation (1-22) become computationally too expensive for calculations lasting longer than a few collision times. More information must be discarded if one wishes to investigate longer time- (and length-) scale phenomena. This reduction is achieved by taking velocity moments of Boltzmann's equation. The six-dimensional μ space description is reduced to a three-dimensional configuration ($\mathbf{x}$) space description. The details of the velocity distribution is reduced to a mean velocity and a spread of velocity (the temperature). As with the hierarchy of kinetic equations, each moment equation has a term involving higher moments, so further approximations are needed to obtain a closed set of equations. Generally, closure is obtained by assuming an equation of state for the pressure $p : p = p(\rho, T)$, where ρ is mass density and T is temperature.

The zeroth-moment equation, which describes mass conservation, is obtained by multiplying Eq. (1-22) by the particle mass and integrating over velocity space:

$$\int d\mathbf{v}\, m\left\{\frac{\partial f}{\partial t}+\mathbf{v}\cdot\frac{\partial f}{\partial \mathbf{x}}+\frac{\mathbf{F}}{m}\cdot\frac{\partial f}{\partial \mathbf{v}}\right\}=\int d\mathbf{v}\, m\left(\frac{\partial f}{\partial t}\right)_c \tag{1-24}$$

From the definition of f, it follows that

$$\int d\mathbf{v}\, mf = mn = \rho \qquad \int d\mathbf{v}\, m\mathbf{v}f = \rho\mathbf{u} \tag{1-25}$$

where ρ is the mass density and $\mathbf{u}$ is the mean (fluid) velocity. If particle numbers are conserved, the right-hand side of Eq. (1-24) is zero. Using Eq. (1-25) and the necessary condition that $f \to 0$ as $\mathbf{v} \to \infty$, the left-hand side may be integrated to give

$$\frac{\partial \rho}{\partial t}+\nabla\cdot\rho\mathbf{u}=0 \tag{1-26}$$

Multiplying Eq. (1-22) by $m\mathbf{v}$ and integrating over velocity space gives the first moment (momentum conservation) equation:

$$\frac{\partial}{\partial t}\rho\mathbf{u}+\nabla\cdot[\rho\mathbf{u}\mathbf{u}+\mathbf{P}]-n\mathbf{F}=\int d\mathbf{v}\, m\mathbf{v}\left(\frac{\partial f}{\partial t}\right)_c \tag{1-27}$$

where the pressure tensor $\mathbf{P}$ is defined by

$$\mathbf{P}=\int d\mathbf{v}\, mf(\mathbf{v}-\mathbf{u})(\mathbf{v}-\mathbf{u}) \tag{1-28}$$

In many applications, the pressure tensor is well approximated by a scalar pressure, $\mathbf{P} \approx p\mathbf{I}$, where p satisfies the ideal gas law $p = \rho RT = nk_BT$. Using this approximation and combining Eqs. (1-26) and (1-27) gives the equation of motion which forms the basis of the collision-dominated particle models:

$$\rho\frac{d\mathbf{u}}{dt}=\rho\left(\frac{\partial \mathbf{u}}{\partial t}+\mathbf{u}\cdot\nabla\mathbf{u}\right)=-\nabla p+n\mathbf{F}+\int d\mathbf{v}\, m\mathbf{v}\left(\frac{\partial f}{\partial t}\right)_c \tag{1-29}$$

The electrons (or holes, as appropriate) in the collision-dominated semiconductor model are regarded as a charged fluid, and the computer particles correspond to elements of that fluid. Particles have attributes charge, effective mass, and position. The rate of change of their positions is given by velocity where the velocity, derived from Eq. (1-29), is treated as a field quantity. On timescales long compared with the electron-lattice relaxation time τ, the collision term in Eq. (1-29) may be approximated by the simple model

$$\int dv\, m^*\mathbf{v}\left(\frac{\partial f}{\partial t}\right)_c \simeq \int d\mathbf{v}\, m^*\mathbf{v}\,\frac{f_0-f}{\tau}=-\frac{m^*n\mathbf{u}}{\tau} \tag{1-30}$$

and the inertial term $\rho\, d\mathbf{u}/dt$ may be neglected. The equilibrium (Fermi) distribution f_0 is symmetric in $\mathbf{v}$ and so leads to no contribution to the integral in Eq. (1-30). Thus, substituting for p and $\mathbf{F}$, Eq. (1-29) becomes

$$0=-k_BT\nabla n+nq\mathbf{E}-\frac{m^*n\mathbf{u}}{\tau}$$

i.e.,
$$\mathbf{u} = -\frac{D}{n}\nabla n + \mu\mathbf{E} \tag{1-31}$$

where D = diffusion coefficient and μ = mobility (cf. Dekker, 1958, Chapter 13).

The mathematical model for the vortex collision-dominated fluid simulation also arises from Eqs. (1-26) and (1-29). If these equations are summed over all species of particles comprising the fluid then the collision terms in Eq. (1-29) for each species, which give the rate of change of momentum per unit volume due to collisions with other species, sum to zero. With the additional assumptions of incompressibility (ρ = constant) and absence of electromagnetic and gravitational body forces ($\mathbf{F} = 0$), we obtain the fluid description

$$\nabla\cdot\mathbf{u} = 0 \qquad \rho\frac{d\mathbf{u}}{dt} = -\nabla p \tag{1-32}$$

For two-dimensional flows (in the x–y plane, say), we introduce the stream function $\boldsymbol{\psi}$ and vorticity vector $\boldsymbol{\omega}$, both of which only have components in the z direction. These are related to the velocity field $\mathbf{u}$ by

$$\mathbf{u} = \operatorname{curl}\boldsymbol{\psi} \tag{1-33}$$
$$\boldsymbol{\omega} = \operatorname{curl}\mathbf{u} \tag{1-34}$$

Vorticity is a conserved attribute of a fluid-element particle, as may be seen by taking the curl of the equation of motion, Eq. (1-32).

The equations of motion for particles in the vortex model are simply $d\mathbf{x}/dt = \mathbf{u}$. In component form Eq. (1-33) becomes

$$u_x = \frac{\partial\psi_z}{\partial y} \qquad u_y = -\frac{\partial\psi_z}{\partial x} \tag{1-35}$$

and, substituting in Eq. (1-34), one obtains the field equation

$$\nabla^2\psi_z = -\omega_z \tag{1-36}$$

A mathematically similar model to the vortex model for fluids is obtained for the guiding-center motion of electrons perpendicular to a strong magnetic field. The guiding-center equations are obtained by separating the fast-timescale Larmor orbits around the magnetic field lines from the slower drift motions. Mathematically the separation is performed by expanding in the smallness parameter $\varepsilon = m_e/e$. In the absence of collisions, the lowest-order approximation ($\varepsilon = 0$) to Eq. (1-29) becomes

$$\mathbf{F} = \mathbf{E} + \mathbf{u}\times\mathbf{B} = 0 \tag{1-37}$$

Assuming uniformity along the magnetic field, taking the field $\mathbf{B} = (0, 0, B_z)$ and writing $\mathbf{E} = -\nabla\phi$ gives

$$u_x = -\frac{1}{B_z}\frac{\partial\phi}{\partial y} \qquad u_y = \frac{1}{B_z}\frac{\partial\phi}{\partial x} \tag{1-38}$$

The potential is related to the charge density ρ by Poisson's equation:

$$\nabla^2 \phi = -\frac{\rho}{\epsilon_0} \tag{1-39}$$

and

$$\rho = -e(n - n_0) \tag{1-40}$$

where e = magnitude of the electronic charge
n = density of electron guiding centers
n_0 = background density

Comparing Eqs. (1-35) and (1-36) with Eqs. (1-38) and (1-39), we see that the electrostatic potential ϕ plays the same role as the stream function ψ_z, and that the charge density plays the role of the vorticity density ω_z. Hence, with appropriate scaling, the same model applies to incompressible inviscid fluids and electron flows in high magnetic fields!

Perhaps the earliest modeling of vorticity by particles was that of Abernathy and Kronauer (1962), using a PP method. The formation of "vortex streets" was successfully observed. A two-dimensional PM model was used by Levy and Hockney (1968) in the analogous electron flow problem. A similar hydrodynamic model has been described by Christiansen (1973*a*,*b*) and used extensively in the study of the interaction between finite-area vortex structures (Christiansen and Zabusky, 1973).

Further cases of phenomena described by fluid equations are amenable to particle simulation. Compressible fluids may be treated by using a fractional timestep method to solve the equation of motion: Changes in velocity field arising from pressure gradients are computed with a fixed configuration, then fluid elements (particles) are moved to transport mass and momentum (Harlow, 1964). Magnetofluids may be treated similarly, with particles having magnetic-flux related attributes in addition to the position, mass, momentum, and entropy constants (McCrory *et al.*, 1977).

1-5 PARTICLE MODELS

We identify three principal types of particle simulation model; the particle-particle (PP) model, the particle-mesh (PM), model and the particle-particle–particle-mesh (the PPPM or P^{3}M) model. The PP model uses the action at a distance formulation of the force law, the PM model treats the force as a field quantity—approximating it on a mesh—and the P^{3}M model is a hybrid of the PP and PM models. The choice of model is dictated partly by the physics of the phenomenon under investigation and partly by consideration of computational costs.

1-5-1 The Particle-Particle Method

The PP method is conceptually and computationally the simplest. The state of the physical system at some time t is described by the set of particle positions and velocities $\{\mathbf{x}_i(t), \mathbf{v}_i(t); i = 1, N_p\}$. The timestep loop updates these values using the forces of interaction and equations of motion to obtain the state of the system at a slightly later time $t+DT$, as follows:

Timestep loop of PP method

1. Compute forces.

   Clear force accumulators

   for $i = 1$ *to* N_p *do*

   $\mathbf{F}_i := 0$

   Accumulate forces

   for $i = 1$ *to* $N_p - 1$ *do*

   for $j = i+1$ *to* N_p *do*

   Find force $\mathbf{F}_{ij}$ of particle j on particle i

   $\mathbf{F}_i := \mathbf{F}_i + \mathbf{F}_{ij}$

   $\mathbf{F}_j := \mathbf{F}_j - \mathbf{F}_{ij}$

2. Integrate equations of motion.

   for $i = 1$ *to* N_p *do*

   $$\mathbf{v}_i^{\text{new}} := \mathbf{v}_i^{\text{old}} + \frac{\mathbf{F}_i}{m_i} DT$$

   $$\mathbf{x}_i^{\text{new}} := \mathbf{x}_i^{\text{old}} + \mathbf{v}_i DT$$

3. Update time counter.

   $t := t + DT$

Repeated application of the timestep loop is used to follow the temporal evolution of the system.

Programming the PP timestep loop is a trivial exercise. The reason that such a simple scheme is used only in a few instances becomes apparent when we consider

Table 1-2 Timing estimates for PP calculations

Typical nominal times, circa 1978, are for one-floating-point arithmetic operation (including allowances for necessary housekeeping operations) on a range of available computers. N_p is the number of particles.

Type of computer	Nominal time for 1 operation	CPU time ($=10N_p^2 \times 1$ *op.*) $N_p = 10^2$	$N_p = 10^3$	$N_p = 10^5$
Mini	100 μs	10 s	$\frac{1}{4}$ h	~1 year
Medium-sized	10 μs	1 s	$1\frac{1}{2}$ min	~1 month
Large scientific	1 μs	0.1 s	10 s	~1 day
Large parallel/ vector processor	0.1 μs	0.01 s	1 s	~1 h

the computational costs. Take, for example, a system of point-charged particles where the action-at-a-distance expression for forces is given by Eq. (1-2). Assume that +, −, ÷, and × each count as one floating-point operation and $(\ )^{3/2}$ count as three. Multiplication by $m_i(q_i q_j/4\pi\epsilon_0)$ and DT can be avoided by careful choice of units (see Chapter 2) and so we shall omit these operations from our count. Working through the PP timestep loop we obtain the following count of operations:

Calculation		*Operations count*
clear force accumulator		$3N_p$
do for $N_p(N_p-1)/2$ particle pairs		
compute $\mathbf{x}_i-\mathbf{x}_j$	3	
compute $\lvert\mathbf{x}_i-\mathbf{x}_j\rvert^3$	8	
compute $\mathbf{F}_{ij}$	3	
update $\mathbf{F}_i$ and $\mathbf{F}_j$	6	
	20	$10N_p(N_p-1)$
update velocities and positions		$6N_p$
	Total	$10N_p^2-N_p$

Table 1-2 gives estimates of the central processor unit (CPU) time needed per timestep of a PP calculation for different numbers of particles. The medium-sized computer is typical of those available locally at British universities and the large scientific one corresponds to the backup facilities of the Regional Computer Centres. Large parallel/vector processors are facilities available at certain national establishments. The times per floating-point operation are nominal and make a reasonable allowance for other necessary housekeeping operations. They may be taken as a reasonable basis for conservatively estimating the viability of a proposed calculation. Undoubtedly, very carefully written code will correspond to nominal times less than those in Table 1-2. The times apply to computers circa 1978.

Typically, a few thousand timesteps are needed to obtain useful results from a computer experiment, so the direct summation PP method is viable only for systems of up to approximately a thousand particles if forces are long ranged, such as is the case in the study of stellar clusters. If, however, interparticle forces are short-ranged then the dependence of the operations count on the square of particle numbers is replaced by dependence on the number of particles multiplied by the number of neighbors which are close enough to a particle to contribute significantly to the force it experiences. The Lennard-Jones model of atomic liquids is one instance of a system with short-ranged forces to which the PP model has been extensively applied.

1-5-2 The Particle-Mesh Method

The PM method exploits the force-at-a-point formulation and a field equation for the potential [for example, Eqs. (1-5) and (1-7), respectively]. The result is a much faster, but generally less accurate, force calculation than is obtained using the PP method.

Field quantities, which in the physical system pervade all space, are represented approximately by values on a regular array of mesh points. Differential operators, such as the laplacian ∇^2, are replaced by finite-difference approximations on the mesh. Potentials and forces at particle positions are obtained by interpolating on the array of mesh-defined values. Mesh-defined densities are obtained by the opposite process of assigning the particle attributes (e.g., charge) to nearby mesh points in order to create the mesh-defined values (e.g., of charge density).

The timestep loop of the PM method differs from that of the PP method only in the calculation of the forces. The PM force calculation corresponding to the charged-particle PP example given above consists of three steps:

1. Assign charge to mesh.
2. Solve Poisson's equation (Eq. 1-7) on the mesh.
3. Compute forces from the mesh-defined potential and interpolate forces at particle positions.

Steps (1) and (3) have operation counts proportional to the number of particles N_p. The operation count for step (2) depends on the number of mesh points N. Hence, the operations count is given by

$$\text{Operations count} = \alpha N_p + \beta(N)$$

where the constant α and function β depend on the particular form of PM scheme being used. These various forms will be investigated further in Chapter 5. For the present, we shall take some characteristic values in order to make a comparison with the PP method:

$$\alpha = 20 \qquad \beta = 5N^3 \log_2 N^3 \text{ for an } (N \times N \times N) \text{ mesh}$$

Taking $N = 32$, $N_p = 10^5$, and a machine with a nominal CPU time per floating-point operation of 1 μs, we obtain

$$\begin{aligned} \text{CPU time} &= (20 \times 10^5 + 5 \times 32^3 \times 15) \times 10^{-6} \text{ seconds} \\ &\simeq 4\tfrac{1}{2} \text{ seconds} \\ \text{cf.} \quad &\sim 1 \text{ day for the PP method!} \end{aligned}$$

The enormous speed gain of the PM over the PP method is obtained at the cost of a loss of resolution in the potential and force fields. Only those field variations which have wavelength longer than the spacing of the mesh can be accurately represented by mesh values. The potential and force fields of a single point charge (or mass) are poorly represented on the mesh for distances less than the mesh spacing H. However, the nature of these errors is such that the inaccurate representation of the fields of a point charge may be interpreted as an accurate representation of the fields of a finite-sized charge cloud whose width is of the order of the mesh spacing H.

The limited resolution of the mesh proves unacceptable for studying correlated systems: A mesh fine enough to resolve close encounters of particles would have so many mesh points that the CPU time needed to solve for the potential would exceed the timestep cycle time for the PP method. The opposite is true for uncorrelated and collisional (including collision-dominated) systems: The limited mesh resolution introduces the properties required to destroy the unphysical correlations between the limited number of computational superparticles. Provided that the mesh spacing is smaller than the wavelengths of importance in the physical system (for example, the Debye length in plasmas and the spiral arm width in galaxies) and the number of computer superparticles in each cell of the mesh is large enough (say ~ 10) to maintain low levels of fluctuations, then the PM model will accurately represent the mathematical models of uncorrelated and collisional physical systems.

1-5-3 The Particle-Particle–Particle-Mesh Method

We have seen that the PP method can be used for small systems with long-range forces or for large systems where the forces of interaction are nonzero for only a few interparticles distances. The PM method, on the other hand, is computationally fast, but can only handle smoothly varying forces. The P^3M method combines the advantages of the PP and PM methods and enables large correlated systems with long-range forces to be simulated.

The trick used in the P^3M method is to split the interparticle forces into two parts

$$\mathbf{F}_{ij} = \mathbf{F}_{ij}^{\mathrm{sr}} + \mathbf{F}_{ij}^{\mathrm{m}} \tag{1-41}$$

where the rapidly varying short range part $\mathbf{F}_{ij}^{\mathrm{sr}}$ is nonzero for only a few interparticle distances and the slowly varying part $\mathbf{F}_{ij}^{\mathrm{m}}$ is sufficiently smooth to be accurately represented on a mesh. The PP method is used to find the total short-range contribution to the force on each particle and the PM method is used to find the total slowly varying force contribution. The addition of the total short-range and total slowly varying force contributions gives the total forces on each particle which is used to update the velocities. The resulting P^3M scheme falls between the PP and PM method in that it can represent close encounters as accurately as the PP method and calculate long-range forces as rapidly as the PM method.

The splitting of the forces (Eq. 1-41) is demonstrated most readily for inverse-square-law forces. For instance, consider uniformly charged spheres with total charge q and radius $a/2$. From Gauss' theorem (Reitz and Milford, 1962, Chapter 2) and Newton's third law (action and reaction equal and opposite) it follows that the force between two such spheres separated by a distance $r \geqslant a$ is given by

$$F(r) = \frac{q^2}{4\pi\epsilon_0} \frac{1}{r^2} \qquad r \geqslant a \tag{1-42}$$

and directed along the line joining the centers of the spheres. A little more detailed calculation reveals that for $r \leqslant a$

$$F(r) = \frac{q^2}{4\pi\epsilon_0 a^2}\left(\frac{8r}{a} - \frac{9r^2}{a^2} + \frac{2r^4}{a^4}\right) \qquad r \leqslant a \tag{1-43}$$

Thus, one possible splitting of the force is to set the mesh part of the force to $F(r)$ and the short-range part (nonzero only for $r < a$) to the difference between the inverse-square-law force and $F(r)$. The question of splitting the force is considered in more detail in Chapter 8.

The operations count for P^3M timestep cycle consists of the part $\alpha N_p + \beta(N)$ from the PM part and integration of the equations of motion, and a part proportional to $N_p N_n$ for the PP part. The number of neighbors, N_n, is given approximately by $4\pi a^3 n_0/3$, where n_0 is the mean density of particles. Thus, the total operations count for the P^3M model may be written

$$\text{Operations count} = \alpha N_p + \beta(N) + \gamma N_n N_p \tag{1-44}$$

The correct choices of N, N_p, and a cause the operations count to scale linearly with particular number. Typically, a timestep cycle CPU time of approximately one second per thousand particles is obtained on large scientific computers. For ten thousand particles, this gives a cycle time of ten seconds, compared with one of approximately twenty minutes for the same situation using the PP method.

CHAPTER

TWO

A ONE-DIMENSIONAL PLASMA MODEL

In this chapter, we shall use the problem of simulating electrostatic waves in a collisionless plasma to introduce the basic elements of the particle-mesh (PM) scheme and to illustrate the process of designing simulation models. The steps we shall go through follow those shown in Fig. 1-1, although we shall leave consideration of the simulation program until the next chapter. The PM scheme we shall obtain is the one-dimensional NGP scheme. Generalizations of this basic scheme, together with detailed analyses of the properties of the PM simulation model, are given in Chapters 5 and 7.

2-1 THE PHYSICAL SYSTEM

Plasmas are hot, ionized gases composed of ions, electrons, and neutral atoms. A great wealth of physical processes, involving a wide range of length and timescales, arise from interactions between the plasma constituents and the electromagnetic fields. The mathematical model of a plasma we shall consider here is an idealized theoretical model in that it retains only a limited subset of the phenomena occurring in a real plasma.

The equations describing the idealized plasma model are the Vlasov equation describing the electron distribution f,

$$\frac{df}{dt} = \frac{\partial f}{\partial t} + \mathbf{v} \cdot \frac{\partial f}{\partial \mathbf{x}} + \frac{\mathbf{F}}{m} \cdot \frac{\partial f}{\partial \mathbf{v}} = 0 \tag{2-1}$$

and the electrostatic force-at-a-point expressions for the potential ϕ, electric field

E, and force **F**,

$$\nabla^2 \phi = -\rho/\epsilon_0 \qquad \mathbf{F} = q\mathbf{E} = -q\nabla\phi \tag{2-2}$$

The positive ions are treated as a fixed, neutralizing, background charge density ρ_0, and the total charge density ρ is given by

$$\rho(\mathbf{x}) = q \int f \, d\mathbf{v} + \rho_0 \tag{2-3}$$

where $q(= -e)$ is the electron charge. Assumptions implicit in Eqs. (2-1) to (2-3) are that there is no charge exchange (total electron charge conserved), that the electron gas is collisionless (i.e., kinetic energy $\gg$ potential energy), that the disturbances from equilibrium are parallel to the electric field (electrostatic approximation), and that velocities are much less than the speed of light. The model is appropriate for studying collective electron oscillations on a length scale much greater than the average electron spacing and on timescales much less than the electron collision time and the timescale of ion motions (cf. Chapter 1).

The one-dimensional model is obtained by assuming that quantities such as charge density, potential, and electric field depend only upon one spatial coordinate, say x. In the other two directions, y and z, the plasma is assumed to be uniform and of infinite extent. Immediate simplification follows from the uniformity: The y and z positions and velocities become ignorable coordinates, and Eqs. (2-1) to (2-3) reduce to

$$\frac{\partial f}{\partial t} + v\frac{\partial f}{\partial x} + \frac{F}{m}\frac{\partial f}{\partial v} = 0 \tag{2-4}$$

$$\frac{d^2\phi}{dx^2} = -\frac{\rho}{\epsilon_0} \qquad E = -\frac{d\phi}{dx} \qquad F = qE \tag{2-5}$$

$$\rho(x) = q \int f(x, v, t) \, dv + \rho_0 \tag{2-6}$$

Although Eqs. (2-4) to (2-6) represent a great simplification of the complete description of a plasma, they still present a formidable task for mathematical analysis. In linearized theory (see, for example, Stix, 1962) variables are expanded about equilibrium solutions (i.e., $f = f_0 + f_1$, $E = E_0 + E_1$, etc.) and only terms to first order are retained. The first-order equations are Fourier-transformed in space and Laplace-transformed in time, yielding the description of disturbances as a superposition of harmonics $\sim e^{i(kx-\omega t)}$. Eliminating first-order variables from the linearized versions of Eqs. (2-4) to (2-6) yield a consistency condition (the dispersion relation) relating ω and k:

$$1 = -\frac{\omega_p^2}{kn_0}\left\{P\int_{-\infty}^{\infty} \frac{\partial f_0/\partial v}{\omega - kv}\,dv - \frac{i\pi}{|k|}\frac{\partial f_0}{\partial v}\bigg|_{\omega/k}\right\} \tag{2-7}$$

where the P indicates the principal part of the integral.

The essential features of Eq. (2-7) may be obtained by assuming some simple form for f_0 in order to evaluate the integral. For instance, if we take an f_0 whose first three moments correspond to a Maxwellian distribution of density n_0 and thermal velocity v_T:

$$f_0 = \begin{cases} \dfrac{n_0}{2\sqrt{3}v_T} & |v| < \sqrt{3}v_T \\ 0 & \text{otherwise} \end{cases} \tag{2-8}$$

then Eq. (2-7) becomes

$$1 \simeq \frac{\omega_p^2}{(\omega^2 - 3k^2 v_T^2)} + \frac{i\pi\omega_p^2}{n_0 k|k|} \frac{\partial f_0}{\partial v}\bigg|_{\omega/k} \tag{2-9}$$

If we let $\omega = \omega_r + i\omega_i$, then for small growth rates $\omega_i \ll \omega_r$, Eq. (2-9) yields the warm plasma approximation to (2-7):

$$\omega_r^2 \simeq \omega_p^2(1 + 3k^2\lambda_D^2) \qquad \omega_i \simeq \frac{\pi}{2} \frac{\omega_p^2 \omega_r}{n_0 k|k|} \frac{\partial f_0}{\partial v}\bigg|_{\omega/k} \tag{2-10}$$

where $\lambda_D = v_T/\omega_p$ is the Debye length. It is clear from the real part of the frequency that the characteristic frequencies and wavelength of electrostatic plasma oscillations are the plasma frequency ω_p and Debye length λ_D, respectively, where

$$\omega_p = \sqrt{\frac{nq^2}{\epsilon_0 m_e}} \qquad \lambda_D = \sqrt{\frac{\epsilon_0 k_B T}{nq^2}} \tag{2-11}$$

The imaginary part of the frequency describes Landau damping. If f_0 has only a single maximum, then all waves are damped; if not, then waves whose phase velocity ω/k have the same sign as $\partial f_0/\partial v|_{\omega/k}$ have exponentially growing amplitudes.

Mathematical theory can be extended into the weakly nonlinear regime by evaluating the changes to f_0 due to f_1 (Drummond and Pines, 1964; Bernstein and Engelmann, 1966), but to study the full nonlinear effects of velocity space instabilities, we must turn to computer simulation.

2-2 DISCRETIZATION OF THE MATHEMATICAL MODEL

The first step from the mathematical to simulation model is to approximate the mathematical equations, Eqs. (2-4) to (2-6), by the algebraic equations required for numerical computations. The simulation model we shall arrive at will be the Nearest-Grid-Point (NGP) particle-mesh (PM) model. To see why the PM model makes good sense, we shall go through the process of the discretization in some detail.

2-2-1 The Superparticle Equations

The first stage of the discretization is to replace Eq. (2-4) by its characteristic equations. If we imagine that $x-v$ phase space is divided into a regular array of infinitesimal cells of volume $d\tau = dx\,dv$, where $d\tau$ is sufficiently small for not more than one electron to occupy it, then $f(x, v, t)\,d\tau$ gives the probability that the cell at (x, v) is occupied at time t. Given that there is an electron in the cell at time t, then it follows that there will be one in the cell at (x', v') at time t', where (x', v') are related to (x, v) by the electron equations of motion

$$x' = x + \int_t^{t'} v\,dt \qquad v' = v + \int_t^{t'} \frac{qE}{m}\,dt \tag{2-12}$$

Pursuing this reasoning, we can show generally that

$$f(x', v', t') = f(x, v, t) \tag{2-13}$$

where (x', v') are related to (x, v) by Eq. (2-12). Equations (2-12) and (2-13) are simply a restatement of Eq. (2-4), as may be demonstrated by Taylor-expanding Eq. (2-13) and taking the limit as $(t' - t)$ tends to zero. The equation of motion (2-12) is the characteristic equation of Eq. (2-4), or, equivalently, f is conserved along electron trajectories, Eq. (2-13). Thus, if we knew the values of f for each infinitesimal cell in phase space at time t, we could map f forward to any later time by integrating the equations of motion.

Clearly, to map f for every infinitesimal cell is computationally impracticable. Instead, we take a sample of points $\{x_i, v_i;\quad i = 1, N_p\}$ where each point represents an element i of phase fluid corresponding to $N_s = \int_i f\,dx\,dv$ plasma electrons per unit area in the y-z plane. The orbits through phase space of the sample points are given by

$$\frac{dx_i}{dt} = v_i \qquad M\frac{dv_i}{dt} = F(x_i) \tag{2-14}$$

where $M = N_s m_e$, and m_e is the electron mass. The physically appealing superparticle interpretation of the sample points introduced in the previous chapter follows naturally from Eq. (2-14).

If the density of superparticles (sample points) is sufficiently great then no explicit reference to the distribution function is required in the discrete model. Approximations to the moments of f are constructed directly from the coordinates of the superparticles. For instance, velocity moments of f may be approximated by

$$\int v^n f\,dv \simeq \frac{N_s}{\lambda}\int_{x-\lambda/2}^{x+\lambda/2} dx' \int v^n \tilde{f}\,dv \tag{2-15}$$

$$= \frac{N_s}{\lambda}\sum_i v_i^n \tag{2-16}$$

where $\tilde{f} = \sum_{i=1}^{N_p} \delta(x - x_i)\delta(v - v_i)$ is the distribution of superparticles and the sum over i in Eq. (2-16) is over all superparticles in the neighborhood $[x - (\lambda/2), x + (\lambda/2)]$ of the point x at which the moment is evaluated. The local average over the neighborhood is necessary if we are to represent a smooth distribution function f: If averaging were not performed, then the superparticles would behave as strongly correlated charged particles rather than as samples of a continuous smoothly varying phase fluid (cf. Sec. 1-4).

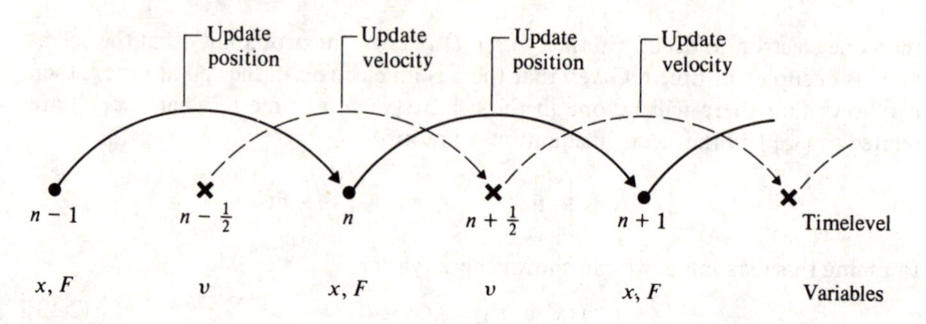

Figure 2-1 The equations of motion are integrated forwards in time using the leapfrog scheme. Positions at timelevel $n-1$ are updated using velocities at timelevel $n-\frac{1}{2}$, velocities at timelevel $n-\frac{1}{2}$ are updated using forces at timelevel n, and so forth.

The approximation Eq. (2-15) gives some indication as to what density of superparticles is sufficient. The characteristic wavelength of electrostatic oscillations is the Debye length λ_D [cf. Eq. (2-10)]. To represent adequately these oscillations we require that the averaging length be less than or the same order as λ_D. The number of superparticles in the averaging length $\lambda \sim \lambda_D$ must be large enough (say ~ 10) for statistical fluctuations of the moments to be small, from which we conclude that a suitable sample density is one where the number of superparticles in a Debye length is large.

The final stage of discretization of the Vlasov equation (2-4) is to replace the continuous time variables in Eq. (2-14) by a discrete set of timelevels separated in time by a small time interval, the timestep DT. The most commonly used scheme is the finite-difference approximation leapfrog scheme, so called because it leapfrogs positions and velocities forward in time (cf. Fig. 2-1). Positions and fields are defined at integral timelevels ($t = 0, DT, 2DT, 3DT \cdots$) and velocities are defined at half-integral timelevels ($t = \frac{1}{2}DT, \frac{3}{2}DT \cdots$). The leapfrog finite-difference approximations to Eq. (2-14) are

$$\begin{aligned} x_i^{n+1} - x_i^n &= v_i^{n+1/2} DT \\ v_i^{n+1/2} - v_i^{n-1/2} &= \frac{F(x_i^n) DT}{m_i} \end{aligned} \qquad (2\text{-}17)$$

The properties of Eqs. (2-17) are investigated in detail in Chapter 4. Here we shall just note that the timestep DT must be chosen to allow plasma oscillations (frequency $\sim \omega_p$) to be represented, i.e., $\omega_p DT \ll 2$.

2-2-2 The Field Equations

The region of space (i.e., the range of x values) spanned by the simulation model (which by necessity is finite) is known as the computational box. Boundary conditions are specified at the surfaces $x = 0$ and $x = L$ of the box—for example, if the surfaces were earthed metal plates, then we would require $\phi = 0$ at the surfaces. Given the potentials at the surfaces and the charge density distribution within the computational box, Poisson's equation, Eq. (2-5), completely specifies the potential. More appropriate boundary conditions for the present example are periodic conditions.

$$\phi(x) = \phi(x+L) \tag{2-18}$$

Periodic conditions yield the same dispersion relation, Eq. (2-7), as for the infinite system, except that the continuum of wavenumbers is replaced by a discrete set $k = 2\pi l/L$ whose wavelengths λ are such that $L = l\lambda$; $l =$ integer.

To get the Eqs. (2-5) for the potential and electric field into a suitable form for numerical solution we again use finite differences. The derivative is defined as

$$\frac{d}{dx} f(x) = \lim_{h\to 0} \left\{ \frac{f(x+h/2) - f(x-h/2)}{h} \right\} \tag{2-19}$$

In the finite-difference approximation the limit is not taken to zero, but to some small quantity H which gives an acceptable compromise between accuracy and computational cost. For the plasma model $H \leqslant \lambda_D$ is a suitable choice. Charge density, potential, and electric fields are represented by a set of values spaced at regular intervals H throughout the computational box (cf. the representation of continuous logarithm function by regularly spaced values in a log table). The points at which values are recorded are the mesh (or grid) points (see Fig. 2-2). Mesh points lie at the centers of cells of width H. If the origin is taken at mesh point 0, the position of mesh point p is at $x_p = pH$. An integral number of cell widths H fit into the computational box length L. For periodic boundary conditions, the number of cells is equal to the number of grid points N_g.

$$L = N_g H \tag{2-20}$$

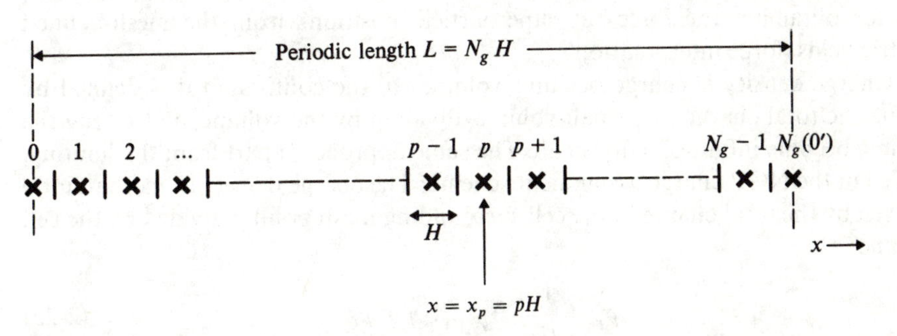

Figure 2-2 The computational box $x \in [0, L]$ is split into cells of width H. At the center of each cell is a mesh point at which values of charge density potential and electric field are stored. The box may be imagined to periodically repeat: Mesh point N_g is a periodic image of mesh point 0, point N_g+1 is a periodic image of point 1, etc.

Derivatives are replaced by finite differences on the mesh as follows:

$$\left.\frac{d^2\phi}{dx^2}\right|_{x_p} = \left.\frac{d}{dx}\left(\frac{d\phi}{dx}\right)\right|_{x_p}$$

$$\simeq \frac{1}{H}\left(\left.\frac{d\phi}{dx}\right|_{x_p+H/2} - \left.\frac{d\phi}{dx}\right|_{x_p-H/2}\right)$$

$$\simeq \frac{1}{H}\left(\frac{\phi(x_{p+1})-\phi(x_p)}{H} - \frac{\phi(x_p)-\phi(x_{p-1})}{H}\right)$$

$$= \frac{\phi(x_{p+1})-2\phi(x_p)+\phi(x_{p-1})}{H^2} \tag{2-21}$$

and

$$\left.\frac{d\phi}{dx}\right|_{x_p} \simeq \frac{(\phi(x_{p+1})-\phi(x_{p-1}))}{2H} \tag{2-22}$$

The resulting finite-difference field equations are

$$\frac{\phi_{p+1}-2\phi_p+\phi_{p-1}}{H^2} = -\frac{\rho_p}{\epsilon_0} \tag{2-23}$$

$$E_p = \frac{\phi_{p-1}-\phi_{p+1}}{2H} \tag{2-24}$$

where the more compact labeling of values by integer subscripts, $\phi_p \equiv \phi(x_p)$, etc., is used.

2-2-3 Charge Assignment and Force Interpolation

To complete the discrete model we require prescriptions for obtaining the charge density at mesh points from the distribution of superparticles (charge assignment) and for obtaining the forces at superparticle positions from the mesh-defined electric fields (force interpolation).

Charge density is charge per unit volume: In the continuum it is defined by taking the total charge in a small volume, dividing by the volume, and letting the volume become infinitesimally small. The same approach, apart from the limiting, is used in the NGP charge assignment scheme: The charge density at mesh point p is given by the total charge in the cell surrounding mesh point p divided by the cell volume:

$$\rho_p = \frac{1}{H} \sum_{\substack{\text{particles} \\ i \text{ in cell } p}} N_s q + \rho_0 \tag{2-25}$$

The NGP charge assignment scheme, Eq. (2-25), is identical to the prescription for constructing approximate moments of the distribution function from the superparticle coordinates. Substituting Eq. (2-15) into Eq. (2-6) and setting $\lambda = H$ gives

$$\rho(x) \simeq \frac{qN_s}{H} \int_{x-H/2}^{x+H/2} dx' \int_{-\infty}^{\infty} \tilde{f}\, dv + \rho_0 \tag{2-26}$$

$$= \frac{qN_s}{H} \int_{x-H/2}^{x+H/2} dx'\, n(x') + \rho_0 \tag{2-27}$$

$n(x') = \Sigma_{i=1}^{N_p} \delta(x' - x_i)$ is the density of superparticle centers. Performing the integral over x' and setting $x = x_p$ recovers Eq. (2-25). However, an alternative form which will prove useful when we analyze charge assignment in more detail (Chapter 5) is obtained by introducing the charge assignment function W:

$$W(x) = \begin{cases} 1 & |x| < H/2 \text{ or } x = H/2 \\ 0 & \text{otherwise} \end{cases} \tag{2-28}$$

Using Eq. (2-28) allows us to rewrite Eq. (2-27) as

$$\rho(x) = \frac{qN_s}{H} \int_0^L W(x' - x) n(x')\, dx' + \rho_0 \tag{2-29}$$

Values of ρ_p are obtained by sampling $\rho(x)$ at mesh points:

$$\rho_p = \rho(x)|_{x = x_p}$$

The NGP force interpolation scheme treats the force in a similar way to charge assignment. The force field in a cell is given by the value at the mesh point at the center of that cell. Thus the force on some particle i at x_i is

$$F(x_i) = N_s q E(x_p) \qquad x_p - H/2 < x_i \leqslant x_p + H/2 \tag{2-30}$$

This also may be rewritten in terms of the NGP charge assignment function:

$$F(x_i) = N_s q \sum_{p=0}^{N_g - 1} W(x_i - x_p) E_p \tag{2-31}$$

Note that here the sum is over mesh points, whereas in charge assignment the sum is over particles. Periodic boundary conditions are handled in Eq. (2-31) by treating mesh point N_g as mesh point 0.

2-2-4 The Discrete Model

Collecting together the results of the previous three subsections, we obtain the equations of the NGP particle-mesh model of a one-dimensional electron plasma:

1. *Charge assignment*

$$\rho_p^n = \frac{qN_s}{H} \sum_{i=1}^{N_p} W(x_i^n - x_p) + \rho_0 \tag{2-32}$$

2. *Field equations*

$$\frac{\phi_{p-1}^n - 2\phi_p^n + \phi_{p+1}^n}{H^2} = -\frac{\rho_p^n}{\epsilon_0} \tag{2-33}$$

$$E_p^n = \frac{\phi_{p-1}^n - \phi_{p+1}^n}{2H} \tag{2-34}$$

3. *Force interpolation*

$$F_i^n = F(x_i^n) = N_s q \sum_{p=0}^{N_g - 1} W(x_i^n - x_p) E_p^n \tag{2-35}$$

4. *Equations of motion*

$$\frac{v_i^{n+1/2} - v_i^{n-1/2}}{DT} = \frac{F(x_i^n)}{N_s m_e} \tag{2-36}$$

$$\frac{x_i^{n+1} - x_i^n}{DT} = v_i^{n+1/2} \tag{2-37}$$

Periodic boundary conditions are applied to the potentials and to the particle positions. Equations (2-32) to (2-37) together with the boundary conditions completely define the evolution of the discrete system. Given the state of the system $\{x_i^n; v_i^{n-1/2}; i = 1, N_p\}$, successively solving Eqs. (2-32) to (2-37) for $\{\rho_p^n\}$, $\{\phi_p^n\}$, $\{E_p^n\}$, $\{F_i^n\}$, $\{v_i^{n+1/2}\}$, $\{x_i^{n+1}\}$, respectively, gives the state of the system $\{x_i^{n+1}, v_i^{n+1/2}; i = 1, N_p\}$ at time DT later.

The parameters H, DT, L, N_p appearing in the discrete model must satisfy constraints

1. $H \gtrsim \lambda_D$
2. $\omega_p DT \ll 2$
3. $L \gg \lambda_D$
4. $N_p \lambda_D \gg L$

in order that plasma waves may be adequately represented and that the model is collisionless. The further requirement that there are a large number of superparticles in the range of velocities near the phase velocity of unstable waves will generally be met if the number of particles per Debye length (constraint 4) is satisfied.

More quantitative statements on the constraints and on the quality of representation of the physical system can be obtained by applying the methods of analysis used for the mathematical model to the discrete model. We shall defer such considerations until later (Chapters 5 and 7).

2-3 NUMERICAL ALGORITHMS

The discrete model is translated into a computer program by first devising a sequence of operations (numerical algorithm) which when performed solve for the required unknowns of the discrete-model equations, then expressing the sequence as a series of computer instructions. A measure of the quality of algorithms is the number of arithmetic operations required to solve a given set of algebraic equations.

2-3-1 Dimensionless Units

The algebraic equations of concern here are Eqs. (2-32) to (2-37). In these equations are a number of multiplying constants which by a careful choice of dimensionless units can be eliminated.

The natural units of length and time in the discretized system of equations are the discretization intervals: the cell width H and timestep DT. If we denote dimensionless quantities by primes then

$$x' = x/H \quad : \quad \text{length} \tag{2-38}$$

$$t' = t/DT \quad : \quad \text{time} \tag{2-39}$$

If, in addition, we measure velocity in units of cell widths per timestep and acceleration in units of cell widths per timestep per timestep

$$v' = vDT/H \quad : \quad \text{velocity} \tag{2-40}$$

$$a' = aDT^2/H \quad : \quad \text{acceleration} \tag{2-41}$$

the equations of motion (2-36) and (2-37) become

$$v_i'^{n+1/2} - v_i'^{n-1/2} = a_i'^n \tag{2-42}$$

$$x_i'^{n+1} - x_i'^n = v_i'^{n+1/2} \tag{2-43}$$

where

$$a_i'^n = \frac{F_i^n DT^2}{N_s m_e H} \tag{2-44}$$

Relating the dimensionless particle acceleration $a_i'^n$ to the mesh-defined electric field using Eq. (2-35)

$$a_i'^n = \sum_p W(x_i'^n - p) \frac{qDT^2}{m_e H} E_p^n \tag{2-45}$$

suggests the choice of dimensionless electric field units

$$E_p'^n = \frac{qDT^2}{m_e H} E_p^n \tag{2-46}$$

giving

$$a_i'^n = \sum_p W(x_i'^n - p) E_p'^n \tag{2-47}$$

The natural choice of units for potential and charge density follow from removing constants from Eqs. (2-34) and (2-33):

$$E_p'^n = \phi_{p+1}'^n - \phi_{p-1}'^n \tag{2-48}$$

$$\phi_{p-1}'^n - 2\phi_p'^n + \phi_{p+1}'^n = \rho_p'^n \tag{2-49}$$

yielding
$$\phi_p'^n = -\frac{qDT^2}{2m_eH^2}\phi_p^n \tag{2-50}$$

and
$$\rho_p'^n = \frac{qDT^2}{2m_e\epsilon_0}\rho_p^n \tag{2-51}$$

Equation (2-32) becomes
$$\rho_p'^n = \frac{\omega_p^2 DT^2}{2}\left[\sum_{i=1}^{N_p}\frac{W(x_i'^n - p)}{N_c} - 1\right] \tag{2-52}$$

where $N_c = \rho_0 H/N_s|q|$ is the average number of superparticles per cell (assuming total charge neutrality). The first term in the bracket on the right-hand side of Eq. (2-52) shows clearly how the graininess of the particle model feeds into the dynamics of particles if N_c is small.

2-3-2 Charge Assignment

Equation (2-52) provides a trivial example of the problems in devising numerical algorithms. If it were used to construct in turn ρ_0', ρ_1', ρ_2', etc. then the computation of $\{\rho_p'; p = 0, N_g - 1\}$ would require approximately $N_g N_p$ tests, N_p additions, and N_g multiplications.

A much more economical procedure is to sweep through the particle coordinates once, simultaneously accumulating all the charge densities:

Algorithm

1. Initialize charge density accumulators:

   for $p = 1$ *to* N_g *do*

   $\rho_p := -N_c$

2. Accumulate charge density:

   for $j = 1$ *to* N_p *do*

   $p := \text{nint}(x_j')$ (locate nearest mesh point)

   $\rho_p := \rho_p + 1$ (increment charge density)

3. Scale charge densities:

   for $p = 1$ *to* N_g *do*

   $\rho_p' := \frac{\omega_p^2 DT^2}{2N_c} \times \rho_p$

This algorithm requires N_p additions and N_g multiplications. There are two points in the algorithm to note particularly: First, in steps 1 and 3, the periodic image mesh point N_g is used rather than mesh point 0 to facilitate the mapping of the algorithm to FORTRAN and second, in step 2, periodic boundary conditions on particle coordinates are applied such that their coordinates x lie in the range

$0.5 \leqslant x < N_g + 0.5$. In practice, steps 2 and 3 are merged (cf. Fig. 3-10) by adding the scaled charge increment rather than unity in step 2.

2-3-3 Poisson's Equation

The solution of Eq. (2-49) for the potential values presents a more difficult problem than that faced for charge assignment. Dropping the primes from the dimensionless quantities, the set of equations to be solved is

$$\begin{aligned}
\phi_{N_g} - 2\phi_1 + \phi_2 &= \rho_1 \\
\phi_1 - 2\phi_2 + \phi_3 &= \rho_2 \\
&\vdots \\
\phi_{N_g-2} - 2\phi_{N_g-1} + \phi_{N_g} &= \rho_{N_g-1} \\
\phi_{N_g-1} - 2\phi_{N_g} + \phi_1 &= \rho_{N_g}
\end{aligned} \tag{2-53}$$

where periodicity has been used to bring the indices in the first and last equation into the range $[1, N_g]$.

There are two conditions to be satisfied for the set of equations to have a unique solution. The first is that the system is charge-neutral. Each potential value appears in three separate equations in Eqs. (2-53), once with a multiplier -2 and twice with a multiplier $+1$. Adding all the equations together gives a zero sum on the left and the total charge on the right. Hence for consistency

$$\sum_{p=1}^{N_g} \rho_p = 0 \qquad \text{charge neutrality} \tag{2-54}$$

The second condition is that a reference potential must be chosen. If we construct a set of values $\{\psi_p = \phi_p + c;\ p \in [1, N_g]\}$ where c is a constant then $\{\psi_p\}$ is also a solution of Eq. (2-53). Taking the constant c to be such that $\phi_{N_g} = 0$ reduces Eq. (2-53) to

$$\begin{aligned}
-2\phi_1 + \phi_2 \qquad\qquad &= \rho_1 \\
\phi_1 - 2\phi_2 + \phi_3 \qquad &= \rho_2 \\
\phi_2 - 2\phi_3 + \phi_4 \qquad &= \rho_3 \\
\phi_3 - 2\phi_4 + \phi_5 &= \rho_4 \\
&\vdots \\
\phi_{N_g-2} - 2\phi_{N_g-1} &= \rho_{N_g-1} \\
\phi_1 \qquad\qquad + \phi_{N_g-1} &= \rho_{N_g}
\end{aligned} \tag{2-55}$$

If we multiply the first equation of (2-55) by one, the second by two, the third by three, etc., and add all the resulting equations together we obtain

$$N_g \phi_1 = \sum_{p=1}^{N_g} p\rho_p \tag{2-56}$$

Equation (2-56) gives the value ϕ_1. The first equation of Eqs. (2-55) then gives ϕ_2, the second ϕ_3, and so forth until the solution is found.

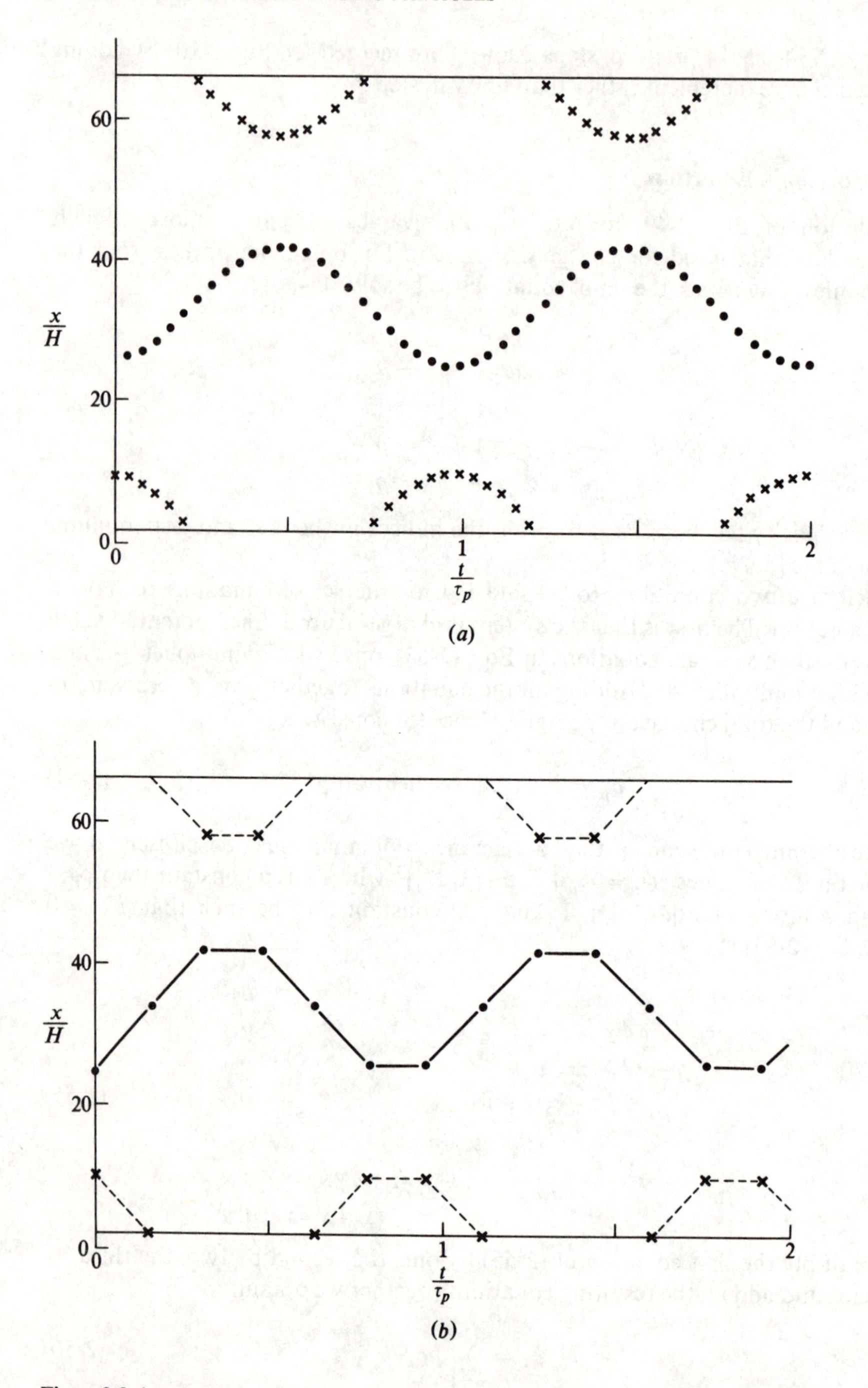

Figure 2-3 $(x-t)$ space-time plots of the oscillation of a pair of electrons separated by $L/4$, for different values of the timestep. (a) $\omega_p DT = 0.25$ (stable and accurate). (b) $\omega_p DT = 1.0$ (stable but inaccurate). (c) $\omega_p DT = 2.25$ (unstable). (d) Total energy for cases (a) and (c).

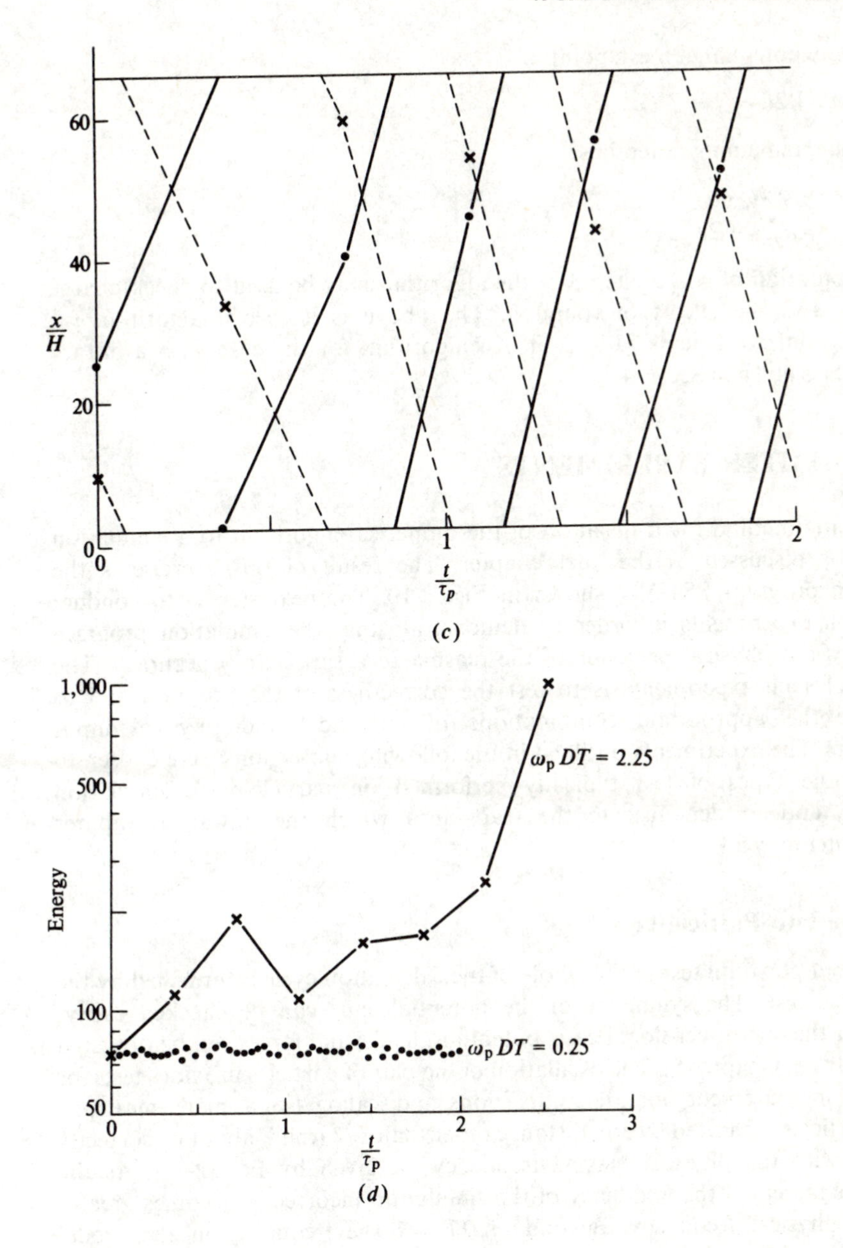

(*c*)

(*d*)

Algorithm

1. Compute potential at mesh point 1:

$\phi_1 := 0$

for $p = 1$ *to* N_g *do*

$\phi_1 := \phi_1 + p\rho_p$

$\phi_1 := \phi_1 / N_g$

2. Compute potential at mesh point 2:

   $\phi_2 := \rho_1 + 2\phi_1$

3. Compute remaining potentials:

   for $p = 3$ *to* N_g *do*

   $\phi_p := \rho_{p-1} + 2\phi_{p-1} - \phi_{p-2}$

The computation of ϕ_{N_g} in step 3 of the algorithm may be used to check charge neutrality and the effects of roundoff. The above is a special algorithm for the tridiagonal coefficients $[1, -2, 1]$. An algorithm for the case with arbitrary coefficients is given in Sec. 6-4-1.

2-4 COMPUTER EXPERIMENTS

A systematic method for translation of the numerical algorithm to a simulation program is discussed in the next chapter. The result of this exercise is the simulation program ES1D1V shown in Fig. 3-10. The next step is to conduct verification experiments in order to demonstrate that the simulation program reproduces the physical behavior of the plasma to a satisfactory accuracy. The purpose of such experiments is to test the correctness of the program and to determine the appropriate combinations of numerical and physical input parameters. The experiments outlined in the following subsections were chosen to illustrate the types of test typically performed on particle-mesh simulation programs, and to demonstrate the traps into which the unwary computer experimenter may fall.

2-4-1 The Two-Particle Test

A simple but powerful test of the whole of the calculation cycle is furnished by the two-particle test. The symmetry of the potential, etc., can be checked easily; indeed, for the one-dimensional case, potentials, fields, and forces can be obtained analytically (cf. Chapter 5). The oscillation of the pair of particles provides tests for the energy measurement, boundary conditions, and scaling. For a small timestep, a pair of particles separated by any distance other than $L/2$ (think about it) correctly oscillates with the physical plasma frequency ω_p given by Eq. (2-11). As the timestep is increased the frequency of the simulation incorrectly becomes greater than the physical frequency. Beyond $\omega_p DT = 2$ the frequency in the model becomes complex, and the time integration becomes unstable (cf. Chapter 4). The computed solution then bears no relation to the physical oscillation.

The default data for the one-dimensional plasma program shown in Fig. 3-10 is the two-particle oscillation data. Readers wishing to experiment with the one-dimensional plasma model will find it instructive to plot position, velocity, kinetic energy, potential energy, and total energy as functions of time for 50 timesteps using the default data. The influence of the choice of timestep on the stability and accuracy may be demonstrated by performing the two-particle test for the following values of timestep:

1. $\omega_p DT = 0.25$ (The default value.) A stable and accurate timestep. The plasma period is correctly reproduced with a period $\tau_p = 2\pi/\omega_p = 25.1$ timesteps (see Fig. 2-3*a*).
2. $\omega_p DT = 1.0$ A stable but inaccurate timestep. The model oscillates with a period of 6 timesteps, instead of the correct physical value $\tau_p = 2\pi/\omega_p = 6.28$ timesteps (see Fig. 2-3*b*).
3. $\omega_p DT = 2.25$ An unstable timestep since $\omega_p DT > 2$. The computed results are nonsense and the exponential growth of the numerically unstable solution will cause arithmetic overflow after only a few timesteps unless provision is

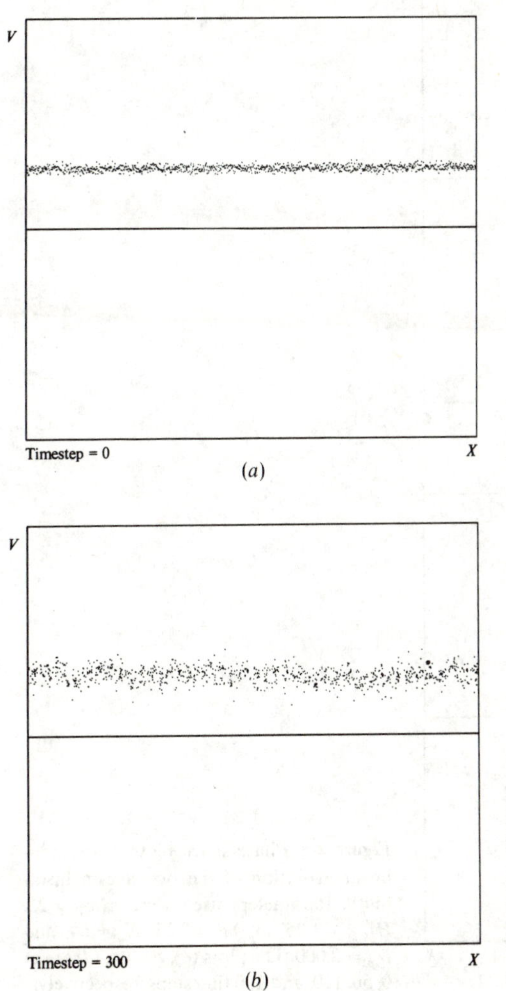

Figure 2-4 $(x-v)$ phase-space plots obtained for a single cold electron beam ($H = 22.5\lambda_D$) using the program shown in Fig. 3-10. The initial spread of velocity in (*a*) is increased by nonphysical instabilities to the stable final state shown in (*b*).

made in the program. In the program given in Fig. 3-10 the calculation is automatically closed down if any particle moves more than about $2L$ per timestep. This occurs after 7 timesteps (see Fig. 2-3*c*). The instability also shows up clearly in the lack of conservation of total energy (see Fig. 2-3*d*), compared with that obtained in case 1 above.

2-4-2 Wave Dispersion

A test of the collective properties of the plasma simulation is provided by measuring the relationship between the frequency and wavelength of small-

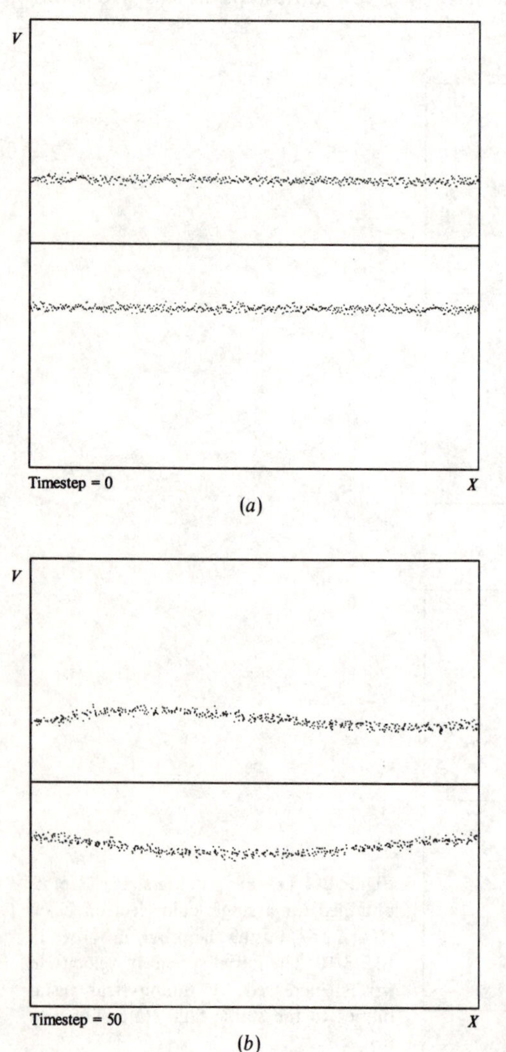

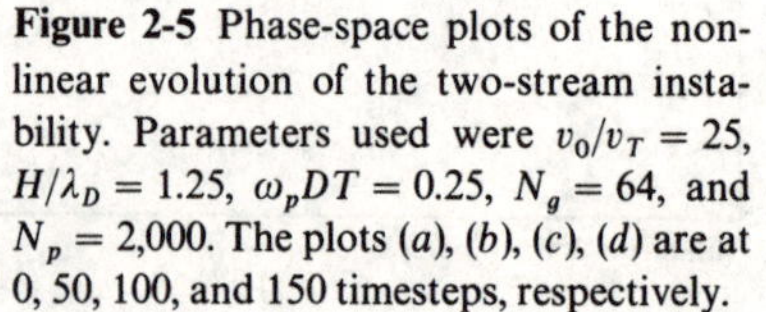

Figure 2-5 Phase-space plots of the non-linear evolution of the two-stream instability. Parameters used were $v_0/v_T = 25$, $H/\lambda_D = 1.25$, $\omega_p DT = 0.25$, $N_g = 64$, and $N_p = 2{,}000$. The plots (*a*), (*b*), (*c*), (*d*) are at 0, 50, 100, and 150 timesteps, respectively.

amplitude waves. Linearized theory predicts the relationship (2-9) for the continuum, although numerical effects can considerably modify the behavior in simulation models (see Chapter 7). In addition, the simulation model includes nonlinear effects which considerably complicate the simple relationship $\omega = \omega(k)$ predicted by linear theory.

The measurement of the frequency-wavelength relationship for a particular mode may be performed as follows: Initialize the program with a thermal distribution of particles. Excite the mode of interest by modulating the particle velocities with the required wavelength, then measure the time evolution of the potential harmonic of the same wavelength. For the weakly, damped, long-wavelength harmonics, the dominant frequency can usually be identified from the

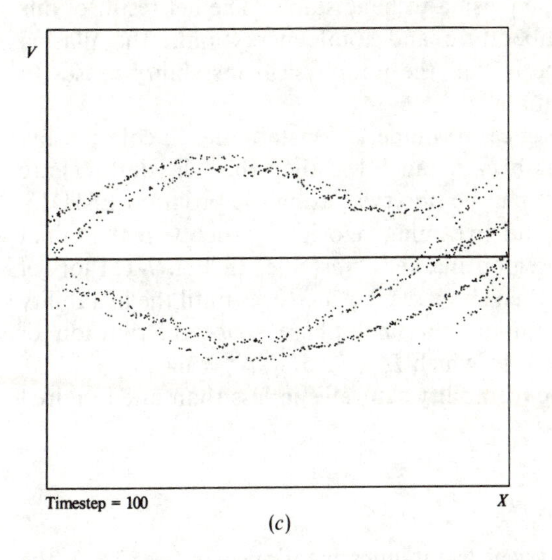

(c)

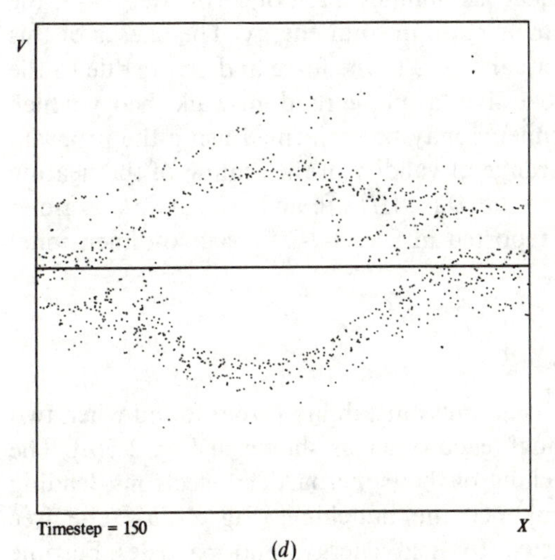

(d)

amplitude-time record after the initial transient has decayed. A more systematic method of finding the dominant frequency is to use transform methods on a segment of the amplitude-time record. For further details of time correlation measurement the reader is referred to Bracewell (1965).

2-4-3 The Cold Plasma

The mathematical limit of zero temperature which is often used to simplify the analysis of plasma phenomena is not accessible to the computer experiment. Equally, this useful mathematical fiction cannot be produced in the laboratory. If one attempts to set up a simulation with a very small plasma temperature, then numerical mode coupling causes the plasma to be unstable: The net result of this instability is to increase the temperature and total energy until the plasma becomes stable. The final outcome is that the nonphysical instability ceases to grow and saturates at a finite amplitude.

Further discussion of and references to numerical instabilities in cold plasma (i.e., plasmas where the Debye length λ_D is much less than the cell width H) are given in Sec. 7-3. These instabilities can be observed using the program ES1D1V by selecting a thermal velocity v_T and streaming velocity v_0, such that the initial state of the simulation lies in the region marked "unstable" in Fig. 7-1. Plots of kinetic, potential, and total energy then show secular increases until the instability stabilizes. Figure 2-4 shows the initial and final phase space distribution of particles for a simulation experiment in which $H = 22.5\lambda_D$, $N_g = 64$, $N_p = 1{,}000$, $\omega_p DT = 0.25$. For this example the instability saturates in less than one hundred timesteps.

2-4-4 Energy Conservation

Even in the absence of the nonphysical instabilities mentioned in Sec. 2-4-3, the simulation would still show a secular increase in total energy. The causes of this increase are fluctuations in the numerical errors in the force and errors due to the finite size of the timestep. Section 7-6-1 gives a simple random-walk theory which describes the energy increase. Experiments may be performed using the program ES1D1V (Fig. 3-10) to examine the range of validity of the scaling of the heating rate [Eq. (7-140)] with the parameters $(\omega_p DT)$, (H/λ_D), and $(n_0\lambda_D)$, in a manner similar to the empirical correlation reported in Sec. 9-2-2 for a two-dimensional electrostatic simulation.

2-4-5 The Two-Stream Instability

The simplest physical electrostatic plasma microinstability is that found when two equal streams of electrons flow through each other as shown in Fig. 2-5(a). The interaction of the beams causes bunching of the (superparticle) electrons, leading to potential wells which further enhance the bunching (Fig. 2-5b). The free streaming energy is rapidly converted to field energy, and particles become trapped in the potential wells (Fig. 2-5*c*). As time progresses the energy becomes concentrated in one long wavelength vortex in x-v phase space (Fig. 2-5*d*).

Further information on the linear theory of two-stream instabilities can be found in most standard texts on plasma physics (e.g., Clemmow and Dougherty, 1969). For those interested in experimenting with two-stream instabilities, we suggest the implementation of a one-dimensional simulation program. In addition to the equal beam case, the one-dimensional electrostatic model can be used for unequal beams, for ion–electron and ion–ion instabilities (using a reduced ion–electron mass ratio to limit the amount of computer time needed).

CHAPTER

THREE

THE SIMULATION PROGRAM

3-1 INTRODUCTION

Features common to the majority of computer programs used for computer experiments are their large size and their complexity. A typical program is several thousand card images long and is the accumulation of man-years rather than man-hours of effort. Clearly, careful planning and programing are needed if the effort invested is to lead to useful physical results.

Although we are primarily concerned here with particle models, the approach to writing simulation programs we discuss in this chapter finds much wider application. This approach may be summarized as "top-downwards" program planning followed by top-downwards programming, testing, and recursive refinement. Immediate benefits of this method of programming are that the evolving program provides its own "testbed," enabling it to be systematically tested as it is being developed and that the natural outcome is a well-structured program.

The program planning stage we have already met: Fig. 1-1 summarizes the steps from the physical phenomenon to the numerical algorithms. In Chapter 2 these steps were followed through in some detail for the one-dimensional NGP plasma model we shall use for our example later in this chapter. Consequently, we shall focus here on questions of planning which are centered on the transformation of the simulation model into a computer program.

The principal objective of a simulation program is to provide a means for obtaining physical results from the simulation model. For simple calculations which require less than, say, two hundred card images of coding, the steps to this objective require little more than the straightforward translation of the algorithms into computer coding. However, the relatively large and complex programs employed in multidimensional particle simulations demand a clear appreciation of the interaction of the computer, the program, and the program user. For this reason, we digress a little in the following section in order to provide a suitable context for the material presented in Secs. 3-3 and 3-4.

Section 3-3 contains a brief description of a programing system, the OLYMPUS Programming System (Roberts, 1974; Christiansen and Roberts, 1974; Hughes et al., 1974*a,b*, 1975; Roberts, 1975; Hughes, 1980*a,b*; Eastwood and Roberts, 1979), which embodies the structured top-downwards approach to programming we advocate. Section 3-4 shows the methodology put into practice for the one-dimensional plasma simulation program, ES1D1V.

3-2 PROGRAM REQUIREMENTS AND SPECIFICATION

3-2-1 User's Requirements

We first look at the simulation program from the point of view of the user. Typically, he will be a physicist, chemist, or engineer. His primary interest is the physical aspect of the simulation rather than the numerical and computational. He should be free to concentrate on his speciality, and for this he should be able to call upon and use a large selection of previously constructed and tested programs. He should not have to spend unnecessary time and effort either in finding out in detail how each individual program operates or in programming his own version. However, he must be able to establish readily the precise scientific assumptions adopted in any program he uses, and be able to vary or extend them to suit his needs. Similarly, he must be aware of the limitations of the numerical schemes employed in order that he, or his colleagues, may modify or replace them in applications where they prove inadequate.

The major difference between large scientific programs and most large programs in other applications is their variability. A large simulation program is rarely run more than once in the same form. The user is always pushing forward into some previously unexplored area of investigation. He calls upon many variations of scientific assumptions, initial conditions, configurations, and numerical schemes, as well as changes in output and program control. As far as he, the user, is concerned, simulation programs should have a clear standard modular structure and the support of an extensive library of previously verified programs (or program modules).

The user's requirements of a simulation program may be summarized by the conditions that the program is easy to read, easy to use, and easy to adapt. His overall view of the program is influenced by how well these conditions are met. At the topmost level, he can treat the program as a "black box" which will produce physical results from the input data he provides. At the next level, the program is seen to consist of the principal control structure, the principal data, and a set of ("black box") modules. The principal control structure defines the flow through the program, the principal data defines the state of the physical system, and the modules (comprising sequences of instructions) act as high-level operators acting on the principal data. For example, the timestep loop of the model described in the last chapter may be viewed as a high-level operator whose action is to map the state of the system forward in time by the timestep *DT*. Each module, when looked at in more detail, may be repeatedly split into control structure, data, and operators until individual variables and instructions are reached.

The conflict between the user requirements—programs which are easy to read, use and adapt—and computer requirements—data storage and operations minimization—is largely avoided by systematic top-downwards programming. User requirements impinge largely on the topmost levels where instructions are executed infrequently, whereas computer constraints are mainly concerned with the bottommost level of mapping the often-executed parts of the algorithms into program instructions.

In practice, the user requirements are often neglected. Programs tend to develop from a nucleus around which program architecture, control structure, and data organization evolve in a haphazard fashion. The dual purpose of high-level languages, that of making the code intelligible to both man and machine, is generally not used to full advantage. Often, documentation is sparse and is not updated as the program is being developed. Indeed, many programs become unusable because unrecorded changes destroy the validity of both the program development tests and the documentation. These problems largely disappear when

Table 3-1 Specifications for systematic programming

Specification	Action
Portability	Use a high-level language. Segregate system-dependent and -independent parts.
Immutability	Make the SPF invariant. Authentication code for checking.
Flexibility	Design flexibility into program control. Use edit files and system-independent editor.
Architecture	Use universal control structure. Orientate operator and operand organization to user.
Control	Use universal control structure. Build in switches and diagnostic points.
Diagnostics	Use standard dumping utilities.
Notation	Initial letter conventions. Mnemonics.
Layout	Standard indentations, ruled lines, etc.
Documentation	Indexes, cross references, decimal numbering. Headings and subheadings.
Construction	Top downwards from skeleton.
Testing	Top downwards using diagnostic dumping.
Automatic program generation	Code and documentation from master indexes. Automatic resetting of layout and labels.

the OLYMPUS Programming System based on the specifications listed in Table 3-1 is employed.

3-2-2 Program Specifications

The first three items listed in Table 3-1 refer principally to the topmost level, the complete program. Specification of the architecture, control, and diagnostics becomes relevant at the next level down when the framework into which the various program modules are to be fitted is considered. Notation, layout, and documentation specifications are important for the clarity both of the overall structure and of the individual modules. The last three specifications pertain to the efficient creation of programs within the scheme defined by the other nine.

Portability is an essential feature of any large simulation program. Often programs are developed on one machine, and then transferred to a more powerful machine for production runs. Programs are interchanged between laboratories having different computer installations. Even programs developed and run on a particular computer at a particular location eventually have to move onto a new machine or die when the computer is superseded.

The minimum requirement for portability is that the program is written, as far as possible, in a widely available high-level language. Here, we assume FORTRAN (N.C.C., 1970) since all present-day scientific computers are FORTRAN-orientated, although other languages, such as ALGOL 68 (Wijngaarden et al., 1976; Lindsey and Meulen, 1971) and PASCAL (Jensen and Wirth, 1978), reflect better the structured approach to programming. If system-dependent coding is necessary then it should be segregated from the machine-independent part to minimize the difficulty in transferring the program. For instance, system-dependent features such as channel conventions and supervisor calls can be avoided by symbolic channel numbers and by using high-level language calls to a suite of utility subprograms, respectively.

Immutability is necessary if a program user is to have confidence in the results he obtains and *flexibility* is needed to allow him to adapt the program to new situations. Immutability and portability can be obtained by establishing a system-independent standard program file (SPF) whose invariance can be tested by a code number generated by a checking program (Eastwood and Roberts, 1979). Flexibility can be introduced by separate edit files (Fig. 3-1) and by anticipation in the design of the program control structure.

All the programs for the simulation models treated in this book are alike in that they have an initialization stage followed by a timestepping loop and a closedown stage. They have control subprograms directing the flow of the calculation and output routines to transmit information to the user. If the various simulation programs are made to have in detail the same *architecture* then those programs become much easier to understand and use. This standardization of architecture can be extended to the operators (subprograms) and operands (data) of the program, as will be shown in Sec. 3-3.

Standardization of program architecture enables most of the *control* to

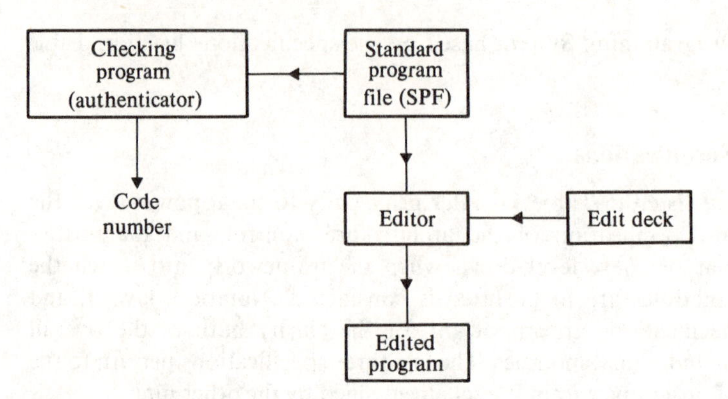

Figure 3-1 Portable and immutable programs can be obtained by defining a system-independent program file (SPF) whose invariance can be checked by a hash-code-generating checking program. Flexibility is introduced by an edit deck and editor which produces a temporary edited program for execution. (*After Roberts, 1975, reproduced by permission of Atom.*)

become part of a universal application-independent skeleton program from which programs for particular applications may be developed. Logical switches and *diagnostic* points in the skeleton program automatically provide a high degree of flexibility in application programs developed in this manner. If, as is the case for OLYMPUS, simple diagnostic utility subprograms are included in the skeleton program, then the skeleton program provides an excellent environment for efficient program development.

The similarity of architecture and control structure of applications programs originating from a universal skeleton program is reinforced if standards for *notation*, *layout*, and *documentation* are also established. The result is programs which are more readable and hence more easy to understand, use, and adapt. Although use of extensive documentation and conventions for layout and notation may appear at first sight to be a waste of time, the opposite is generally true. Notation and layout conventions reduce the frequency of programming errors and documentation helps maintain well-structured coding. Moreover, if the conventions are embedded in a programming system, then software utilities can be developed to *automatically generate* many of the housekeeping sections of programs and reset to standard form manually typed coding.

3-3 THE OLYMPUS PROGRAMMING SYSTEM

OLYMPUS is a FORTRAN-based programming system developed according to the specifications listed in Table 3-1. We use this particular system for illustration because (1) it embodies and promotes good programming practice, (2) it is widely used, particularly in the plasma research community, and (3) it is freely available from the Computer Physics Communications Library, Belfast.

There are four aspects to the OLYMPUS system. First and foremost,

OLYMPUS is a methodology to efficiently create well-structured programs. The other three aspects are designed to help the programmer achieve the objectives of the methodology: software to provide a system-independent environment, conventions to clarify program and data structure, and further software to partially automate program construction. The methodology and conventions can be applied to programs (not necessarily FORTRAN) irrespective of the presence of the software, although the software greatly simplifies matters. We will limit ourselves in this section to a brief outline of the OLYMPUS conventions and software. The methodology we will meet in the context of the one-dimensional plasma example in the next section.

OLYMPUS programs are written in Standard FORTRAN to meet the needs of portability, ease of use, and compatibility with most existing scientific programs and program modules. An OLYMPUS program for a particular application is defined in a system-independent format known as the Standard Program File (SPF). The immutable SPF contains documentation, common blocks, FORTRAN subprograms, and test data. The OLYMPUS SPF processor converts the SPF into the form required by a particular installation and links in from the OLYMPUS libraries applications independent (but perhaps machine dependent) program modules needed to create an executable program.

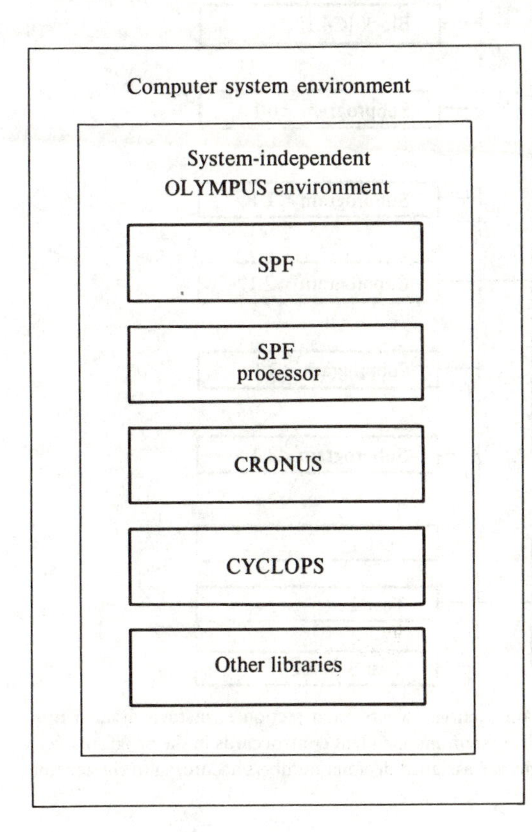

Figure 3-2 The OLYMPUS software provides a system-independent environment for the user's program. Applications programs in SPF form can be transferred to different installations and run without modification.

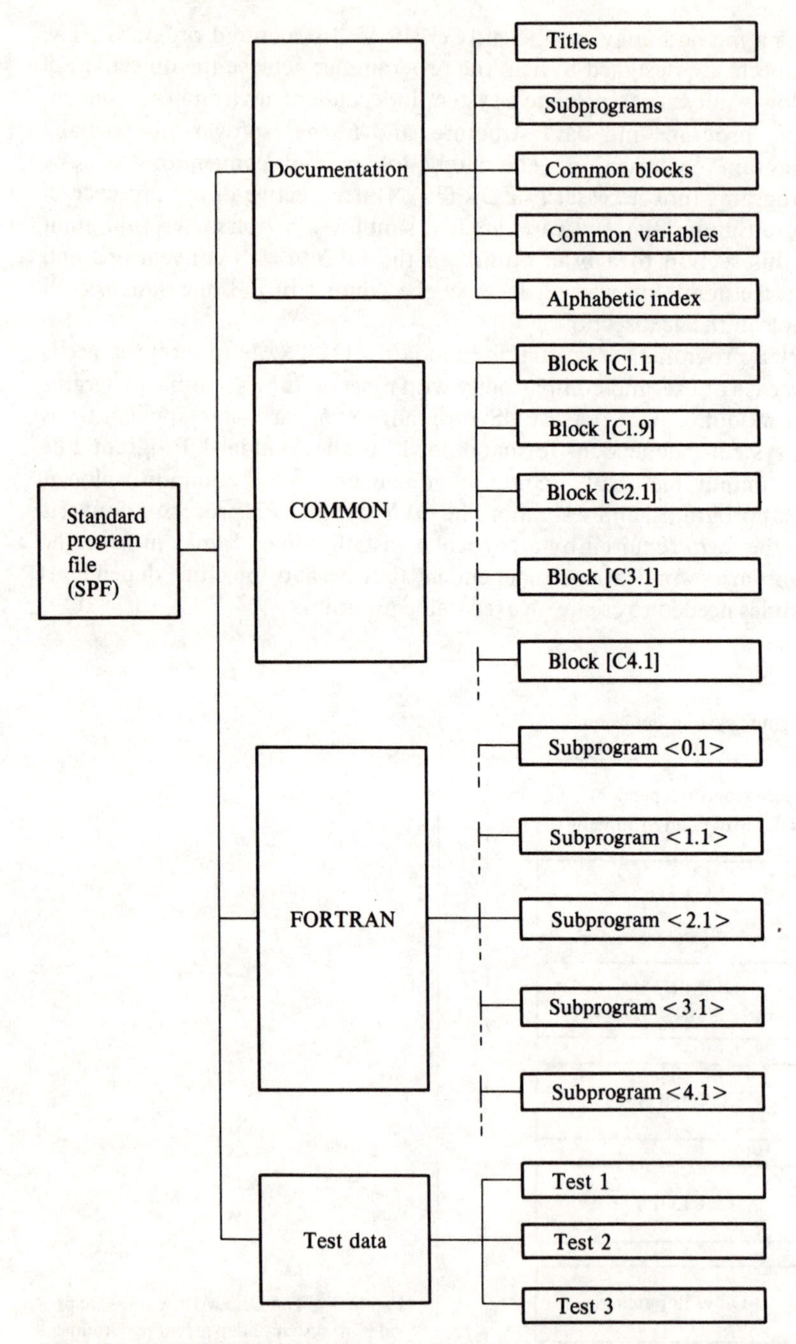

Figure 3-3 The SPF is divided into four main sections, where each section is in turn divided into modules. Sections and modules are preceded by system-independent control cards in the program deck. Common blocks and FORTRAN subprograms are assigned decimal numbers according to the scheme given in Table 3.2.

The simple, and largely FORTRAN, software of the OLYMPUS system provides an interface between the SPF and the computer system. SPF programs can be freely interchanged between computers where the OLYMPUS environment is established. This is shown schematically in Fig. 3-2. The box labeled "CRONUS" in Fig. 3-2 refers to the universal skeleton program which defines the basic architecture for all OLYMPUS programs and that labeled "CYCLOPS" refers to the library of OLYMPUS utilities designed to speed up program development. "Other libraries" refers both to the OLYMPUS software used to automate program construction and to the growing body of available applications programs and program modules.

Figure 3-3 shows the structure of the standard program file (SPF). The SPF is divided into four main sections: documentation, common blocks, FORTRAN subprograms, and test data. Each section is divided into subsections. The main sections and subsections are headed by system-independent control statements beginning "C/" and "C/ MODULE," respectively. These control statements enable the SPF processors to convert the SPF into the forms required for execution on particular installations.

The "Documentation" section contains a title module identifying the programs and a set of modules giving indexes of subroutines, of common blocks, of variables in each common block and of all common variables listed in alphabetic order. Each module of the "COMMON" section contains a documented common block and each module of the FORTRAN section contains a FORTRAN subprogram. Subprograms are regarded as operators and the data in the labeled common blocks are the operands. As far as possible, each subprogram and each

Table 3-2 The OLYMPUS division of subroutines into classes and common blocks into groups

Instructions		
		0 Control
		1 Prologue
		2 Calculation
	Classes of subprograms	3 Output
		4 Epilogue
		5 Diagnostics
		U Utilities
Data		
		1 General OLYMPUS data
		2 Physical problem
	Groups of labeled common blocks	3 Numerical scheme
		4 Housekeeping
		5 I/O and diagnostics
		6 Text manipulation

Source: Roberts (1974); courtesy of *Computer Phys. Commun.*

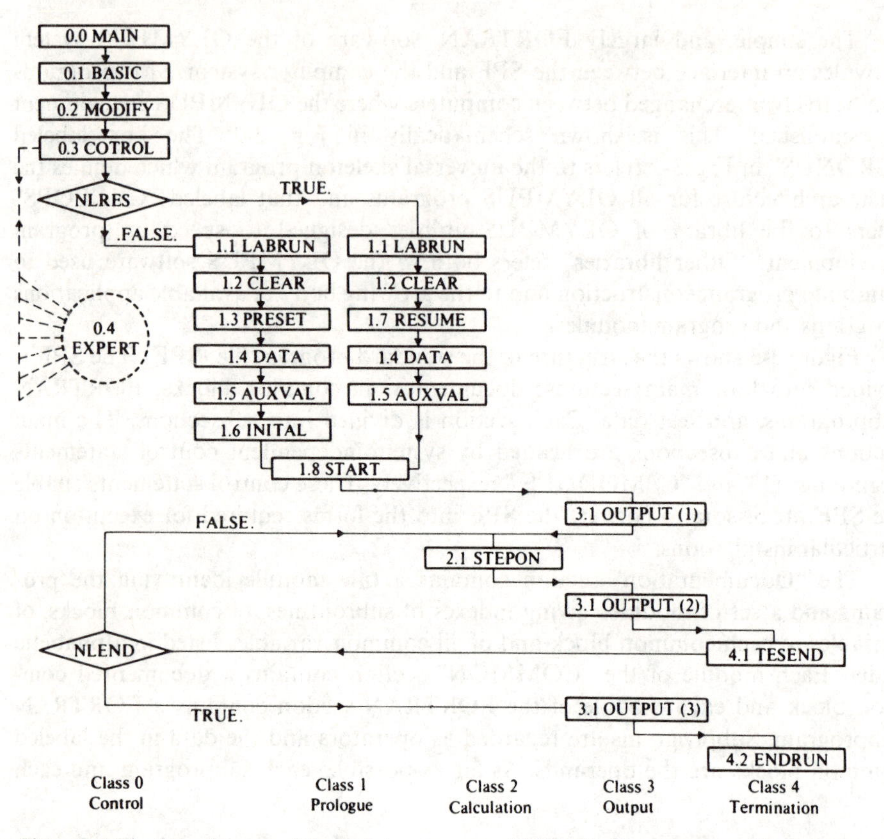

Figure 3-4 The flow diagram of the skeleton program, CRONUS. The purposes of the subprograms are listed in Table 3-3. (*After Christiansen and Roberts, 1974, courtesy of Computer Physics Communications.*)

common block has a well-defined purpose which is reflected by decimal numbers ascribed to them. Subprograms are divided into classes and are given decimal numbers ⟨class.subprogram⟩. Common blocks are divided into groups and are labeled [group.block]. Classes of subprograms and groups of common blocks are listed in Table 3-2.

The standard OLYMPUS skeleton SPF, CRONUS, sets the basic architecture for all OLYMPUS programs. CRONUS (Fig. 3-4 and Table 3-3) contains the universal OLYMPUS control ([C1.1] COMBAS) and diagnostic ([C1.9] COMDDP) common blocks, the universal (Class 0) control subprograms (⟨0.0⟩ – ⟨0.3⟩), the standard diagnostic structure (Fig. 3-6), plus a set of dummy subprograms with standard names and numbering. A listing of the standard form of the control subprogram, ⟨0.3⟩ COTROL, is shown in Fig. 3-5. In any "real" OLYMPUS program the dummies in CRONUS are replaced, as necessary, by the SPF versions with the same name and decimal numbering. Thus, the programmer would supply his own version of ⟨2.1⟩ STEPON which controls

Table 3-3 PROGRAM CRONUS: subprogram names and identification numbers

Name	No.	Dummy arguments	Purpose	Status
		Class 0 Main Control		
(MAIN)	0.0		Fortran main program	P
BASIC	0.1		Initialize basic control data	P
MODIFY	0.2		Modify basic data if required	D
COTROL	0.3		Control the run	P
EXPERT	0.4	KCLASS, KSUB, KPOINT	Modify standard operation of program	M
		Class 1 Prologue		
LABRUN	1.1		Label the run	D
CLEAR	1.2		Clear variables and arrays	D
PRESET	1.3		Set default values	D
DATA	1.4		Define data specific to run	D
AUXVAL	1.5		Set auxiliary values	D
INITAL	1.6		Define physical initial conditions	D
RESUME	1.7		Resume run from previous record	D,M
START	1.8		Start the calculation	D
		Class 2 Calculation		
STEPON	2.1		Step on the calculation	D
		Class 3 Output		
OUTPUT	3.1	K	Control the output	D
		Class 4 Epilogue		
TESEND	4.1		Test for completion of run	M
ENDRUN	4.2		Terminate the run	M
		Class 5 Diagnostics		
REPORT	5.1	KCLASS, KSUB, KPOINT	Control the diagnostics	M
CLIST	5.2	KGROUP, KBLOCK	Print COMMON variables	D
ARRAYS	5.3	KGROUP, KBLOCK	Print COMMON arrays	D

Status codes

D Dummies; replace by "real" versions provided by the programer.

M Can be left in, or modified by the programer as required.

P Permanent control routine; should not require alteration.

Source: Christiansen and Roberts (1974); courtesy of *Computer Phys. Commun.*

the timestep loop and calls in other routines, ⟨2.2⟩, ⟨2.3⟩, etc., to do the actual work. ⟨3.1⟩ OUTPUT organizes the output, and so forth.

The disciplined approach adopted in the OLYMPUS system extends to the notation, layout, and documentation of the subprograms and common blocks.

```
C/ MODULE COS3
C
      SUBROUTINE COTROL
C
C 0.3  Control the run
C
C     Version 2a          1/8/73      KVR/JPC        Culham
C
C/ INSERT COMBAS
C/ INSERT COMDDP
C-----------------------------------------------------------------------
C
      DATA ICLASS,ISUB/0,3/
      CALLMESAGE(48H    0.3 ENTER RUN CONTROL                         )
C
C-----------------------------------------------------------------------
CL              1.        Prologue
C
      IF(NLRES) GO TO 170
C
C               A.         New run
C
CL                    1.1      Label the run
  110 CALL LABRUN
                                   CALL EXPERT(ICLASS,ISUB,1 )
C
CL                    1.2      Clear variables and arrays
  120 CALL CLEAR
                                   CALL EXPERT(ICLASS,ISUB,2 )
C
CL                    1.3      Set default values
  130 CALL PRESET
                                   CALL EXPERT(ICLASS,ISUB,3 )
C
CL                    1.4      Define data specific to run
  140 CALL DATA
                                   CALL EXPERT(ICLASS,ISUB,4 )
C
CL                    1.5      Set auxiliary values
  150 CALL AUXVAL
                                   CALL EXPERT(ICLASS,ISUB,5 )
C
CL                    1.6      Define physical initial conditions
  160 CALL INITAL
                                   CALL EXPERT(ICLASS,ISUB,6 )
      GO TO 180
C
C               B.         Resume a previous run
C
CL                    1.7      Pick up record, modify required parameters
  170 CONTINUE
                                   CALL EXPERT(ICLASS,ISUB,7 )
C     Label the continuation run
      CALL LABRUN
```

Figure 3-5 The default form of subprogram ⟨0.3⟩ COTROL of CRONUS. (*From Christiansen and Roberts, 1974, courtesy of Computer Physics Communications.*)

```
                                        CALL EXPERT(ICLASS,ISUB,8 )
C     Clear variables and arrays
         CALL CLEAR
                                        CALL EXPERT(ICLASS,ISUB,9 )
C     Pick up record and print details
         CALL RESUME

                                      CALL EXPERT(ICLASS,ISUB,10)
C     Read any new data needed
         CALL DATA
                                      CALL EXPERT(ICLASS,ISUB,11)
C     Modify auxiliary variables as required
         CALL AUXVAL
                                      CALL EXPERT(ICLASS,ISUB,12)
C
C               C.        Preliminary operations
C
CL                  1.8      Start or restart the run
  180    CALL START
                                      CALL EXPERT(ICLASS,ISUB,13)
C     Initial output
         CALL OUTPUT(1)
                                      CALL EXPERT(ICLASS,ISUB,14)
C
C
C-----------------------------------------------------------------------
CL              2.        Calculation
C
CL                  2.1      Step on the calculation
  210    CALL STEPON
                                      CALL EXPERT(ICLASS,ISUB,15)
C
C-----------------------------------------------------------------------
CL              3.        Output
C
CL                  3.1      Periodic production of output
  310    CALL OUTPUT(2)
                                      CALL EXPERT(ICLASS,ISUB,16)
C
C-----------------------------------------------------------------------
CL              4.        Epilogue
C
CL                  4.1      Test for completion of run
  410    CALL TESEND
                                      CALL EXPERT(ICLASS,ISUB,17)
         IF(.NOT.NLEND) GO TO 210
C
C     Final output
         CALL OUTPUT(3)
                                      CALL EXPERT(ICLASS,ISUB,18)
C
CL                  4.2      Terminate the run
  420    CALL ENDRUN
C
         RETURN
         END
```

Table 3-4 The OLYMPUS initial letter convention

	Real, complex	Integer (and Hollerith)	Logical
Subprogram dummy arguments	*P*	*K*	*KL*
Common variable and array names	*A–H, O, Q–Y*	*L, M, N*	*LL, ML, NL*
Local variable and array names	*Z*	*I*	*IL*
Loop indexes		*J*	

Source: Christiansen and Roberts (1974); courtesy of *Computer Phys. Commun.*

Table 3-5 The CYCLOPS utility subroutines.
They are designed for four functions (1) output, (2) array manipulation, (3) timing, and (4) diagnostics

No.	Name	Arguments	Use
1	RVAR	KNAME, PVALUE	Print real variable and value.
2	IVAR	KNAME, KVALUE	Print integer variable and value.
3	HVAR	KNAME, KVALUE	Print hollerith variable and value.
4	LVAR	KNAME, KLVAL	Print logical variable and value.
5	RARRAY	KNAME, PA, KDIM	Print real array and values.
6	IARRAY	KNAME, KA, KDIM	Print integer array and values.
7	HARRAY	KNAME, KA, KDIM	Print hollerith array and values.
8	LARRAY	KNAME, KLA, KDIM	Print logical array and values.
9	RARAY2	KNAME, PA, KDIMX, KX, KY	Print doubly subscripted real array.
10	RESETR	PA, KDIM, PVALUE	Reset real array to specified value.
11	RESETI	KA, KDIM, KVALUE	Reset integer array to specified value.
12	RESETH	KA, KDIM, KVALUE	Reset hollerith array to specified value.
13	RESETL	KLA, KDIM, KLVAL	Reset logical array to specified value.
14	SCALER	PA, KDIM, PC	Scale a real array by real value.
15	SCALEI	KA, KDIM, KC	Scale an integer array by integer value.
16	COPYR	PA1, K1, PA2, K2, KDIM	Copy one real array into another.
17	COPYI	KA1, K1, KA2, K2, KDIM	Copy one integer array into another.
18	SIGNR	PA, KDIM	Change the sign of real array.
19	SIGNI	KA, KDIM	Change the sign of integer array.
20	PAGE		Fetch new page on output channel.
21	BLINES	K	Print K blank lines.
22	MESAGE	KMESS	Print 48 character message.
23	REPTHD	KCLASS, KSUB, KPOINT	Print heading for diagnostic report.
24	RUNTIM		Update CPU time and print it.
25	DAYTIM		Print date and time.
26	JOBTIM	PTIME	Fetch allocated jobtime.
27	DUMCOM	KCLASS, KSUB, KPOINT	Dump selected common block.

Source: Christiansen and Roberts (1974); courtesy of *Computer Phys. Commun.*

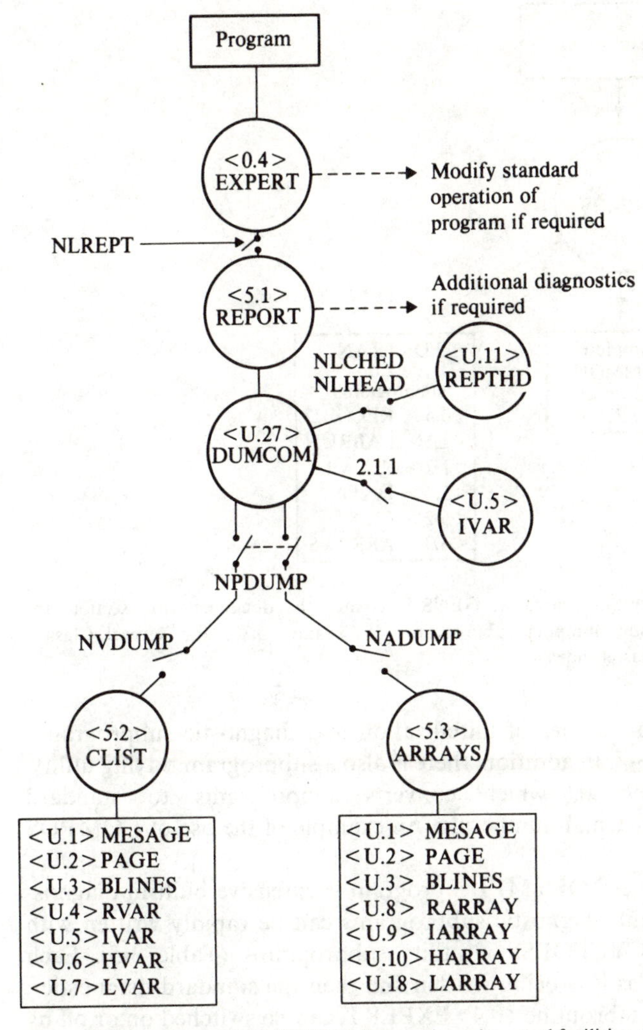

Figure 3-6 The standard OLYMPUS diagnostic and control facilities. (*After Christiansen and Roberts, 1974, courtesy of Computer Physics Communications.*)

Subprograms are divided into decimally numbered sections and subsections. Mnemonic names, the use of symbolic rather than arithmetic notation, and the initial letter conventions summarized in Table 3-4 are used to make the meaning of the code clearer. Ruled lines delineating sections and differing indentations of headings, subheadings, comments, and statements are employed to make the program more readable.

The clearly defined structure of OLYMPUS programs permits much of the tedious labor of programming to be eliminated. The automatic program generator, GENSIS (Hughes, 1980*b*), generates from a compact definition of the data structure (the master index, cf. Fig. 3-8), documentation and common sections of

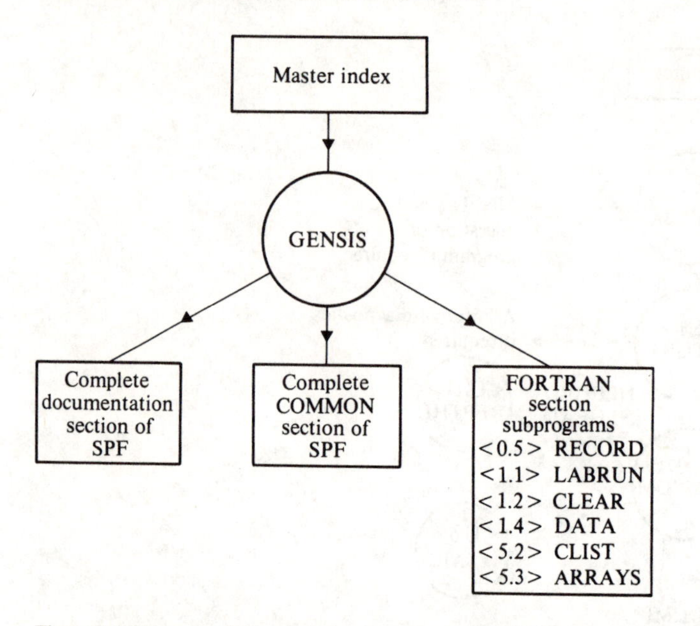

Figure 3-7 The automatic program generator, GENSIS, creates the documentation section, the common section, the diagnostic dumping subprograms ⟨5.2⟩ and ⟨5.3⟩, and several Class 1 subprograms from a compact master index.

the SPF, together with a number of initialization and diagnostic subprograms. This is illustrated in Fig. 3-7. In addition, there is also a subprogram tidying utility, COMPOS (Hughes, 1980*a*), which converts subprograms to standard OLYMPUS layout and decimal numbering. An example of the use of COMPOS is shown in Fig. 3-9.

An integral part of every OLYMPUS program is extensive built-in diagnostics. Application-dependent diagnostic subprograms can be rapidly written with the aid of the library, CYCLOPS, of utility subprograms (Table 3-5). Each program is instrumented as it is being written and then the standard diagnostics, accessed through calls to subroutine ⟨0.4⟩ EXPERT, can be switched on or off by means of logical switching variables in the common block [C1.9] COMDPP.

3-4 THE PROGRAM ES1D1V

The planning stages for ES1D1V, going through the steps from the physical phenomenon to numerical algorithm, can be summarized by lists of (1) the differential equations, (2) the discrete equations and scaling factors, and (3) the numerical algorithm steps. These lists, together with the standard structure and variables provided by OLYMPUS, give the starting point for the top-downwards and recursive refinement methodology of the OLYMPUS system.

The first step in the program development is to determine whether a suitable library program or program modules are available for the problem to be tackled.

If so, then these should provide the starting point for, or constraints on, the development of the program. For instance, if an appropriate module for solving Poisson's equation were available, then list (4) containing tables of the subprogram names, common block names, and common variable names would be added to the lists (1) to (3) mentioned above. By this means, unnecessary conflicts of names can be avoided.

```
PROGRAM ES1D1V
TITLE 1-D ELECTROSTATIC PLASMA MODEL
AUTHOR J.W.Eastwood and R.W.Hockney
VERSION 1   RWH/JWE                   Reading Univ.      1979

INDEX OF SUBPROGRAMS
2.2 POTSOL Solve for the potential

ARRAY DIMENSIONS
MAX1=256,NGMAX
MAX2=2048,NPMAX

INDEX OF COMMON VARIABLES

S 2.1 Principal Physical Quantities

COMMON/COMPPQ/
R EPSO *Free Space Permittivity (SI)
R ELMASS *Electron Mass (SI)
R ELCHAG *Electron Charge (SI)
R API Pi
R BOLTZK *Boltzmann's Constant (SI)
R DEBLNO Initial Debye Length (SI)
R BLMETR *Box length (SI)
R DENBAK *Background Density (SI)
R FREQPO Initial Plasma Frequency (SI)
R TODEGK *Initial Temperature (Deg K)
R VTHERO Initial Thermal Velocity (SI)
RA ELPOT(MAX1) Electric Potential
RA CHARG(MAX1) Charge Density
RA XPOS(MAX2) Particle Positions
RA VEL(MAX2) Particle Velocities
R VDRIFT Electron Drift Velocity (SI)

S 2.2 Secondary Physical Quantities

COMMON/COMSPQ/
R BOXLEN Box Length
R DELCHG Charge Increment Per Particle
R QBAKG Background Charge Density
R ENKIN Kinetic Energy
R ENPOT Potential Energy
```

Figure 3-8 A section of the GENSIS input for the program ES1D1V. The output is included in the program listing in Fig. 3-10.

3-4-1 The Program Control Structure

Once the lists are drawn up, and we suggest that the reader follows through this exercise by referring to Chapter 2, our attention can be turned to planning the

```
SUBROUTINE INITAL
*
* 1.6   Set Physical Initial Conditions
*
*  Note: XPOS and VEL are in dimensionless units
*INSERT COMBAS,COMDDP,COMPPQ,COMSPQ
*INSERT COMNUM,COMSCA,COMHOK,COMOUT
DATA  ICLASS,ISUB/1,6/
CALL EXPERT
IF(NLOMT1(ISUB)) RETURN
C
C--------------------------------------------------------------------
CL              1.          Two Particle Test
C
         IF(NCASE.NE.1) GO TO 10
         XPOS(1) = X2PT1
         XPOS(2) = X2PT2
         VEL(1)  = V2PT1
         VEL(2)  = V2PT2
         RETURN
#Drifting Thermal Distrbn.,Uniform Density
  10    CONTINUE
IF(NCASE.NE.2) GO TO 5
##Load Positions
CALL RANPOS(XPOS,1,NP,XMIN,XMAX)
##Load Velocities
ZVDR = VDRIFT / SCVEL
ZVTH = VTHERO / SCVEL
CALL THERDI(VEL,1,NP,ZVDR,ZVTH)
RETURN
#Two Warm Streams
5
IF(NCASE.NE.3) GO TO 15
CALL RANPOS(XPOS,1,NP,XMIN,XMAX)
ZVDR1 = VDR1 / SCVEL
ZVDR2 = VDR2 / SCVEL
ZVTH1 = SQRT(BOLTZK * TST1 / ELMASS) / SCVEL
ZVTH2 = SQRT(BOLTZK * TST2 / ELMASS) / SCVEL
CALL THERDI(VEL,1,NST1,ZVDR1,ZVTH1)
CALL THERDI(VEL,NST2,NP,ZVDR2,ZVTH2)
RETURN
#Other Cases
C
15
CALL EXPERT
*-----------
RETURN
       END
```

Figure 3-9 An example of the effect of the compositor program COMPOS. (*a*) shows the input to and (*b*) shows the output from COMPOS.

```
C
      SUBROUTINE INITAL
C
C 1.6  Set Physical Initial Conditions
C
C    Note: XPOS and VEL are in dimensionless units
C/ INSERT COMBAS
C/ INSERT COMDDP
C/ INSERT COMPPQ
C/ INSERT COMSPQ
C/ INSERT COMNUM
C/ INSERT COMSCA
C/ INSERT COMHOK
C/ INSERT COMOUT
     DATA       ICLASS,    ISUB/1,    6/
                                      CALL EXPERT(ICLASS,ISUB,1)
      IF(NLOMT1(ISUB)) RETURN
C
C-----------------------------------------------------------------------
CL          1.          Two Particle Test
C
      IF(NCASE.NE.1) GO TO 200
      XPOS(1) = X2PT1
      XPOS(2) = X2PT2
      VEL(1) = V2PT1
      VEL(2) = V2PT2
      RETURN
C
C-----------------------------------------------------------------------
CL          2.          Drifting Thermal Distrbn.,Uniform Density
C
  200 CONTINUE
      IF(NCASE.NE.2) GO TO 300
C
CL          2.1         Load Positions
      CALL RANPOS(XPOS,1,NP,XMIN,XMAX)
C
CL          2.2         Load Velocities
      ZVDR = VDRIFT / SCVEL
      ZVTH = VTHERO / SCVEL
      CALL THERDI(VEL,1,NP,ZVDR,ZVTH)
      RETURN
C
C-----------------------------------------------------------------------
CL          3.          Two Warm Streams
C
  300 CONTINUE
      IF(NCASE.NE.3) GO TO 400
      CALL RANPOS(XPOS,1,NP,XMIN,XMAX)
      ZVDR1 = VDR1 / SCVEL
      ZVDR2 = VDR2 / SCVEL
      ZVTH1 = SQRT(BOLTZK * TST1 / ELMASS) / SCVEL
      ZVTH2 = SQRT(BOLTZK * TST2 / ELMASS) / SCVEL
      CALL THERDI(VEL,1,NST1,ZVDR1,ZVTH1)
      CALL THERDI(VEL,NST2,NP,ZVDR2,ZVTH2)
      RETURN
C
C-----------------------------------------------------------------------
CL          4.          Other Cases
C
  400 CONTINUE
                                      CALL EXPERT(ICLASS,ISUB,2)
C-----------------------------------------------------------------------
      RETURN
      END
```

overall program structure. Since ES1D1V is a straightforward initial-value–boundary-value calculation, the control structure (see Fig. 3-5) and principal subprogram names used in CRONUS (Fig. 3-4) will become the flow diagram of ES1D1V. As the program is developed, the dummy subprograms of CRONUS (cf. Table 3-3) are progressively replaced by and supplemented with the working subprograms of ES1D1V. In larger programs, a number of the CRONUS subprograms, such as ⟨2.1⟩ STEPON, would simply become control subprograms. STEPON would call a sequence of subprograms ⟨2.2⟩, ⟨2.3⟩, etc. which would do the numerical computations: In the present simple example (cf. Fig. 3-10) the majority of the calculation is performed within ⟨2.1⟩ STEPON, with only the standard diagnostic subprogram, ⟨0.4⟩ EXPERT, and potential solver subprogram, ⟨2.2⟩ POTSOL, being called by it.

3-4-2 The Master Index

The third stage of the program development is to sweep through the lists (1) to (4) again to obtain a master index of the data. The first level of the data organization is provided by the OLYMPUS common block groups listed in Table 3-2. In each of these groups common blocks and then the variables used within the program must be defined.

The universal-control common blocks [C1.1] COMBAS and [C1.9] COMDDP are already defined in CRONUS. COMBAS contains basic OLYMPUS data and COMDDP contains diagnostic and development parameters. Variables in groups 2 to 4 correspond almost exactly to the variables introduced by the mathematical model (list 1), the discretization (list 2), and the numerical algorithms (list 3), respectively.

List (1) [cf. Eqs. (3-5) to (3-17)] introduces physical constants and variables. An appropriate grouping of these quantities is achieved by introducing, for example, the common blocks [C2.1] COMPPQ for principal and [C2.2] COMSPQ for secondary physical quantities. Discretization brings in cell widths, timestep, mesh points, superparticles, etc., and the scaling factors: Hence common blocks [C3.1] COMNUM and [C3.2] COMSCA. Further blocks that are likely to be needed are ones for housekeeping variables ([C4.1] COMHOK) and output ([C5.1] COMOUT). In larger programs, the choice of common blocks would be influenced by the need for modularity: Secondary physical, numerical, and housekeeping variables used for intercommunication between a group of subprograms performing a single well-defined task (i.e., within a module) would logically be assigned their own common blocks to minimize chances of accidental corruption.

The variables and arrays of the group 1 common blocks, [C1.1] COMBAS and [C1.9] COMDDP, are defined and initialized in the dummy program CRONUS. Their name, type, dimension, purpose, and default values are summarized in Table 3-6. The first quantity in [C1.1] COMBAS, ALTIME, is set up by a call to the utility subprogram ⟨U.17⟩ JOBTIM already built into the Class 0 subprogram ⟨0.0⟩ MAIN of CRONUS (Fig. 3-4). Its value is the CPU time remaining at the start of the "go" step. It may be used in ⟨4.1⟩ TESEND in

Table 3-6 Master indexes for the standard OLYMPUS common blocks [C1.1] COMBAS and [C1.9] COMDDP

Block	Name	Type	Dimension	Purpose	Preset value
[C1.1] COMBAS	ALTIME	R		CPU time allocated for job (s)	Supervisor Call
	CPTIME	R		CPU time used so far (s)	0.0
	NLEDGE	I		Channel for restart records	30
	NLEND	L		.TRUE. if run to be terminated	.FALSE.
	NLRES	L		.TRUE. if run being resumed	.FALSE.
	NONLIN	I		Channel for on-line input/output	1
	NOUT	I		Current output channel	NPRINT
	NPRINT	I		Channel for printed output	99
	NREAD	I		Channel for card input (or equivalent)	97
	NREC	I		Current record number	1
	NRESUM	I		Resume from record on this channel	NLEDGE
	NSTEP	I		Current step number	0
	STIME	R		Start time (s)	0.0
	LABEL1	IA	12	Labels used to describe the run, set by program in LABRUN	
	LABEL2	IA	12		
	LABEL3	IA	12		
	LABEL4	IA	12		
	LABEL5	IA	12	Labels available to programmer	Blank
	LABEL6	IA	12		
	LABEL7	IA	12	Labels reserved for system use	
	LABEL8	IA	12		
	NDIARY	I		Channel for diary	NPUNCH
	NIN	I		Current input channel	NREAD
	NPUNCH	I		Channel for punched card output, (or equivalent)	98
	NRUN	I		Maximum number of steps	1
[C1.9] COMDDP	MAXDUM	I		Maximum dimension of dump arrays	20
	MXDUMP	I		Actual dimension of dump arrays	10
	NADUMP	IA	20	Codes for array dumps	0
	NCLASS	I		Most recent class reported	0
	NPDUMP	IA	20	Codes for dumping points	0
	NPOINT	I		Most recent point reported	0
	NSUB	I		Most recent subprogram reported	1
	NVDUMP	IA	20	Codes for dumping variables	0
	NLCHED	L		.TRUE. if report heads need for central class	.FALSE.
	NLHEAD	LA	9	.TRUE. if report heads needed for class 1–9	.FALSE.
	NLOMT1	LA	50	.TRUE. if subprogram in class 1 to be omitted	.FALSE.
	NLOMT2	LA	50	.TRUE. if subprogram in class 2 to be omitted	.FALSE.
	NLOMT3	LA	50	.TRUE. if subprogram in class 3 to be omitted	.FALSE.
	NLREPT	L		.TRUE. if any reports required	.FALSE.

Source: Christiansen and Roberts (1974); courtesy of *Computer Phys. Commun.*

conjunction with CPTIME (whose value is updated and printed by ⟨U.12⟩ RUNTIM) to close down a run if the CPU time is being exhausted before the planned number of calculation steps is completed. A third timing variable, STIME, is included in [C1.1] to store the reference zero time needed on some machines. The purpose of the logical switches NLEND and NLRES should be clear from Fig. 3-4, and the various input and output channels are self-explanatory—NREAD, NOUT, and NPRINT being the ones of principal interest to us here. The current step number, NSTEP, is incremented by one in ⟨2.1⟩ STEPON, and NRUN contains the maximum number of steps. The "LABELn" arrays are used to store 48 character labels.

Elements of [C1.9] are all concerned with the extensive development and diagnostic aids built into CRONUS. MAXDUM and MXDUMP are symbolic dimensions used for clearing and dumping the dump arrays NADUMP, NPDUMP, and NVDUMP. Unless we plan to have more than ten dumps in any one run we need not concern ourselves with them. NCLASS, NSUB, and NPOINT are tracer variables recording the most recent class, subprogram, and point in the subprogram from which ⟨0.4⟩ EXPERT was called. Logical arrays NLOMT1, NLOMT2, NLOMT3 provide the facility for switching off chosen subprograms. For instance, if we want to switch off ⟨1.6⟩ INITAL then we would set (via input to ⟨1.4⟩ DATA) NLOMT1(6) = .TRUE..

The uses of the remaining elements of [C1.9] are summarized by Fig. 3-6. NLREPT is the "master switch" on reports. If any reports are needed it must be set to .TRUE.. ⟨5.1⟩ REPORT calls the dumping utility ⟨U.27⟩ DUMCOM. If NLCHED is set to .TRUE., then messages of the form

CLASS = 0, SUBPROGRAM = 3, POINT = 6

are printed at each call of ⟨5.1⟩ REPORT (this example corresponds to the call to EXPERT at point 6 in ⟨0.3⟩ COTROL). The array NLHEAD provides a similar facility for other classes of subprograms. The call to ⟨0.4⟩ EXPERT from point 1 in ⟨2.1⟩ STEPON causes

STEP = (value of NSTEP)
****

to be printed if NLREPT = .TRUE..

Points at which selected variables and arrays are to be dumped are specified by packing codes into the array NPDUMP. For example, suppose we want two dumps, the first at (class.subprog.point) = 0.3.6 and the second at 3.1.4, then we set

NPDUMP(1) = 306 (3-1)

NPDUMP(2) = 30104 (3-2)

class subprogram point

The information to be dumped at these points is specified by packing codes into the corresponding elements of NVDUMP and NADUMP. For instance, to dump all variables in block [C2.1], all group 3 blocks and block [C4.1] at dumping point

1, we would set

$$\text{NVDUMP(1)} = 213041 \qquad (3\text{-}3)$$

block [C2.1] block [C4.1]

zero denotes dumping all group 3 variables

The same code format is used in NADUMP. The code 100 is used to denote "dump everything." Thus, if we wanted to dump all arrays at point 2 we would set

$$\text{NADUMP(2)} = 100 \qquad (3\text{-}4)$$

In general, up to four requests, each of two digits, can be packed into each element of NVDUMP and NADUMP. Expression (3-3) causes ⟨U.27⟩ DUMCOM to issue the calls

```
CALL CLIST (2,1)
CALL CLIST (3,0)
CALL CLIST (4,1)
```

while (3-4) leads to

```
CALL ARRAYS (0,0)
```

The subprograms ⟨5.2⟩ CLIST and ⟨5.3⟩ ARRAYS are automatically generated from a master index of variables by the Automatic Program Generator, GENSIS, and the remainder of the standard diagnostics facilities are provided in the skeleton program CRONUS. Thus, once a master index is formed, the OLYMPUS software immediately provides an environment for testing each subprogram as it is written. Even in the absence of GENSIS, the CYCLOPS utilities (Fig. 3-5) and the skeleton forms in CRONUS reduce the process of writing ⟨5.2⟩ CLIST and ⟨5.3⟩ ARRAYS to a simple mechanical process once the master index is established.

The next stage of refinement of the master index of the data is achieved by again bringing together the evolving index and the lists which summarize the planning stages. This time we correlate the detailed physical, numerical, and algorithmic descriptions with the member common blocks of the appropriate groups in order to identify and name constants, variables, and arrays which are required.

The mathematical description of the one-dimensional electron plasma we obtained in Chapter 2 is summarized by the following equations:

$$\frac{dx}{dt} = v \qquad \frac{dv}{dt} = a \qquad (3\text{-}5)\quad (3\text{-}6)$$

$$a = -\frac{q}{m}\frac{d\phi}{dx} \qquad \frac{d^2\phi}{dx^2} = -\frac{\rho}{\epsilon_0} \qquad (3\text{-}7)\quad (3\text{-}8)$$

$$\rho = \rho_0 + qn_e = -qn_0 + qn_e \qquad (3\text{-}9)$$

The particle coordinates $\{x\}$ and the potentials $\{\phi\}$ are subject to periodic boundary conditions over the computational box of length L:

$$x_{\min} \leqslant x \leqslant x_{\max} \tag{3-10}$$

$$\phi(x) = \phi(x+L) \tag{3-11}$$

where

$$x_{\max} = x_{\min} + L \tag{3-12}$$

The initial configuration is usually characterized by bulk quantities: the background density n_0 and the distribution of electron velocities. The distribution of velocities is generally described by the first two velocity moments; the drift velocity v_d and the root-mean-squared (thermal) velocity v_T, where v_T is related to the temperature T by

$$v_T = \sqrt{\frac{k_B T}{m_e}} \tag{3-13}$$

where k_B is Boltzmann's constant. Length and time scales are characterized by the Debye length λ_D and plasma frequency ω_p, respectively:

$$\lambda_D = \sqrt{\frac{\epsilon_0 k_B T}{nq^2}} \qquad \omega_p = \sqrt{\frac{nq^2}{\epsilon_0 m_e}} \qquad \text{(3-14)} \quad \text{(3-15)}$$

The total kinetic energy per electron $\mathcal{T}$ is given by

$$\mathcal{T} = \tfrac{1}{2} m_e v_T^2 = \tfrac{1}{2} k_B T \tag{3-16}$$

and the total potential energy per electron V is given by

$$V = \frac{1}{2n_0 L} \int dx\, \rho\phi \tag{3-17}$$

Equations (3-5) to (3-17) constitute list (1). We infill [C2.1] COMPPQ and [C2.2] COMSPQ by methodically transferring variables and constants from these equations to the master index sections for the two common blocks, together with information describing the meaning and type of the variables. The same procedure is followed for the group 3 and 4 common blocks. Equations (2-32) to (2-37) give the discrete algebraic description of the plasma. The numerical parameters block [C3.1] COMNUM contains the numerical quantities introduced into the mathematical equations by discretization, the block [C3.2] COMSCA contains scaling factors, and so forth.

The result of this exercise is the initial version of the master index input data for the automatic program generator, GENSIS. A section of this input data is shown in Fig. 3-8. The dummy array dimensions MAX1 and MAX2 are given numerical values in common blocks and symbolic names (NGMAX and NPMAX) for use in the code itself. NGMAX and NPMAX are, of course, housekeeping quantities and so are included in [C1.4] COMHOK. The asterisks against the descriptions of a number of the entries are to denote quantities subject to user specification. GENSIS creates from the input the indexes, common blocks, and a number of the subprograms shown in Fig. 3-10.

Figure 3-10 The SPF listing of the program ES1D1V.

```
C/ PACKAGE
C/ PROGRAM ES1D1V
C/ DOCUMENTATION
C/ MODULE TITLE
C
C
C
C
C                        ********************
C                        *                  *
C                        *    E S 1 D 1 V   *
C                        *                  *
C                        ********************
C
C
C                        1-D ELECTROSTATIC PLASMA MODEL
C
C                        BY
C
C                        J.W.Eastwood and R.W.Hockney
C
C
C                                              Computational Physics Group
C                                              Computer Science Department
C                                              University of Reading
C                                              Reading,Berkshire
C                                              England
C
C
C
C      This is a basic version of a one-dimensional
C      NGP particle-mesh plasma model.Its purpose is to
C      illustrate the top-downwards approach to writing
C      Particle-Mesh simulation programs using the
C      OLYMPUS programming system.
C
C      NOTE(1): OLYMPUS common variables and unchanged
C      CRONUS subprograms are not included in indexes.
C
C      NOTE(2):The output routines,together with the
C      appropriate common variables,are omitted from
C      this version of the program.
C
C
C      VERSION  1   RWH/JWE                Reading Univ.     1979
C
C/ MODULE INDSUB
C-----------------------------------------------------------------------
CL                             INDEX OF SUBPROGRAMS
C
C VERSION 1   RWH/JWE                 Reading Univ.  1979
CL                                PROLOGUE            CLASS 1
C  RANPOS              Load Random Positions                        1.9
C  THERDI              Load Thermal Velocity Distribution           1.10
CL                                CALCULATION         CLASS 2
C  POTSOL              Solve for the potential                      2.2
C
C/ MODULE INDCOM
C-----------------------------------------------------------------------
CL                             INDEX OF COMMON BLOCKS
```

```
C
C VERSION 1   RWH/JWE                 Reading Univ.    1979
C
CL                     1.       GENERAL OLYMPUS DATA
C   COMBAS             BASIC SYSTEM PARAMETERS                        C1.1
C   COMDDP             DEVELOPMENT AND DIAGNOSTIC PARAMETER           C1.9
C
CL                     2.       PHYSICAL PROBLEM
C   COMPPQ             Principal Physical Quantities                  C2.1
C   COMSPQ             Secondary Physical Quantities                  C2.2
C
CL                     3.       NUMERICAL SCHEME
C   COMNUM             Numerical Parameters                           C3.1
C   COMSCA             Scaling Factors                                C3.2
C
CL                     4.       HOUSEKEEPING
C   COMHOK             Housekeeping                                   C4.1
C
CL                     5.       I/O AND DIAGNOSTICS
C   COMOUT             Output Variables                               C5.1
C
C/ MODULE INDVAR
C-----------------------------------------------------------------------
CL                          ALPHABETIC INDEX OF COMMON VARIABLES
C
C VERSION 1   RWH/JWE                 Reading Univ.    1979
C
C   API                Pi                                          R  2.1
C   BLMETR             *Box length (SI)                            R  2.1
C   BOLTZK             *Boltzmann's Constant (SI)                  R  2.1
C   BOXLEN             Box Length                                  R  2.2
C   CHARG(MAX1)        Charge Density                              RA 2.1
C   DEBLNO             Initial Debye Length (SI)                   R  2.1
C   DELCHG             Charge Increment Per Particle               R  2.2
C   DENBAK             *Background Density (SI)                    R  2.1
C   DT                 *Timestep                                   R  3.1
C   ELCHAG             *Electron Charge (SI)                       R  2.1
C   ELMASS             *Electron Mass (SI)                         R  2.1
C   ELPOT(MAX1)        Electric Potential                          RA 2.1
C   ENKIN              Kinetic Energy                              R  2.2
C   ENPOT              Potential Energy                            R  2.2
C   ENTOT              Total Energy                                R  2.2
C   EPS0               *Free Space Permittivity (SI)               R  2.1
C   FACPE              Dimensionless PE to KE Units                R  3.2
C   FREQP0             Initial Plasma Frequency (SI)               R  2.1
C   FRPDT              *Plasma Freq. * Timestep                    R  3.1
C   H                  Cell Width                                  R  3.1
C   NCASE              *Select Initial Conditions                  I  4.1
C   NG                 *Number Of Grid Points                      I  3.1
C   NGDIM              *Dimension of Grid Arrays                   I  4.1
C   NGMAX              *Dump Limit Of Mesh Arrays                  I  4.1
C   NGP1               NG + 1                                      I  3.1
C   NGP2               NG + 2                                      I  3.1
C   NOPSEL             *Select Output Sequence                     I  5.1
C   NP                 *Number Of Particles                        I  3.1
C   NPDIM              *Dimension of Particle Arrays               I  4.1
C   NPMAX              *Dump Limit Of Particle Arrays              I  4.1
C   NS1                *Output Frequency Selector                  I  5.1
C   NS2                *  ''        ''       ''                    I  5.1
```

```
C  NS3                  *  "      "      "                          I  5.1
C  NS4                  *  "      "      "                          I  5.1
C  NS5                  *  "      "      "                          I  5.1
C  NST1                 *No. of particles in stream 1 (case 3)      I  2.2
C  NST2                 *No. of particles in stream 2 (case 3)      I  2.2
C  QBAKG                Background Charge Density                   R  2.2
C  RNC                  No. of Particles Per Cell                   R  3.1
C  SCELF                Electric Fields To SI                       R  3.2
C  SCENG                Energies To SI                              R  3.2
C  SCPOS                Positions To SI                             R  3.2
C  SCPOT                Potentials To SI                            R  3.2
C  SCVEL                Velocities To SI                            R  3.2
C  TODEGK               *Initial Temperature (Deg K)                R  2.1
C  TDEGK                Electron Temperature                        R  2.2
C  TOLRNC               *POTSOL Error Limit                         R  3.1
C  TOTMOM               Total Momentum                              R  2.2
C  TST1                 *Temperature,stream 1 (case 3)              R  2.2
C  TST2                 *Temperature,stream 2 (case 3)              R  2.2
C  V2PT1                *VEL(1) for 2 particle test                 R  2.2
C  V2PT2                *VEL(2) for 2 particle test                 R  2.2
C  VDR1                 *Drift velocity,stream 1 (case 3)           R  2.2
C  VDR2                 *Drift velocity,stream 2 (case 3)           R  2.2
C  VDRIFT               Electron Drift Velocity (SI)                R  2.1
C  VEL(MAX2)            Particle Velocities                         RA 2.1
C  VTHERO               Initial Thermal Velocity (SI)               R  2.1
C  X2PT1                *XPOS(1) for 2 particle test                R  2.2
C  X2PT2                *XPOS(2) for 2 particle test                R  2.2
C  XMAX                 Maximum Particle Position                   R  2.2
C  XMIN                 Minimum Particle Position                   R  2.2
C  XPOS(MAX2)           Particle Positions                          RA 2.1
C
C/ MODULE INDBLK
C----------------------------------------------------------------------
CL                                INDEX OF COMMON VARIABLES
C
C VERSION 1   RWH/JWE                     Reading Univ.     1979
C
CL                     C2.1      Principal Physical Quantities
C VERSION 1   RWH/JWE                     Reading Univ.     1979
C       COMMON/COMPPQ/
C
C  API                  Pi                                          R  2.1
C  BLMETR               *Box length (SI)                            R  2.1
C  BOLTZK               *Boltzmann's Constant (SI)                  R  2.1
C  CHARG(MAX1)          Charge Density                              RA 2.1
C  DEBLNO               Initial Debye Length (SI)                   R  2.1
C  DENBAK               *Background Density (SI)                    R  2.1
C  ELCHAG               *Electron Charge (SI)                       R  2.1
C  ELMASS               *Electron Mass (SI)                         R  2.1
C  ELPOT(MAX1)          Electric Potential                          RA 2.1
C  EPSO                 *Free Space Permittivity (SI)               R  2.1
C  FREQPO               Initial Plasma Frequency (SI)               R  2.1
C  TODEGK               *Initial Temperature (Deg K)                R  2.1
C  VDRIFT               Electron Drift Velocity (SI)                R  2.1
C  VEL(MAX2)            Particle Velocities                         RA 2.1
C  VTHERO               Initial Thermal Velocity (SI)               R  2.1
C  XPOS(MAX2)           Particle Positions                          RA 2.1
C
CL                     C2.2      Secondary Physical Quantities
```

```
C VERSION 1   RWH/JWE                   Reading Univ.     1979
C       COMMON/COMSPQ/
C
C  BOXLEN             Box Length                                  R  2.2
C  DELCHG             Charge Increment Per Particle               R  2.2
C  ENKIN              Kinetic Energy                              R  2.2
C  ENPOT              Potential Energy                            R  2.2
C  ENTOT              Total Energy                                R  2.2
C  QBAKG              Background Charge Density                   R  2.2
C  TDEGK              Electron Temperature                        R  2.2
C  TOTMOM             Total Momentum                              R  2.2
C  TST1               *Temperature,stream 1 (case 3)              R  2.2
C  TST2               *Temperature,stream 2 (case 3)              R  2.2
C  V2PT1              *VEL(1) for 2 particle test                 R  2.2
C  V2PT2              *VEL(2) for 2 particle test                 R  2.2
C  VDR1               *Drift velocity,stream 1 (case 3)           R  2.2
C  VDR2               *Drift velocity,stream 2 (case 3)           R  2.2
C  X2PT1              *XPOS(1) for 2 particle test                R  2.2
C  X2PT2              *XPOS(2) for 2 particle test                R  2.2
C  XMAX               Maximum Particle Position                   R  2.2
C  XMIN               Minimum Particle Position                   R  2.2
C  NST1               *No. of particles in stream 1 (case 3)      I  2.2
C  NST2               *No. of particles in stream 2 (case 3)      I  2.2
C
CL                     C3.1     Numerical Parameters
C VERSION 1   RWH/JWE                   Reading Univ.     1979
C       COMMON/COMNUM/
C
C  DT                 *Timestep                                   R  3.1
C  FRPDT              *Plasma Freq. * Timestep                    R  3.1
C  H                  Cell Width                                  R  3.1
C  RNC                No. of Particles Per Cell                   R  3.1
C  TOLRNC             *POTSOL Error Limit                         R  3.1
C  NG                 *Number Of Grid Points                      I  3.1
C  NGP1               NG + 1                                      I  3.1
C  NGP2               NG + 2                                      I  3.1
C  NP                 *Number Of Particles                        I  3.1
C
CL                     C3:2     Scaling Factors
C VERSION 1   RWH/JWE                   Reading Univ.     1979
C       COMMON/COMSCA/
C
C  FACPE              Dimensionless PE to KE Units                R  3.2
C  SCELF              Electric Fields To SI                       R  3.2
C  SCENG              Energies To SI                              R  3.2
C  SCPOS              Positions To SI                             R  3.2
C  SCPOT              Potentials To SI                            R  3.2
C  SCVEL              Velocities To SI                            R  3.2
C
CL                     C4.1     Housekeeping
C VERSION 1   RWH/JWE                   Reading Univ.     1979
C       COMMON/COMHOK/
C
C  NCASE              *Select Initial Conditions                  I  4.1
C  NGDIM              *Dimension of Grid Arrays                   I  4.1
C  NGMAX              *Dump Limit Of Mesh Arrays                  I  4.1
C  NPDIM              *Dimension of Particle Arrays               I  4.1
C  NPMAX              *Dump Limit Of Particle Arrays              I  4.1
C
```

```
CL                 C5.1     Output Variables
C VERSION 1   RWH/JWE                Reading Univ.     1979
C      COMMON/COMOUT/
C
C  NOPSEL          *Select Output Sequence                         I  5.1
C  NS1             *Output Frequency Selector                      I  5.1
C  NS2             *  ''        ''        ''                       I  5.1
C  NS3             *  ''        ''        ''                       I  5.1
C  NS4             *  ''        ''        ''                       I  5.1
C  NS5             *  ''        ''        ''                       I  5.1
C
C/ COMMON
C/ MODULE COMPPQ
C-----------------------------------------------------------------------
CL                 C2.1     Principal Physical Quantities
C VERSION 1   RWH/JWE                Reading Univ.    1979
       COMMON/COMPPQ/
     R  API    ,  BLMETR,  BOLTZK,  CHARG ,  DEBLNO,  DENBAK,
     R  ELCHAG,   ELMASS,  ELPOT ,  EPS0  ,  FREQP0,  TODEGK,
     R  VDRIFT,   VEL   ,  VTHERO,  XPOS
       DIMENSION
     R  CHARG(256),        ELPOT(256),        VEL(2048),
     R  XPOS(2048)
C/ MODULE COMSPQ
C-----------------------------------------------------------------------
CL                 C2.2     Secondary Physical Quantities
C VERSION 1   RWH/JWE                Reading Univ.   1979
       COMMON/COMSPQ/
     R  BOXLEN,   DELCHG,  ENKIN ,  ENPOT ,  ENTOT ,  QBAKG ,
     R  TDEGK ,   TOTMOM,  TST1  ,  TST2  ,  V2PT1 ,  V2PT2 ,
     R  VDR1  ,   VDR2  ,  X2PT1 ,  X2PT2 ,  XMAX  ,  XMIN  ,
     I  NST1  ,   NST2
C/ MODULE COMNUM
C-----------------------------------------------------------------------
CL                 C3.1     Numerical Parameters
C VERSION 1   RWH/JWE                Reading Univ.    1979
       COMMON/COMNUM/
     R  DT    ,   FRPDT ,  H     ,  RNC   ,  TOLRNC,
     I  NG    ,   NGP1  ,  NGP2  ,  NP
C/ MODULE COMSCA
C-----------------------------------------------------------------------
CL                 C3.2     Scaling Factors
C VERSION 1   RWH/JWE                Reading Univ.     1979
       COMMON/COMSCA/
     R  FACPE ,   SCELF ,  SCENG ,  SCPOS ,  SCPOT ,  SCVEL
C/ MODULE COMHOK
C-----------------------------------------------------------------------
CL                 C4.1     Housekeeping
C VERSION 1   RWH/JWE                Reading Univ.     1979
       COMMON/COMHOK/
     I  NCASE ,   NGDIM ,  NGMAX ,  NPDIM ,  NPMAX
C/ MODULE COMOUT
C-----------------------------------------------------------------------
CL                 C5.1     Output Variables
C VERSION 1   RWH/JWE                Reading Univ.     1979
       COMMON/COMOUT/
     I  NOPSEL,   NS1   ,  NS2   ,  NS3   ,  NS4   ,  NS5
C
C/ FORTRAN
```

```
C
C/ MODULE C1S1
      SUBROUTINE LABRUN
C
C 1.1  LABEL THE RUN
C
C VERSION 1   RWH/JWE                    Reading Univ.     1979
C
C/ INSERT COMBAS
C-----------------------------------------------------------------------
C
C    READ LABELS
      READ(NREAD,9900)LABEL1
      READ(NREAD,9900)LABEL2
      READ(NREAD,9900)LABEL3
      READ(NREAD,9900)LABEL4
C
C    WRITE HEADING
      CALL BLINES(8)
      WRITE(NOUT,9901)
      WRITE(NOUT,9902)
      CALL BLINES(4)
C
C    WRITE LABELS
      WRITE(NOUT,9903) LABEL1
      CALL BLINES(1)
      WRITE(NOUT,9903) LABEL2
      CALL BLINES(1)
      WRITE(NOUT,9903) LABEL3
      CALL BLINES(1)
      WRITE(NOUT,9903) LABEL4
      CALL BLINES(1)
C
      RETURN
 9900 FORMAT(12A4)
 9901 FORMAT(50X,27HP R O G R A M   E S 1 D 1 V  )
 9902 FORMAT(50X,27(1H*))
 9903 FORMAT(12X,12A4)
      END
C
C/ MODULE C1S2
      SUBROUTINE CLEAR
C
C 1.2  CLEAR VARIABLES AND ARRAYS
C
C VERSION 1   RWH/JWE                    Reading Univ.     1979
C
     COMMON/COMPPQ/ R21(4620)
     COMMON/COMSPQ/ R22(18),I22(2)
     COMMON/COMNUM/ R31(5),I31(4)
     COMMON/COMSCA/ R32(6)
     COMMON/COMHOK/ I41(5)
     COMMON/COMOUT/ I51(6)
C-----------------------------------------------------------------------
C
C
      ZEROS=0.0
C
C    BLOCK COMPPQ
```

```
      CALL RESETR(R21,4620,ZEROS)
C     BLOCK COMSPQ
      CALL RESETR(R22,18,ZEROS)
      CALL RESETI(I22,2,0)
C     BLOCK COMNUM
      CALL RESETR(R31,5,ZEROS)
      CALL RESETI(I31,4,0)
C     BLOCK COMSCA
      CALL RESETR(R32,6,ZEROS)
C     BLOCK COMHOK
      CALL RESETI(I41,5,0)
C     BLOCK COMOUT
      CALL RESETI(I51,6,0)
C
      RETURN
      END
C
C/ MODULE C1S3
      SUBROUTINE PRESET
C
C 1.3  Set Default Values
C
C/ INSERT COMBAS
C/ INSERT COMDDP
C/ INSERT COMPPQ
C/ INSERT COMSPQ
C/ INSERT COMNUM
C/ INSERT COMSCA
C/ INSERT COMHOK
C/ INSERT COMOUT
C
C-----------------------------------------------------------------
CL              1.          Olympus Data
C
      NRUN = 50
C
C-----------------------------------------------------------------
CL              2.          Physical System
C
C     These values give results in units of plasma freq and debye length
C
CL                   2.1        Physical constants
      API = 3.14159265
      BOLTZK = 1.0
      ELCHAG = -1.0
      ELMASS = 1.0
      EPS0 = 1.0
C
CL                   2.2        Physical State
      BLMETR = 64.0
      DENBAK = 1.0
      TODEGK = 1.0
      VDRIFT = 0.0
C
CL                   2.3        Secondary Physical Quantities
      XMIN = 2.0
C
CL                   2.4        Particle Distribution Info
      X2PT1 = 10.0
```

```
      X2PT2 = 26.0
      V2PT2 = 0.0
C
      TST1 = 0.01
      TST2 = 0.01
      VDR1 = 1.0
      VDR2 = -1.0
      NST1 = 1000
      NST2 = 1000
C
C-----------------------------------------------------------------------
CL              3.           Numerical Parameters
C
      NG = 64
      NP = 2
      FRPDT = 0.25
      TOLRNC = 1.0E-5
C
C-----------------------------------------------------------------------
CL              4.           Housekeeping
C
      NGMAX = 70
      NPMAX = 2
      NGDIM = 250
      NPDIM = 2048
      NCASE = 1
C
C-----------------------------------------------------------------------
CL              5.           Output Control
C
      NOPSEL = 1
      NS1 = 25
      NS2 = 1
      NS3 = 1
      NS4 = 1
      NS5 = 1
C-----------------------------------------------------------------------
      RETURN
      END
C
C/ MODULE C1S4
      SUBROUTINE DATA
C
C 1.4  DEFINE DATA SPECIFIC TO RUN
C
C/ INSERT COMBAS
C/ INSERT COMDDP
C/ INSERT COMPPQ
C/ INSERT COMSPQ
C/ INSERT COMNUM
C/ INSERT COMSCA
C/ INSERT COMHOK
C/ INSERT COMOUT
C-----------------------------------------------------------------------
     NAMELIST/NEWRUN/
    . NDIARY,   NIN   ,   NLEDGE,   NONLIN,   NOUT  ,   NPRINT,
    . NPUNCH,   NRUN  ,   NADUMP,   NCLASS,   NPDUMP,   NPOINT,
    . NSUB  ,   NVDUMP,   NLCHED,   NLHEAD,   NLOMT1,   NLOMT2,
    . NLOMT3,   BLMETR,   BOLTZK,   DENBAK,   ELCHAG,   ELMASS,
```

```
     .  EPS0   ,   TODEGK,   TST1   ,   TST2   ,   V2PT1 ,   V2PT2 ,
     .  VDR1   ,   VDR2   ,  X2PT1 ,   X2PT2 ,   NST1   ,   NST2   ,
     .  DT     ,   FRPDT  ,  TOLRNC,   NG     ,   NP     ,   NCASE ,
     .  NGDIM ,   NGMAX ,   NPDIM ,   NPMAX ,   NOPSEL,   NS1    ,
     .  NS2    ,   NS3    ,   NS4    ,   NS5
C-----------------------------------------------------------------------
      NAMELIST/RESET/
     .  NDIARY,   NIN    ,   NLEDGE,   NONLIN,   NOUT   ,   NPRINT,
     .  NPUNCH,   NRUN   ,   NADUMP,   NCLASS,   NPDUMP,   NPOINT,
     .  NSUB   ,   NVDUMP,   NLCHED,   NLHEAD,   NLOMT1,   NLOMT2,
     .  NLOMT3
C-----------------------------------------------------------------------
CL              1.          New Run
C
        IF(NLRES) GO TO 200
        READ(NREAD,NEWRUN)
        WRITE(NOUT,NEWRUN)
        RETURN
C
C-----------------------------------------------------------------------
CL              2.          Run Restarted
C
  200   CONTINUE
        READ(NREAD,RESET)
        WRITE(NOUT,RESET)
C
        RETURN
        END
C
C/ MODULE C1S5
        SUBROUTINE AUXVAL
C
C 1.5  Set Auxiliary Values
C
C/ INSERT COMBAS
C/ INSERT COMDDP
C/ INSERT COMPPQ
C/ INSERT COMSPQ
C/ INSERT COMNUM
C/ INSERT COMSCA
C/ INSERT COMHOK
C/ INSERT COMOUT
C
C-----------------------------------------------------------------------
CL              1.          Physical System
C
C
CL                  1.1       Principal Physical Quantities
        DEBLN0 = SQRT(EPS0*BOLTZK*TODEGK/(DENBAK*ELCHAG**2))
        FREQP0 = SQRT(DENBAK*ELCHAG**2/(EPS0*ELMASS))
        VTHER0 = FREQP0 * DEBLN0
C
CL                  1.2       Secondary Physical Quantities
        IF(NCASE.EQ.3) NP = NST1 + NST2
        BOXLEN = NG
        DELCHG = FRPDT**2 * BOXLEN / (2.0*FLOAT(NP))
        QBAKG  = -FRPDT**2 / 2.0
        XMAX   = XMIN + BOXLEN
C
```

```
C------------------------------------------------------------------------
CL             2.           Numerical Parameters
C
      NGP1 = NG + 1
      NGP2 = NG + 2
      DT   = FRPDT / FREQP0
      H    = BLMETR / BOXLEN
      RNC  = FLOAT(NP) / BOXLEN
C
C------------------------------------------------------------------------
CL             3.           Scaling Factors
C
      SCPOS = H
      SCVEL = H / DT
      SCELF = ELCHAG * DT**2 / (ELMASS*H)
      SCPOT = -2.0 * ELMASS / ELCHAG * SCVEL**2
      SCENG = ELMASS * SCVEL**2 / (8.0*FLOAT(NP))
      FACPE = -16.0 *RNC / FRPDT**2
C
C------------------------------------------------------------------------
CL             4.           Initialisation Checks
C
      IF(NG.LE.NGDIM) GO TO 400
      CALLMESAGE(48HNG TOO LARGE - PROGRAM HALTED                    )
      STOP
C
  400 IF(NP.LE.NPDIM) GO TO 401
      CALLMESAGE(48HNP TOO LARGE - PROGRAM HALTED                    )
      STOP
C
  401 IF(FRPDT.LT.2.0) GO TO 402
      CALLMESAGE(48HWARNING - DT STABILITY LIMIT EXCEEDED            )
      RETURN
C
  402 IF(FRPDT.LE.1.0) GO TO 403
      CALLMESAGE(48HWARNING - DT TOO LARGE FOR ACCURACY              )
  403 CONTINUE
C
C------------------------------------------------------------------------
      RETURN
      END
C
C/ MODULE C1S6
      SUBROUTINE INITAL
C
C 1.6  Set Physical Initial Conditions
C
C     Note: XPOS and VEL are in dimensionless units
C/ INSERT COMBAS
C/ INSERT COMDDP
C/ INSERT COMPPQ
C/ INSERT COMSPQ
C/ INSERT COMNUM
C/ INSERT COMSCA
C/ INSERT COMHOK
C/ INSERT COMOUT
     DATA      ICLASS,  ISUB/1,   6/
                                  CALL EXPERT(ICLASS,ISUB,1)
      IF(NLOMT1(ISUB)) RETURN
```

```
C
C-----------------------------------------------------------------------
CL              1.          Two Particle Test
C
         IF(NCASE.NE.1) GO TO 200
         XPOS(1) = X2PT1
         XPOS(2) = X2PT2
         VEL(1) = V2PT1
         VEL(2) = V2PT2
         RETURN
C
C-----------------------------------------------------------------------
CL              2.          Drifting Thermal Distribution,Uniform Density
C
  200    CONTINUE
         IF(NCASE.NE.2) GO TO 300
         CALL RANPOS(XPOS,1,NP,XMIN,XMAX)
         ZVDR = VDRIFT / SCVEL
         ZVTH = VTHERO / SCVEL
         CALL THERDI(VEL,1,NP,ZVDR,ZVTH)
         RETURN
C
C-----------------------------------------------------------------------
CL              3.          Two Warm Streams
C
  300    CONTINUE
         IF(NCASE.NE.3) GO TO 400
         INST2 = NST1 + 1
         CALL RANPOS(XPOS,1,NST1,XMIN,XMAX)
         CALL RANPOS(XPOS,INST2,NP,XMIN,XMAX)
         ZVDR1 = VDR1 / SCVEL
         ZVDR2 = VDR2 / SCVEL
         ZVTH1 = SQRT(BOLTZK * TST1 / ELMASS) / SCVEL
         ZVTH2 = SQRT(BOLTZK * TST2 / ELMASS) / SCVEL
         CALL THERDI(VEL,1,NST1,ZVDR1,ZVTH1)
         CALL THERDI(VEL,INST2,NP,ZVDR2,ZVTH2)
         RETURN
C
C-----------------------------------------------------------------------
CL              4.          Other cases
C
  400    CONTINUE
                                       CALL EXPERT(ICLASS,ISUB,2)
C-----------------------------------------------------------------------
         RETURN
         END
C
C/ MODULE C1S8
         SUBROUTINE START
C
C 1.8  Start The Calculation
C
C/ INSERT COMBAS
C/ INSERT COMDDP
C/ INSERT COMPPQ
C/ INSERT COMSPQ
C/ INSERT COMNUM
C/ INSERT COMSCA
C/ INSERT COMHOK
```

```
C/ INSERT COMOUT
C-----------------------------------------------------------------------
      DATA        ICLASS,   ISUB/1,   8/
                                      CALL EXPERT(ICLASS,ISUB,1)
        IF(NLOMT1(ISUB)) RETURN
C
C-----------------------------------------------------------------------
CL              1.          Ensure Position BC Are Satisfied
C
        DO 104 JPART = 1,NP
C
  100   IF(XPOS(JPART).GE.XMIN) GO TO 101
        XPOS(JPART) = XPOS(JPART) + BOXLEN
        GO TO 100
  101   CONTINUE
C
  102   IF(XPOS(JPART).LT.XMAX) GO TO 103
        XPOS(JPART) = XPOS(JPART) - BOXLEN
        GO TO 102
  103   CONTINUE
C
  104   CONTINUE
C-----------------------------------------------------------------------
        RETURN
        END
C
C/ MODULE C1S9
        SUBROUTINE RANPOS(POS,KFRST,KLAST,PMIN,PMAX)
C
C 1.9  Load Randomised Cell Distribution
C      Note: the function RND returns random number on interval (0.,1.)
C
C-----------------------------------------------------------------------
C     POS      Coordinate Array
C     KFRST    Address Of First Element
C     KLAST    Address Of Last Element
C     PMIN     Minimum Value Of POS Elements
C     PMAX     Maximim Value Of POS Elements
C-----------------------------------------------------------------------
C
      DIMENSION   POS(1)
C
C-----------------------------------------------------------------------
CL              1.          Initialise
C
        ZINC = (PMAX-PMIN) / FLOAT(KLAST-KFRST+1)
        ZCEL = PMIN
        ZSEED = RND(127)
C
C-----------------------------------------------------------------------
CL              2.          Load Positions
C
        DO 200 JPAR = KFRST,KLAST
        POS(JPAR) = ZCEL + ZINC * RND(0)
        ZCEL = ZCEL + ZINC
  200   CONTINUE
C
C-----------------------------------------------------------------------
        RETURN
```

```
      END
C
C/ MODULE C1S10
      SUBROUTINE THERDI(PVEL,KFRST,KLAST,PDRIFT,PTHERM)
C
C 1.10 Load Thermal Distribution
C
C     Velocities are scaled to give exact first two moments
C     Gaussian is truncated at 3 * PTHERM
C-----------------------------------------------------------------------
C     PVEL      Coordinate Array
C     KFRST     Address Of First Element
C     KLAST     Address Of Last Element
C     PDRIFT    Drift Velocity
C     PTHERM    Thermal Velocity
C-----------------------------------------------------------------------
      DIMENSION   PVEL(1)
      DATA        ZRMIN/1.1109E-2/,   ZTUPI/6.283185/
C
C-----------------------------------------------------------------------
CL              1.          Load Unscaled Gaussian
C
      ZSEED = RND(271)
C
      DO 101 JPART = KFRST,KLAST
  100 ZRAND = RND(0)
      IF(ZRAND.LT.ZRMIN) GO TO 100
      PVEL(JPART) = SQRT(-2.0*ALOG(ZRAND))
  101 CONTINUE
C
      DO 102 JPART = KFRST,KLAST
      ZARG = ZTUPI * RND(0)
      PVEL(JPART) = PVEL(JPART) * SIN(ZARG)
  102 CONTINUE
C
C-----------------------------------------------------------------------
CL              2.          Check First Two Moments
C
      ZMEANV = 0.0
      ZMVSQ = 0.0
C
      DO 200 JPART = KFRST,KLAST
      ZMEANV = ZMEANV + PVEL(JPART)
      ZMVSQ = ZMVSQ + PVEL(JPART)**2
  200 CONTINUE
C
      ZNVALS = KLAST - KFRST + 1
      ZMEANV = ZMEANV / ZNVALS
      ZMVSQ = ZMVSQ / ZNVALS - ZMEANV**2
      ZRMSV = SQRT(ZMVSQ)
C
C-----------------------------------------------------------------------
CL              3.          Scale and Shift Distribution
C
      DO 300 JPART = KFRST,KLAST
      PVEL(JPART) = (PVEL(JPART) - ZMEANV) * PTHERM / ZRMSV + PDRIFT
  300 CONTINUE
C-----------------------------------------------------------------------
      RETURN
```

```
         END
C
C/ MODULE C2S1
         SUBROUTINE STEPON
C
C 2.1  Advance One Timestep
C
C/ INSERT COMBAS
C/ INSERT COMDDP
C/ INSERT COMPPQ
C/ INSERT COMSPQ
C/ INSERT COMNUM
C/ INSERT COMSCA
C/ INSERT COMHOK
C/ INSERT COMOUT
C    Version 1  JWE/RWH
C-----------------------------------------------------------------------
       DATA       ICLASS,   ISUB/2,   1/
                                      CALL EXPERT(ICLASS,ISUB,1)
         IF(NLOMT2(ISUB)) RETURN
         NSTEP = NSTEP + 1
C
C-----------------------------------------------------------------------
CL              1.          Assign Charge
C
C
CL                   1.1      Reset Charge Accumulators
         DO 110 JMESH = 2,NGP1
  110    CHARG(JMESH) = QBAKG
C
CL                   1.2      Assign Charge
         DO 120 JPART = 1,NP
         ICEL = XPOS(JPART)
  120    CHARG(ICEL) = CHARG(ICEL) + DELCHG
C
C-----------------------------------------------------------------------
CL              2.          Obtain Potentials and Potential Energy
C
C
CL                   2.1      Solve Poissons Equation
         CALL POTSOL
C
CL                   2.2      Find P.E.
         ENPOT = 0.0
         DO 220 JMESH = 2,NGP1
  220    ENPOT = ENPOT + CHARG(JMESH) * ELPOT(JMESH)
         ENPOT = ENPOT * FACPE
C
C-----------------------------------------------------------------------
CL              3.          Accelerate Particles and Find Energies
C
         ENKIN = 0.0
         DO 300 JPART = 1,NP
         ICEL = XPOS(JPART)
         ICM1 = ICEL - 1
         ICP1 = ICEL + 1
         ZACCN = ELPOT(ICP1) - ELPOT(ICM1)
         ZVOLD = VEL(JPART)
         VEL(JPART) = ZVOLD + ZACCN
```

```
      ENKIN = ENKIN + (ZVOLD+VEL(JPART))**2
  300 CONTINUE
C
      ENTOT = ENKIN + ENPOT
C
C-----------------------------------------------------------------------
CL            4.          Move Particles
C
      DO 421 JPART = 1,NP
C
CL                4.1     Move
      XPOS(JPART) = XPOS(JPART) + VEL(JPART)
C
CL                4.2     Apply B.C.
      IF(XPOS(JPART).LT.XMAX) GO TO 420
      XPOS(JPART) = XPOS(JPART) - BOXLEN
      IF(XPOS(JPART).GE.XMAX) GO TO 500
      GO TO 421
C
  420 IF(XPOS(JPART).GE.XMIN) GO TO 421
      XPOS(JPART) = XPOS(JPART) + BOXLEN
      IF(XPOS(JPART).LT.XMIN) GO TO 500
  421 CONTINUE
      RETURN
C
C-----------------------------------------------------------------------
CL            5.          Velocity Runaway Failure
C
  500 CONTINUE
      CALLMESAGE(48HVELOCITIES TOO LARGE - RUN TERMINATING           )
      NLEND = .TRUE.
      RETURN
      END
C
C/ MODULE C2S2
      SUBROUTINE POTSOL
C
C 2.2  Solve for Potential in 1-D Periodic System
C
C/ INSERT COMBAS
C/ INSERT COMDDP
C/ INSERT COMPPQ
C/ INSERT COMSPQ
C/ INSERT COMNUM
C/ INSERT COMSCA
C/ INSERT COMHOK
C/ INSERT COMOUT
C     Version 1  JWE/RWH
C-----------------------------------------------------------------------
     LOGICAL     ILREDO
     DATA        ICLASS,   ISUB/2,   2/
                                     CALL EXPERT(ICLASS,ISUB,1)
      IF(NLOMT2(ISUB)) RETURN
C
C-----------------------------------------------------------------------
CL            1.          Obtain Potentials at End Points
C
      ILREDO = .TRUE.
  100 ZPOT1 = 0.0
```

```
C
      DO 101 JMESH = 2,NGP1
  101 ZPOT1 = ZPOT1 + FLOAT(JMESH-NGP2) * CHARG(JMESH)
      ZPOT1 = ZPOT1 / BOXLEN
C
      ELPOT(2) = ZPOT1
      ELPOT(NGP1) = 0.0
C
CL              1.1     Add Periodic Repeats
      ELPOT(1) = 0.0
      ELPOT(NGP2) = ZPOT1
C
C----------------------------------------------------------------------
CL            2.        Compute Interior Points
C
      DO 200 JMESH = 3,NG
      ELPOT(JMESH) = CHARG(JMESH-1)
     + +2.0 * ELPOT(JMESH-1) - ELPOT(JMESH-2)
  200 CONTINUE
C
C----------------------------------------------------------------------
CL            3.        Consistency Check and Fixup
C
      ZERROR = CHARG(NG)+2.0*ELPOT(NG)-ELPOT(NG-1)
      IF(ABS(ZERROR).LT.TOLRNC) GO TO 323
C
CL              3.1     Warning Message
      CALLMESAGE(48HINCONSISTENT POTENTIALS                         )
      CALL IVAR(8HNSTEP   ,NSTEP)
      CALL RVAR(8HZERROR  ,ZERROR)
C
      IF(ILREDO) GO TO 320
                                   CALL EXPERT(ICLASS,ISUB,2)
      CALLMESAGE(48HCHARGE INCONSISTENCY IN C2S2                    )
      STOP
C
CL              3.2     Charge Neutralising Fixup
  320 ILREDO = .FALSE.
      ZTOTCH = 0.0
C
      DO 321 JMESH = 2,NGP1
  321 ZTOTCH = ZTOTCH + CHARG(JMESH)
      ZTOTCH = ZTOTCH / FLOAT(NG)
      CALL RVAR(8HZTOTCH  ,ZTOTCH)
C
      DO 322 JMESH = 2,NGP1
  322 CHARG(JMESH) = CHARG(JMESH) - ZTOTCH
      GO TO 100
  323 CONTINUE
C----------------------------------------------------------------------
      RETURN
      END
C
C/ MODULE C3S1
      SUBROUTINE OUTPUT(K)
C
C 3.1  Control The Output
C
C/ INSERT COMBAS
```

```
C/ INSERT COMDDP
C/ INSERT COMPPQ
C/ INSERT COMSPQ
C/ INSERT COMNUM
C/ INSERT COMSCA
C/ INSERT COMHOK
C/ INSERT COMOUT
       DATA      ICLASS,   ISUB/3,    1/
                                          CALL EXPERT(ICLASS,ISUB,1)
       IF(NLOMT3(ISUB)) RETURN
C-----------------------------------------------------------------------
       GO TO (100,200,300),K
C
C-----------------------------------------------------------------------
CL             1.          Initialisation Output
C
  100  CONTINUE
C
CL                 1.1      General Output
C
CL                 1.2      Output selection 1
       IF(NOPSEL.NE.1) GO TO 130
       RETURN
C
CL                 1.3      Output Selection 2
  130  CONTINUE
       IF(NOPSEL.NE.2) GO TO 140
       RETURN
C
CL                 1.4      Output Selection 3
  140  CONTINUE
       IF(NOPSEL.NE.3) GO TO 150
       RETURN
C
CL                 1.5      Further Output Selections
  150  CONTINUE
                                          CALL EXPERT(ICLASS,ISUB,2)
       RETURN
C
C-----------------------------------------------------------------------
CL             2.          Periodic Output
C
  200  CONTINUE
C
CL                 2.1      General Output
C
CL                 2.2      Output Selection 1
C
CL                 2.3      Output Selection 2
C
CL                 2.4      Output Selection 3
C
CL                 2.5      Further Output Selections
                                          CALL EXPERT(ICLASS,ISUB,3)
C
C-----------------------------------------------------------------------
CL             3.          Final Output
C
  300  CONTINUE
```

```
CL                  3.1      General Output
C
CL                  3.2      Output Selection 1
C
CL                  3.3      output Selection 2
C
CL                  3.4      Output Selection 3
C
CL                  3.5      Further Output Selections
                                        CALL EXPERT(ICLASS,ISUB,4)
C-----------------------------------------------------------------------
      RETURN
      END
C
C/ MODULE C5S2
      SUBROUTINE CLIST(KGROUP,KBLOCK)
C
C 5.2  PRINT COMMON VARIABLES
C
C VERSION 1   RWH/JWE                  Reading Univ.     1979
C
C/ INSERT COMBAS
C/ INSERT COMDDP
C/ INSERT COMPPQ
C/ INSERT COMSPQ
C/ INSERT COMNUM
C/ INSERT COMSCA
C/ INSERT COMHOK
C/ INSERT COMOUT
C-----------------------------------------------------------------------
C
      IF(KGROUP.EQ.0)GO TO 100
      IF((KGROUP.LT.0).OR.(KGROUP.GT.9))RETURN
      GO TO(100,200,300,400,500,999,999,999,999),KGROUP
C
C-----------------------------------------------------------------------
CL                1.           GENERAL OLYMPUS DATA
C
  100 CONTINUE
      IF(KBLOCK.EQ.0)GO TO 101
      GO TO(110,999,999,999,999,999,999,999,190),KBLOCK
  101 CONTINUE
C
CL                  1.1      BLOCK COMBAS
  110 CONTINUE
      CALL PAGE
      CALLMESAGE(48H BLOCK COMBAS                                   )
      CALL BLINES(1)
      CALL IVAR(8HNDIARY  ,NDIARY)
      CALL IVAR(8HNIN     ,NIN   )
      CALL IVAR(8HNLEDGE  ,NLEDGE)
      CALL IVAR(8HNONLIN  ,NONLIN)
      CALL IVAR(8HNOUT    ,NOUT  )
      CALL IVAR(8HNPRINT  ,NPRINT)
      CALL IVAR(8HNPUNCH  ,NPUNCH)
      CALL IVAR(8HNREAD   ,NREAD )
      CALL IVAR(8HNREC    ,NREC  )
      CALL IVAR(8HNRESUM  ,NRESUM)
```

```
      CALL IVAR(8HNRUN    ,NRUN  )
      CALL IVAR(8HNSTEP   ,NSTEP )
      CALL LVAR(8HNLEND   ,NLEND )
      CALL LVAR(8HNLRES   ,NLRES )
      IF(KBLOCK.NE.0)RETURN
C
CL                1.9     BLOCK COMDDP
  190 CONTINUE
      CALL PAGE
      CALLMESAGE(48H BLOCK COMDDP                                   )
      CALL BLINES(1)
      CALL IVAR(8HMAXDUM  ,MAXDUM)
      CALL IVAR(8HMXDUMP  ,MXDUMP)
      CALL IVAR(8HNCLASS  ,NCLASS)
      CALL IVAR(8HNPOINT  ,NPOINT)
      CALL IVAR(8HNSUB    ,NSUB  )
      CALL LVAR(8HNLCHED  ,NLCHED)
      CALL LVAR(8HNLREPT  ,NLREPT)
      IF((KGROUP.NE.0).OR.(KBLOCK.NE.0))RETURN
C
C-----------------------------------------------------------------------
CL              2.          PHYSICAL PROBLEM
C
  200 CONTINUE
      IF(KBLOCK.EQ.0)GO TO 201
      GO TO(210,220,999,999,999,999,999,999,999),KBLOCK
  201 CONTINUE
C
CL                2.1     BLOCK COMPPQ
  210 CONTINUE
      CALL PAGE
      CALLMESAGE(48H BLOCK COMPPQ                                   )
      CALL BLINES(1)
      CALL RVAR(8HAPI     ,API   )
      CALL RVAR(8HBLMETR  ,BLMETR)
      CALL RVAR(8HBOLTZK  ,BOLTZK)
      CALL RVAR(8HDEBLNO  ,DEBLNO)
      CALL RVAR(8HDENBAK  ,DENBAK)
      CALL RVAR(8HELCHAG  ,ELCHAG)
      CALL RVAR(8HELMASS  ,ELMASS)
      CALL RVAR(8HEPS0    ,EPS0  )
      CALL RVAR(8HFREQP0  ,FREQP0)
      CALL RVAR(8HTODEGK  ,TODEGK)
      CALL RVAR(8HVDRIFT  ,VDRIFT)
      CALL RVAR(8HVTHERO  ,VTHERO)
      IF(KBLOCK.NE.0)RETURN
C
CL                2.2     BLOCK COMSPQ
  220 CONTINUE
      CALL PAGE
      CALLMESAGE(48H BLOCK COMSPQ                                   )
      CALL BLINES(1)
      CALL RVAR(8HBOXLEN  ,BOXLEN)
      CALL RVAR(8HDELCHG  ,DELCHG)
      CALL RVAR(8HENKIN   ,ENKIN )
      CALL RVAR(8HENPOT   ,ENPOT )
      CALL RVAR(8HENTOT   ,ENTOT )
      CALL RVAR(8HQBAKG   ,QBAKG )
      CALL RVAR(8HTDEGK   ,TDEGK )
```

```
      CALL RVAR(8HTOTMOM  ,TOTMOM)
      CALL RVAR(8HTST1    ,TST1  )
      CALL RVAR(8HTST2    ,TST2  )
      CALL RVAR(8HV2PT1   ,V2PT1 )
      CALL RVAR(8HV2PT2   ,V2PT2 )
      CALL RVAR(8HVDR1    ,VDR1  )
      CALL RVAR(8HVDR2    ,VDR2  )
      CALL RVAR(8HX2PT1   ,X2PT1 )
      CALL RVAR(8HX2PT2   ,X2PT2 )
      CALL RVAR(8HXMAX    ,XMAX  )
      CALL RVAR(8HXMIN    ,XMIN  )
      CALL IVAR(8HNST1    ,NST1  )
      CALL IVAR(8HNST2    ,NST2  )
      IF((KGROUP.NE.0).OR.(KBLOCK.NE.0))RETURN
C
C-----------------------------------------------------------------------
CL              3.          NUMERICAL SCHEME
C
  300 CONTINUE
      IF(KBLOCK.EQ.0)GO TO 301
      GO TO(310,320,999,999,999,999,999,999,999),KBLOCK
  301 CONTINUE
C
CL                  3.1     BLOCK COMNUM
  310 CONTINUE
      CALL PAGE
      CALLMESAGE(48H BLOCK COMNUM                                   )
      CALL BLINES(1)
      CALL RVAR(8HDT      ,DT    )
      CALL RVAR(8HFRPDT   ,FRPDT )
      CALL RVAR(8HH       ,H     )
      CALL RVAR(8HRNC     ,RNC   )
      CALL RVAR(8HTOLRNC  ,TOLRNC)
      CALL IVAR(8HNG      ,NG    )
      CALL IVAR(8HNGP1    ,NGP1  )
      CALL IVAR(8HNGP2    ,NGP2  )
      CALL IVAR(8HNP      ,NP    )
      IF(KBLOCK.NE.0)RETURN
C
CL                  3.2     BLOCK COMSCA
  320 CONTINUE
      CALL PAGE
      CALLMESAGE(48H BLOCK COMSCA                                   )
      CALL BLINES(1)
      CALL RVAR(8HFACPE   ,FACPE )
      CALL RVAR(8HSCELF   ,SCELF )
      CALL RVAR(8HSCENG   ,SCENG )
      CALL RVAR(8HSCPOS   ,SCPOS )
      CALL RVAR(8HSCPOT   ,SCPOT )
      CALL RVAR(8HSCVEL   ,SCVEL )
      IF((KGROUP.NE.0).OR.(KBLOCK.NE.0))RETURN
C
C-----------------------------------------------------------------------
CL              4.          HOUSEKEEPING
C
  400 CONTINUE
      IF(KBLOCK.EQ.0)GO TO 401
      GO TO(410,999,999,999,999,999,999,999,999),KBLOCK
  401 CONTINUE
```

```
C
CL                 4.1     BLOCK COMHOK
  410    CONTINUE
         CALL PAGE
         CALLMESAGE(48H BLOCK COMHOK                                        )
         CALL BLINES(1)
         CALL IVAR(8HNCASE   ,NCASE )
         CALL IVAR(8HNGDIM   ,NGDIM )
         CALL IVAR(8HNGMAX   ,NGMAX )
         CALL IVAR(8HNPDIM   ,NPDIM )
         CALL IVAR(8HNPMAX   ,NPMAX )
         IF((KGROUP.NE.0).OR.(KBLOCK.NE.0))RETURN
C
C----------------------------------------------------------------------
CL              5.         I/O AND DIAGNOSTICS
C
  500    CONTINUE
         IF(KBLOCK.EQ.0)GO TO 501
         GO TO(510,999,999,999,999,999,999,999,999),KBLOCK
  501    CONTINUE
C
CL                 5.1     BLOCK COMOUT
  510    CONTINUE
         CALL PAGE
         CALLMESAGE(48H BLOCK COMOUT                                        )
         CALL BLINES(1)
         CALL IVAR(8HNOPSEL  ,NOPSEL)
         CALL IVAR(8HNS1     ,NS1   )
         CALL IVAR(8HNS2     ,NS2   )
         CALL IVAR(8HNS3     ,NS3   )
         CALL IVAR(8HNS4     ,NS4   )
         CALL IVAR(8HNS5     ,NS5   )
         IF((KGROUP.NE.0).OR.(KBLOCK.NE.0))RETURN
C
C----------------------------------------------------------------------
C
  999    CONTINUE
         RETURN
         END
C
C/ MODULE C5S3
         SUBROUTINE ARRAYS(KGROUP,KBLOCK)
C
C 5.3  PRINT COMMON ARRAYS
C
C VERSION 1   RWH/JWE                 Reading Univ.     1979
C
C/ INSERT COMBAS
C/ INSERT COMDDP
C/ INSERT COMPPQ
C/ INSERT COMSPQ
C/ INSERT COMNUM
C/ INSERT COMSCA
C/ INSERT COMHOK
C/ INSERT COMOUT
C----------------------------------------------------------------------
C
         IF(KGROUP.EQ.0)GO TO 100
         IF((KGROUP.LT.0).OR.(KGROUP.GT.9))RETURN
```

```
      GO TO(100,200,300,400,500,999,999,999,999),KGROUP
C
C-----------------------------------------------------------------------
CL            1.          GENERAL OLYMPUS DATA
C
  100 CONTINUE
      IF(KBLOCK.EQ.0)GO TO 101
      GO TO(110,999,999,999,999,999,999,999,190),KBLOCK
  101 CONTINUE
C
CL                1.1      BLOCK COMBAS
  110 CONTINUE
      CALL PAGE
      CALL BLINES(1)
      CALLMESAGE(48H BLOCK COMBAS                                   )
      CALL BLINES(1)
      CALL HARRAY(8HLABEL1  ,LABEL1,12)
      CALL HARRAY(8HLABEL2  ,LABEL2,12)
      CALL HARRAY(8HLABEL3  ,LABEL3,12)
      CALL HARRAY(8HLABEL4  ,LABEL4,12)
      CALL HARRAY(8HLABEL5  ,LABEL5,12)
      CALL HARRAY(8HLABEL6  ,LABEL6,12)
      CALL HARRAY(8HLABEL7  ,LABEL7,12)
      CALL HARRAY(8HLABEL8  ,LABEL8,12)
      IF(KBLOCK.NE.0)RETURN
C
CL                1.9      BLOCK COMDDP
  190 CONTINUE
      CALL PAGE
      CALL BLINES(1)
      CALLMESAGE(48H BLOCK COMDDP                                   )
      CALL BLINES(1)
      CALL IARRAY(8HNADUMP  ,NADUMP,20)
      CALL IARRAY(8HNPDUMP  ,NPDUMP,20)
      CALL IARRAY(8HNVDUMP  ,NVDUMP,20)
      CALL LARRAY(8HNLHEAD  ,NLHEAD,9)
      CALL LARRAY(8HNLOMT1  ,NLOMT1,50)
      CALL LARRAY(8HNLOMT2  ,NLOMT2,50)
      CALL LARRAY(8HNLOMT3  ,NLOMT3,50)
      IF((KGROUP.NE.0).OR.(KBLOCK.NE.0))RETURN
C
C-----------------------------------------------------------------------
CL            2.          PHYSICAL PROBLEM
C
  200 CONTINUE
      IF(KBLOCK.EQ.0)GO TO 201
      GO TO(210,999,999,999,999,999,999,999,999),KBLOCK
  201 CONTINUE
C
CL                2.1      BLOCK COMPPQ
  210 CONTINUE
      CALL PAGE
      CALL BLINES(1)
      CALLMESAGE(48H BLOCK COMPPQ                                   )
      CALL BLINES(1)
      CALL RARRAY(8HCHARG   ,CHARG ,NGMAX)
      CALL RARRAY(8HELPOT   ,ELPOT ,NGMAX)
      CALL RARRAY(8HVEL     ,VEL   ,NPMAX)
      CALL RARRAY(8HXPOS    ,XPOS  ,NPMAX)
```

```
      IF(KBLOCK.NE.0)RETURN
C
C-----------------------------------------------------------------------
CL              3.          NUMERICAL SCHEME
C
  300   CONTINUE
        IF(KBLOCK.EQ.0)GO TO 301
        GO TO(999,999,999,999,999,999,999,999,999),KBLOCK
  301   CONTINUE
C
C-----------------------------------------------------------------------
CL              4.          HOUSEKEEPING
C
  400   CONTINUE
        IF(KBLOCK.EQ.0)GO TO 401
        GO TO(999,999,999,999,999,999,999,999,999),KBLOCK
  401   CONTINUE
C
C-----------------------------------------------------------------------
CL              5.          I/O AND DIAGNOSTICS
C
  500   CONTINUE
        IF(KBLOCK.EQ.0)GO TO 501
        GO TO(999,999,999,999,999,999,999,999,999),KBLOCK
  501   CONTINUE
C
C-----------------------------------------------------------------------
C
  999   CONTINUE
        RETURN
        END
C
C/ TEST DATA
C/ MODULE TEST1
1-D Plasma Model
Two Particle Test
R.W.Hockney and J.W.Eastwood
Reading University 1979
C/ NAMELIST NEWRUN
   NLREPT = T,
   NPDUMP(1)=313,NADUMP(1)=100,NVDUMP(1)=100,
   NPDUMP(2)=30101,NADUMP(2)=21,NVDUMP(2)=22,
C/ END NAMELIST
C/ END PROGRAM
C/ END PACKAGE
```

3-4-3 Class 1: The Prologue Subprograms

Once GENSIS has been run using the initial attempt at the master index, the skeleton program CRONUS begins to take on the identity of the plasma program ES1D1V: The evolving program now contains the operands, the diagnostic and development facilities (Fig. 3-6), the Class 0 (control) subprograms and initialization subprograms ⟨1.1⟩ LABRUN, ⟨1.2⟩ CLEAR and ⟨1.4⟩ DATA. Any alterations to these routines required as the program is being developed are performed by updating the master index and rerunning GENSIS. The remaining Class 1 subprograms are still in skeleton form.

Subprograms ⟨1.3⟩, ⟨1.4⟩ and ⟨1.5⟩ form a standard data initialization structure. Nonzero principal variables and constants are assigned default values in subroutine ⟨1.3⟩ PRESET. The set of principal data should be complete in that the program should execute a standard test run without any further data input. The default values of those principal parameters which are subject to user specification can be altered for specific runs by including the names and new values of the quantities to be changed in the namelists read by ⟨1.4⟩ DATA. Dependent quantities are calculated in ⟨1.5⟩ AUXVAL from the principal quantities initialized by ⟨1.3⟩ PRESET and ⟨1.4⟩ DATA.

Subprograms ⟨1.3⟩ and ⟨1.5⟩ provide particularly simple examples of the top-downwards approach to coding. The first stage is to give the subprograms access to common variables by including the SPF control cards.

C/ INSERT COMnnn

where "COMnnn" denotes the block name (e.g., COMBAS). On execution of the program, the SPF processor will replace these cards by copies of the relevant common blocks from the OLYMPUS libraries (in the case of COMBAS or COMDDP) or from the SPF deck (for all other blocks), as appropriate. The next two stages are to divide the subprograms into sections (which reflect the groups of data) and subsections (corresponding to the nature of the variables in each common block). By convention, sections and subsections are headed by a decimally numbered heading and sections are separated by ruled lines (cf. Fig. 3-10). Once these are included the executable statements can be inserted in the appropriate places. A benefit of this approach is that the evolving program remains tidy as it is being developed. There is always a well-defined place for new variables to be inserted if and when they are required.

Subprogram ⟨1.6⟩ INITAL sets the physical initial conditions. It is developed along the same lines as those described above for ⟨1.3⟩ PRESET and ⟨1.5⟩ AUXVAL. However, in this case there is a further level in the recursive refinement. The executable statements may be divided into two groups: those concerned with control and those concerned with the computation of initial positions and velocity. The former are written first to provide a scaffolding into which the latter can be inserted, just as CRONUS provides the control structure for the standard-named OLYMPUS subprograms to be inserted. A further point to note (cf. Fig. 3-10) is that the control structure in ⟨1.6⟩ is designed so that further options can be readily added, either in an ad-hoc fashion by providing a version of ⟨0.4⟩ EXPERT which sets positions and velocity when called with arguments (1, 6, 2), or permanently by adding further sections at the end. In both instances all previous runs of the program are repeatable without further program modification.

The CRONUS subprogram ⟨1.7⟩ RESUME is left in dummy form in ES1D1V. The final CRONUS prologue subprogram, ⟨1.8⟩ START, is generally used to cope with timestepping schemes which are not selfstarting (e.g., the main calculation loop requires information at more than the present timelevel to advance by one step). In ES1D1V it is used only for consistency checks. The remaining Class 1 subprograms (see Fig. 3-10) are used to load particle positions and velocities.

3-4-4 Calculation and Output Subprograms

The main calculation (Class 2) and output (Class 3) subprograms may be developed in parallel once initial versions of the Class 1 routines have been checked out using the standard OLYMPUS dumping facilities. The principal control routines of these two classes are ⟨2.1⟩ STEPON and ⟨3.1⟩ OUTPUT, respectively.

Subroutine ⟨2.1⟩ STEPON evolves in the same manner as described earlier for the Class 1 subprograms. The dummy version is first given access to common variables using the C/ INSERT control cards. The standard switching and diagnostic structure

```
      DATA ICLASS, ISUB/2,1/
                                  CALL EXPERT (ICLASS, ISUB, 1)
      IF (NLOMT2(ISUB)) RETURN
```

is then added, together with the arithmetic expression

```
NSTEP = NSTEP+1
```

which updates the timestep counter. Next, the section and subsection headings are defined and the statements which control program flow through these sections are added. Finally, the coding which does the work plus the call to the lower-level routine, ⟨2.2⟩ POTSOL, are added.

In larger programs, the coding which does the actual calculation would be almost totally moved from ⟨2.1⟩ STEPON to lower-level subprograms. In such cases, the final stages of development of ⟨2.1⟩ STEPON would be to insert calls to the relevant subprograms and to include these subprograms initially in dummy form to maintain the evolving program in executable form. Furthermore, the structuring of the program need not stop there. It is often desirable in large programs for the subprograms called from STEPON themselves to be control routines for a module of the calculation. If the module needs further common blocks and/or variables, then these can be added to the master index and initialized in the standard Class 1 initialization structure.

Since the development of ⟨3.1⟩ OUTPUT and its associated Class 3 routines follows the same pattern as that used for the Class 1 and Class 2 routines, we shall not pursue them further here. In the program listing (Fig. 3-10), ⟨3.1⟩ OUTPUT is left in partially written form to illustrate the control structure into which calls to specific routines can be added as required.

3-5 FINAL REMARKS

Simulation programs are the apparatus of a computational scientist. As with their laboratory counterparts, the larger the programs become, the greater is the need for systematic management of their development. We have used the OLYMPUS system to formalize the management for ES1D1V, although OLYMPUS is not essential for the concepts we have tried to convey. In the absence of OLYMPUS, the same principal stages would be followed but with the tedious bookkeeping work being performed manually.

The basic steps of program development may be summarized by the following work schedule.

1. Follow through the planning steps (Fig. 1-1) thoroughly before beginning the programming. Make summary lists of the planning stages to provide ready cross-referencing. If a properly tested program is available for the problem to be solved, then use it.
2. Plan the overall program architecture, using CRONUS as a model. Identify modular parts and if modules exist incorporate them into the program.
3. Plan the overall data organization, using OLYMPUS groups and program modularity as a guide. Start to build up a master index of variables.
4. Write and test the topmost level of the program, initially using dummy subprograms for those at lower levels in the program tree. Include at this stage program flow control switches, diagnostic test points and diagnostic output. Quantities in [C1.1] COMBAS and [C1.9] COMDDP (Table 3-6) illustrate the nature of variables required at this stage.
5. Develop the first level of initialization (Class 1) subprograms, adding principal physical, numerical, and housekeeping variables into the appropriate places in the data organization and recording their name, type, and description in the master index. The dummy subprograms are infilled by adding, in turn, the non-executable parts, the switch and diagnostic point, sections and subsections, the control structure, and finally the executable statements which do the actual calculation and/or calls to lower-level subprograms.
6. Repeat (5) for the first level of the main calculation (Class 2), output (Class 3), and final calculation (Class 4) subprograms, extending the master index of variables and Class 1 routines as needed.
7. Proceed to the next level of subprograms, using the diagnostic facilities to test the new coding at each stage of development. Introduce further common blocks, initialization subprograms, etc., for separate modules of the calculation.
8. Repeat (7), refining and extending previously developed parts of the program as

necessary until a fully tested working program is obtained. At all stages ensure that the master indexes, diagnostic facilities, common blocks, and comments are amended to agree with each other and with the executable coding.

Implicit in the work schedule are test runs of the partially developed program at each stage of its evolution. The simplest and often most useful test is to hand check the complete dump from test runs where few particles and coarse meshes are used. Such checks, however, are very laborious and not foolproof, so it is always worthwhile to use known properties of particle models to help verify the program correctness: It is clear from elementary statistical arguments (Bayes' theorem) that the more independent checks there are, the more likely is the program to be bug-free.

Three important classes of program verifications which prove valuable for particle models are *consistency* tests, *symmetry* tests, and *convergence* tests. The conservation of total momentum and the requirement of charge neutrality in periodic systems are examples of consistency tests. Errors in the total momentum and total charge should be no greater than roundoff error. Symmetry tests check the correctness of the form rather than the value of quantities. For instance, such tests may be applied to a configuration containing one or two particles to check the symmetries of charge density, potential, and forces. For periodic and isolated systems, the displacement invariance of these quantities provides further symmetry checks. Moreover, the method of images (Reitz and Milford, 1962, Chapter 3) allows displacement invariance checks to be extended to models with Neumann or Dirichlet boundary conditions. Convergence tests are used to check whether numerical solutions tend towards known analytic solutions as cell size and timestep are reduced. The potential of a single particle and the oscillation of a pair of particles provide the basis for two such tests, although for collisionless systems it should be remembered that collective behavior (where $N_p \gg 1$) is of interest and the "best" pair force may not give the best overall approximation (cf. Chapter 7).

CHAPTER

FOUR

TIME INTEGRATION SCHEMES

4-1 INTRODUCTION

The particles in most particle simulations move according to Newton's laws of motion (see Table 1-1).

$$\frac{d\mathbf{x}}{dt} = \mathbf{v} \tag{4-1}$$

$$m\frac{d\mathbf{v}}{dt} = \mathbf{F} \tag{4-2}$$

The form **F** takes depends on the forces involved. For instance, the force **F** for a particle with charge q moving in electric field **E** and magnetic field **B** is the sum of the electrical force $q\mathbf{E}$ and the Lorentz force $q\mathbf{v} \times \mathbf{B}$.

In the steps from the mathematical description of physical phenomena to the simulation program, the continuous differential equations (4-1) and (4-2) are replaced by linear algebraic relationships. Continuous functions **x** and **v** are replaced by values at discrete time intervals. One example of discretization which we have already met (see Chapter 2) is the leapfrog scheme:

$$\frac{\mathbf{x}^{n+1} - \mathbf{x}^n}{DT} = \mathbf{v}^{n+1/2} \tag{4-3}$$

$$m\frac{(\mathbf{v}^{n+1/2} - \mathbf{v}^{n-1/2})}{DT} = F(\mathbf{x}^n) \tag{4-4}$$

where DT is the timestep and the superscript designates the time level ($t = nDT$, etc.).

Equations (4-3) and (4-4) represent just one possible choice. There are many ways in which Eqs. (4-1) and (4-2) can be approximated by discrete analogs, although generally the discrete approximations are linear multistep equations relating values at discrete timelevels:

$$\sum_{i=0}^{k} a_{k-i}\mathbf{x}^{n+k-i} = \frac{DT^2}{m}\sum_{i=0}^{k} b_{k-i}\mathbf{F}^{n+k-i} \tag{4-5}$$

Discrete equations of this form are classified under two headings depending on the value of b_k. If b_k is zero, the scheme is *explicit* and $\mathbf{x}^n$ can be solved for directly in terms of known quantities. If b_k is nonzero the scheme is *implicit* and, unless $\mathbf{F}$ is a simple (such as linear) function of $\mathbf{x}$, $\mathbf{x}^n$ must be found iteratively.

It is clear from Eq. (4-5) that we require some criteria to help us evaluate and choose schemes from among the almost limitless possibilities. These criteria we consider below under the headings:

consistency
accuracy
stability
efficiency

The discussion of these criteria given below is concerned with practical guidance for the integration of the equations of motion most commonly encountered in simulations using particles. For a thorough mathematical treatment of the numerical analysis of integrating ordinary differential equations, of which these are a special case, the reader is referred to other texts; for example Henrici (1962), Daniel and Moore (1970), or most texts on numerical analysis.

4-2 CONSISTENCY

The first requirement for any algebraic approximation to a differential equation is that it should tend towards the differential equation in the limit of infinitesimal timestep. Thus, for example, we would say that the discrete equation

$$\frac{x^{n+1} - x^n}{DT} = v^n \tag{4-6}$$

is consistent with the differential equation

$$\frac{dx}{dt} = v \tag{4-7}$$

since

$$x^{n+1} - x^n = x(t^n + DT) - x(t^n) \tag{4-8}$$

$$= \left[x(t^n) + \left.\frac{dx}{dt}\right|_{t=t^n} DT + \cdots \right] - x(t^n)$$

gives

$$\lim_{DT\to 0}\left[\frac{x(t^n+DT) - x(t^n)}{DT}\right] = \frac{dx}{dt} = v \tag{4-9}$$

A second desired consistency property of the discrete approximation to the differential equation is that it reflects the same time symmetry. The differential equations (4-1) and (4-2) are time-reversible, i.e., if a particle is integrated forwards in time in a given force field, and then time (and velocity) are reversed, the particle will retrace its path and return to its starting point.

Time-reversible difference approximations are obtained by defining time-centered derivatives. Equations (4-3) and (4-4) are examples with time-centered difference approximations: the difference $(x^{n+1}-x^n)$ is centered about time $t^{n+1/2}$ and the difference $(v^{n+1/2}-v^{n-1/2})$ is centered about t^n. Equation (4-6) contains a difference approximation which is not properly time-centered; $(x^{n+1}-x^n)$ is centered about $t^{n+1/2}$, whereas the right-hand side, v^n, is defined at time t^n.

Unfortunately, it is not always practicable to employ properly time-centered schemes, as generally they lead to implicit equations in variables at the new timelevel. Explicit difference equations introduce function values at the new timelevel only through the derivative term, allowing new values to be obtained by simple evaluation of an expression depending only on old values. In implicit schemes new values appear in both the derivative and the driving term. Consequently, implicit schemes require the inversion of a set of equations which may be nonlinear. For example, an implicit approximation to the equations of motion

$$\frac{d\mathbf{x}}{dt} = \mathbf{v} \tag{4-10}$$

$$\frac{d\mathbf{v}}{dt} = \mathbf{v} \times \mathbf{\Omega}(x, t) \tag{4-11}$$

is

$$(\mathbf{x}^{n+1}-\mathbf{x}^n) = \tfrac{1}{2}(\mathbf{v}^{n+1}+\mathbf{v}^n)DT \tag{4-12}$$

$$(\mathbf{v}^{n+1}-\mathbf{v}^n) = \tfrac{1}{2}(\mathbf{v}^{n+1}\times\mathbf{\Omega}^{n+1}+\mathbf{v}^n\times\mathbf{\Omega}^n)DT \tag{4-13}$$

Unless Ω is a simple algebraic function of $\mathbf{x}$, Eqs. (4-12) anc (4-13) cannot be rearranged to give new values (timelevel $n+1$) in terms of old values (timelevel n). The cost of iteratively solving the implicit equations is acceptable for small numbers of particles but not for the large ensembles used in many particle–mesh models. Hence explicit schemes which are not properly time-centered are sometimes employed.

4-3 ACCURACY

Accuracy and stability of difference schemes are related to the requirement that the deviation of the computed values of positions and velocities from values given by the solution of the differential equation is small.

Accuracy is concerned with local errors. Local errors arise from two sources. First are the roundoff errors resulting from the finite wordlength of numbers within a computer, and second are the truncation errors caused by representing continuous variables by discrete sets of values. A third possible source of errors in calculations is the blunder—this we assume you will avoid by employing the systematic approach to programing described in Chapter 3, but, if not, then it is dealt with under the heading of consistency!

Generally, roundoff errors are much smaller than truncation errors, and provided the scheme is stable (see next section) they can usually be ignored. Truncation errors, which arise from the representation of continuous quantities by discrete sets of values, may be described in either of two ways: in terms of sampling (cf. Chapter 5), where the errors manifest themselves as aliases in the frequency domain, or in terms of the difference between the differential and algebraic equations. In the latter case, a measure of the smallness of truncation errors is given by the order of the difference scheme. The order p is a statement that truncation errors are proportional to $(DT)^P$ for small DT.

The leapfrog approximation, Eqs. (4-3) and (4-4), is a second-order accurate approximation to the equations of motion, Eqs. (4-1) and (4-2). Eliminating $\mathbf{v}$ from Eqs. (4-3) and (4-4), we obtain (in one dimension)

$$\frac{x^{n+1}-2x^n+x^{n-1}}{DT^2}=\frac{F(x^n)}{m} \tag{4-14}$$

Let X be the solution to the differential equations (4-1) and (4-2)

$$\frac{d^2X}{dt^2}=F/m \tag{4-15}$$

then we define the truncation error at timelevel n, δ^n, by substituting X into Eq. (4-14):

$$\frac{X^{n+1}-2X^n+X^{n-1}}{DT^2}=\frac{F(X^n)}{m}-\delta^n \tag{4-16}$$

Taylor-expanding X^{n+1} and X^{n-1} about $X^n = X(t^n)$ allows us to write Eq. (4-16) as

$$\frac{d^2X}{dt^2}+\frac{DT^2}{12}\frac{d^4X}{dt^4}+\text{h.o.t.}=\frac{F(X^n)}{m}-\delta^n \tag{4-17}$$

where the derivatives are to be evaluated at $t = t^n$. Subtracting Eq. (4-17) from Eq. (4-15) then gives the desired result:

$$\delta^n=-\frac{DT^2}{12}\frac{d^4X}{dt^4}+\text{h.o.t.} \tag{4-18}$$

Namely, leapfrog is second-order accurate ($\delta^n \propto DT^2$).

4-4 STABILITY

Stability is concerned with the propagation of errors. Even if truncation and roundoff errors are very small, a scheme will be of little value if the effects of the small errors grow rapidly with time. Instability arises from the nonphysical solution of the discretized equations. If the discrete equations have solutions which grow much more rapidly than the correct solution of the differential equations, then even a very small roundoff error is certain eventually to seed that solution and render the numerical results meaningless.

A numerical method is said to be stable if a small error at any stage does not lead to a larger cumulative error. To get a quantitative measure of stability, we must obtain the equation describing the evolution of errors and then investigate how the solutions of that equation behave. Let us again take the leapfrog scheme as an example.

Expressing the leapfrog scheme in terms of positions only, we have

$$x^{n+1} - 2x^n + x^{n-1} = \frac{F(x^n)}{m} DT^2 \tag{4-19}$$

If we were given $x^0 = X^0$, $x^1 = X^1$ as initial conditions, then in the absence of roundoff error, Eq. (4-19) would give the solution values $X^2, X^3, \ldots,$ where

$$X^2 - 2X^1 + X^0 = \frac{F(X^1)}{m} DT^2 \tag{4-20}$$

$$X^3 - 2X^2 + X^1 = \frac{F(X^2)}{m} DT^2 \tag{4-21}$$

and so forth. However, the inexact floating-point arithmetic performed in computers causes the exact-solution values $X^2, X^3, X^4, \ldots$ of the difference equations to be replaced by approximations $x^2, x^3, x^4, \ldots,$ where

$$x^2 - 2X^1 + X^0 = \frac{F(X^1)}{m} DT^2 \tag{4-22}$$

$$x^3 - 2x^2 + X^1 = \frac{F(x^2)}{m} DT^2 \tag{4-23}$$

etc. The roundoff error in the position at timelevel 2 is

$$\epsilon^2 = x^2 - X^2 \tag{4-24}$$

Roundoff errors will be made at every arithmetic operation. Our present concern is to see how the error at a particular timelevel propagates, so let us see how ϵ^2 affects the solution at later timelevels in the absence of any further roundoff.

Let

$$\epsilon^n = x^n - X^n \tag{4-25}$$

be the error in the solution of Eq. (4-19) due to roundoff error at timelevel 2; then from Eqs. (4-23) and (4-25) we have

$$(X^3 + \epsilon^3) - 2(X^2 + \epsilon^2) + X^1 = \frac{F(X^2 + \epsilon^2)}{m} DT^2 \tag{4-26}$$

Subtracting Eq. (4-21) gives

$$\epsilon^3 - 2\epsilon^2 = \frac{F(X^2 + \epsilon^2)}{m} DT^2 - \frac{F(X^2)}{m} DT^2 \tag{4-27}$$

We assume roundoff is small, so that the right-hand side of Eq. (4-27) may be Taylor-expanded about X^2 to give

$$\epsilon^3 - 2\epsilon^2 \simeq \epsilon^2\left(\frac{\partial F}{\partial x}\bigg|_{x=X^2}\frac{DT^2}{m}\right) \tag{4-28}$$

Similarly,

$$\epsilon^4 - 2\epsilon^3 + \epsilon^2 = \epsilon^3\left(\frac{\partial F}{\partial x}\bigg|_{x=X^3}\frac{DT^2}{m}\right) \tag{4-29}$$

and at timelevel n

$$\epsilon^{n+1} - 2\epsilon^n + \epsilon^{n-1} = \epsilon^n\left(\frac{\partial F}{\partial x}\bigg|_{x=X^n}\frac{DT^2}{m}\right) \tag{4-30}$$

Equation (4-30) is as difficult to solve as the original equation, Eq. (4-19), unless $\partial F/\partial x$ = constant. Since we are looking for the worst case (we are only interested in whether or not a scheme is unstable, not how unstable it is) we replace $(\partial F/\partial x)|_{x=X^n}$ by its maximum negative value $-|\partial F/\partial x|_{\max}$ to obtain the error propagation equation:

$$\epsilon^{n+1} - 2\epsilon^n + \epsilon^{n-1} = -\frac{1}{m}\left|\frac{\partial F}{\partial x}\right|_{\max} DT^2\epsilon^n \tag{4-31}$$

$$= -\Omega^2 DT^2 \epsilon^n \tag{4-32}$$

The modulus and negative sign, i.e., $\partial F/\partial x \to -|\partial F/\partial x|_{\max}$, arise because we assume that the solution to Eq. (4-19) is an oscillatory solution. In general, it need not be. For an unbounded (nonoscillatory) solution to Eq. (4-19), we would have to ask the question: "Does the propagated error ϵ^n grow more slowly than X^n as n increases?" Stability analysis for a bounded solution (oscillatory) only needs an answer to the much simpler question: "Does ϵ^n grow as n increases?" We already know that X^n is never greater than amplitude of the oscillation.

Equation (4-32) is linear in ϵ and so will have solutions of the form $\epsilon^n = \lambda^n = \exp(i\omega nDT)$, where the superscript n on λ indicates a power (square, cube, etc.). Substituting this trial solution into Eq. (4-32) and dividing through by λ^{n-1} gives a quadratic in λ

$$\lambda^2 - 2\lambda + 1 = -(\Omega DT)^2\lambda \tag{4-33}$$

yielding the two-characteristic solution functions λ_+ and λ_-:

$$\lambda_\pm = 1 - \frac{(\Omega DT)^2}{2} \pm \left[\frac{(\Omega DT)^2}{2}\right]\left[1 - \frac{4}{(\Omega DT)^2}\right]^{1/2} \tag{4-34}$$

The general solution to Eq. (4-32) may be written as

$$\epsilon^n = a\lambda_+^n + b\lambda_-^n \tag{4-35}$$

where the constants a and b are determined by the initial errors. In general, the constants a and b, and the characteristic solutions λ_+ and λ_-, are complex quantities.

The characteristic solutions, λ_+ and λ_-. completely describe the behavior of solutions of the error propagation equation (4-32). If $|\lambda_+|$ and $|\lambda_-|$ are less than or equal to unity then an error incurred at a particular step will remain small and the scheme is stable. However, if either $|\lambda_+|$ or $|\lambda_-|$ is greater than unity, then propagating errors grow exponentially and will eventually swamp the solution. The criterion for stability thus becomes

For a time integration scheme to be numerically stable, the characteristic solutions λ of its error propagation equation must lie in or on the unit circle, i.e., $|\lambda| \leqslant 1$.

The question of timestep dependence of stability reduces to the problem of determining the locus of the characteristic solutions (or roots) λ in the complex λ plane as the timestep varies from zero to infinity. If any root should cross the unit circle then the numerical scheme has become unstable. The value of timestep for which this occurs is the stability limit for the difference scheme.

4-4-1 The Root Locus Method

The root locus method exploits the complex λ plane representation in investigating the stability of numerical schemes. In this method, the locus of the roots is followed as a function of the "gain" parameter Γ. The principal advantage of this approach is that the general behavior of a given scheme can often be found without explicitly solving for the roots λ.

Returning to the leapfrog example, we have from Eq. (4-33)

$$\frac{(\lambda-1)(\lambda-1)}{\cdot\lambda} = -\Gamma \tag{4-36}$$

where

$$\Gamma = (\Omega DT)^2 \tag{4-37}$$

and Γ can take values $0 \leqslant \Gamma \leqslant \infty$. The gain parameter for the general multistep scheme, Eq. (4-5), can be expressed as

$$\frac{\rho(\lambda)}{\sigma(\lambda)} = \frac{\prod_{i=0}^{i=k} (\lambda-\alpha_i)}{\prod_{i=0}^{i=k} (\lambda-\beta_i)} = -\Gamma \tag{4-38}$$

where

$$\rho(\lambda) = \sum_{i=0}^{k} a_{k-i}\lambda^{k-i} \qquad \sigma(\lambda) = \sum_{i=0}^{k} b_{k-i}\lambda^{k-i}$$

and

$$\Gamma = (\Omega DT)^2 \qquad \Omega = \left(\frac{1}{m}\left|\frac{\partial F}{\partial x}\right|_{\max}\right)^{1/2}$$

It follows from Eq. (4-38) that the roots for $\Gamma = 0$ (small DT) occur at the zeros α_i of $\rho(\lambda)$ and that the roots for $\Gamma = \infty$ (large DT) occur at the zeros β_i of $\sigma(\lambda)$ or at infinity. The general multistep scheme is consistent if $\rho(\lambda)$ has a double root at $\lambda = +1$, and is accurate for the solution of the Newton equation if these roots traverse the unit circle as Γ increases from zero.

For the leapfrog scheme [Eq. (4-36)] the two extremes of Γ give

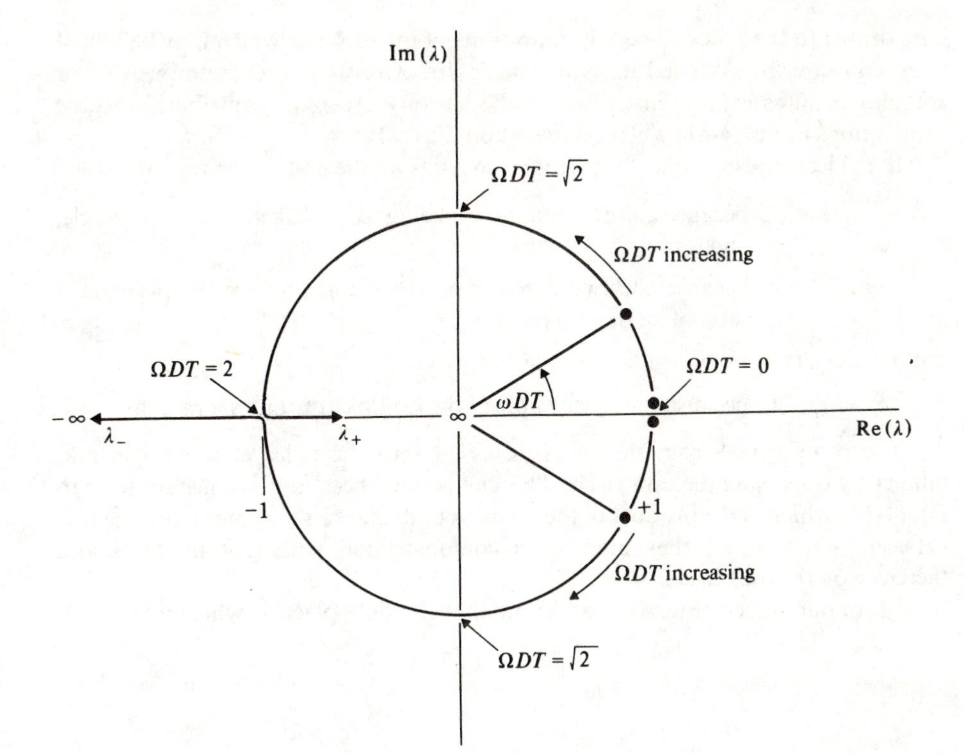

Figure 4-1 Changes of the roots λ_+ and λ_- of the leapfrog difference equation as ΩDT increases from zero to infinity. At $\Omega DT = 0$, both λ_+ and λ_- are unity. As ΩDT increases, λ_+ and λ_- migrate anticlockwise and clockwise, respectively, around the unit circle until they again become degenerate at $\lambda_+ = \lambda_- = -1$, $\Omega DT = 2$. Beyond $\Omega DT = 2$, λ_- goes to $-\infty$ and λ_+ goes to zero, both along the real axis. The behavior of λ_+ and λ_- may be related to the frequency of a leapfrog harmonic oscillator by setting $\lambda = e^{i\omega DT}$ (see Sec. 4-6 and Fig. 4-3).

$$\begin{aligned} &\lambda_+ = \lambda_- = 1 && \text{when } \Gamma = 0 \\ &\lambda_+ = 0 \quad \lambda_- = -\infty && \text{when } \Gamma = \infty \end{aligned} \tag{4-39}$$

These points are marked in the complex λ plane shown in Fig. 4-1 by the symbols 0 and ∞. The next problem is to trace the path between the extremes. Taking the real part ϕ and imaginary part ψ of the logarithm of Eq. (4-38) we obtain:

$$\phi = \sum_i \log|(\lambda - \alpha_i)| - \sum_i \log|(\lambda - \beta_i)| = \log \Gamma \tag{4-40}$$

$$\psi = \sum_i \arg(\lambda - \alpha_i) - \sum_i \arg(\lambda - \beta_i) = \pi \tag{4-41}$$

The second equation, (4-41), proves to be the more useful. It states that the root λ moves about the complex plane in such a way that the sum of the angles subtended by lines drawn to the zeros α_i, minus the sum of the angles subtended by lines drawn to the poles β_i, equals π. Sections of the real axis which may be loci of roots can now be sketched in. Note that pairs of roots (whether double-real or complex conjugates) and any roots to the left of λ give zero contribution to the summations in Eq. (4-41) when evaluated on the real axis.

In the leapfrog example, there can be no roots on the real axis in the ranges

$+1 < \lambda_r$, because all roots and poles, being to the left, subtend zero angle, and

$0 < \lambda_r \leqslant +1$, because the two zeros at $\alpha = +1$ contribute π, each giving a total of 2π or zero radians;

but roots exist for

$-\infty < \lambda_r < 0$, because the simple pole of the origin contributes π radians.

The roots travel from the zeros at $\lambda = +1$ to the poles at the origin and infinity by traversing the unit circle. This can be seen because the constant term in Eq. (4-34), which is the product of the roots, is unity. Since the roots cannot be real between $+1$ and -1 they must be a conjugate pair with unit modulus and therefore on the unit circle.

From our sketch (Fig. 4-1), we know that the only place at which a root can

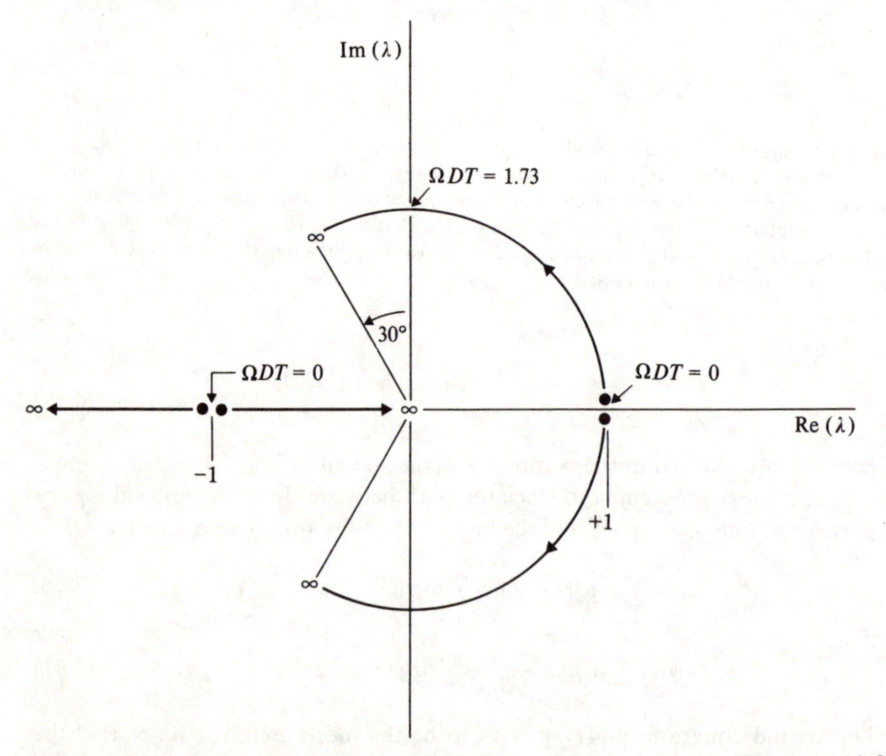

Figure 4-2 The root locus diagram for the five-timelevel explicit scheme (4-43). The zero and infinity symbols on the loci of the roots refer to values of Γ.

leave the unit circle is where the real axis crosses the unit circle, that is to say, at $\lambda = -1$. Substituting this value into the original equation (4-36) we immediately obtain the stability boundary:

$$-\Gamma = -(\Omega DT)^2 = -4 \qquad \text{or} \qquad \Omega DT = 2 \tag{4-42}$$

We note that, in the case of the leapfrog scheme, we are approximating a second-order differential equation by a second-order difference equation. This is usually desirable because the general solution of both the differential and difference equations may be expressed as a linear combination of the same number (in this case two) of characteristic functions [cf. Eq. (4-35)]. Each characteristic solution of the difference equation is an approximation to a characteristic solution of the differential equation.

More accurate difference approximations necessarily involve reference to more timelevels. This causes the number of characteristic solutions to the difference equation to exceed the number of characteristic solutions of the physical problem, and it is these nonphysical parasitic solutions which often go unstable first, leading to more severe timestep restrictions than those arising for the lower-order approximations.

As an example of such a higher-order scheme let us consider the five-timelevel scheme defined by

$$\frac{x^{n+2}-2x^n+x^{n-2}}{4DT^2} = \tfrac{1}{3}[F^{n+1}+F^n+F^{n-1}]+O(DT)^4 \tag{4-43}$$

This scheme is explicit, since F^{n+2} does not occur on the right-hand side and is accurate to order DT^4 compared with the leapfrog scheme, which is order DT^2.

The error propagation equation for Eq. (4-43) is

$$\epsilon^{n+2}-2\epsilon^n+\epsilon^{n-2} = -\frac{4(\Omega DT)^2}{3}[\epsilon^{n+1}+\epsilon^n+\epsilon^{n-1}] \tag{4-44}$$

which leads to the characteristic equation

$$\lambda^4-2\lambda^2+1 = -\Gamma[\lambda^3+\lambda^2+\lambda] \tag{4-45}$$

where

$$\Gamma = 4(\Omega DT)^2/3 \tag{4-46}$$

We now have a higher-order algebraic equation for the roots. Although the full root locus for this case can be found fairly easily (cf. Fig. 4-2) it is useful now to discuss an alternative method for determining a stability boundary which can be applied routinely to higher-order equations.

In all the difference schemes examined in this chapter we will find that the stability boundary occurs when $\lambda = -1$. It is therefore a useful first hypothesis to make this assumption and attempt to prove that a root does indeed leave the unit circle at this point. This may be done by making the expansion $\lambda = -(1+\beta)$ and proving that there is a valid solution for Γ in the range $0 \leqslant \Gamma \leqslant \infty$, when $\beta > 0$.

Substituting $\lambda = -(1+\beta)$ in Eq. (4-45) gives to lowest order in β the valid solution for Γ

$$\Gamma = 4\beta^2 \tag{4-47}$$

The stability boundary occurs when $\beta = 0$, which by Eq. (4-46) is when $DT = 0$. Hence we may conclude that the five-level scheme, although more accurate than leapfrog, is unstable for any DT greater than zero.

The root locus diagram for the five-level scheme [Eq. (4-43)] is shown in Fig. 4-2. The root which goes immediately unstable for nonzero DT is from the zeros at $\lambda = -1$ which were introduced by differencing over twice the timestep on the left-hand side of Eq. (4-43). These roots do not appear in the simpler leapfrog scheme and are parasitic roots. We leave the determination of this root locus diagram as an exercise for the reader.

One final point worth noting from the root locus method is the practical measure of accuracy it provides. For oscillatory problems the nonparasitic roots of the difference equations are approximations to the roots $\lambda = \exp(\pm i\Omega DT)$ of the differential equation. Taking four timesteps per period ($\Omega DT = \pi/2 \simeq 1.57$) gives $\lambda = i$. Substituting $\lambda = i$ in Eqs. (4-36) and (4-45) yields $\Omega DT = \sqrt{2} \simeq 1.41$ for leapfrog and $\Omega DT = \sqrt{3} \simeq 1.73$ for the five-level scheme. Both methods differ from the exact value by the same amount. Thus, although Eq. (4-43) is more accurate than leapfrog for small DT, the advantage is rapidly lost as DT is increased [neglecting also, of course, the fact that Eq. (4-43) is unconditionally unstable!].

4-4-2 The Amplification Matrix

A valuable method for obtaining the characteristic equation of time integration schemes which are discrete approximations to a set of differential equations is to formulate the eigenvalue problem in terms of the amplification matrix **G**. To illustrate the use of the amplification matrix, let us again go through the stability analysis for the leapfrog scheme

$$x^{n+1} - x^n = v^{n+1/2} DT \tag{4-48}$$

$$v^{n+1/2} - v^{n-1/2} = \frac{F^n}{m} DT \tag{4-49}$$

Error propagation equations are obtained by letting $x = X + \epsilon_x$ and $v = V + \epsilon_v$, assuming ϵ is small so that $F(X^n + \epsilon_x^n)$ may be Taylor-expanded, and then subtracting the roundoff-free solutions from the approximate solutions to give

$$\epsilon_x^{n+1} - \epsilon_x^n = \epsilon_v^{n+1/2} DT \tag{4-50}$$

$$\epsilon_v^{n+1/2} - \epsilon_v^{n-1/2} = \frac{dF^n/dx}{m} DT\epsilon_x^n = -\Omega^2 DT \epsilon_x^n \tag{4-51}$$

If we define the error vector ϵ as

$$\epsilon^n = \begin{bmatrix} \epsilon_x^n \\ \epsilon_v^{n-1/2} \end{bmatrix} \tag{4-52}$$

then we may write Eqs. (4-50) and (4-51) as a matrix equation

$$\epsilon^{n+1}-\epsilon^{n}=\begin{bmatrix}0, & DT\\ 0, & 0\end{bmatrix}\epsilon^{n+1}+\begin{bmatrix}0, & 0\\ -\Omega^2 DT, & 0\end{bmatrix}\epsilon^{n} \tag{4-53}$$

Collecting terms, we have

$$\begin{bmatrix}1, & -DT\\ 0, & 1\end{bmatrix}\epsilon^{n+1}=\begin{bmatrix}1, & 0\\ -\Omega^2 DT, & 1\end{bmatrix}\epsilon^{n} \tag{4-54}$$

and multiplying by $\begin{bmatrix}1, & DT\\ 0, & 1\end{bmatrix}$, the inverse of the left-hand side matrix,

$$\epsilon^{n+1}=\begin{bmatrix}1-\Omega^2 DT^2, & DT\\ -\Omega^2 DT, & 1\end{bmatrix}\epsilon^{n} \tag{4-55}$$

i.e.,

$$\epsilon^{n+1}=G\epsilon^{n} \tag{4-56}$$

where G is the amplification matrix.

A matrix equation of form of Eq. (4-56) can be obtained for any time integration scheme. If we make the similarity transformation which diagonalizes G then Eq. (4-56) becomes

$$(S^{-1}\epsilon^{n+1})=(S^{-1}GS)(S^{-1}\epsilon^{n}) \tag{4-57}$$

or

$$\epsilon'^{n+1}=G'\epsilon'^{n} \tag{4-58}$$

where $\epsilon'=S^{-1}\epsilon$.

G' is a diagonal matrix, where the values on the diagonal are the eigenvalues of G:

$$G'=\begin{bmatrix}\lambda_1 & & & & & \\ & \lambda_2 & & 0 & & \\ & & \lambda_3 & & & \\ 0 & & & \cdot & & \\ & & & & \cdot & \\ & & & & & \cdot\end{bmatrix} \tag{4-59}$$

The error at timelevel $n+1$ may be related to the original error seed ϵ'^{1} by repeated application of Eq. (4-58):

$$\epsilon'^{n+1}=G'\epsilon'^{n}=(G')^2\epsilon'^{n-1}=\cdots=(G')^n\epsilon'^{1}\cdots \tag{4-60}$$

where

$$(G')^n=\begin{bmatrix}\lambda_1^n & & & & \\ & \lambda_2^n & & 0 & \\ & & \cdot & & \\ 0 & & & \cdot & \\ & & & & \cdot\end{bmatrix} \tag{4-61}$$

It follows from Eqs. (4-60) and (4-61) that if any of the eigenvalues of G lie outside the unit circle (i.e., $|\lambda|>1$) then the original seed error ϵ'^{1} will be amplified. The error eventually grows exponentially, increasing approximately as $|\lambda|^n_{max}$, where $|\lambda|_{max}$ is the eigenvalue of G with the greatest magnitude.

A quantitative stability criterion in terms of the amplification matrix G may be stated as:

A time integration scheme is stable if the eigenvalues of its amplification matrix lie on or within the unit circle; i.e.,

$$|\lambda|_{max} \leqslant 1 \tag{4-62}$$

The eigenvalues of G are given by roots of the equation

$$\det(G-\lambda I) = 0 \tag{4-63}$$

Thus, for the leapfrog scheme

$$\det(G-\lambda I) = \begin{vmatrix} 1-\Omega^2 DT^2-\lambda, & DT \\ -\Omega^2 DT, & 1-\lambda \end{vmatrix} = 0 \tag{4-64}$$

giving
$$\lambda^2-(2-\Omega^2 DT^2)\lambda+1 = 0 \tag{4-65}$$

which is the same as we obtained earlier [cf. Eq. (4-33)].

In general, when deriving error propagation equations, we obtain matrix equations of the form

$$A\epsilon^{n+1} = B\epsilon^n \tag{4-66}$$

giving the amplification matrix $G = A^{-1}B$. However, the inversion of matrix A can be avoided by using the result that

$$\det(G-\lambda I) = \det(A^{-1}(B-\lambda A))$$

$$= \frac{1}{\det A}\det(B-\lambda A) = 0 \tag{4-67}$$

$$\Rightarrow \det(B-\lambda A) = 0 \tag{4-68}$$

Evaluating Eq. (4-68) to find $\{\lambda\}$ is usually much easier than inverting A, as may be seen in the examples discussed in Sec. 4-7.

4-5 EFFICIENCY

Efficiency is particularly important in particle–mesh simulations because of the large number (up to a million!) of particles used. The two factors of concern in evaluating the cost of a scheme are storage and time. Storage limitations favor schemes which involve as few timelevels as possible in the process of advancing positions and velocities by one timestep. Time limitations point to schemes with small numbers of operations per particle per timestep.

The requirements for stability generally restrict timestep DT to be small compared to the characteristic periods of the system. The compromise between accuracy and efficiency can be altered in two ways—either by using a higher-order scheme and larger timestep, or by using a low-order scheme and smaller timestep. The former approach suffers because (1) the timestep is limited by natural frequencies of the system, (2) higher-order schemes often have more restrictive stability limits on the timestep, and (3) high-order schemes need force values at several timelevels. Usually, the best compromise between accuracy, stability, and efficiency in many-body calculations is found by using simple second-order accurate schemes (such as leapfrog) and adjusting the timestep accordingly.

An example of the waste which can arise if the constraints of efficiency, stability, and accuracy are not carefully used is the following integration scheme:

$$x^{n+1} = x^n + v^n DT + \frac{DT^2}{6}\left[4\frac{F^n}{m} - \frac{F^{n-1}}{m}\right] \tag{4-69}$$

$$v^{n+1} = v^n + \frac{DT}{6}\left[2\frac{F^{n+1}}{m} + 5\frac{F^n}{m} - \frac{F^{n-1}}{m}\right] \tag{4-70}$$

It has been suggested that Eqs. (4-69) and (4-70) provide a superior alternative to the leapfrog scheme. They involve roughly four times more operations per particle per timestep than leapfrog, and require the accelerations at the two previous timelevels to be retained. However, if we difference Eq. (4-69) for two successive timelevels

$$x^n - x^{n-1} = v^{n-1} DT + \frac{DT^2}{6}\left[4\frac{F^{n-1}}{m} - \frac{F^{n-2}}{m}\right] \tag{4-71}$$

$$x^{n+1} - x^n = v^n DT + \frac{DT^2}{6}\left[\frac{4F^n}{m} - \frac{F^{n-1}}{m}\right] \tag{4-72}$$

we find

$$x^{n+1} - 2x^n + x^{n-1} = (v^n - v^{n-1})DT + \frac{DT^2}{6}\left[4\frac{F^n}{m} - 5\frac{F^{n-1}}{m} + \frac{F^{n-2}}{m}\right] \tag{4-73}$$

Eliminating $(v^n - v^{n-1})$ from Eqs. (4-70) and (4-73) gives

$$x^{n+1} - 2x^n + x^{n-1} = \frac{F^n DT^2}{m} \tag{4-74}$$

Equation (4-74) is identical to that obtained by eliminating velocity from the leapfrog equations! Hence Eqs. (4-69) and (4-70) are equivalent to leapfrog, but require roughly four times more work to give the same answer!

4-6 THE LEAPFROG HARMONIC OSCILLATOR

The equation of motion of a particle oscillating in a parabolic potential well is

$$\frac{d^2 X}{dt^2} = -\Omega^2 X \tag{4-75}$$

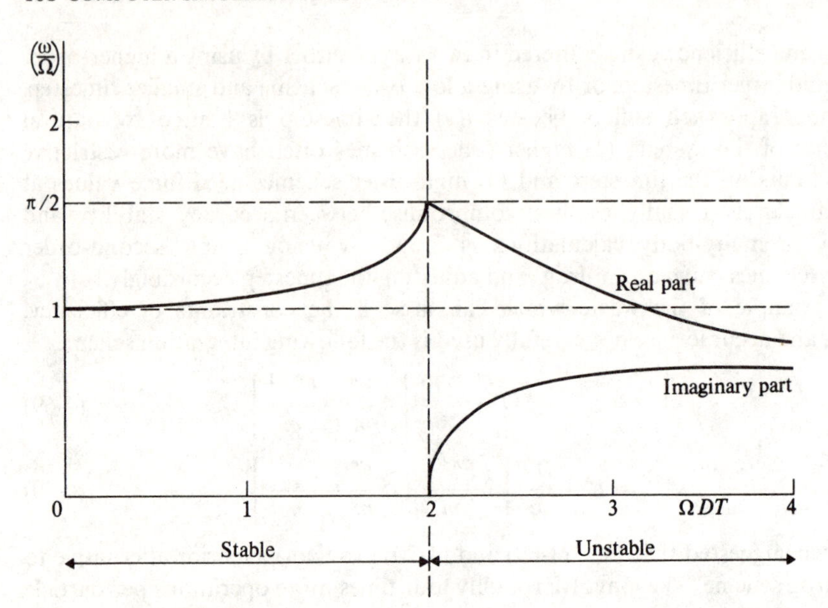

Figure 4-3 The oscillation frequency ω of the leapfrog harmonic oscillator as a function of timestep. Beyond the stability limit, the frequency becomes complex (corresponding to an exponential growth of the oscillation amplitude). Ω is the correct frequency of the harmonic oscillator.

The solution to Eq. (4-75) is a sinusoidal variation of the particle position X, with frequency Ω. In general

$$X = a\cos\Omega t + b\sin\Omega t \tag{4-76}$$

where constants a and b are determined by the initial conditions.

The leapfrog approximation to Eq. (4-75) gives

$$x^{n+1} - 2x^n + x^{n-1} = -\Omega^2 DT^2 x^n \tag{4-77}$$

If we look for solutions to Eq. (4-77) of the form $x = x_0 \exp(i\omega t)$ we find

$$e^{i\omega DT} - 2 + e^{-i\omega DT} = -\Omega^2 DT^2 \tag{4-78}$$

giving

$$\sin\frac{\omega DT}{2} = \pm\frac{\Omega DT}{2} \tag{4-79}$$

Thus, the general solution to Eq. (4-77) may be written

$$x^n = a\cos\omega t^n + b\sin\omega t^n \tag{4-80}$$

where $t^n = nDT$, i.e., the leapfrog approximation, Eq. (4-77), also gives a sinusoidal variation of particle positions, but with the modified frequency ω given by Eq.

(4-79). The frequency ω increases with timestep until the stability limit $\Omega DT = 2$ is reached. At the stability limit there are two timesteps per period—successive x values are alternately positive and negative. It follows from Eq. (4-79) that beyond the stability limit ω becomes complex. Substituting $\omega = \omega_r + i\omega_i$ into Eq. (4-79) shows that the real part of the frequency continues to give two timesteps per period (i.e., $\omega_r DT = \pi$) and the imaginary part is given by

$$\cosh \frac{\omega_i DT}{2} = \pm \frac{\Omega DT}{2} \tag{4-81}$$

The timestep dependence of the frequency of the leapfrog harmonic oscillator is sketched in Fig. 4-3. Examples of the motion of a particle calculated using the leapfrog harmonic scheme are shown in Fig. 4-4 for a range of timesteps. Beyond $\Omega DT = 2$, the amplitude of the oscillations grows exponentially.

The differential harmonic oscillator, Eq. (4-75), has the energy constant

$$E = (V)^2 + \Omega^2 (X)^2 = \text{constant} \tag{4-82}$$

where $V = dX/dt$. The leapfrog oscillator has an analogous energy constant. From Eq. (4-77) we have

$$v^{n+1/2} - v^{n-1/2} = -\Omega^2 DT x^n \tag{4-83}$$

where

$$v^{n+1/2} = (x^{n+1} - x^n)/DT \tag{4-84}$$

Multiplying Eq. (4-83) by $v^{n+1/2} + v^{n-1/2}$ and rearranging yields

$$E'^n = (v^{n+1/2})^2 + \Omega^2 x^n x^{n+1} = \text{constant} \tag{4-85}$$

Conservation of E' can be used to interpret the growth of oscillation amplitude when ΩDT exceeds two. It also enables the growth of total energy in particle–mesh simulations to be understood. If a random force term ADT^2 (representing errors in the particle–mesh force calculation) is added to the right-hand side of Eq. (4-77), then it can be shown by simple random-walk arguments that

$$E'^n = E'^0 + \langle A^2 \rangle n DT^2 \tag{4-86}$$

It is possible to do better than the finite-difference approximation, Eq. (4-77). Comparing Eqs. (4-76) and (4-80), we see that Eq. (4-77) gives the correct form of solution, but with the wrong frequency. Thus, by adjusting the frequency Ω to some other frequency Ω' on the right-hand side of Eq. (4-77) it should be possible to get solutions of the difference equation which exactly match the correct solution. Differencing values of the solution [Eq. (4-76)] of Eq. (4-75) gives

$$X^{n+1} - 2X^n + X^{n-1} = -\left(\Omega \frac{\sin \dfrac{\Omega DT}{2}}{\dfrac{\Omega DT}{2}} \right)^2 DT^2 X^n \tag{4-87}$$

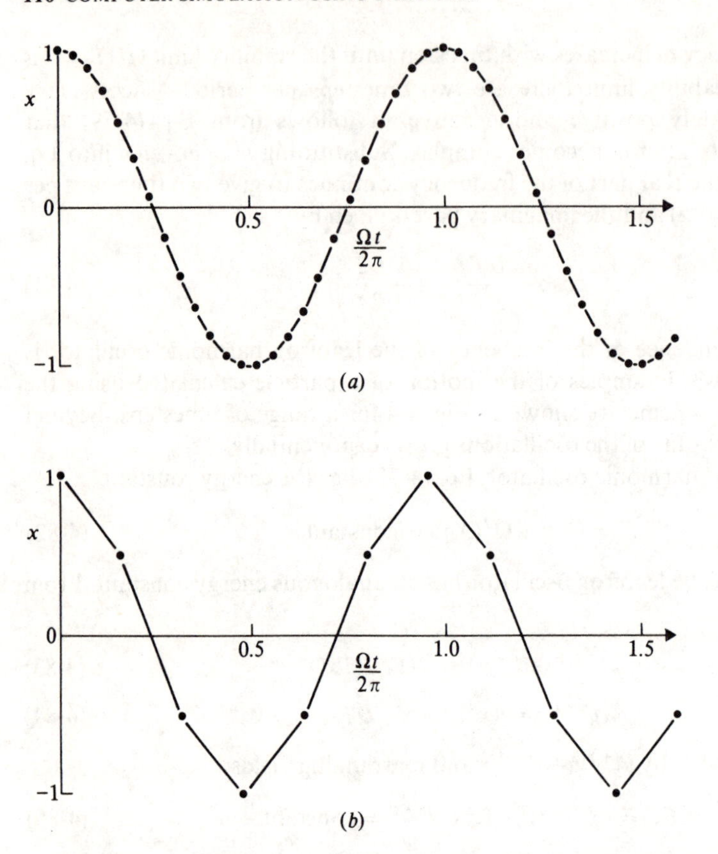

Figure 4-4 Position as a function of time of particle moving in a parabolic potential well according to the leapfrog equations of motion. Initial conditions used were $x^0 = 1$, $v^0 = 0$ ($\Rightarrow v^{-1/2} = \Omega^2 x_0 DT/2$). Crosses mark the position of the particles at each timestep. Timesteps used were (*a*) $\Omega DT = \frac{1}{4}$, (*b*) $\Omega DT = 1$, (*c*) $\Omega DT = 1.5$, and (*d*) $\Omega DT = 2$.

Hence, by replacing Ω by

$$\Omega' = \Omega \left(\frac{\sin \dfrac{\Omega DT}{2}}{\dfrac{\Omega DT}{2}} \right) \tag{4-88}$$

in Eq. (4-77) we obtain a difference equation which reproduces the exact solution values. In addition, the modified version remains stable for all DT. Unfortunately, for forces which are not linear functions of x, the frequency correction cannot be so simply made, although similar correction schemes can be derived to overcome problems with rapidly varying short-range forces encountered in molecular dynamics and galaxy clustering simulations. Finally, we note that an alternative interpretation of Eq. (4-87) is to regard it as an algebraic relationship between

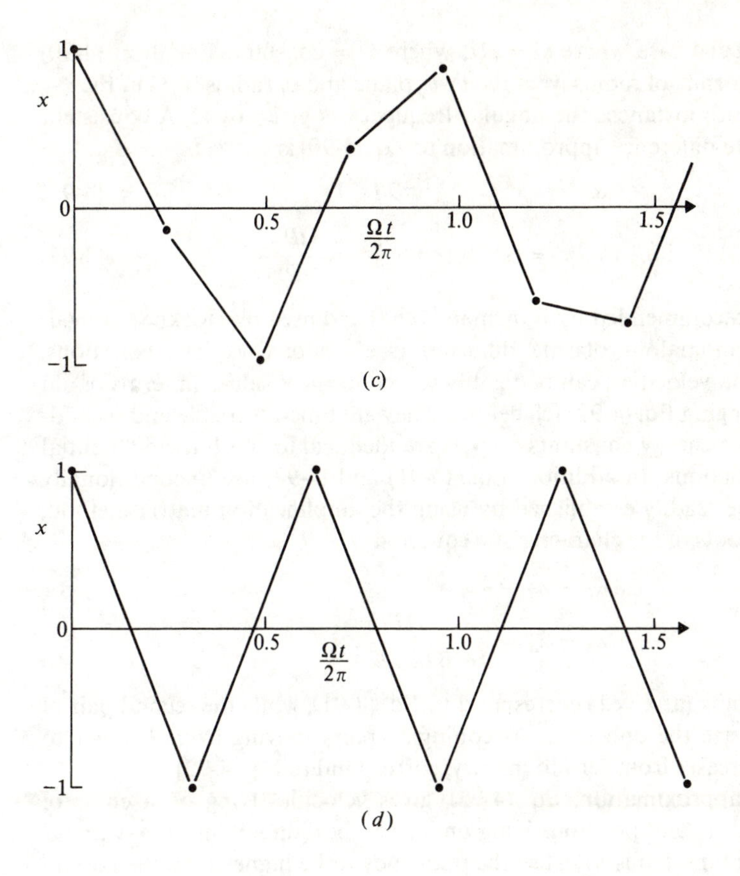

samples of the solution of Eq. (4-75) taken at time intervals DT. In Chapter 5 we shall show that this sampling interpretation proves invaluable when analyzing the errors introduced in spatial discretization.

4-7 EXAMPLES OF INTEGRATION SCHEMES

4-7-1 Lorentz Force Integrators

In the presence of a magnetic field $\mathbf{B}$ the force on a particle of charge q with velocity $\mathbf{v}$ is given by the Lorentz force:

$$\mathbf{F} = q\mathbf{v} \times \mathbf{B} \tag{4-89}$$

Positions and velocities are obtained by integrating

$$\frac{d\mathbf{x}}{dt} = \mathbf{v} \qquad \frac{d\mathbf{v}}{dt} = \mathbf{v} \times \boldsymbol{\Omega} \tag{4-90}$$

where $\boldsymbol{\Omega} = q\mathbf{B}/m$ is the cyclotron (or gyro) frequency.

Consider first the case where $\mathbf{\Omega} = \Omega\hat{\mathbf{z}}$, where Ω = constant. Equation (4-90) describes circular orbits of radius $|\mathbf{v}|$ in the v_x-v_y plane and of radius $|\mathbf{v}|/\Omega$ in the x-y plane, where in both instances the angular frequency is given by Ω. A consistent time-centered finite-difference approximation to Eq. (4-90) is

$$\mathbf{x}^{n+1} - \mathbf{x}^n = \mathbf{v}^{n+1/2} DT \tag{4-91}$$

$$\mathbf{v}^{n+1/2} - \mathbf{v}^{n-1/2} = (\mathbf{v}^{n+1/2} + \mathbf{v}^{n-1/2}) \times \frac{\mathbf{\Omega} DT}{2} \tag{4-92}$$

This scheme was recommended by Buneman (1967) and used by Hockney (1966*a*) in the study of anomalous plasma diffusion (see Chapter 9). The equations, although implicit in velocities, can be readily solved for new values in terms of old by suitably rearranging Eq. (4-92) (cf. below). They are time-reversible and second-order accurate. The energy constants $(=|\mathbf{v}|^2)$ are identical for both the differential and difference equations. In addition, Eqs. (4-91) and (4-92) are unconditionally stable: this may be readily established by using the amplification matrix method, yielding the four roots of the characteristic equation

$$\begin{aligned} \lambda_1 &= \lambda_2 = 1 \\ \lambda_3 &= \lambda_4^* = \frac{1 + i\Omega DT/2}{1 - i\Omega DT/2} \end{aligned} \tag{4-93}$$

The first pair of roots (at $\lambda = 1$) correspond to Eq. (4-91), while the second pair of roots, which traverse the unit circle as conjugate pairs moving from $\lambda = +1$ to $\lambda = -1$ as DT increases from zero to infinity, correspond to Eq. (4-92).

The discrete approximation, Eq. (4-92), gives velocities lying on a circle of radius $|\mathbf{v}|$ in the $v_x - v_y$ and positions lying on a circle of radius R' in the x-y plane. The effect of finite timestep is to cause the frequency to be higher than the correct frequency Ω (cf. Sec. 4-6) and the radius R' to differ from the cyclotron radius $R = |\mathbf{v}|/\Omega$. The frequency ω of the orbits described by Eq. (4-92) may be found by setting λ_3 or λ_4 to $\exp(i\omega DT)$ in Eq. (4-93), yielding

$$\tan\frac{\omega DT}{2} = \pm\frac{\Omega DT}{2} \tag{4-94}$$

The radius R' may be found from Eqs. (4-91) and (4-92) using Eq. (4-94) and the result that $|\mathbf{v}|$ = constant:

$$R' = \frac{|\mathbf{v}|}{\Omega}\sec\frac{\omega DT}{2} \tag{4-95}$$

As in the case of the leapfrog harmonic oscillator, the error in the frequency can be eliminated by adjusting the frequency appearing in the difference equations (see Hockney, 1966*a*, pp. 93–98). In this case, Eq. (4-94) tells us that we must replace $\Omega DT/2$ in Eq. (4-92) by $\tan(\Omega DT/2)$.

The results for $\mathbf{\Omega}$ = constant immediately generalize to a variable $\mathbf{\Omega}$ by replacing $\mathbf{\Omega}$ by $\mathbf{\Omega}^n = \mathbf{\Omega}(\mathbf{x}^n)$ in Eq. (4-92). In addition, if there is an electric field present, then the leapfrog approximation for the electric field force can be combined with the Lorentz force integrator to give the scheme

$$\mathbf{x}^{n+1} - \mathbf{x}^n = \mathbf{v}^{n+1/2} DT \tag{4-96}$$

$$\mathbf{v}^{n+1/2} - \mathbf{v}^{n-1/2} = \frac{q\mathbf{E}^n}{m} DT + (\mathbf{v}^{n+1/2} + \mathbf{v}^{n-1/2}) \times \frac{\mathbf{\Omega}^n DT}{2} \tag{4-97}$$

The components of Eqs. (4-96) and (4-97) parallel to the magnetic field are simply the leapfrog scheme, and so must satisfy the leapfrog stability criterion. The components perpendicular to the magnetic field are unconditionally stable and have the interesting property that for large timestep they tend towards the adiabatic drift equation:

$$\bar{\mathbf{v}}^n = \frac{(\mathbf{v}^{n+1/2} + \mathbf{v}^{n-1/2})}{2} = \frac{\mathbf{E}^n \times \mathbf{B}^n}{|B^n|^2} + O(DT^{-1}) \tag{4-98}$$

This property was used by Levy and Hockney (1968) in the study of crossed-field electron beams, and has been extensively exploited by Birdsall and his coworkers (Birdsall and Langdon, 1981).

The electric acceleration terms and the rotational (Lorentz force) terms in Eq. (4-97) can be separated by introducing two intermediate velocities $\mathbf{v}_1^*$ and $\mathbf{v}_2^*$:

$$\mathbf{v}_1^* = \mathbf{v}^{n-1/2} + \frac{q}{m} \mathbf{E}^n \frac{DT}{2} \tag{4-99}$$

$$\mathbf{v}_2^* = \mathbf{v}_1^* + (\mathbf{v}_2^* + \mathbf{v}_1^*) \times \frac{\mathbf{\Omega}^n DT}{2} \tag{4-100}$$

$$\mathbf{v}^{n+1/2} = \mathbf{v}_2^* + \frac{q}{m} \mathbf{E}^n \frac{DT}{2} \tag{4-101}$$

Equation (4-100) may further be factorized by taking its cross product with $\mathbf{\Omega} DT/2$ and eliminating the triple cross-product term involving $\mathbf{v}_2^*$ to yield

$$\mathbf{v}_2^* = \mathbf{v}_1^* + \frac{2}{1 + \left(\frac{\Omega DT}{2}\right)^2} \mathbf{v}_3^* \times \mathbf{\Omega} \tag{4-102}$$

where

$$\mathbf{v}_3^* = \mathbf{v}_1^* + \mathbf{v}_1^* \times \frac{\mathbf{\Omega} DT}{2} \tag{4-103}$$

The factorization equations [Eqs. (4-99), (4-103), (4-102), and (4-101)] form the basis of Boris' CYLRAD algorithm (Boris, 1970).

The frequency correction factor introduced for the constant $\mathbf{\Omega}$ case can also be incorporated into the more general form [Eq. (4-97)]. If we assume that $\mathbf{E}$ and $\mathbf{\Omega}$ are approximately constant over a timestep, then the equation of motion

$$\dot{\mathbf{v}} = \frac{q\mathbf{E}}{m} + \mathbf{v} \times \mathbf{\Omega} \tag{4-104}$$

can be integrated over that small time interval. Decomposing the velocity into components parallel ($\mathbf{v}_{\parallel}$) and perpendicular ($\mathbf{v}_{\perp}$) to the magnetic field, and expressing $\mathbf{v}_{\perp}$ as the sum of a drift velocity $\mathbf{u}$ and a cyclotron velocity $\mathbf{w}$ gives

$$\mathbf{v}_{\parallel} = \mathbf{v}_{\parallel}^{0} + \frac{q}{m}\mathbf{E}_{\parallel} t \tag{4-105}$$

$$\mathbf{v}_{\perp} = \mathbf{u} + \mathbf{w} \tag{4-106}$$

$$\mathbf{u} = \frac{q}{m}\frac{\mathbf{E}\times\mathbf{\Omega}}{|\mathbf{\Omega}|^2} \tag{4-107}$$

$$\dot{\mathbf{w}} = \mathbf{w}\times\mathbf{\Omega} \tag{4-108}$$

Solving Eq. (4-108) over the time interval $[(n-\frac{1}{2})DT, (n+\frac{1}{2})DT]$ and eliminating $\mathbf{w}$ and $\mathbf{u}$ using Eqs. (4-106) and (4-109) yields

$$\left(\mathbf{v}_{\perp}^{n+1/2} - \frac{q}{m}\frac{\mathbf{E}^n\times\mathbf{\Omega}^n}{|\mathbf{\Omega}^n|^2}\right) = \left(\mathbf{v}_{\perp}^{n-1/2} - \frac{q}{m}\frac{\mathbf{E}^n\times\mathbf{\Omega}^n}{|\mathbf{\Omega}^n|^2}\right)\cos\Omega^n DT + \left(\mathbf{v}_{\perp}^{n-1/2} - \frac{q}{m}\frac{\mathbf{E}^n\times\mathbf{\Omega}^n}{|\mathbf{\Omega}^n|^2}\right)\times\mathbf{\Omega}^n\,\frac{\sin\Omega^n DT}{\Omega^n} \tag{4-109}$$

Rearranging Eq. (4-109) gives

$$\mathbf{v}_{\perp}^{n+1/2} = \mathbf{v}_{\perp}^{n-1/2} + \alpha\left[\frac{q\mathbf{E}^n}{m}DT + (\mathbf{v}_{\perp}^{n+1/2} + \mathbf{v}_{\perp}^{n-1/2})\times\frac{\mathbf{\Omega}^n DT}{2}\right] \tag{4-110}$$

which is identical to Eq. (4-97) apart from the correction factor

$$\alpha = \frac{\tan\dfrac{\Omega DT}{2}}{\dfrac{\Omega DT}{2}} \tag{4-111}$$

4-7-2 Viscous Force Integrators

In particle models where the inertia is ignored, Newton's laws of motion are replaced by first-order equations describing decays to a steady state. Examples of such simulations include the collision-dominated semiconductor model, the vortex fluid model, and the guiding center plasma model (cf. Sec. 1-4 and Table 1-1). In such cases, the velocity is proportional to the driving term

$$\frac{d\mathbf{x}}{dt} = \mathbf{f}(\mathbf{x}) \tag{4-112}$$

For $\mathbf{f}$ linear, we have

$$\frac{d\mathbf{x}}{dt} = -\alpha\mathbf{x} \qquad \alpha > 0 \tag{4-113}$$

One possible finite difference approximation to Eq. (4-113) is

$$\mathbf{x}^{n+1} - \mathbf{x}^{n-1} = -2\alpha \mathbf{x}^n DT \tag{4-114}$$

Equation (4-114) is explicit, consistent, time-centered, and reversible, and second-order accurate. However, it leads to an error propagation equation

$$\epsilon^{n+1} - \epsilon^{n-1} = -2\alpha DT \epsilon^n \tag{4-115}$$

whose characteristic equation has roots

$$\lambda_{\pm} = -\alpha DT \pm (1 + \alpha^2 DT^2)^{1/2} \tag{4-116}$$

Since $\lambda_- < -1$, the scheme is unconditionally unstable. This is another example where the nonphysical parasitic solution rapidly swamps the physically meaningful decaying solution. The unstable parasitic root has arisen because the first-order differential equation (4-112) has been approximated by a higher- (second-) order difference equation (4-114).

A more suitable approximation to Eq. (4-114) is the second-order-accurate implicit scheme

$$\mathbf{x}^{n+1} - \mathbf{x}^n = -\alpha \frac{(\mathbf{x}^{n+1} + \mathbf{x}^n)}{2} DT \tag{4-117}$$

Because this scheme refers to only two timelevels, it is a first-order difference approximation to the first-order differential equation. Equation (4-117) yields the characteristic equation

$$\lambda = \frac{1 - \dfrac{\alpha DT}{2}}{1 + \dfrac{\alpha DT}{2}} \tag{4-118}$$

implying $|\lambda| \leqslant 1$ for all DT, i.e., the scheme is unconditionally stable. The disadvantage of this scheme is that for nonlinear driving terms the implicit

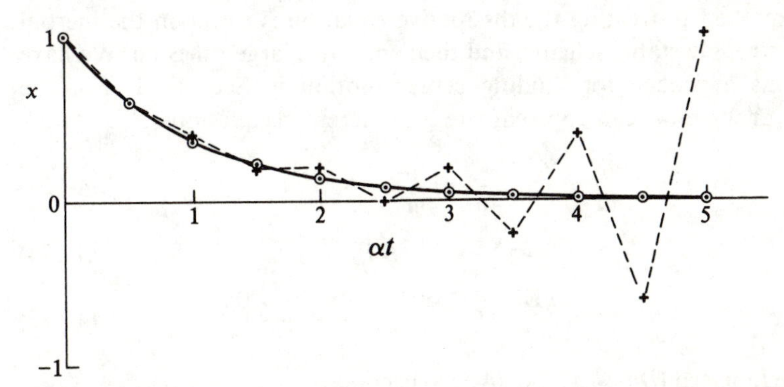

Figure 4-5 Integration of the equation $\dot{x} = -\alpha x$. The continuous curve gives the correct solution $e^{-\alpha t}$, the crosses joined by straight lines give the solution to Eq. (4-114), and the circled points correspond to the solution of Eq. (4-117). Initial conditions used were $x_0 = 1$, and the timestep was such that $\alpha DT = \frac{1}{2}$. Starting values for Eq. (4-114) were obtained using Eq. (4-117).

difference equation

$$\mathbf{x}^{n+1} - \mathbf{x}^n = [f(\mathbf{x}^{n+1}) + f(\mathbf{x}^n)]\frac{DT}{2} \tag{4-119}$$

must be solved at each timelevel: usually this makes such a scheme uneconomical for many body calculations.

Figure 4-5 compares the results of two time-centered and second-order-accurate schemes: the explicit scheme defined by Eq. (4-114) and the implicit scheme defined by Eq. (4-117). The implicit scheme can be seen to decay exponentially to zero like the exact solution. The initial decay of the explicit scheme, however, is swamped by the exponentially growing oscillatory parasitic solution after only seven timesteps, and at the end of the calculation the explicit solution bears no relation to the solution of the differential equation. The explicit scheme results are nonsense because it is numerically unstable, even though it is time-centered and second-order-accurate.

Parasitic solutions are avoided by making the characteristic equation the same order as the differential equation. The only consistent explicit difference approximation to Eq. (4-112) and Eq. (4-113) of this form is the Euler scheme:

$$\mathbf{x}^{n+1} - \mathbf{x}^n = -\alpha \mathbf{x}^n DT \tag{4-120}$$

whose characteristic equation is

$$\lambda = 1 - \alpha DT \tag{4-121}$$

giving the stability criterion

$$\alpha DT < 2 \tag{4-122}$$

Note that, as for the frequency in the oscillatory cases, the decay rate in the linear case can be adjusted to improve the finite-difference approximations. For the cases described here, the correction factors can be found by setting $\lambda = \exp(-at)$ for the physical root of the characteristic equation and then finding α' for use in the difference equation such that $a = \alpha$ in the characteristic equation.

A further method of treating the dissipative equation is to retain the inertial term, obtain a strongly stable scheme, and then employ a large timestep. We have already met this approach for guiding center motion in Sec. 4-7-1. For the semiconductor problem we can approximate the differential equations

$$\frac{d\mathbf{x}}{dt} = \mathbf{v} \qquad \frac{d\mathbf{v}}{dt} = \frac{q\mathbf{E}}{m^*} - \frac{\mathbf{v}}{\tau} \tag{4-123}$$

by

$$\mathbf{x}^{n+1} = \mathbf{x}^n + \mathbf{v}^{n+1/2} DT \tag{4-124}$$

$$\mathbf{v}^{n+1/2} - \mathbf{v}^{n-1/2} = \frac{q\mathbf{E}^n}{m^*} DT - \frac{(\mathbf{v}^{n+1/2} + \mathbf{v}^{n-1/2})}{2} \frac{DT}{\tau} \tag{4-125}$$

where, for large timestep ($DT \gg \tau$), Eq. (4-125) becomes

$$\frac{(\mathbf{v}^{n+1/2} + \mathbf{v}^{n-1/2})}{2} = \bar{\mathbf{v}}^n = \frac{q\tau \mathbf{E}^n}{m^*} = \mu \mathbf{E}^n \tag{4-126}$$

which is simply the desired mobility approximation.

A mathematically equivalent problem to Eq. (4-123) is the equation of motion of the galaxies in comoving coordinates:

$$\dot{V}' + 2HV' = F' \tag{4-127}$$

where H is the Hubble factor and F' is the corrected gravitational force. The scheme given by Eqs. (4-124) and (4-125) was used to approximate Eq. (4-127) in the simulations described in Sec. 11-4-3.

4-7-3 Low-Storage Runge–Kutta Schemes

One of the most popular schemes for integrating ordinary differential equations is the explicit Runge–Kutta method (Henrici, 1962, Chapter 2). This term describes a family of noniterative one-step methods that can be made of high order and which do not require a knowledge of any past values of the solution. The implicit Runge–Kutta methods of Butcher (1964), although more accurate, generally are less suitable for large systems of particles because, unless the forces vary linearly with distance (an unlikely special case), the nonlinear equations for the new values of the solution must be solved by iteration.

Runge–Kutta schemes define the velocity and position at the same value of time, can change the timestep without difficulty, and generally have a smaller truncation error than the finite-difference schemes discussed earlier in this chapter. Their main practical drawback is the need to provide storage for several auxiliary vectors, each of length equal to the number of variables being integrated. In contrast, the new values in the simpler leapfrog finite-difference scheme may overwrite the old, and no auxiliary storage is required. The extra storage required by the Runge–Kutta method may be minimized by a careful choice of the constants in the scheme, or by a careful arrangement of the arithmetic. Methods of minimizing the storage are discussed by Williamson (1980) and by Shampine (1979). We now describe a fourth- and a third-order scheme which are designed to minimize storage.

All the time integration problems considered in this chapter are special cases of the general problem

$$\frac{d\mathbf{x}}{dt} = f(\mathbf{x}) \tag{4-128a}$$

where the $\mathbf{x}$ is a vector of length N containing all the variables being integrated, and $f(\mathbf{x})$ is a function of these variables which is usually nonlinear. If this function is an explicit function of time, this case is included by adding t to the list of variables in the vector $\mathbf{x}$ and including the extra equation

$$\frac{dt}{dt} = 1 \tag{4-128b}$$

In our case $f(\mathbf{x})$ will not usually be an explicit function of time, except in such cases as an imposed time-dependent magnetic field, and we will assume it is not in estimating the storage requirement. In the case of Newton's laws, Eqs. (4-1) and (4-2), the vector $\mathbf{x}$ is

$$\mathbf{x} = \{x_i, y_i, z_i, v_{xi}, v_{yi}, v_{zi};\ i = 1, \ldots, N_p\} \tag{4-129a}$$

and

$$N = 6N_p$$

where i is the particle number and N_p the total number of particles. On the other hand, in the case of viscous force motion, Eq. (4-112), the velocities need not be stored, hence

$$\mathbf{x} = \{x_i, y_i, z_i;\ i = 1, \ldots, N_p\} \tag{4-129b}$$

and

$$N = 3N_p$$

In both cases, $\mathbf{x}$ is an implicit function of time through the change in value of the particle coordinates.

The classic fourth-order Runge–Kutta scheme (Runge, 1895; Kutta, 1901) evaluates the gradient at four places in the neighbourhood of the solution from

$$\begin{aligned}
\mathbf{k}_1 &= DT\, f(\mathbf{x}^n) \\
\mathbf{k}_2 &= DT\, f(\mathbf{x}^n + \tfrac{1}{2}\mathbf{k}_1) \\
\mathbf{k}_3 &= DT\, f(\mathbf{x}^n + \tfrac{1}{2}\mathbf{k}_2) \\
\mathbf{k}_4 &= DT\, f(\mathbf{x}^n + \mathbf{k}_3)
\end{aligned} \tag{4-130a}$$

and finally obtains the value $\mathbf{x}$ at the new timelevel from a weighted average of these values

$$\mathbf{x}^{n+1} = \mathbf{x}^n + \tfrac{1}{6}(\mathbf{k}_1 + 2\mathbf{k}_2 + 2\mathbf{k}_3 + \mathbf{k}_4) \tag{4-130b}$$

The truncation error is proportional to DT^4. The storage required by this method would appear to be $4N$. One vector of length N may be used to store both the vectors $\mathbf{x}^n$ and $\mathbf{x}^{n+1}$ because the latter may overwrite the former during the evaluation of Eq. (4-130b). In addition, three auxiliary vectors are apparently required; one to accumulate the sum of the vectors $\mathbf{k}_i$ in Eq. (4-130b), and a second and third to contain respectively the argument and value of the function $f(\mathbf{x})$ during the successive evaluations of $\mathbf{k}_i$ in Eq. (4-130a).

However, Blum (1962) has shown how the storage may be reduced to $3N$ by the following rearrangement of the calculation

$$\begin{aligned}
\mathbf{z}_0 &= \mathbf{x}^n \\
\mathbf{p}_0 &= DT\, f(\mathbf{z}_0) \\
\mathbf{z}_1 &= \mathbf{z}_0 + \mathbf{p}_0/2 \\
\mathbf{q}_1 &= \mathbf{p}_0 \\
\mathbf{p}_1 &= DT\, f(\mathbf{z}_1) \\
\mathbf{z}_2 &= \mathbf{z}_1 + \mathbf{p}_1/2 - \mathbf{q}_1/2 \\
\mathbf{q}_2 &= \mathbf{q}_1/6 \\
\mathbf{p}_2 &= DT\, f(\mathbf{z}_2) - \mathbf{p}_1/2 \\
\mathbf{z}_3 &= \mathbf{z}_2 + \mathbf{p}_2 \\
\mathbf{q}_3 &= \mathbf{q}_2 - \mathbf{p}_2 \\
\mathbf{p}_3 &= DT\, f(\mathbf{z}_3) + 2\mathbf{p}_2
\end{aligned} \tag{4-131}$$

and finally

$$\mathbf{x}^{n+1} = \mathbf{z}_4 = \mathbf{z}_3 + \mathbf{q}_3 + \mathbf{p}_3/6$$

The Eqs. (4-131) are algebraically equivalent to Eqs. (4-130), but now it is obvious that only three vectors are required—one for all $\mathbf{z}_i$, one for all $\mathbf{q}_i$, and one for all $\mathbf{p}_i$. The evaluation takes place in the sequence: $\mathbf{z}_{i+1}$ which overwrites $\mathbf{z}_i$, $\mathbf{q}_{i+1}$ which overwrites $\mathbf{q}_i$, and $\mathbf{p}_{i+1}$ which overwrites $\mathbf{p}_i$; for $i = 0, 1, 2, 3$. The total storage required for the integration of Newton's equations in three dimensions is therefore $18N_p$, compared with $6N_p$ for the leapfrog scheme. Earlier, Gill (1951) showed how a different set of constants in the Runge–Kutta scheme leads to a fourth-order method which requires only $3N$ storage. Later Fyfe (1966) proved that any fourth-order Runge–Kutta scheme can be organized to require only $3N$ storage.

If the number of particles is large, say more than 10,000, it is likely to be unacceptable to use a scheme which requires three times the minimum storage of N. For the Newton equations in three dimensions, this means more than 180,000 numbers or 0.72 megabytes in 32-bit precision. However, Williamson (1980) has discussed a class of second-, third-, and fourth-order schemes which require a total storage of $2N$. He recommends the following third-order scheme

$$\begin{aligned}
\mathbf{z}_0 &= \mathbf{x}^n \\
\mathbf{q}_1 &= DT\, f(\mathbf{z}_0) \\
\mathbf{z}_1 &= \mathbf{z}_0 + \tfrac{1}{3}\mathbf{q}_1 \\
\mathbf{q}_2 &= DT\, f(\mathbf{z}_1) + \tfrac{5}{9}\mathbf{q}_1 \\
\mathbf{z}_2 &= \mathbf{z}_1 + \tfrac{15}{16}\mathbf{q}_2 \\
\mathbf{q}_3 &= DT\, f(\mathbf{z}_2) - \tfrac{153}{128}\mathbf{q}_2
\end{aligned} \tag{4-132}$$

and finally

$$\mathbf{x}^{n+1} = \mathbf{z}_3 = \mathbf{z}_2 + \tfrac{8}{15}\mathbf{q}_3$$

During the evaluation of Eqs. (4-132) it is apparent that only two vectors need be stored—one for all $\mathbf{q}_i$ and one for all $\mathbf{z}_i$—since $\mathbf{q}_{i+1}$ may overwrite $\mathbf{q}_i$, and $\mathbf{z}_{i+1}$ may overwrite $\mathbf{z}_i$. If the scheme is applied to the viscous-force equations (4-112), the storage required is $6N_p$, which is the same as that required if the inertial term is introduced as in Eqs. (4-123) to (4-125). However the Runge–Kutta scheme has a truncation error proportional to DT^3 compared to DT^2 for the inertial method.

Williamson's scheme (4-132) has been applied to the two-dimensional guiding-center (or vortex) equations for a pair of particles moving in a plane. These may be conveniently expressed in complex notation by

$$\frac{dr}{dt} = i\frac{r}{|r|^2} \tag{4-133}$$

where the complex variable r represents the separation of the particles. The correct orbit is a circle with the frequency varying inversely as the square of the radius. Using a timestep of $\frac{1}{8}$ of a period, the relative error in the radius was 0.001 percent after integrating over one period. When applied to a simulation with 10,000 particles, the above third-order Runge–Kutta scheme conserves the potential energy, which should be a constant, to 0.25 percent over a calculation of 350 timesteps.

CHAPTER

FIVE

THE PARTICLE-MESH FORCE CALCULATION

5-1 INTRODUCTION

The simplest example of the particle-mesh calculations, the NGP (nearest grid point) scheme was employed in Chapters 2 and 3 to illustrate program planning and development. In this chapter we investigate how that basic scheme is generalized to higher-order accuracy and to higher dimensions. To avoid ambiguity, we will use a charged system for our discussion, although the comments apply equally well to other systems for which the particle-mesh scheme is used (see Chapter 1).

Recall from Chapter 1 that the four principal steps of the particle-mesh calculation are:

1. Assign charge to the mesh.
2. Solve the field equation on the mesh.
3. Calculate the mesh-defined force field.
4. Interpolate to find forces on the particles.

where the forces found at step 4 are used to integrate the equations of motion (cf. Chapter 4). Each stage introduces errors. In step 1 the charge distribution of particles whose positions vary continuously throughout the computational box is replaced by a finite set of charge density values. Clearly, the finite set of values are only able to represent a density distribution which is in some sense "smooth"; this concept we shall clarify later (Secs. 5-3-2 and 5-6-2). Steps 2 and 3 introduce truncation errors by replacing differentials by differences (or integrals by sums). The final step, step 4, makes its contribution through interpolation errors. The net result is an approximate force whose errors further degrade the time integration through the acceleration (for collision-dominated schemes, the velocity) term in the equations of motion.

The important quantity in measuring the quality of a particular force calculation scheme is the overall error in the force rather than errors in the component parts, for it is only the total force which affects the dynamics of the system. Quite large numerical errors in each of the steps of the force calculation are acceptable if their net outcome is to provide an accurate force. In order to see how to obtain such cancellation of errors, we need to understand the nature of the errors introduced at each step and how those errors change in response to simple modifications of the numerical schemes.

We use two complementary (and to some extent equivalent) methods to investigate particle-mesh algorithms. First, in x space, we examine schemes by ordering, using the cell size H as a smallness parameter. Concepts such as smoothness, truncation, and angular anisotropy are used to characterize the nature of the errors. In the second method, the operations in the steps of the calculation cycle are represented by their Fourier transforms; there, concepts more familiar to signal processing—sampling, aliasing, convolutions—are used to shed light on the nature of the approximations. The first approach is used primarily to identify how variations of individual elements affect the quality and cost of the approximation, whereas the second, more powerful method can be used to optimally combine those elements.

The next three sections of this chapter involve only the use of x space analysis. The final section (Sec. 5-6) introduces the transform space description. Later (Chapters 7 and 8) we shall employ the transform space description in combination with methods of mathematical physics to show how a comprehensive description of the "physics" of the numerical model may be obtained and how that "physics" may be optimized.

5-2 FORCES IN ONE DIMENSION

We begin our analysis with the simple periodic system; a one-dimensional (i.e., slab-symmetric, cf. Chapter 2) system containing two equal and opposite charges. Despite the simplicity of the system, the results we obtain have bearing on the many-body system since the forces are additive and more general boundary conditions are equivalent to image charges.

5-2-1 The Continuous System

If we have a continuous periodic system of period length L, with a charge q at $x = x_1 = x'$ and a charge $-q$ at $x = x_2 = -x'$ then the potential is given by the solution of

$$\frac{d^2\phi}{dx^2} = -\frac{q}{\epsilon_0}(\delta(x-x')-\delta(x+x')) \tag{5-1}$$

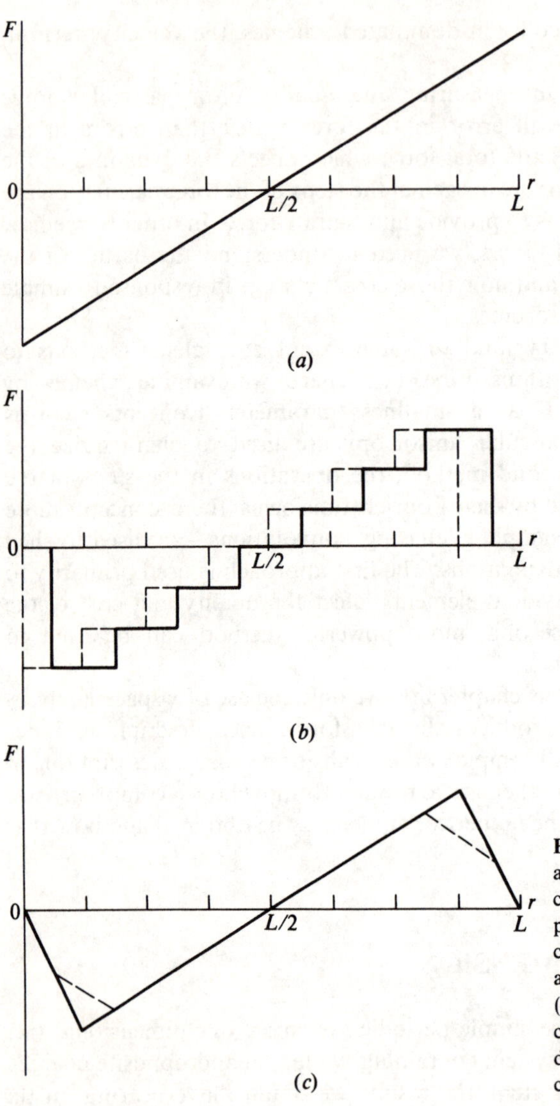

Figure 5-1 The force F on a charge q as a function of its separation r from a charge $-q$ in a one-dimensional periodic system of period length L (8 cells) for (*a*) a continuous system, (*b*) an NGP system, (*c*) a CIC system. In (*b*) and (*c*) the solid lines are for the charge $-q$ at a mesh point and the dashed line is for the charge $-q$ at a cell boundary.

where
$$\phi(x) = \phi(x+L) \tag{5-2}$$

and δ is the Dirac δ function (see Appendix A). The solution to Eqs. (5-1) and (5-2) is easily shown to be

$$\phi(x) = \begin{cases} \dfrac{q}{\epsilon_0 L}(L-2x')x & |x| \leqslant x' \\ \dfrac{q}{\epsilon_0 L}(L-2x)x' & x' \leqslant |x| \leqslant \dfrac{L}{2} \end{cases} \tag{5-3}$$

where the arbitrary constant in the potential is set to make the average potential zero. The electric field

$$E = -\frac{d\phi}{dx} \tag{5-4}$$

takes its mean value at $|x| = x'$

$$E = -\frac{q}{2\epsilon_0 L}(L-4x') \qquad |x| = x' \tag{5-5}$$

yielding respectively the forces $F_e(x_1)$ and $F_e(x_2)$ on the charged particles at positions $x_1 = x'$ and $x_2 = -x'$

$$F_e(x_1) = -F_e(x_2) = \begin{cases} -\dfrac{q^2}{2\epsilon_0 L}(L-4x') & |x'| \neq 0 \\ 0 & |x'| = 0 \text{ and } L/2 \end{cases} \tag{5-6}$$

Figure 5-1(*a*) shows the force on particle 1 as a function of the particle separation $r = 2x'$.

5-2-2 The NGP Scheme

The NGP force calculation (see Chapter 2) which corresponds most closely to the continuous case is that where both particles lie on mesh points. If particles 1 and 2 have charges q and $-q$ and lie on mesh points at $x_1 = x_{p'}$ and $x_2 = x_{-p'} = -x_{p'}$, respectively, then the mesh-defined charge density is given by

$$\rho_p \equiv \rho(x_p) = \frac{q}{H}[\delta_{p,\,p'} - \delta_{p,\,-p'}] \tag{5-7}$$

where $\delta_{p,\,p'}$ is the Kronecker delta (see Sec. A-4-4) and $x_p = pH$ is the position of the mesh point labeled by the integer p.

The mesh-defined potential is given by the solution of the set of algebraic equations

$$\phi_{p-1} - 2\phi_p + \phi_{p+1} = -\frac{\rho_p H^2}{\epsilon_0} \tag{5-8}$$

subject to the boundary conditions

$$\phi_p = \phi_{p+N_g} \tag{5-9}$$

where N_g is the number of mesh points in a periodic length, L.

The solution to Eqs. (5-8) and (5-9) is

$$\phi_p = \begin{cases} \dfrac{q}{\epsilon_0 L}(L-2x_{p'})x_p & |x_p| \leqslant x_{p'} \\ \dfrac{q}{\epsilon_0 L}(L-2x_p)x_{p'} & x_{p'} \leqslant |x_p| < L/2 \end{cases} \tag{5-10}$$

and calculating the electric field using

$$E_p = -\left[\frac{\phi_{p+1} - \phi_{p-1}}{2H}\right] \tag{5-11}$$

gives the forces on particles 1 and 2, F_1 and F_2, as

$$F_1 = -F_2 = \begin{cases} \dfrac{q^2}{2\epsilon_0 L}(L - 4x_{p'}) & x_{p'} \neq 0 \\ 0 & x_{p'} = 0 \end{cases} \tag{5-12}$$

Comparing Eqs. (5-3) and (5-10) and Eqs. (5-6) and (5-12) reveals the apparently remarkable result that the three-point approximation to the laplacian and the finite-difference approximation to the derivative give exactly the same result as the differential equations, provided that particles lie on mesh points! A little further reflection shows that this result is to be expected. The finite-difference approximations to the first and second derivatives to the potential are second-order accurate, with leading error terms proportional to the third and fourth derivatives of the potential, respectively. By Taylor expansion,

$$\begin{aligned} E_p = \frac{\phi(x_p - H) - \phi(x_p + H)}{2H} &= -\left.\frac{d\phi}{dx}\right|_{x_p} - \left.\frac{H^2}{6}\frac{d^3\phi}{dx^3}\right|_{x_p} + \text{h.o.t.} \\ \frac{\phi(x_p - H) - 2\phi(x_p) + \phi(x_p + H)}{H^2} &= \left.\frac{d^2\phi}{dx^2}\right|_{x_p} + \left.\frac{H^2}{12}\frac{d^4\phi}{dx^4}\right|_{x_p} + \text{h.o.t.} \end{aligned} \tag{5-13}$$

The potential of a single isolated charge in one dimension (the Green's function) is piecewise linear for the differential problem. This gives zero third and higher derivatives and consequently exactly satisfies Eqs. (5-8) and (5-11). Since any solution to those equations can be written as linear superpositions of scaled and shifted Green's functions for the isolated charge problem, it follows that the solution to the differential and difference equations should be identical.

The zero truncation error for a simple finite-difference approximation to the laplacian is special to one dimension. In two and three dimensions the Green's function solutions are logarithmic and reciprocal functions of distance, respectively, and so will have all derivatives nonzero. However, by employing a Green's function formulation rather than the Poisson equation for the potential, the property of exact potentials at mesh points can be recovered in higher dimensions, although in certain cases the computational cost of such an approach can be considerable.

The NGP scheme assigns all the charge from a given particle to its nearest grid point and takes as the value of the force on a particle the mesh-computed value at the nearest grid point. Therefore, the only values the interparticle forces can take are those given by Eq. (5-12). Thus, as the interparticle separation is varied the interparticle force changes discontinuously as particles cross cell boundaries. This is illustrated in Fig. 5-1(*b*). In addition, the force depends on the positions of the particles with respect to the mesh as well as their separation: If the particle separations are held constant and the particles are displaced with respect to the mesh, then the interparticle force fluctuates with a period given by the cell width H. It is this loss of displacement invariance which has the most damaging effect on the physical reality of simulation results. In the NGP case, the largest fluctuations in the interparticle forces are for particles separated by less than H, where the forces switch from zero to their maximum value as a charge crosses a cell boundary.

5-2-3 The CIC Scheme

The errors in the interparticle forces in the one-dimensional NGP model arise solely from the crudeness of the charge assignment and force interpolation scheme. A better approximation to the force, but one which is more costly in terms of number of arithmetic operations per particle per timestep, can be obtained by replacing nearest grid point assignment by a scheme involving the two nearest neighbors. This more accurate scheme is known as the "cloud-in-cell" scheme or CIC (Birdsall and Fuss, 1969), a name which arose from the physical interpretation of the assignment of charge (cf. Sec. 5-4).

To demonstrate the improvement in accuracy brought about by assigning charge to and interpolating force from two mesh points rather than one, we again consider the interparticle forces. First, we take the case where particle 2 is at a mesh point (say mesh point 0 at $x = 0$) and particle 1 lies at position x between mesh point p and $p+1$ where we choose $p > 1$. We ignore for the present any forces particle 1 may exert upon itself. The correct value of the force F_e exerted on particle 1 by particle 2 is obtained from Eq. (5-6) with particle separation $2x' = x$:

$$F_e(x) = \frac{-q^2}{2\epsilon_0 L}(L-2x) \tag{5-14}$$

The (NGP) forces defined at mesh points p and $p+1$ are

$$F(x_p) = -\frac{q^2}{2\epsilon_0 L}(L-2x_p) \tag{5-15}$$

$$F(x_{p+1}) = -\frac{q^2}{2\epsilon_0 L}(L-2x_{p+1}) \tag{5-16}$$

If, instead of taking the NGP force value, we take a linear combination of the two nearest values we obtain

$$F(x) = \alpha F(x_p) + (1-\alpha)F(x_{p+1}) \tag{5-17}$$

where α is a function of $x - x_p$. The error in the force on particle 1, ϵ_1, is now given by

$$\begin{aligned}\epsilon_1 &= F_e - F \\ &= \frac{q^2}{\epsilon_0 L}[(x-x_{p+1}) - \alpha(x_p - x_{p+1})]\end{aligned} \tag{5-18}$$

Thus, if we choose

$$\alpha = \frac{x_{p+1} - x}{x_{p+1} - x_p} = 1 - \frac{x - x_p}{H} \tag{5-19}$$

we obtain $\epsilon_1 = 0$. Substituting Eq. (5-19) for α into Eq. (5-17) gives, after a little rearrangement,

$$F = \left(1 - \frac{x - x_p}{H}\right) F(x_p) + \left(1 + \frac{x - x_{p+1}}{H}\right) F(x_{p+1}) \tag{5-20}$$

Defining the CIC force interpolation function

$$W(x - x_p) = \begin{cases} 1 - \dfrac{|x - x_p|}{H} & |x - x_p| \leqslant H \\ 0 & \text{otherwise} \end{cases} \tag{5-21}$$

enables Eq. (5-20) to be written in the more general form

$$F(x) = \sum_p W(x - x_p) F(x_p) \tag{5-22}$$

where the sum p is taken over all mesh points. Equation (5-22) is of exactly the same form as that we obtained for the NGP scheme in Chapter 2 [Eq. (2-31)].

A similar exercise may be performed for the CIC charge assignment scheme. Consider the case where particle 1 is at a position between mesh points p and $p+1$, but now particle 2 lies at x between mesh points 0 and 1. If NGP charge assignment were used, then the force on particle 1 due to particle 2 would be insensitive to displacements of particle 2 in the ranges $x \in [0, H/2]$ and $x \in [H/2, H]$, but would change discontinuously at $x = H/2$. However, if the charge $-q$ from particle 2 is shared between the two grid points, 0 and 1 at x_0 and x_1, nearest to it

$$\rho(x_0) = \frac{(-q)}{H} \beta \tag{5-23}$$

$$\rho(x_1) = \frac{(-q)}{H} (1 - \beta) \tag{5-24}$$

then the error in the force can again be reduced to zero (assuming mesh points involved in assignment and interpolation are distinct) by choosing $\beta = 1 - x/H$. This choice of β is equivalent to using a charge assignment function identical to the force interpolation function given by Eq. (5-21):

$$\rho(x_0) = \frac{(-q)}{H} W(x - x_0) \tag{5-25}$$

$$\rho(x_1) = \frac{(-q)}{H} W(x - x_1) \tag{5-26}$$

For an ensemble of N_p charged particles, Eqs. (5-25) and (5-26) generalize to give

$$\rho(x_p) = \frac{1}{H} \sum_{i=1}^{N_p} q_i W(x_i - x_p) \tag{5-27}$$

where x_i and q_i are the position and charge of particle i, respectively. Alternatively, Eq. (5-27) may be expressed in terms of the density $n(x)$ of particle coordinates, in which case the expression for charge assignment obtained for the NGP scheme [Eq. (2-29)] is recovered.

The CIC scheme, in which Eqs. (5-27) and (5-22) are used for charge assignment and force interpolation and Eq. (5-21) defines the charge-assignment–force-interpolation function, yields the interparticle force shown in Fig. 5-1(*c*). In comparing Figs. 5-1(*a*)–(*c*) note that

1. CIC gives a much smoother force than the NGP scheme (piecewise linear rather than piecewise constant).
2. CIC reduces the amplitude of fluctuations in the interparticle forces as the particles are displaced with respect to the mesh.
3. The residual errors in the force are spatially more localized for the CIC scheme.

The exact suppression of force fluctuations for particle separations greater than two cell widths is special to the one-dimensional case, although the three points as noted above still hold true in two and three dimensions.

5-2-4 Mixed Schemes

The NGP and CIC schemes both employ charge assignment functions which are the same as their force interpolation functions. A result of this symmetry is that they conserve momentum, i.e., the forces between a pair of particles are equal and opposite, and the force of a particle upon itself (the self force) is zero (cf. Sec. 5-3-3).

A possible variation is to use a charge assignment function which is different from the force interpolation function. In such cases, symmetry still causes the force between pairs of particles to be equal and opposite, but no longer ensures that the self force is zero. At best, the presence of the self force presents a nonphysical restriction on the timestep and, at worst, it is disastrous.

To illustrate the problems introduced by the self force we shall evaluate the self force of an isolated charge of charge q for cases where the CIC and NGP schemes are used for charge assignment and force interpolation and vice versa. In both instances we shall assume that the mesh-defined potentials and electric field are given by Eqs. (5-8) and (5-11), respectively.

Solving Eq. (5-8) for an isolated unit charge at mesh point 0, taking $\phi_0 = 0$, gives

$$\phi_p = -\frac{q}{2\epsilon_0} |x_p| \tag{5-28}$$

For a charge q at position x, where $0 \leqslant x \leqslant H/2$, the potential obtained when using CIC charge assignment is given by the superposition of potentials as given by Eq. (5-28) for charge $q(1-x/H)$ at mesh point zero and charge qx/H at mesh point 1. Following through this calculation gives for a charge in cell 0 (i.e., $|x| \leqslant H/2$)

$$\phi_p = -\frac{q}{2\epsilon_0}|x-x_p| \tag{5-29}$$

Hence, by Eqs. (5-11), (2-28), and (5-22) we obtain the self force of a particle of charge q in cell 0

$$F_{\text{self}} = -\frac{q^2}{2\epsilon_0 H}x \tag{5-30}$$

for the CIC charge assignment, NGP force interpolation scheme. Equation (5-30) describes the force for a simple harmonic oscillator of frequency

$$\omega_{\text{self}} = \left(\frac{q^2}{2\epsilon_0 mH}\right)^{1/2} \tag{5-31}$$

provided that the particle remains in the region $|x| \leqslant H/2$. To ensure stability of the leapfrog time integration scheme we require $\omega_{\text{self}}DT < 2$ in addition to any constraints on the timestep imposed by the physics of the many-body system. For typical choices of parameters in collisionless systems the timestep limit imposed by ω_{self} is usually less severe than that imposed by the natural physical frequencies (cf. Chapter 7). However, the ω_{self} timestep limit cannot be totally ignored as it sets a limit on the ratio DT^2/H.

If we repeat the self force calculation for NGP charge assignment and CIC force interpolation we find that the self force is now given by

$$F_{\text{self}} = \frac{q^2}{2\epsilon_0 H}x \tag{5-32}$$

Thus, whereas the CIC/NGP force calculation introduces relatively benign self force oscillations, the NGP/CIC self force is of opposite sign, giving rise to exponentially unstable motion even in the limit of zero timestep. Since such a situation is clearly unacceptable in a numerical scheme, we must restrict the force interpolation to an order the same as or lower than the charge assignment. Furthermore, if we want to ensure exact momentum conservation, then we require identical charge assignment and force interpolation functions (cf. Sec. 5-3-3).

5-3 THE HIERARCHY OF CHARGE ASSIGNMENT SCHEMES

The numerical errors arising from the operations of charge assignment and force interpolation cause the interparticle forces to be incorrect in magnitude and to fluctuate under displacements. Increasing the number of mesh points used in assignment and interpolation allows the undesirable numerical effects to be weakened but at the expense of greater computational cost. From an operational viewpoint, it is important to minimize errors for the given number of mesh points used, or equivalently to obtain the best quality–cost compromise.

A hierarchy of charge-assignment–force-interpolation functions, ordered in terms of cost and quality, can be obtained by applying the following criteria:

1. At particle separations large compared with the mesh spacing, the fluctuations should become negligible (i.e., spatial localization of errors).
2. The charge assigned to the mesh from a particle and the force interpolated to a particle from the mesh should smoothly vary as the particle moves across the mesh (i.e., fluctuation of spatially localized errors should be small).
3. Momentum should be conserved (i.e., avoid problems arising from self forces).

The first criterion enables sets of constraint equations restricting the forms of charge assignment functions to be derived from expansions in terms of mesh-size-related smallness parameters. The second criterion reduces the set of assignment functions to an hierarchy whose first two members turn out to be the NGP and CIC scheme. The third criterion causes the force interpolation function to be identical to the charge assignment function, although an alternative criterion, energy conservation, which can be substituted for criterion 3 leads to different schemes (cf. Sec. 5-5).

5-3-1 The Long-Range Constraints

Criterion 1 is applied by considering the potential or field due to a single unit charge. The results generalize to other charge values and many body systems by linearity and superposition. We consider first a one-dimensional scheme where charge is assigned from the particle to the m mesh points nearest to it.

Let $W_p(x)$ be the fraction of charge assigned to mesh point p at position x_p from a particle at position x and let $G(x'-x_p)$ (the Green's function) be the potential at position x' due to a unit charge at mesh point p, then charge assignment leads to a potential at point x' given by

$$\phi(x') = \sum_{p=1}^{m} W_p(x)G(x'-x_p) \tag{5-33}$$

$W_p(x)[= W(x-x_p)]$ is simply the charge assignment function, special cases of which were considered in the previous section. The correct potential ϕ_c due to a unit charge at position x is a function only of the distance $|x'-x|$, whereas the approximate potential is a function of both $|x'-x|$ and $(x-x_p)$, the position of x with respect to the mesh points. According to criterion 2 (and to the physics of the situation!), it is the dependence on $(x-x_p)$ which should be suppressed.

Taylor-expanding $G(x'-x_p)$ about $(x'-x)$ in Eq. (5-33) gives

$$\phi(x') = \sum_{p=1}^{m} W_p(x)G(x'-x) + \sum_{p=1}^{m} W_p(x) \sum_{n=1}^{\infty} \frac{\Delta_p^n}{n!} \frac{d^n G(x-x')}{dx^n} \tag{5-34}$$

$$\Delta_p = x - x_p \tag{5-35}$$

The requirement of charge conservation gives the first constraint on the choice of the form of W:

$$\sum_{p=1}^{m} W_p(x) = 1 \tag{5-36}$$

If charge is assigned to only one mesh point ($m = 1$) then Eq. (5-36) completely specifies the assignment scheme—that is, the NGP scheme. However, if m is greater than one, then the extra freedoms can be used to weaken the mesh dependence of $\phi(x')$: This is done by requiring successively higher-order terms (ordered by the smallness parameter, $\Delta_p/|x-x'|$) in Eq. (5-34) to be mesh-independent, that is, we demand that

$$\sum_{p=1}^{m} W_p \Delta_p^n = \text{constant} \tag{5-37}$$

for increasingly large n values as the number of mesh points m involved in the assignment of charge is increased. Since both ϕ_c and G are even functions of their arguments, Eq. (5-37) must be further restricted to give the constraint equations

$$\sum_{p=1}^{m} W_p \Delta_p^n = \begin{cases} 0 & n \text{ odd} \\ \text{constant} & n \text{ even} \end{cases} \tag{5-38}$$

For $m = 2$, only the zeroth order constraint [Eq. (5-36)] and the first-order constraint [$n = 1$ in Eq. (5-38)] can be satisfied, giving the leading error term in Eq. (5-34) of order $(\Delta_p)^2$. Using Eq. (5-35), the constraint equations become

$$W_1 + W_2 = 1 \tag{5-39}$$

$$W_1 x_1 + W_2 x_2 = x \tag{5-40}$$

where all W_p for $p \neq 1$ or 2 are zero and $x_1 \leqslant x \leqslant x_2$. Solving Eqs. (5-39) and (5-40) gives

$$W_1(x) = 1 - \frac{(x-x_1)}{H} \tag{5-41}$$

$$W_2(x) = 1 + \frac{(x-x_2)}{H} \tag{5-42}$$

where H (the cell width) is the spacing of the mesh points. Using the displacement invariance property of the charge assignment function

$$W_p(x) = W(x-x_p) \tag{5-43}$$

enables Eqs. (5-41) and (5-42) to be combined to give

$$W(x) = \begin{cases} 1 - \dfrac{|x|}{H} & |x| < H \\ 0 & \text{otherwise} \end{cases} \tag{5-44}$$

which is, of course, the CIC charge assignment function.

Increasing m to three allows the constraint equations (Eq. 5-38) to be satisfied up to $n = 2$

$$W_1 + W_2 + W_3 = 1 \tag{5-45}$$

$$W_1 x_1 + W_2 x_2 + W_3 x_3 = x \tag{5-46}$$

$$W_1 x_1^2 + W_2 x_2^2 + W_3 x_3^2 = C + x^2 \tag{5-47}$$

where
$$W_p = 0 \quad \text{for} \quad p \neq 1, 2 \text{ or } 3 \tag{5-48}$$

and $-H/2 \leqslant x - x_2 < H/2$. For this case, Eq. (5-34) becomes

$$\phi(x') = G(x'-x) + \frac{C}{2}\frac{d^2G(x'-x)}{dx^2} + O(\Delta^3) \tag{5-49}$$

Solving Eqs. (5-45) to (5-47) gives

$$W_1(x) = (X^2 - HX + C)/2H^2 \tag{5-50}$$

$$W_2(x) = 1 - (X^2 + C)/H^2 \tag{5-51}$$

$$W_3(x) = (X^2 + HX + C)/2H^2 \tag{5-52}$$

where $X = x - x_2$ and $-H/2 \leqslant X < H/2$. Equations (5-50) to (5-52) describe charge assignment in the form which gives the basis for the algorithm for charge assignment. Using the displacement invariance property [Eq. (5-43)] and Eq. (5-48), Eqs. (5-50) to (5-52) can be rewritten, as was done for the CIC scheme [Eq. (5-44)], in terms of a single piecewise polynomial charge assignment function:

$$W(x) = \begin{cases} (x^2 + 3Hx + 2H^2 + C)/2H^2 & -3H/2 \leqslant x < -H/2 \\ 1 - (x^2 + C)/H^2 & -H/2 \leqslant x < H/2 \\ (x^2 - 3Hx + 2H^2 + C)/2H^2 & H/2 \leqslant x < 3H/2 \\ 0 & \text{otherwise} \end{cases} \tag{5-53}$$

If C is set to zero in Eq. (5-53), then the mesh-calculated potential [Eq. (5-49)] will represent the correct potential ϕ_c to within $O(\Delta^3)$ provided that G is set equal to ϕ_c. If, however, C is chosen nonzero, then G must be chosen to satisfy

$$\frac{C}{2}\frac{d^2G}{dx^2} + G = \phi_c \tag{5-54}$$

to offset the reshaping effects of charge assignment on the potential and retain second-order accuracy. In practice, it turns out to be more advantageous to reshape the interparticle forces and accept the errors in the potential (see Chapters 7 and 8).

Similar long-range constraint equations, ordered in terms of the smallness parameter of the form $\Delta/|\mathbf{x} - \mathbf{x}'|$, are obtained in two and three dimensions by generalizing Eq. (5-33) and its expansion. Replacing x by $\mathbf{x}$ [$= (x, y)$ in two and (x, y, z) in three dimensions], etc., gives

$$\phi(\mathbf{x}') = \sum_{\mathbf{p}} W_{\mathbf{p}}(\mathbf{x}) G(\mathbf{x}' - \mathbf{x}_{\mathbf{p}}) \tag{5-55}$$

The vector **p** is now a pair (p_1, p_2) or triplet (p_1, p_2, p_3) of integers labeling the mesh point **p** at position $\mathbf{x_p}$ and the sum is taken over a number m of mesh points neighboring the source point **x**.

The expansion of Eq. (5-55) in two dimensions is

$$\phi(\mathbf{x}') = \sum_{\mathbf{p}} W_{\mathbf{p}}(\mathbf{x}) \sum_{r,s=0}^{\infty} \frac{\Delta_1^r \Delta_2^s}{r!s!} \frac{\partial^{r+s} G}{\partial x^r \partial y^s} \tag{5-56}$$

and in three dimensions

$$\phi(\mathbf{x}') = \sum_{\mathbf{p}} W_{\mathbf{p}}(\mathbf{x}) \sum_{r,s,t=0}^{\infty} \frac{\Delta_1^r \Delta_2^s \Delta_3^t}{r!s!t!} \frac{\partial^{r+s+t} G}{\partial x^r \partial y^s \partial z^t} \tag{5-57}$$

where
$$\mathbf{\Delta} = (\Delta_1, \Delta_2, \Delta_3) = \mathbf{x} - \mathbf{x_p} \tag{5-58}$$

If, as in the one-dimensional case, we only make the minimal restriction on G that $G(\mathbf{x}) = G(|\mathbf{x}|)$, we obtain from Eq. (5-56) or (5-57) the hierarchy of constraints on the charge assignment function:

$$\sum_{\mathbf{p}} W_{\mathbf{p}}(\mathbf{x}) = 1 \qquad \text{(charge conservation)} \tag{5-59}$$

$$\sum_{\mathbf{p}} W_{\mathbf{p}}(\mathbf{x})\Delta_i = 0 \qquad \text{(first-order)} \tag{5-60}$$

$$\sum_{\mathbf{p}} W_{\mathbf{p}}(\mathbf{x})\Delta_i\Delta_j = C\delta_{ij} \qquad \text{(second-order)} \tag{5-61}$$

$$\sum_{\mathbf{p}} W_{\mathbf{p}}(\mathbf{x})\Delta_i\Delta_j\Delta_k = 0 \qquad \text{(third-order)} \tag{5-62}$$

and so forth.

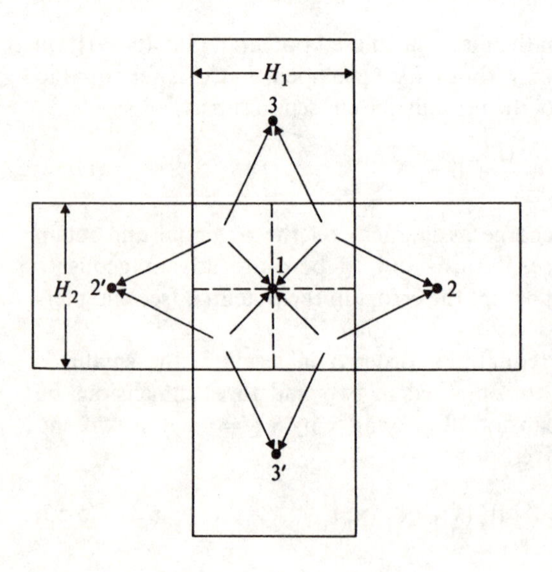

Figure 5-2 The two-dimensional three-point charge-sharing scheme assigns charge to the nearest grid point (labeled 1) and the next nearest grid points in the east-west direction (labeled 2 or 2′) and in the north-south direction (labeled 3 or 3′). The choice of the next nearest points is determined by the cell quadrant in which the particle lies.

Equations (5-59) through (5-62) are applied in the same manner as their one-dimensional counterpart. For example, in two-dimensions the lowest-order charge assignment function, satisfying only Eq. (5-59), is obtained by assigning all the charge to the nearest grid point:

$$W(\mathbf{x}) = \begin{cases} 1 & -\mathbf{H}/2 \leqslant \mathbf{x} < \mathbf{H}/2 \\ 0 & \text{otherwise} \end{cases} \tag{5-63}$$

$\mathbf{H} = (H_1, H_2)$ is the vector describing the cell dimensions of the mesh used. Extending assignment to the three nearest grid point to the source charge enables both the charge conservation and first-order constraint to be satisfied. Employing the labeling scheme shown in Fig. 5-2 the constraint equations become

$$W_1 + W_2 + W_3 = 1 \tag{5-64}$$

$$W_1 x_1 + W_2 x_2 + W_3 x_3 = x \tag{5-65}$$

$$W_1 y_1 + W_2 y_2 + W_3 y_3 = y \tag{5-66}$$

giving

$$W_1 = 1 - \frac{|x|}{H_1} - \frac{|y|}{H_2} \tag{5-67}$$

$$W_2 = \frac{|x|}{H_1} \tag{5-68}$$

$$W_3 = \frac{|y|}{H_2} \tag{5-69}$$

for $-\mathbf{H}/2 \leqslant \mathbf{x} < \mathbf{H}/2$.

Displacement invariance can be used to combine Eqs. (5-67) to (5-69) to form the charge assignment function:

$$W(\mathbf{x}) = \begin{cases} 1 - \dfrac{|x|}{H_1} - \dfrac{|y|}{H_2} & -\dfrac{\mathbf{H}}{2} \leqslant \mathbf{x} < \dfrac{\mathbf{H}}{2} \\ 1 - \dfrac{|x|}{H_1} & x = \dfrac{H_1}{2} \text{ or } \dfrac{H_1}{2} < |x| < H_1 \quad -\dfrac{H_2}{2} \leqslant y < \dfrac{H_2}{2} \\ 1 - \dfrac{|y|}{H_2} & -\dfrac{H_1}{2} \leqslant x < \dfrac{H_1}{2} \quad y = \dfrac{H_2}{2} \text{ or } \dfrac{H_2}{2} < |y| < H_2 \\ 0 & \text{otherwise} \end{cases} \tag{5-70}$$

Increasing the number of mesh points used to four still only permits the constraints up to first order to be satisfied, but introduces an extra degree of freedom in the choice of W. For instance, if we prescribe a four-point scheme which may be written in the product form

$$W(x, y) = P(x)P(y) \tag{5-71}$$

then satisfying the first-order constraint gives the charge assignment function

$$W(\mathbf{x}) = \begin{cases} \left(1 - \dfrac{|x|}{H_1}\right)\left(1 - \dfrac{|y|}{H_2}\right) & -\mathbf{H} < \mathbf{x} < \mathbf{H} \\ 0 & \text{otherwise} \end{cases} \tag{5-72}$$

Eq. (5-72) gives the two-dimensional version of the CIC charge assignment function.

The list of examples of two-dimensional assignment schemes can be extended indefinitely: Schemes involving five, six, seven, and more mesh points may be devised, allowing progressively higher-order constraints to be satisfied or further degrees of freedom to be added. The problem with many such schemes is that considerable computational effort is expended in deciding where the odd points are to be placed unless there is a regular invariant pattern of points, as in the four-point, symmetrical five-point, or nine-point assignment schemes. Of the nine-point schemes, perhaps the most useful is the second-order accurate version which is a product form whose factors are given by Eq. (5-53).

Three-dimensional schemes are obtained by repeating the two-dimensional analysis with a third component added. The three-dimensional version of the two-dimensional three-point scheme involves four points, and so forth. Since this exercise adds nothing further apart from straightforward algebra we shall not pursue it further here.

The expansion schemes, Eqs. (5-56) and (5-57), may be interpreted physically as a multipole expansion. The $O(\Delta)$ terms correspond to the dipole moment, the $O(\Delta^2)$ terms correspond to the quadrupole moment, and so forth. This relationship can be readily seen by substituting $G = \log r/2\pi\epsilon_0$ in Eq. (5-56) or $G = (4\pi\epsilon_0 r)^{-1}$ in Eq. (5-57). The distant potential and fields of a distribution of charges which are assigned using a zeroth-order scheme (NGP) will have a leading error term proportional to the dipole moment. In a first-order scheme (e.g., CIC), the leading error is quadrupole in nature, and so forth. The multipole interpretation becomes apparent when one inspects the charge assignment scheme examples. For instance, the three-point scheme, Eqs. (5-67) to (5-69), may be seen as a unit charge at mesh point 1, together with a dipole in the N-S direction with moment $H_2 y$ and a dipole in the E-W direction with moment $H_1 x$. Similar interpretations follow for other examples.

In two dimensions, the result of an error which is dipolar in nature is to give an error in the potential which varies like $\cos\theta$. For a quadrupole error (i.e., for a first-order scheme), the error varies like $\cos 2\theta$, and so forth for higher-order schemes, where the angle θ increases from 0 to 2π as the source charge is encircled in an anticlockwise direction. As mentioned above, this can be shown by setting $G = \log r/2\pi\epsilon_0$ in Eq. (5-56), or more straightforwardly by setting $G = (\log z)/2\pi\epsilon_0$ and using the complex variable equivalent of Eq. (5-56) (cf. Eastwood and Hockney, 1974)

$$\begin{aligned}\psi &= \frac{1}{2\pi\epsilon_0}\sum_p W_p(z)\log(z-z_p)\\ &= \frac{1}{2\pi\epsilon_0}\left[\log z - \sum_p \sum_{n=1}^{\infty} \frac{W_p}{n}\left(\frac{z_p}{z}\right)^n\right]\end{aligned} \tag{5-73}$$

The real part of Eq. (5-73) gives the potential ϕ. z is the usual complex variable, $z = x+iy$, p is a complex index, $p = p_1+ip_2$, labeling mesh points, and the origin is set at the source charge position.

The error in the potential is given by

$$\operatorname{Re}[\epsilon] = \operatorname{Re}\left[\sum_{n=1}^{\infty} \epsilon_n\right] = -\operatorname{Re}\left[\frac{1}{2\pi\epsilon_0}\sum_{n=1}^{\infty}\left(\frac{1}{nz^n}\sum_p W_p z_p^n\right)\right] \tag{5-74}$$

For the NGP scheme, the leading error is ϵ_1, giving

$$\epsilon_1 = -\frac{1}{2\pi\epsilon_0}\frac{z_1}{z} = -\frac{1}{2\pi\epsilon_0}\frac{r_1 e^{i\theta_1}}{re^{i\theta}} \tag{5-75}$$

$$\therefore \qquad \operatorname{Re}[\epsilon_1] = -\frac{r_1}{2\pi\epsilon_0 r}\cos(\theta_1-\theta) \tag{5-76}$$

i.e., NGP gives an error which is dipolar in nature with extrema along the line joining the source charge to the nearest grid point. For first-order schemes, the leading error term is

$$\epsilon_2 = -\frac{1}{4\pi\epsilon_0}\sum_p \frac{W_p z_p^2}{z^2} \tag{5-77}$$

which gives, for example, the error for the three-point and the CIC scheme for a charge midway between neighbouring mesh points of

$$\operatorname{Re}[\epsilon_2] = -\frac{1}{4\pi\epsilon_0 r^2} r_1^2 \cos 2\theta \tag{5-78}$$

The net result of these errors is to give a directional dependence, or angular anisotropy, to the potential. This nonphysical angular dependence of potential, and also of the fields, is further compounded by the errors arising from approximating the field equations (see Sec. 5-4). An alternative approach to the multipole expansion viewpoint is given by Rosen et al. (1970) and Kruer et al. (1973).

5-3-2 The Smoothness Constraints

The constraint equations on the form of the charge assignment function generate a hierarchy of schemes where increasingly rapid decay of errors with distance from the source charge is obtained as more mesh points are employed in the assignment scheme. However, the higher-order schemes, despite giving more localized errors, may not prove superior to lower-order schemes if the fluctuations of the force for small particle separations is large. The second criterion, the smoothness constraint, is concerned with the reduction of these fluctuations.

In x space, the concept of smoothness is quantified in terms of the continuity of the derivatives of the approximated function. An example of this is furnished by the forces plotted in Fig. 5-1: The lowest-order (NGP) scheme gives forces which are discontinuous in value, while the CIC gives forces continuous in value but discontinuous in their first derivative. Further improvements in smoothness are obtained by first demanding continuity in both value and first derivative of the force and then, in addition, the continuity of the second derivative, and so forth. The same remarks apply for the mesh-defined charge density and for the potential. In k space, the smoothness is reflected in the rapidity of the decrease of amplitude of the harmonics of the Fourier-transformed function as wavenumber k increases (cf. Sec. 5-6).

The order of smoothness obtainable is determined by the number of mesh points involved. In one dimension, the highest-order polynomial that can be uniquely fitted to n mesh values is of order $n-1$, and therefore the highest derivative which can be made continuous is the $(n-1)$th derivative (the $n=0$ derivative is the function and the $n=-1$ derivative is the integral of the function). Consequently, the best charge-assignment–force-interpolation functions according to the smoothness criterion are, for one and two mesh points, respectively, the NGP and CIC functions [Eqs. (2-28) and (5-21)].

Continuity of value and derivative is obtained by the TSC ("triangular shaped cloud") scheme, which in one dimension employs three mesh points and has an assignment–interpolation function which is piecewise quadratic. To illustrate how the smoothness criterion may be applied, we shall outline its use in deriving the TSC charge assignment (inverse interpolation) and interpolation functions.

Consider the situation depicted in Fig. 5-3. If the particle at position x carries a unit charge, then the fractions (W_{-1}, W_0, W_1) of the charge assigned to mesh points $(-1, 0, -1)$ satisfy

$$W_{-1} + W_0 + W_1 = 1 \tag{5-79}$$

where each of the fractions are quadratic functions of the position x. It follows from symmetry that

$$W_0(x) = ax^2 + b \tag{5-80}$$

and

$$W_{-1}(-x) = W_1(x) = cx^2 + dx + e \tag{5-81}$$

The smoothness criterion applied to charge assignment requires both the amounts

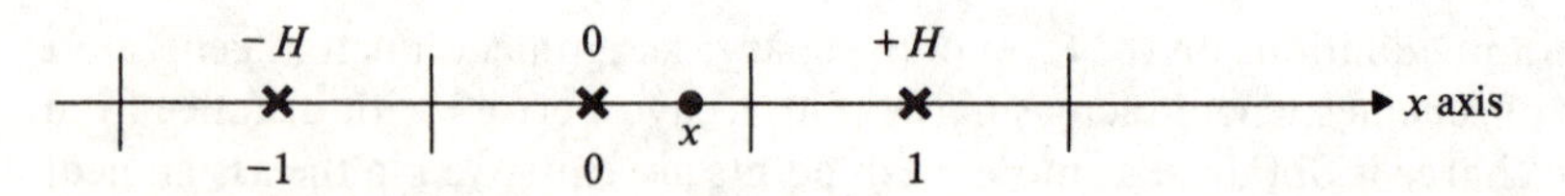

Figure 5-3 In the one-dimensional three-point scheme, assignment from a charge at position x, where $|x| \leqslant H/2$, is to the three neighboring mesh points $(-1, 0, 1)$ at $(-H, 0, H)$, respectively. The vertical bars indicate cell boundaries.

of charge assigned to each point and their derivatives to vary continuously. The polynomial form, Eqs. (5-80) and (5-81), guarantees this for $|x| < H/2$, so we only need to consider changes at $x = H/2$ or $-H/2$. As the particle crosses the cell boundary at $x = H/2$, the three nearest mesh points become $(0, 1, 2)$, so we require for continuity of value

$$W_{-1}(H/2) = 0 \tag{5-82}$$

$$W_0(H/2) = W_1(H/2) \tag{5-83}$$

and for continuity of derivative

$$\frac{d}{dx} W_{-1}(H/2) = 0 \tag{5-84}$$

$$\frac{d}{dx} W_0(H/2) = -\frac{d}{dx} W_1(H/2) \tag{5-85}$$

Solving Eqs. (5-79) to (5-85) gives the TSC scheme:

$$W_0(x) = \tfrac{3}{4} - \left(\frac{x}{H}\right)^2 \tag{5-86}$$

$$W_{-1}(-x) = W_1(x) = \tfrac{1}{2}\left(\tfrac{1}{2} + \frac{x}{H}\right)^2 \tag{5-87}$$

Using displacement invariance, Eqs. (5-86) and (5-87) combine to give the TSC assignment function:

$$W(x) = \begin{cases} \tfrac{3}{4} - \left(\dfrac{x}{H}\right)^2 & |x| \leqslant \dfrac{H}{2} \\ \tfrac{1}{2}\left(\tfrac{3}{2} - \dfrac{|x|}{H}\right)^2 & \dfrac{H}{2} \leqslant |x| \leqslant \dfrac{3H}{2} \\ 0 & \text{otherwise} \end{cases} \tag{5-88}$$

Comparing Eq. (5-88) with Eq. (5-53), we see that the TSC assignment function is the special case, $C = H^2/4$, of the second-order-accurate, three-point scheme obtained by using the long-range constraints.

One consequence of using the TSC assignment scheme is a spreading of the charge. It follows from Eqs. (5-86) and (5-87) that a unit charge at $x = 0$ leads to the fractional charges $(\frac{1}{8}, \frac{3}{4}, \frac{1}{8})$ being assigned to mesh points $(-1, 0, +1)$, respectively. In some, but not all, circumstances this may be regarded as a favorable increase in the width of a finite-sized particle obtained in conjunction with the higher order of smoothness. More commonly, it is desirable to obtain the smoothness properties independently of the spreading, an end achieved by following assignment by a reshaping step.

To illuminate the parallel with the discussion of Sec. 5-3-1, we take again the case of the potential of a unit charge. If we let the potential at mesh point p due to a unit charge at mesh point p' be $G_{p-p'}$, then the effect of charge assignment is to give a potential at mesh point p due to a unit charge at $x = 0$ of

$$\phi_p = \tfrac{1}{8}G_{p+1} + \tfrac{3}{4}G_p + \tfrac{1}{8}G_{p-1} \tag{5-89}$$

i.e.,

$$\phi_p = G_p + \frac{H^2}{8}\frac{(G_{p+1} - 2G_p + G_{p-1})}{H^2} \tag{5-90}$$

Equation (5-90) is simply the finite-difference approximation to Eq. (5-54) with $C = (H/2)^2$, i.e., C takes the value which reduces Eq. (5-53) to the TSC assignment function, Eq. (5-88). If we let $G^+_{p-p'}$ be the correct potential at mesh point p due to a unit charge at mesh point p', then Eq. (5-90) can be used to define values of G which offset the spreading effects of charge assignment

$$G^+_p = \left(1 + \frac{H^2}{8}D^2\right)G_p \tag{5-91}$$

In Eq. (5-91), D^2 is the second difference operator analogous to the laplacian operator for continuous functions.

Equation (5-89) becomes for a distribution of charges

$$\phi_p = \sum_{p'} G_{p-p'}\rho_{p'}H \tag{5-92}$$

where $(\rho_{p'}H)$ is the charge assigned to mesh point p', or, equivalently, the set of values $\{\rho_p\}$ define the mesh charge density. In Eq. (5-92), values of G have been adjusted in accordance with Eq. (5-91) to offset the spreading of the charge. Alternatively, we could specify the correct interaction potential $\phi_{p-p'}$ to give

$$\phi_p = \sum_{p'} G^+_{p-p'}\rho^*_{p'}H \tag{5-93}$$

where Eqs. (5-91) to (5-93) define the corrected charge density values $\{\rho^*_p\}$. Substituting Eq. (5-91) into Eq. (5-93), rearranging the sum, and subtracting Eq. (5-92) gives the relationship between ρ and ρ^*:

$$\left(1 + \frac{H^2}{8}D^2\right)\rho^*_p = \rho_p \tag{5-94}$$

Equations (5-91)–(5-94) give us two alternative ways of correcting the spreading effects of charge assignment. Either we first solve Eq. (5-94) to obtain $\{\rho^*_p\}$ and then solve

$$\nabla^2\phi_p = -\frac{\rho^*_p}{\epsilon_0} \tag{5-95}$$

for the potential, or first solve for the uncorrected potential

$$\nabla^2\phi^*_p = -\frac{\rho_p}{\epsilon_0} \tag{5-96}$$

and then correct the potentials by solving

$$\left(1+\frac{H^2}{8}D^2\right)\phi_p = \phi_p^* \tag{5-97}$$

for the values $\{\phi_p\}$. In both of these approaches, no account is taken of finite-difference errors in the approximations to the laplacians in Eqs. (5-95) and (5-96).

The smoothness criterion, when applied to interpolation, is generally called *spline fitting*. Referring again to Fig. 5-3, if we have some quantity A (for example, the potential or the force field) which has values $(A_{-1}^\dagger, A_0^\dagger, A_1^\dagger)$ at mesh points $(-1, 0, 1)$, respectively, then the interpolated value of A at position x is given by

$$A(x) = W_{-1}A_{-1}^\dagger + W_0 A_0^\dagger + W_1 A_1^\dagger \tag{5-98}$$

Applying symmetry, and continuity of value and derivative of $A(x)$ at $x = H/2$, yields Eqs. (5-80) to (5-85). Prescribing $A(x)$ equal to mesh values for the function $A(x) = $ constant recovers Eq. (5-79) and thence Eq. (5-88) for the interpolation function.

Interpolation leads to spreading similar to that found for assignment. Again, the spreading can be offset by introducing an extra correction step. If we demand that the interpolated function $A(x)$ takes the correct values $A(x_p) = A_p$ at mesh points, then Eq. (5-98) yields the equation for the mesh-defined quantities $\{A_p^\dagger\}$:

$$A_p = \left(1+\frac{H^2}{8}D^2\right)A_p^\dagger \tag{5-99}$$

In the case of the potential, Eq. (5-99) is simply a repeat of the prescription for the charge assignment correction.

For the particle–mesh force calculation with both charge assignment and force interpolation corrections the sequence of steps becomes the following:

1. Assign charge:

$$\rho_p = \frac{q}{H}\sum_{i=1}^{N_p} W(x_i - x_p) \tag{5-100}$$

2. Correct the charge density:

$$L_1 \rho_p^* = \rho_p \tag{5-101}$$

3. Solve for potential and field:

$$\nabla^2 \phi_p = -\rho_p^*/\epsilon_0 \tag{5-102}$$

$$F_p = -q\frac{d\phi_p}{dx} \tag{5-103}$$

4. Correct the forces:

$$L_2 F_p^\dagger = F_p \tag{5-104}$$

5. Interpolate:

$$F(x_i) = \sum_p W(x_i - x_p) F_p^{\dagger} \tag{5-105}$$

In steps 2 and 4, the operators L_1 and L_2 are the "sharpening" operators which offset the spreading due to assignment and interpolation, respectively. The preceding discussion used fitting of mesh values for interactions between point particles. In practice, aiming for a finite sized particle is more realistic, in which case the sharpening operators, L_1 and L_2, will differ from the examples given. Furthermore, no account has yet been taken of the effects of errors in step 3. Although such errors can be included in the analysis of the form outlined, the manipulation becomes exceedingly cumbersome, particularly when dealing with two and three dimensions. As mentioned in the introduction, the overall analysis is best tackled using transform space methods.

To retain the smoothness properties of the one-dimensional charge-assignment–force-interpolation functions, the assignment functions for two-dimensional schemes must take the product form

$$W(\mathbf{x}) = W(x, y) = W(x)W(y) \tag{5-106}$$

where both of the separate functions in the product on the right-hand side of Eq. (5-106) are the one-dimensional assignment functions with the required order of smoothness. Similarly, in three dimensions, the assignment functions must be of the form

$$W(\mathbf{x}) = W(x, y, z) = W(x)W(y)W(z) \tag{5-107}$$

to obtain everywhere continuity of value (first order), continuity of value and first derivative (second order), etc. These product forms have the additional advantage that the factorization enables them to be computed with a relatively small number of arithmetic operations. For example, finding the twenty-seven point charge weights using the three-dimensional TSC scheme gives a real arithmetic operations count of no more than thirty-three multiplications and fifteen additions to obtain all twenty-seven charge weights, i.e., approximately one multiplication and half an addition per mesh point per particle.

The product forms which satisfy successively higher orders of the long-range constraints and of smoothness form the sequence of assignment schemes NGP, CIC, TSC,... involving $1^d, 2^d, 3^d, \ldots$ mesh points where d is the dimensionality. In general these schemes are to be preferred, although in three dimensions the cubic increase in the number of mesh points involved with order can make intermediate schemes the more cost-effective alternatives. The intermediate schemes can be devised to satisfy the smoothness criteria everywhere up to a certain order and along particular axes to higher orders. For example, the three-point scheme [Eq. (5-70)], which is intermediate in computational cost between the NGP and CIC scheme, satisfies the first-order long-range constraint (giving a quadrupole leading error term) but only gives continuity of charge variation and force when particles cross cell boundaries along lines which are parallel to the x axis, the y axis, and the $|x| = |y|$ axes, and pass through mesh points. The largest discontinuous change in the charge assigned to a particular mesh point from a particle with unit charge is $\frac{1}{2}$ for the three-point scheme, compared with 1 for NGP and 0 for CIC.

5-3-3 The Momentum Conservation Constraint

The total momentum, in the absence of roundoff errors, is identically conserved by all particle mesh calculations which have

1. identical charge assignment and force interpolation functions and
2. correctly space-centered difference approximations to derivatives (or in the case of a Green's function formulation, a discrete Green's function with the correct symmetry).

Condition 1 states that if the charge assigned to mesh point $\mathbf{p}$, $\delta\rho(\mathbf{x_p})$, from a particle of charge q at position $\mathbf{x}$ is given by

$$\delta\rho(\mathbf{x_p}) = \frac{q}{V_c} W(\mathbf{x}-\mathbf{x_p}) \tag{5-108}$$

then the force, $\mathbf{F}(\mathbf{x})$, interpolated to that particle is given by

$$\mathbf{F}(\mathbf{x}) = \sum_{\mathbf{p}} qW(\mathbf{x}-\mathbf{x_p})\mathbf{E}(\mathbf{x_p}) \tag{5-109}$$

where $\mathbf{E}(\mathbf{x_p})$ is the mesh-defined electric field at mesh point $\mathbf{p}$ and the sum is over all mesh points. Condition 2 requires that the approximate equations used to solve for the mesh-defined potential and fields can be formally expressed in the form

$$\mathbf{E}(\mathbf{x_p}) = V_c \sum_{\mathbf{p}'} \mathbf{d}(\mathbf{x_p}; \mathbf{x_{p'}})\rho(\mathbf{x_{p'}}) \tag{5-110}$$

where

$$\mathbf{d}(\mathbf{x_p}; \mathbf{x_{p'}}) = -\mathbf{d}(\mathbf{x_{p'}}; \mathbf{x_p}) \tag{5-111}$$

In Eqs. (5-108) and (5-110), V_c is the volume of a mesh cell, for one-dimensional schemes, $V_c = H$, for two, $V_c = H_1 H_2$ and for three, $V_c = H_1 H_2 H_3$. Eq. (5-111) is the requirement that the mesh-defined field components due to a charge at $\mathbf{x} = 0$ are odd functions of x, y, and z.

To show that conditions 1 and 2 are sufficient to ensure momentum conservation we must prove (*a*) that self forces are zero and (*b*) that forces between particles are equal and opposite.

(*a*) *The self force*, $\mathbf{F}_{\text{self}}$:

Combining Eqs. (5-108) to (5-110) for a particle of charge q at position $\mathbf{x}$ gives

$$\mathbf{F}_{\text{self}}(\mathbf{x}) = q^2 \sum_{\mathbf{p},\mathbf{p}'} \mathbf{d}(\mathbf{x_p}; \mathbf{x_{p'}})W(\mathbf{x}-\mathbf{x_p})W(\mathbf{x}-\mathbf{x_{p'}}) \tag{5-112}$$

also interchanging p and p'

$$= q^2 \sum_{\mathbf{p},\mathbf{p}'} \mathbf{d}(\mathbf{x_{p'}}; \mathbf{x_p})W(\mathbf{x}-\mathbf{x_{p'}})W(\mathbf{x}-\mathbf{x_p}) \tag{5-113}$$

Adding Eq. (5-112) to Eq. (5-113) and using Eq. (5-111) yields the desired result that $\mathbf{F}_{\text{self}} = 0$.

(*b*) *Interparticle forces:*

The force on particle 1 of charge q_1 at position $\mathbf{x}_1$ due to particle 2 of charge q_2 at $\mathbf{x}_2$ is, from Eqs. (5-108) to (5-110),

$$\mathbf{F}_{12} = q_2 q_1 \sum_{\mathbf{p},\mathbf{p}'} W(\mathbf{x}_1 - \mathbf{x}_\mathbf{p})\mathbf{d}(\mathbf{x}_\mathbf{p}; \mathbf{x}_{\mathbf{p}'})W(\mathbf{x}_2 - \mathbf{x}_{\mathbf{p}'}) \tag{5-114}$$

Similarly, the force on particle 2 due to particle 1 is

$$\mathbf{F}_{21} = q_1 q_2 \sum_{\mathbf{p},\mathbf{p}'} W(\mathbf{x}_2 - \mathbf{x}_{\mathbf{p}'})d(\mathbf{x}_{\mathbf{p}'}; \mathbf{x}_\mathbf{p})W(\mathbf{x}_1 - \mathbf{x}_\mathbf{p}) \tag{5-115}$$

Adding Eqs. (5-114) and (5-115) gives the result

$$\mathbf{F}_{21} + \mathbf{F}_{12} \equiv 0 \tag{5-116}$$

i.e., forces are equal and opposite.

5-3-4 Cloud and Assignment Function Shapes

The hierarchy of charge assignment schemes obtained by applying the long-range and smoothness criteria can be interpreted physically by ascribing a finite width to the distribution of charge in each particle. The advantage of this interpretation over the multipole expansion description is that it allows, with some minor modifications, the wave-kinetic analysis of the continuous differential system to be brought to bear on the analysis of the simulation models. However, care must be taken when using the finite-sized particle idea, since the shape one obtains depends on the physical quantity used to determine that shape.

The cloud shape and the assignment function shape arise from slightly different interpretations of the mesh-defined charge density. If we use the definition of the mesh-defined charge density used in Sec. 2-2-3, namely that the charge density at a mesh point is the total charge in the cell surrounding that mesh point divided by the cell volume, then the appropriate shape is the cloud shape. Alternatively, we may regard the distribution of overlapping finite-sized particles as leading to a smoothly varying continuous charge density, sample values of which are recorded at mesh points. In this case the appropriate shape is the assignment function shape.

The cloud shape $S(x')$ of a particle with unit charge is its charge density, where x' measures the distances from the center of the particle. The fraction of the charge assigned from a particle of shape S at position x to mesh point p at x_p is given by the overlap of the cloud with cell p, i.e.,

$$W(x - x_p) = W_p(x) = \int_{x_p - H/2}^{x_p + H/2} S(x' - x)\,dx' \tag{5-117}$$

Using the top-hat function Π (see Appendix A) and Eq. (5-117) gives the relationship between the cloud shapes S and the assignment function W:

$$W(x) = \int \Pi\left(\frac{x'}{H}\right) S(x' - x)\,dx' \tag{5-118}$$

If we restrict S to even functions then Eq. (5-118) may be compactly expressed as

$$W(x) = \Pi\left(\frac{x}{H}\right) * S(x) \tag{5-119}$$

Cloud shapes S for the hierarchy of schemes may be obtained using an approach similar to that used for the assignment function in Secs. 5-3-1 and 5-3-2. However, since we have already determined the functions W, we can use Eq. (5-119) to obtain them. Using the definitions of the functions Π and $\wedge$ given in Appendix A, we can rewrite the charge assignment functions for the NGP, CIC, and TSC schemes [Eqs. (2-28), (5-44), and (5-88)], respectively as

$$\text{NGP:}\; W(x) = \Pi\left(\frac{x}{H}\right) \equiv \frac{1}{H}\Pi\left(\frac{x}{H}\right) * \delta\left(\frac{x}{H}\right) \tag{5-120}$$

$$\text{CIC:}\; W(x) = \wedge\left(\frac{x}{H}\right) = \frac{1}{H}\Pi\left(\frac{x}{H}\right) * \Pi\left(\frac{x}{H}\right) \tag{5-121}$$

$$\text{TSC:}\; W(x) = \frac{1}{H}\wedge\left(\frac{x}{H}\right) * \Pi\left(\frac{x}{H}\right) = \frac{1}{H^2}\Pi\left(\frac{x}{H}\right) * \Pi\left(\frac{x}{H}\right) * \Pi\left(\frac{x}{H}\right) \tag{5-122}$$

Therefore, the cloud shapes are

$$\text{NGP:}\; S(x) = \frac{1}{H}\delta\left(\frac{x}{H}\right) = \delta(x) \tag{5-123}$$

$$\text{CIC:}\; S(x) = \frac{1}{H}\Pi\left(\frac{x}{H}\right) \tag{5-124}$$

$$\text{TSC:}\; S(x) = \frac{1}{H}\wedge\left(\frac{x}{H}\right) \tag{5-125}$$

Each successively higher-order assignment function is obtained by convolving the previous assignment function with $(1/H)\Pi(x/H)$, so the sequence summarized in Table 5-1 can be extended indefinitely.

Figure 5-4 shows a pictorial representation of the cloud shape interpretation. The particle carries with it the cloud shape appropriate to the assignment scheme. The area of overlap of the cloud shapes with a cell determines the fraction of the charge assigned to the mesh point in that cell. Figure 5-5 shows the corresponding situation for the assignment function interpretation of assignment. In this case, the

Table 5-1 The hierarchy of assignment schemes of the product form ($H = 1$)

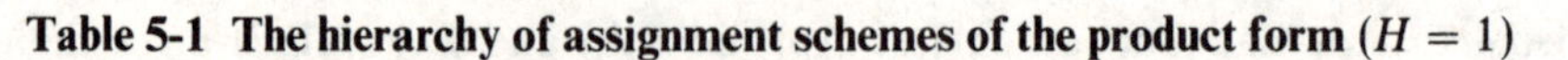

Scheme	Order	Number of points	Cloud shape	Assignment function shape	Force
NGP	0	1^d	δ	Π	Stepwise
CIC	1	2^d	Π	$\Lambda = \Pi * \Pi$	Continuous piecewise linear
TSC	2	3^d	Λ	$\Pi * \Pi * \Pi$	Continuous value and first derivative
PQS	3	4^d	$\Lambda * \Pi$	$\Pi * \Pi * \Pi * \Pi$	Continuous value first and second derivative

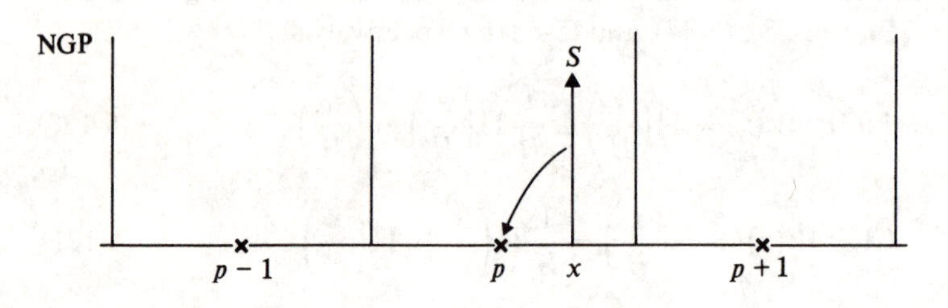

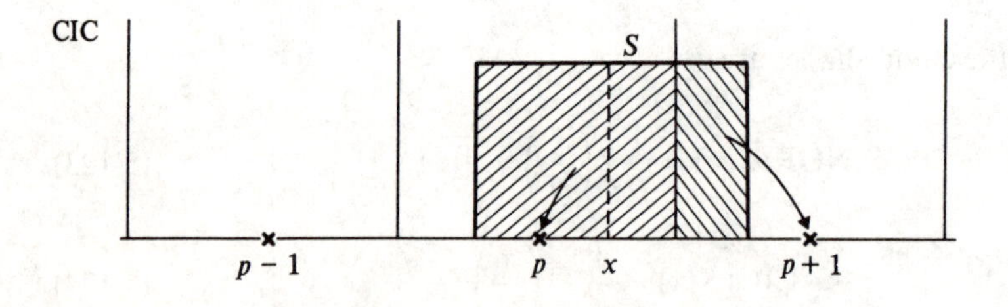

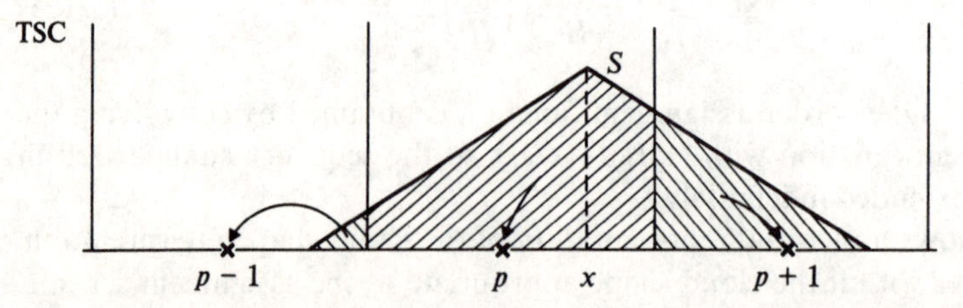

Figure 5-4 The cloud shape interpretation of charge assignment. The fraction of charge assigned from a particle at position x to a given mesh point is equal to the area of overlap of the cloud shape S with the cell containing that point.

particles carry the assignment function, and the value of the function at a mesh point gives the fraction of the charge assigned to that point.

The multidimensional versions of the schemes listed in Table 5-1 have cloud shapes S and assignment function shapes W which are given by products of the corresponding one-dimensional functions for each component. Figure 5-6 shows a pictorial interpretation of the two-dimensional CIC scheme: The product of Π functions gives the particle a shape whose projection is equal to the cell area. The area of overlap of the cloud shape gives the fraction of charge assigned to each of the four mesh points around the particle: It is for this reason that the CIC scheme is sometimes referred to as "area weighting" (Harlow 1964, Morse 1970).

Cloud shapes can also be defined for the two- and three-dimensional schemes which are not of the product form. For example, the three-point assignment

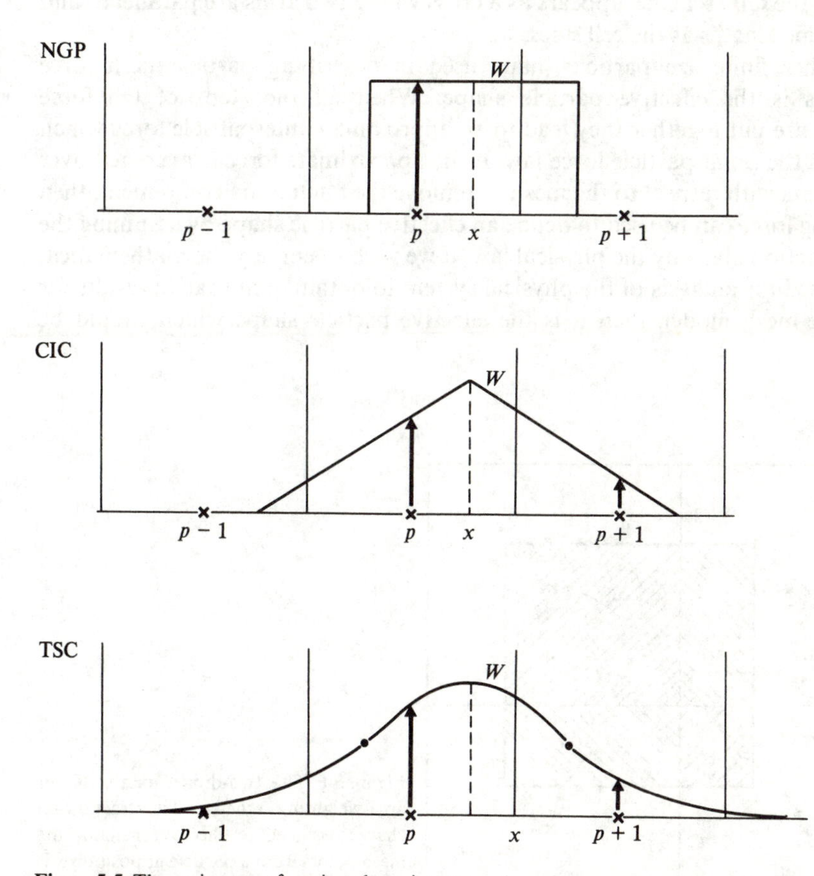

Figure 5-5 The assignment function shape interpretation of charge assignment. The fraction of charge assigned from a particle at position x to a given mesh point is equal to the value of the assignment function W at that point.

scheme, Eq. (5-70), has a cloud shape

$$S(x,y) = \frac{1}{H_1 H_2}\left[\Pi\left(\frac{x}{H_1}\right)\delta\left(\frac{y}{H_2}\right) + \delta\left(\frac{x}{H_1}\right)\Pi\left(\frac{y}{H_2}\right) - \delta\left(\frac{x}{H_1}\right)\delta\left(\frac{y}{H_2}\right)\right] \tag{5-126}$$

and an assignment function

$$W(x,y) = \Pi\left(\frac{x}{H_1}\right)\Pi\left(\frac{y}{H_2}\right) * S(x,y) \tag{5-127}$$

$$= \Lambda\left(\frac{x}{H_1}\right)\Pi\left(\frac{y}{H_2}\right) + \Pi\left(\frac{x}{H_1}\right)\Lambda\left(\frac{y}{H_2}\right) - \Pi\left(\frac{x}{H_1}\right)\Pi\left(\frac{y}{H_2}\right) \tag{5-128}$$

The three-point scheme cloud shape, when plotted in the same manner as shown in Fig. 5-6 for the CIC scheme, appears as a cross whose two arms are parallel to and have the same lengths as the cell sides.

A further finite-size particle shape used in describing particle-mesh force calculations is the effective particle shape. When all the steps of the force calculation are put together they lead to an approximate interparticle force which differs from the point particle force law. If the approximate force is averaged over displacements with respect to the mesh to remove the fluctuating component, then the resulting force can be used to define an effective particle shape by assuming the point interaction given by the physical law. If we wish to employ the mathematical methods used for analysis of the physical system to obtain quantitative results for the particle-mesh model, then it is the effective particle shape which should be used.

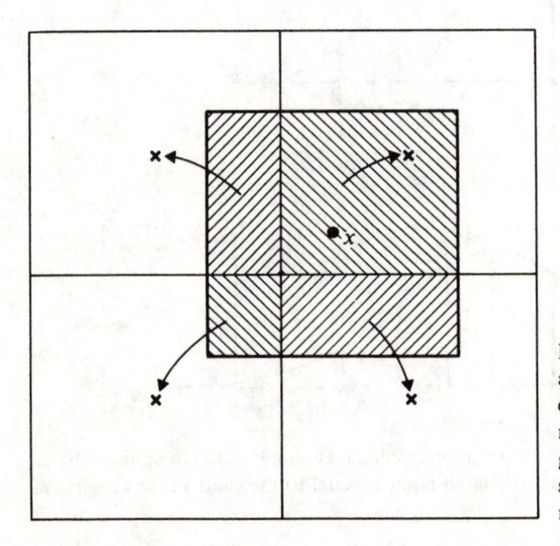

Figure 5-6 The two-dimensional CIC or area-weighting scheme. The fraction of charge assigned to the four neighboring mesh points from a particle at position x is given by the area of overlap of its cloud shape with the cells containing those neighboring mesh points.

5-4 TRUNCATION ERRORS

The truncation errors in the one-dimensional example discussed in Sec. 5-2 were absent except at the source point because of the nature of the solution of the one-dimensional Poisson equation. This absence of truncation errors for simple finite-difference approximations does not occur in two and three dimensions.

The simplest two-dimensional approximation to Poisson's equation uses a five-point approximation to the laplacian, which for a square mesh $(H_1 = H_2 = H)$ may be written

$$\nabla^2\phi \simeq D^2\phi = \frac{\phi_{p_1+1,p_2} - 2\phi_{p_1,p_2} + \phi_{p_1-1,p_2}}{H^2} + \frac{\phi_{p_1,p_2+1} - 2\phi_{p_1,p_2} + \phi_{p_1,p_2-1}}{H^2} \tag{5-129}$$

where mesh point $\mathbf{p} = (p_1, p_2)$ is at position $\mathbf{x_p} = \mathbf{p}H$. Taylor-expanding $D^2\phi$ about the point $\mathbf{x} = \mathbf{x_p}$ gives

$$\nabla^2\phi = D^2\phi - \frac{H^2}{12}\left(\frac{\partial^4\phi}{\partial x^4} + \frac{\partial^4\phi}{\partial y^4}\right) + \text{h.o.t.} \tag{5-130}$$

Thus, the five-point approximation is second-order accurate, the second term on the right-hand side of Eq. (5-130) giving the truncation error δ.

An estimate of the truncation error for a unit (rod) charge at the origin may be obtained by substituting the solution to the differential problem, $\phi = \log r/2\pi\epsilon_0$, into the expression for the truncation error, giving

$$\delta = -H^2 \frac{\cos 4\theta}{2\pi\epsilon_0 r^4} \tag{5-131}$$

where θ is the azimuthal angle. The important point to note from Eq. (5-131) is that the truncation error will lead to an error in the magnitude of the potential which depends both on the distance from the source charge and on the direction. The exact inversion of the five-point operator, Eq. (5-129), for a unit charge (Buneman, 1971) quantifies the $\cos 4\theta$ dependence of the error in the potential, with the largest negative errors along the $\pm x$ and $\pm y$ axes and the largest positive errors along the $|x| = |y|$ axes.

Higher-order difference approximations, e.g., the nine-point approximation, may be used to reduce the truncation error and the resulting angular anisotropy in the potential (Buneman, 1973). However, if the boundary conditions are such that the Green's function formulation of Poisson's equation takes the displacement invariant form

$$\phi(\mathbf{x_p}) = V_c \sum_{p'} G(\mathbf{x_p} - \mathbf{x_{p'}})\rho(\mathbf{x_{p'}}) \tag{5-132}$$

then the solution of the Green's function formulation involves the same order of computational cost as the difference equations (cf. Chapter 6). If we treat values of

G as adjustable parameters, then the truncation errors in the potential can take whatever value we choose. In particle-mesh calculations, the most sensible choice is that which gives the best approximation to the forces.

The replacement of the gradient of the potential by a finite-difference approximation is a further source of truncation errors. The second-order-accurate centered two-point finite-difference scheme replaces the true electric field

$$\mathbf{E} = -\nabla\phi \tag{5-133}$$

by

$$\mathbf{E}(\mathbf{x_p}) = -\left(\frac{\phi_{p_1+1,p_2}-\phi_{p_1-1,p_2}}{2H}, \frac{\phi_{p_1,p_2+1}-\phi_{p_1,p_2-1}}{2H}\right) \tag{5-134}$$

Taylor-expanding Eq. (5-134) gives the truncation error ϵ in the electric field

$$\epsilon = \frac{H^2}{6}\left(\frac{\partial^3\phi}{\partial x^3}, \frac{\partial^3\phi}{\partial y^3}\right) \tag{5-135}$$

For the case of a unit charge at the origin, Eq. (5-135) becomes

$$\epsilon = \frac{H^2}{6\pi\epsilon_0 r^3}(-\cos 3\theta, \sin 3\theta) \tag{5-136}$$

Thus, to lowest order, the magnitude of the error $|\epsilon|$ is independent of direction. The direction of the error is aligned with the correct field when $\tan 3\theta = \pm\tan\theta$, i.e., when $\theta = n\pi/4$, and is perpendicular to the correct field when $\theta = (2n+1)\pi/8$, where n is an integer.

The next-higher-order approximation to the gradient involves four mesh points for each component. The four-point scheme takes a linear combination of centered difference approximations over the two neighboring points and the two next-nearest neighbors.

$$E_x(\mathbf{x_p}) = -\alpha\frac{(\phi_{p_1+1,p_2}-\phi_{p_1-1,p_2})}{2H} - (1-\alpha)\frac{(\phi_{p_1+2,p_2}-\phi_{p_1-2,p_2})}{4H} \tag{5-137}$$

Setting $\alpha = \frac{4}{3}$ in Eq. (5-137) leads to a cancellation of the $O(H^2)$ term in the truncation error, leaving

$$\epsilon = -\frac{H^4}{30}\left(\frac{\partial^5\phi}{\partial x^5}, \frac{\partial^5\phi}{\partial y^5}\right) \tag{5-138}$$

For the case of a unit charge at the origin, the magnitude of the error is again independent of direction, and is along the direction of the correct field when $\theta = n\pi/4$ or $n\pi/6$ where n is integer.

In principle, truncation errors in the field can be eliminated by using an integral formulation for the field. In practice, such an approach is rarely taken because of the demands on computer storage and time such a calculation would make. The compromise adopted for the P^3M algorithm (Chapter 8) is to use Eq.

(5-137) for the field components, and adjust the Green's function in Eq. (5-132) to compensate for errors in the magnitude of the force.

5-5 ENERGY-CONSERVING SCHEMES

The approach used so far in the analysis of the P-M force calculation has treated the mesh approximations using finite differences. An alternative approach introduced by H. R. Lewis (Lewis, 1970*a,b*) is to use a finite-element method. The results of employing Lewis's approach are the so-called "energy-conserving schemes." The advantages of the finite-element formulation are that it leads to a class of P-M calculations which are optimal for the basis functions chosen and that it can be readily applied to fully electromagnetic systems, awkward geometries, and arbitrarily spaced meshes. However, for the PM and P^3M models treated in this book, it transpires that the optimized finite difference ("Momentum-Conserving") schemes provide the more cost-effective alternative.

The finite-element method uses an integral rather than differential description of the physical system as its starting point. To discretize the system, the functions appearing in the integrals are replaced by a restricted class of functions, where the restricted class has continuity in a sufficient number of derivatives to avoid singularities in the integrand (i.e., the functions are conforming). For quadratic forms of integral, such as those arising for the energy-conserving schemes, discrete linear equations are obtained by applying variational methods to the approximated integral.

The appropriate integral for a system of charged particles interacting according to Coulomb's law is

$$I[\Phi, \mathbf{x}] = \int_{t_0}^{t_1} L\,dt \tag{5-139}$$

where L is the combined particle and field lagrangian (Goldstein, 1959):

$$L = \int d\mathbf{x}_0 d\mathbf{v}_0 f(\mathbf{x}_0, \mathbf{v}_0, 0)[\tfrac{1}{2}mv^2 - q\Phi(\mathbf{x})] + \int d\mathbf{x}' \left(\epsilon_0 \frac{(\nabla\Phi)^2}{2} - \rho_0 \Phi \right) \tag{5-140}$$

The integral I is a functional of the potential Φ and particle coordinates $\mathbf{x}$. Minimizing I recovers the differential equations describing the system. The first integral in Eq. (5-140) is over the computational box and all velocities. $f(\mathbf{x}_0, \mathbf{v}_0, 0)$ is the distribution function at time $t = 0$, and the position $\mathbf{x}$ and velocity $\mathbf{v} = \dot{\mathbf{x}}$ are lagrangian coordinates (corresponding to particle orbits)

$$\mathbf{x} = \mathbf{x}(\mathbf{x}_0, \mathbf{v}_0, t) \qquad \mathbf{v} = \dot{\mathbf{x}}(\mathbf{x}_0, \mathbf{v}_0, t) \tag{5-141}$$

The second integral is taken over the computational box. The quantity ρ_0 is the background charge density.

Variation of the integral, Eq. (5-139), with respect to $\mathbf{x}$ gives

$$\begin{aligned}\delta_{\mathbf{x}} I &= \int dt \int d\mathbf{x}_0 d\mathbf{v}_0 f(\mathbf{x}_0, \mathbf{v}_0, 0)[m\mathbf{v} \cdot \delta\dot{\mathbf{x}} - q\delta\mathbf{x} \cdot \nabla\Phi] \\ &= \int dt \int d\mathbf{x}_0 d\mathbf{v}_0 f(\mathbf{x}_0, \mathbf{v}_0, 0)[-m\dot{\mathbf{v}} - q\nabla\Phi] \cdot \delta\mathbf{x} \\ &\quad + \int d\mathbf{x}_0 d\mathbf{v}_0 f(\mathbf{x}_0, \mathbf{v}_0, 0) m\mathbf{v} \cdot \delta\mathbf{x}\Big|_{t_0}^{t_1} \end{aligned} \tag{5-142}$$

Restricting variations to those which vanish at (arbitrary) times t_0 and t_1 causes the second term in Eq. (5-142) to vanish. Then, since the first term must vanish for all remaining variations $\delta\mathbf{x}$, we obtain

$$m\dot{\mathbf{v}} = -q\nabla\Phi \tag{5-143}$$

i.e., the equation of motion. Variation with respect to Φ gives

$$\begin{aligned}\delta_{\Phi} I = \int dt \{ \int d\mathbf{x}_0 d\mathbf{v}_0 f(\mathbf{x}_0, \mathbf{v}_0, 0)[-q\delta\Phi(x)] \\ + \int d\mathbf{x}'[\epsilon_0 \nabla\phi \cdot \nabla\delta\Phi - \rho_0 \delta\Phi(\mathbf{x}')]\}\end{aligned} \tag{5-144}$$

Using the result $f(\mathbf{x}, \mathbf{v}, t) = f(\mathbf{x}_0, \mathbf{v}_0, 0)$ (see Sec.2-2-1) and boundary conditions on Φ at the surface of the computational box enables Eq. (5-144) to be written

$$\delta_{\phi} I = \int dt \int d\mathbf{x}[-q \int f(\mathbf{x}, \mathbf{v}, t)\, d\mathbf{v} - \epsilon_0 \nabla^2\Phi - \rho_0]\delta\Phi = 0 \tag{5-145}$$

Since Eq. (5-145) is true for any $\delta\phi$ satisfying the boundary conditions, it follows that

$$\nabla^2\Phi = -\frac{1}{\epsilon_0}[\rho_0 + q \int f(\mathbf{x}, \mathbf{v}, t)\, d\mathbf{v}] \tag{5-146}$$

Similar lagrangians can be defined for several species of particles and for more general fields (Lewis, 1970*a*).

To discretize Eq. (5-140), we replace Φ by the approximation

$$\Phi \simeq \phi(\mathbf{x}, t) = \sum_{\mathbf{p}} \phi_{\mathbf{p}}(t) W_{\mathbf{p}}(\mathbf{x}) \tag{5-147}$$

where the sum is taken over all mesh points, the values $\{\phi_{\mathbf{p}}\}$ are the mesh-defined potential values, and $W_p(\mathbf{x})$ are the basis functions. To conform, ϕ must be at least continuous in value. Hence, if Eq. (5-147) represents a piecewise polynomial approximation on a rectangular mesh, $W_{\mathbf{p}}(\mathbf{x})$ must be the CIC or a higher-order assignment function (cf. Secs. 5-3-2 and 5-3-4). Explicit reference to the orbits of N_p particles is obtained by setting

$$f(\mathbf{x}_0, \mathbf{v}_0) = \sum_{i=1}^{N_p} \delta(\mathbf{x}_i - \mathbf{x}_{0i})\delta(\mathbf{v}_i - \mathbf{v}_{0i}) \tag{5-148}$$

Substituting Eqs. (5-147) and (5-148) into Eqs. (5-139) and (5-140) gives the approximated integral

$$\begin{aligned} I = \int dt\, L = \int dt \Bigg\{ \sum_{i=1}^{N_p} \left[\tfrac{1}{2} m v_i^2 - q \sum_{\mathbf{p}} \phi_{\mathbf{p}} W_{\mathbf{p}}(\mathbf{x}_i) \right] \\ + \int d\mathbf{x}' \left[\frac{\epsilon_0}{2} \left(\sum_{\mathbf{p}} \phi_{\mathbf{p}} \nabla W_p(\mathbf{x}') \right)^2 - \rho_0 \sum_{\mathbf{p}} \phi_{\mathbf{p}} W_p(\mathbf{x}') \right] \Bigg\} \end{aligned} \tag{5-149}$$

Following through the variational argument again for Eq. (5-149) leads to the Euler–Lagrange equations

$$\frac{\partial L}{\partial \mathbf{x}_i} - \frac{d}{dt}\frac{\partial L}{\partial \mathbf{v}_i} = 0 \qquad i \in [1, N_p] \tag{5-150}$$

i.e.,
$$m\dot{\mathbf{v}}_i = -q \sum_{\mathbf{p}} \phi_{\mathbf{p}} \nabla W_{\mathbf{p}}(\mathbf{x}_i) \tag{5-151}$$

and
$$\frac{\partial L}{\partial \phi_{\mathbf{p}}} = 0 \tag{5-152}$$

i.e.,
$$\epsilon_0 \sum_{\mathbf{p}'} \phi_{\mathbf{p}'} \int d\mathbf{x}' \nabla W_{\mathbf{p}'}(\mathbf{x}') \cdot \nabla W_{\mathbf{p}}(\mathbf{x}') = \rho_0 \int d\mathbf{x}' W_p(\mathbf{x}') + q \sum_{i=1}^{N_p} W_p(\mathbf{x}_i) \tag{5-153}$$

The reason for the name "energy-conserving schemes" can be seen by taking the dot product of Eq. (5-151) with $\mathbf{v}$ and using the result that ϕ is an implicit function of time:

$$\frac{d}{dt}\frac{mv^2}{2} = -\mathbf{v}\cdot\nabla\left(q\sum_p \phi_p W_p\right) = -\frac{d}{dt}\left(q\sum_p \phi_p W_p\right)$$

i.e.,
$$\frac{mv_i^2}{2} + q\sum_p \phi_p W_p = \text{constant} \tag{5-154}$$

Unfortunately, the exact energy conservation implied by Eq. (5-154) is lost when the time derivatives are replaced by finite-difference approximations (Lewis et al., 1972).

Equations (5-151) and (5-153) completely specify the optimal (in the sense that it minimizes I) energy-conserving scheme once the function W is chosen. For example, in one dimension, the lowest-order conforming choice of W, the CIC assignment function [Eq. (5-121)], gives for Eqs. (5-151) and (5-153)

$$m\dot{v} = q\frac{\phi_p - \phi_{p+1}}{H}\Pi\left(\frac{x - x_{p+1/2}}{H}\right) \tag{5-155}$$

$$\frac{\phi_{p-1} - 2\phi_p + \phi_{p+1}}{H^2} = -\frac{1}{\epsilon_0}\left[\rho_0 + \frac{q}{H}\sum_{i=1}^{N_p} \Lambda\left(\frac{x_i - x_p}{H}\right)\right] \tag{5-156}$$

The right-hand side of Eq. (5-156) is the normal one-dimensional CIC charge assignment. However, the force interpolation expression on the right-hand side of Eq. (5-155) describes NGP force interpolation from a mesh whose mesh points are located at the cell boundaries of the mesh used for assignment. The electric field on the shifted mesh is given by a centered difference over half the distance used in the momentum-conserving NGP scheme (Sec. 5-2-2). It follows from the discussion of Sec. 5-2-3 that this mixed CIC/NGP scheme does not conserve momentum owing to the self forces. An isolated particle placed arbitrarily on a mesh will oscillate about the cell boundaries with the frequency

(Langdon, 1973)

$$\omega_{\text{self}} = \left(\frac{q^2}{\epsilon_0 mH}\right)^{1/2} \tag{5-157}$$

where the factor of $\sqrt{2}$ change in the frequency from that given by Eq. (5-31) arises from the difference in the definition of the electric fields in the two instances.

5-6 TRANSFORM SPACE ANALYSIS

There are four features of Fourier transform or series analysis which make it particularly valuable in the analysis of the PM force calculation: the convolution theorem, the sampling theorem, the compact transform space representation of piecewise polynomial functions, and the link transforms provide with wave and kinetic analysis of the physical system. The application of transform methods to PM models was pioneered by Birdsall and coworkers at Berkeley (Birdsall et al., 1970; Langdon and Birdsall, 1970; Okuda and Birdsall, 1970). In particular, it is work by Langdon (1970*a*, 1970*b*, 1973, 1979*a*, 1979*b*) which has led to the present sound theoretical basis for particle simulation of collisionless systems. The insight that the transform space analysis provides has led to improvements in the design of collisionless models (Chen et al., 1974; Eastwood, 1975, 1976*a*; Eastwood and Brownrigg, 1979) and has made the three-dimensional P^3M algorithm a viable proposition (Eastwood, 1976*b*; Eastwood et al., 1980).

The steps of the PM force calculation are summarized by the following equations.

Charge assignment

$$\rho(\mathbf{x_p}) = \frac{q}{V_c}\sum_{i=1}^{N_p} W(\mathbf{x}_i - \mathbf{x_p}) \tag{5-158}$$

Solve for potential

$$\phi(\mathbf{x_p}) = V_c \sum_{\mathbf{p}'} G(\mathbf{x_p} - \mathbf{x_{p'}})\rho(\mathbf{x_{p'}}) \tag{5-159}$$

Compute forces

$$\mathbf{E}(\mathbf{x_p}) = -\mathbf{D}\phi(\mathbf{x_p}) \tag{5-160}$$

$$\mathbf{F}(\mathbf{x}_i) = q\sum_{\mathbf{p}} W(\mathbf{x}_i - \mathbf{x}_p)\mathbf{E}(\mathbf{x_p}) \tag{5-161}$$

or

$$\mathbf{F}(\mathbf{x}_i) = -q\sum_{\mathbf{p}} \nabla W(\mathbf{x}_i - \mathbf{x_p})\phi(\mathbf{x_p}) \tag{5-162}$$

where the notation follows that used earlier, but with the addition of the symbolic representation $\mathbf{D}$ of potential differencing in Eq. (5-160). In Eq. (5-159), the

Green's function G may be defined to absorb "sharpening" operators (cf. Sec. 5-3-2) which may be included in the calculation cycle. When deriving Fourier transforms of Eqs. (5-158) to (5-162), we shall for reasons of clarity take the case of a one-dimensional infinite system with mesh spacing H for illustration: The generalization to higher dimensions, which involves replacing scalars by vectors and redefining scaling factors where appropriate, will be treated in more detail in Chapters 7 and 8.

5-6-1 Charge Assignment

If we define the density of particle centers in the one-dimensional infinite system by

$$n(x) = \sum_{i=1}^{N_p} \delta(x - x_i) \tag{5-163}$$

where x_i is the position of particle i, and if, as is always true for cases we consider, W is an even function, then charge assignment [Eq. (5-158)] is described by

$$\rho(x_p) = \frac{q}{H} \int n(x') W(x_p - x')\, dx' \tag{5-164}$$

Comparing Eq. (5-164) with those given in Table A-3, we see that the mesh-defined values of charge density are given by particular values, sampled at points $\{x_p = pH\}$, of the convolution of the particle density with the charge assignment function, i.e.,

$$\begin{matrix}\text{CHARGE} \\ \text{ASSIGNMENT}\end{matrix} \equiv \text{CONVOLUTION} + \text{SAMPLING}$$

The continuous charge density quantity

$$\rho'(x) = \frac{q}{H} \int n(x') W(x - x')\, dx' \tag{5-165}$$

whose values at $x = x_p$ give the mesh-defined density can be immediately transformed using the convolution theorem (Table A-3)

$$\rho'(x) \supset \hat{\rho}'(k) = \frac{q}{H} \hat{n}(k) \hat{W}(k) \tag{5-166}$$

The symbol $\supset$ is used to denote "transforms to" or "whose transform is", and the circumflex indicates that the quantity is the Fourier transform (or harmonic) of the x space quantity denoted by the same symbol. We use the same symbol $\supset$ for Fourier transforms, series, and finite Fourier transforms, since the appropriate transform is generally obvious from the context. For instance, in the case of ρ' we have from Eqs. (A-1) and (A-2)

$$\rho'(x) = \int_{-\infty}^{\infty} \frac{dk}{2\pi} \hat{\rho}'(k) e^{ikx} \tag{5-167}$$

$$\hat{\rho}'(k) = \int_{-\infty}^{\infty} dx\, \rho'(x) e^{-ikx} \tag{5-168}$$

The appropriate transform for the set of values defined by Eq. (5-164) is the Fourier series, Eqs. (A-7) and (A-8). The transform of $\{\rho(x_p)\}$ is

$$\hat{\rho}(k) = H \sum_{p=-\infty}^{\infty} \rho(x_p) e^{-ikx_p} \tag{5-169}$$

and the inverse transformation is

$$\rho(x_p) = \int_{k_g} \frac{dk}{2\pi} \hat{\rho}(k) e^{ikx_p} \tag{5-170}$$

where $k_g = 2\pi/H$ is the period length in k space. However, since $\rho'(x_p) = \rho(x_p)$, we can use Eq. (5-167) to obtain

$$\rho(x_p) = \int_{-\infty}^{\infty} \frac{dk}{2\pi} \hat{\rho}'(k) e^{ikx_p} \tag{5-171}$$

Splitting the range of integration in Eq. (5-171) into segments of length k_g and changing variables (i.e., $k \to k + nk_g$, n integer) to bring all the integrals to a single period of length k_g gives

$$\rho(x_p) = \int_{k_g} \frac{dk'}{2\pi} \sum_{n=-\infty}^{\infty} \hat{\rho}'(k') e^{ik'x_p} \tag{5-172}$$

which on comparison with Eq. (5-170) yields

$$\hat{\rho}(k) = \sum_{n=-\infty}^{\infty} \hat{\rho}'(k - nk_g) \tag{5-173}$$

$$= \frac{q}{H} \sum_{n=-\infty}^{\infty} \hat{n}(k - nk_g) \hat{W}(k - nk_g) \tag{5-174}$$

The same result, Eq. (5-173), can be obtained rigorously by taking limits of appropriate finite integrals and sums but, as we shall show below, the manipulation of the transforms is more conveniently performed using generalized functions.

A graphical interpretation of the real part of Eq. (5-173) is shown in Fig. 5-7(*a*). The Fourier transform $\hat{\rho}$ of the set of mesh-defined values is given by summing the contributions of replicas of the transform $\hat{\rho}'$ of the continuous density function, where the replicas are shifted by multiples of k_g $(=2\pi/H)$. The contribution of the term $\hat{\rho}'(k - nk_g)$ to $\hat{\rho}$ is called the "nth alias contribution," and the sum over n in Eq. (5-173) is known as the "alias sum." The same graphical interpretation holds for the imaginary part of Eq. (5-173), the only difference being that the imaginary parts of $\hat{\rho}$ and $\hat{\rho}'$ are odd functions of k, since (real) $\supset$ (hermitian) (cf. Table A-2).

The reason for the name "alias" becomes apparent when one considers a particular harmonic in the principal zone ($|k| \leqslant \pi/H$). The solid curve in Fig. 5-7(*b*)

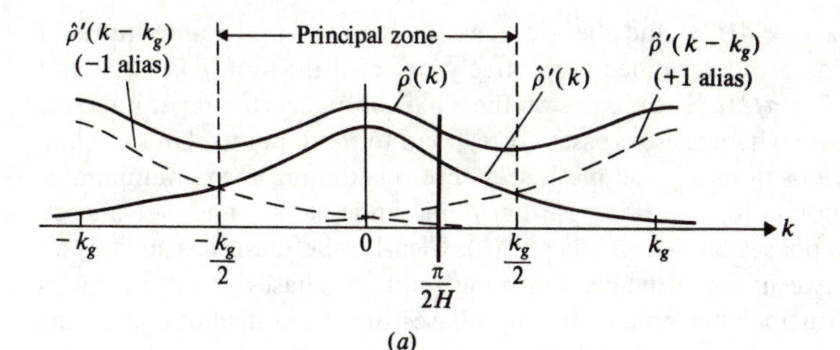

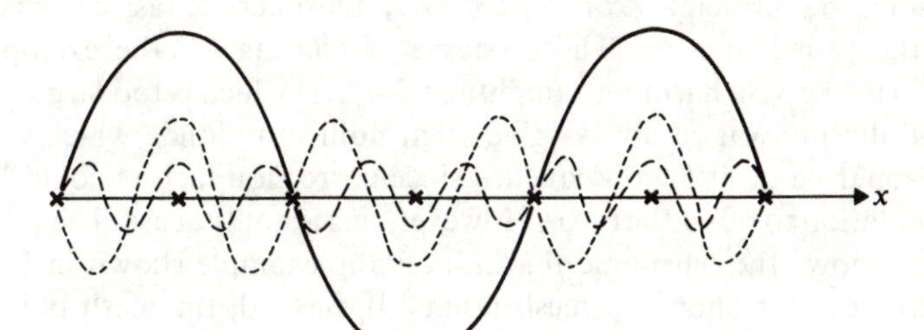

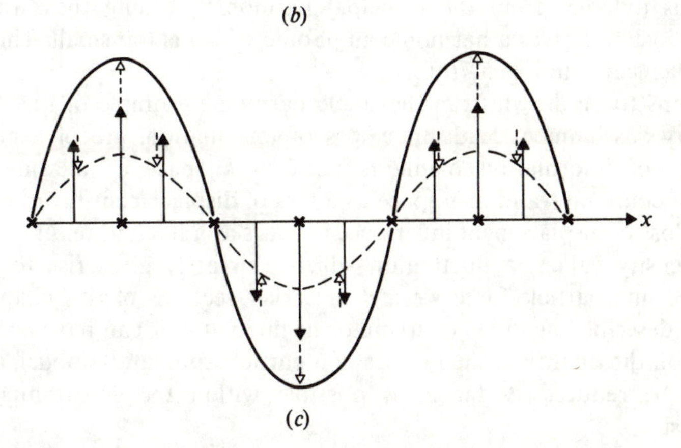

Figure 5-7 (*a*) The Fourier transform $\hat{\rho}$ of the set of charge density values sampled on a mesh of spacing H is equal to the sum of contributions from replicas, spaced at intervals $k_g = 2\pi/H$, of the transform $\hat{\rho}'$ of the continuous density distribution.

(*b*) At a particular wavelength (shown as $k = \pi/2H$) the harmonic $\hat{\rho}$ contains contributions from the principal value of $\hat{\rho}'$ and its aliases.

(*c*) Depending on the relative phases of principal harmonic and aliases, the mesh sees a harmonic amplitude which may be too large or too small.

shows the x space form of a single harmonic of $\hat{\rho}'$ with wavenumber $k = \pi/2H$, or, equivalently, with wavelength $\lambda = 4H$. The amplitude of the sinusoid is given by $|\hat{\rho}'(\pi/2H)|$ and its phase is determined by the relative sizes of the real and imaginary parts of $\hat{\rho}'(\pi/2H)$. The two broken curves are sinusoids corresponding to the $+1$ alias ($\lambda_{+1} = 4H/3$) and the -1 alias ($\lambda_{-1} = 4H/5$). The amplitudes of these two sinusoids are determined respectively by the values of $|\hat{\rho}'(k-k_g)|$ and $|\hat{\rho}'(k+k_g)|$, where $k = \pi/2H$. The crosses on the x axis of Fig. 5-7(b) are at intervals H. Consider the case where these crosses correspond to mesh points. On sampling values of $\rho'(x)$ at positions x_p, the mesh sees zero, maximum, zero, minimum at successive mesh points for the principal harmonic and the $+1$ and -1 aliases, given the relative phases shown in Fig. 5-7(b). Clearly, the mesh has no way of distinguishing between the principal harmonic and its aliases: As far as it is concerned, the principal harmonics and its aliases are all contributions to the harmonic $\hat{\rho}$, whose wavenumber equals the wavenumber of the principal harmonic $\hat{\rho}'$, i.e., in the mesh calculation, harmonics of the charge density whose wavenumbers lie outside the principal zone, $|k| \leqslant \pi/H$, masquerade as, or alias, wavenumbers in the principal zone. The outcome of aliasing for the example shown in Fig. 5-7(b) is to give a harmonic amplitude $\hat{\rho}(\pi/2H)$ which is too large.

If the result of aliasing was to always give a harmonic amplitude which was too large then it would be of little consequence since correction factors could be included in the calculation to offset the error. However, this is not the case. The solid curve in Fig. 5-7(c) shows the harmonic $\hat{\rho}(\pi/2H)$ for the example shown in Fig. 5-7(b) when the crosses correspond to mesh points. If, instead, the mesh points were midway between the cross points, then the sign of the contribution of the $+1$ and -1 aliases is different from the principal harmonic. Adding these three harmonics in this instance gives a harmonic amplitude which is too small. This is shown by the broken curve in Fig. 5-7(c).

The conclusions to be drawn from the single harmonic example of Fig. 5-7, namely that charge assignment leads to a loss of information, are of general validity. The loss of information manifests itself in k space as aliasing—a nonphysical mode coupling—and in x space as a loss of displacement invariance. In Fig. 5-7(c) the loss of displacement invariance appears as a fluctuation in mesh-defined charge density values, a fluctuation which ultimately gives rise to the fluctuations of the interparticle force we met in earlier sections of this chapter. Aliasing, whether described as mode coupling or as fluctuations, can have only a deleterious effect on the quality of the physics of a particle simulation model, and therefore should be reduced as far as is possible within the constraints of computational cost.

The Fourier transformation of the charge assignment step, where both continuous functions and discrete sets of values are involved, can be compactly expressed by employing the generalized function Ш. The sampling function Ш is an infinite row of δ functions spaced at unit interval (see Sec. A-4-5). If a continuous function f is multiplied by Ш then the resulting function $f\dagger$ is an infinite row of impulse functions at unit interval, where the strength of the impulse at $x = n$ is equal to the value of f at $x = n$, i.e.,

$$f(n) = \int_{n-1/2}^{n+1/2} f\dagger(x)\,dx = \int_{n-1/2}^{n+1/2} \text{Ш}(x) f(x)\,dx \qquad (5\text{-}175)$$

Thus, instead of describing the mesh-defined charge density by the set of values $\{\rho(x_p); x_p = pH, p \text{ integer}\}$ we can define the generalized function

$$\rho\dagger(x) = \text{Ш}\left(\frac{x}{H}\right)\rho'(x) \tag{5-176}$$

which comprises an infinite row of impulse functions at interval H and whose strengths give the values of the mesh-defined charge density. The Fourier transform of $\rho\dagger(x)$ is, by the convolution theorem,

$$\hat{\rho}\dagger(k) = H\,\text{Ш}\left(\frac{kH}{2\pi}\right) * \hat{\rho}'(k) \tag{5-177}$$

where $\text{Ш}(x/H) \supset H\,\text{Ш}(kH/2\pi)$ by Eq. (A-29) and the similarity theorem (Table A-4). Explicitly evaluating the convolution in Eq. (5-177) gives

$$\hat{\rho}\dagger(k) = \int_{-\infty}^{\infty} dk' \sum_{n=-\infty}^{\infty} \delta(k - nk_g)\hat{\rho}'(k - k') \tag{5-178}$$

$$= \sum_{n=-\infty}^{\infty} \hat{\rho}'(k - nk_g) = \hat{\rho}(k) \tag{5-179}$$

The equality of $\hat{\rho}$ and $\hat{\rho}\dagger$ can also be established directly, since by definition

$$\hat{\rho}\dagger(k) = \int_{-\infty}^{\infty} dx\, \rho\dagger(x) e^{-ikx} = \int_{-\infty}^{\infty} dx\, \text{Ш}\left(\frac{x}{H}\right)\rho'(x) e^{-ikx} \tag{5-180}$$

which upon evaluation gives

$$\hat{\rho}\dagger(k) = H \sum_p \rho'(x_p) e^{-ikx_p} = H \sum_p \rho(x_p) e^{-ikx_p} = \hat{\rho}(k) \tag{5-181}$$

The loss of information caused by sampling depends on the smoothness of the function being sampled and on the mesh spacing. The definition of smoothness used in Sec. 5-3-2 can be translated to the asymptotic decay of harmonic amplitudes as k increases: A function f, which is continuous in all derivatives up to the nth derivative, has a transform $\hat{f}$ which decays asymptotically as $k^{-(n+1)}$. Another k space viewpoint of smoothness, in terms of the harmonic content of the transformed function beyond wavenumbers $|k| = \pi/H$, provides further insight into the importance of smoothness in choosing assignment functions.

Figure 5-8(a) shows an extreme case of a smooth density distribution, $\rho'(x)$. Its transform is band-limited with a cutoff wavenumber k_c, i.e., $\hat{\rho}'(k) = 0$ for all $|k| \geqslant k_c/2$. Figures 5-8(b–d) illustrate the changes in x and k space as the mesh spacing H is varied. In x space, the arrows indicate the strengths and positions of the impulses of the generalized charge density [cf. (5-176)]. Multiplying $\rho'(x)$ in x space by $\text{Ш}(x/H)$ corresponds to convolving $\hat{\rho}'(k)$ with $H\,\text{Ш}(k/k_g)$ in k space. Convolutions with $H\,\text{Ш}(k/k_g)$ lead to a replication of $\hat{\rho}'(k)$ at intervals $k_g = 2\pi/H$ in k space. Figure 5-8(b) shows the case of oversampling ($k_g > k_c$). Since k_g is greater than k_c, there is no overlap in k space and therefore $\rho'(x)$ can be recovered from the

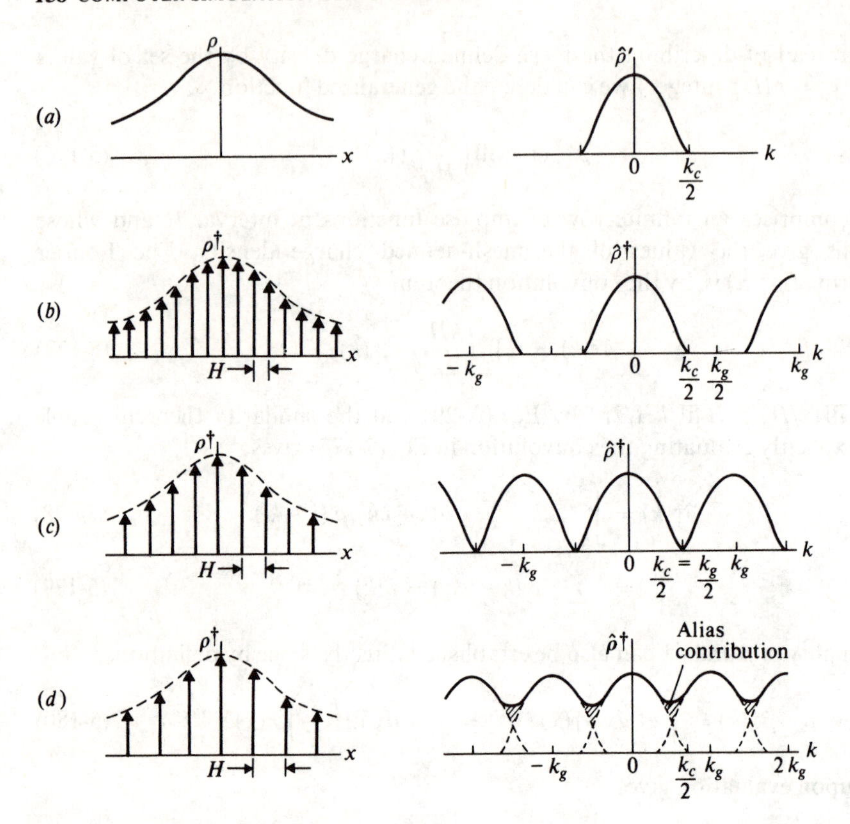

Figure 5-8 (a) The extreme case of a smooth charge distribution ρ' has a transform which is band-limited, $\hat{\rho}'$.

(b) If the mesh spacing H is such that $k_g (= 2\pi/H) > k_c$, then ρ' is oversampled.

(c) If H is such that $k_g = k_c$, then ρ' is critically sampled.

(d) If $k_g < k_c$, ρ' is undersampled. The overlap of the displaced replicas of $\hat{\rho}'$ forming $\hat{\rho}\dagger$ leads to aliasing. In the x space plots (b)–(d), the vertical arrows following the envelope given by ρ' represent impulse functions whose strengths are proportional to the value of ρ' at the position of the arrows.

sample function $\rho\dagger(x)$ by multiplying $\hat{\rho}\dagger(k)$ by $\Pi(k/k_g)$ and performing the inverse transformation. Equivalently, $\rho'(x)$ may be constructed from $\rho\dagger(x)$ by convolution with $1/H \operatorname{sinc}(x/H)[\supset \Pi(k/k_g)]$:

$$\rho'(x) = \frac{1}{H} \operatorname{sinc}\left(\frac{x}{H}\right) * \rho\dagger(x) \tag{5-182}$$

$$= \sum_p \operatorname{sinc}\left(\frac{x - x_p}{H}\right) \rho(x_p) \tag{5-183}$$

As the mesh spacing H is increased towards $H = 2\pi/k_c$, the periodic repeat interval k_g in k space reduces. Figure 5-8(c) shows the critical sampling interval $k_g = k_c$, the largest mesh spacing for which the function $\rho'(x)$ can be recovered from its sample. Beyond the critical point, $\rho'(x)$ is said to be undersampled. It is clear from Fig. 5-8(*d*) that $\rho'(x)$ cannot be uniquely recovered from the undersampled set of values since $\hat{\rho}\dagger$ no longer has a one-to-one correspondence with $\hat{\rho}'$ in the interval $|k| \leqslant k_g/2$. The difference between $\hat{\rho}\dagger$ and $\hat{\rho}'$ in the undersampled case is the alias contribution.

If we could devise a charge assignment scheme which acted as a low pass filter—removing all the harmonics of $\hat{n}$ for $|k| \geqslant \pi/H$—then $\rho'(x)$ would be critically sampled by the mesh and aliases would be completely suppressed. Such schemes do exist, for example, the assignment function $W(x) = (1/H)\,\mathrm{sinc}\,(x/H) \supset \hat{W}(k) = \Pi(k/k_g)$ acts as an ideal low-pass filter: it gives harmonics $\hat{\rho}$ of the mesh-assigned density identical to $q\hat{n}$ for $|k| < \pi/H$ and hence values $\{\rho(x_p)\}$ equal to samples of the best least-squares band-limited function fit to the true charge density $qn(x)$. Indeed, any functions $W(x) \supset \hat{W}(k)$ which are band-limited to the range $|k| < k_g/2$ will totally suppress aliases, and their reshaping effects can be offset (provided $\hat{W}(k) \neq 0$ for $|k| < k_g/2$) for harmonics in the principal zone by applying the sharpening operator $L \supset [\hat{W}(k)]^{-1}$ to the mesh-assigned values. Unfortunately, assignment functions which have band-limited transforms cannot be also space-limited (i.e., nonzero only for $|x| < x_c/2$) and so would involve assignment to all mesh points. Hence, complete alias elimination is computationally impracticable.

Among the assignment functions which employ only a small number of mesh points, those which satisfy both the smoothness and long-range constraints (cf. Table 5-1) provide good approximations to low-pass filters in k space. The lowest order of these is the NGP scheme. Using the viewpoint that charge assignment is convolution followed by sampling, we can interpret the NGP scheme as first taking a running mean (over an interval H) of the true charge density $qn(x)$ to obtain the smoothed density $\rho'(x)$:

$$\rho'(x) = \frac{1}{H}\int_{x-H/2}^{x+H/2} qn(x')\,dx' \tag{5-184}$$

and then sampling $\rho'(x)$ at points $\{x = x_p\}$. Similarly, in the first-order scheme, CIC, one takes a running mean of the running mean

$$\rho'(x) = \frac{1}{H}\int_{x-H/2}^{x+H/2} dx' \,\frac{1}{H}\int_{x'-H/2}^{x'+H/2} qn(x'')\,dx'' \tag{5-185}$$

and then samples the values of this smoothed density to obtain the mesh charge densities. Each successively higher-order scheme in the hierarchy listed in Table 5-1 takes a further running mean before sampling.

The Fourier transforms of the hierarchy of assignment functions NGP, CIC, TSC are found most easily by applying the convolution theorem to Eqs. (5-120)–(5-122). From Eq. (A-18) and the similarity theorem (Table A-4) we have

$$\Pi\left(\frac{x}{H}\right) \supset H\,\mathrm{sinc}\left(\frac{k}{k_g}\right) = H\,\frac{\sin\left(\dfrac{kH}{2}\right)}{\left(\dfrac{kH}{2}\right)} \tag{5-186}$$

Hence, using the convolution theorem we find

$$\text{NGP: } \hat{W}(k) = H \operatorname{sinc}\left(\frac{k}{k_g}\right) \tag{5-187}$$

$$\text{CIC: } \hat{W}(k) = H \operatorname{sinc}^2\left(\frac{k}{k_g}\right) \tag{5-188}$$

$$\text{TSC: } \hat{W}(k) = H \operatorname{sinc}^3\left(\frac{k}{k_g}\right) \tag{5-189}$$

and so forth. The mth order scheme has an assignment function whose transform is given by $H \operatorname{sinc}^{m+1}(k/k_g)$.

A measure of the effectiveness of a charge assignment function, W, in filtering out the nth alias of the wavenumber k (wavelength $\lambda = 2\pi/k$) in the principal zone ($|k| < k_g/2$, $\lambda > 2H$) is furnished by the relative magnitudes of $\hat{W} \supset W$ at the alias and principal wavenumbers. For the mth order scheme in the NGP-CIC-TSC etc. hierarchy this gives

$$\left|\frac{\hat{W}(k-nk_g)}{\hat{W}(k)}\right| = \left(1-n\frac{k_g}{k}\right)^{-(m+1)} = \left(1-\frac{n\lambda}{H}\right)^{-(m+1)} \tag{5-190}$$

i.e., attenuation increases with alias number, wavelength, and order of the scheme.

5-6-2 The Potential Solver

Equation (5-159) becomes for a one-dimensional infinite system

$$\phi(x_p) = H \sum_{p'=-\infty}^{\infty} G(x_p - x_{p'})\rho(x_{p'}) \tag{5-191}$$

Again referring to Table A-3, we recognize this as the convolution sum for a system which is discrete in x space and periodic in k space. Therefore

$$\phi(x_p) \supset \hat{\phi}(k) = \hat{G}(k)\hat{\rho}(k) \tag{5-192}$$

where periodicity implies for n integer

$$\hat{\phi}(k+nk_g) = \hat{\phi}(k) \tag{5-193}$$

The convolution sum, Eq. (5-191), can be expressed as a convolution integral by constructing continuous functions $\phi'(x)$ and $G'(x)$ for the potential and Green's function. Clearly, there are an infinite number of possible continuous functions ϕ' and G' that we could choose, although there is only one choice which does not add spurious information: This special choice is the one where the sets of values $\{\phi(x_p)\}$ and $\{G(x_p)\}$ are those obtained by sampling ϕ' and G' at the critical sampling interval, i.e.,

$$G'(x) \supset \hat{G}'(k) = \Pi\left(\frac{k}{k_g}\right)\hat{G}(k) \tag{5-194}$$

$$\Rightarrow G'(x) = \sum \operatorname{sinc}\left(\frac{x-x_p}{H}\right) G(x_p) \tag{5-195}$$

and similarly for ϕ'. Rewriting Eq. (5-191) in terms of ϕ' and G', and using Eq. (5-176) to express the charge density as a function of the continuous argument x', yields

$$\phi'(x) = \int_{-\infty}^{\infty} dx'\, G'(x-x')\rho\dagger(x') \tag{5-196}$$

which transforms to give

$$\begin{aligned} \phi'(x) \supset \hat{\phi}'(k) &= \hat{G}'(k)\hat{\rho}\dagger(k) \\ &= \Pi\left(\frac{k}{k_g}\right)\hat{\phi}(k) \end{aligned} \tag{5-197}$$

The expression, Eq. (5-196), for the potential is of identical form to the Green's function formulation of Poisson's equation for the physical system. In the continuous system, the Green's function g satisfied

$$\nabla^2 g = -\frac{\delta(\mathbf{x})}{\epsilon_0} \tag{5-198}$$

i.e.,
$$\hat{g}(\mathbf{k}) = \frac{1}{\epsilon_0 k^2} \tag{5-199}$$

where in two and three dimensions $k^2 = \mathbf{k}\cdot\mathbf{k}$. If we specified that $\phi'(x)$ be the best least-squares fit to the exact potential

$$\Phi(x) = \int_{-\infty}^{\infty} dx'\, g(x-x')\rho\dagger(x') \tag{5-200}$$

then we would find that

$$\hat{G}'(k) = \Pi\left(\frac{k}{k_g}\right)\hat{g}(k) \tag{5-201}$$

which corresponds to the influence function $\hat{G}(k) = (1/\epsilon_0 k^2)$ used in the "poor-man's Poisson solver" (Boris and Roberts, 1969). Moreover, if the approximated influence function is set to zero for wavenumbers greater than some cutoff value $\alpha k_g/2$; $0 < \alpha < 1$, then the best least-squares approximation $\phi'(x)$ to $\Phi(x)$ is still obtained by setting the nonzero values of $\hat{G}$ equal $\hat{g}$,

i.e.,
$$\hat{G}'(k) = \Pi\left(\frac{k}{k_g}\right)\Pi\left(\frac{k}{\alpha k_g}\right)\hat{g}(k) \tag{5-202}$$

An alternative criterion, as has been used for example in the simulation of isolated gravitating systems (cf. Chapter 11), is to specify that the values of the potential are correct at mesh points. This gives the prescription for the influence function

$$\hat{G}'(k) = \Pi\left(\frac{k}{k_g}\right) \sum_{n=-\infty}^{\infty} \hat{g}(k-nk_g) \tag{5-203}$$

The influence functions obtained from using finite-difference approximations to the laplacian in Poisson's equation are obtained by directly transforming the difference equations. For instance, the three-point approximation in one dimension [Eq. (5-8)] yields

$$\hat{G}'(k) = \Pi\left(\frac{k}{k_g}\right)\frac{(H/2)^2}{\sin^2\left(\frac{kH}{2}\right)} \tag{5-204}$$

and for the corresponding two-dimensional case [Eq. (5-131)]

$$\hat{G}'(k_1, k_2) = \Pi\left(\frac{k_1}{k_g}\right)\Pi\left(\frac{k_2}{k_g}\right)\frac{(H/2)^2}{\sin^2\left(\frac{k_1 H}{2}\right) + \sin^2\left(\frac{k_2 H}{2}\right)} \tag{5-205}$$

Comparing Eq. (5-204) with Eq. (5-201) we see that the finite-difference approximation causes the high-wavenumber (i.e., $|k|$ approaching $k_g/2$) potential harmonics to be much larger than for the best least-squares approximation, i.e., differencing errors act in the opposite direction to charge assignment errors: Assignment errors attenuate high-wavenumber components whereas differencing errors amplify them. Hence, if a sharpening operator (cf. Sec. 5-3-2) is used on the mesh-assigned charge without taking into account the reshaping effects of the truncation errors in the field equation, one is likely to end up with a poorer approximation than one would obtain if no attempts at correction were made!

5-6-3 Force Interpolation

As may no doubt be deduced from the discussions of charge assignment and the potential solver, the expressions Eqs. (5-160) and (5-161) for computing the force in momentum-conserving schemes and the corresponding expression Eq. (5-162) for energy-conserving schemes may also be recognized as convolutions.

The first step in computing the force for momentum-conserving schemes is to construct the mesh-defined electric field by differencing potential values. The simplest scheme in one dimension, Eq. (5-11), may be expressed as

$$E(x_p) = -H\sum_{p'}\frac{1}{H}\frac{(\delta_{p+1-p'} - \delta_{p-1-p'})}{2H}\phi(x_{p'}) \tag{5-206}$$

where δ is the Kronecker delta [Eq. (A-27)]. The transform of E may be obtained from Eq. (5-206) by using the convolution theorem for discrete quantities [FS(ii) in Table A-3]. The integral form of Eq. (5-206) is obtained by constructing the band-limited continuous function $E'(x)$ from the values $\{E(x_p)\}$ in the same manner employed for G' and ϕ' [Eqs. (5-194) and (5-196)] and rewriting the right-hand side in terms of ϕ':

$$E'(x) = -\int dx'\left[\frac{\delta(x+H-x') - \delta(x-H-x')}{2H}\right]\phi'(x') \tag{5-207}$$

The term in the bracket in Eq. (5-207) is the two-point difference operator approximation to ∇. For the four-point operator, the difference operator D becomes

$$D(x) = \alpha \frac{\delta(x+H) - \delta(x-H)}{2H} + (1-\alpha) \frac{\delta(x+2H) - \delta(x-2H)}{4H} \tag{5-208}$$

cf. Eq. (5-137). In general, we have for one-dimensional schemes

$$E'(x) = -\int dx'\, D(x-x')\phi'(x') \tag{5-209}$$

giving

$$\hat{E}'(k) = -\hat{D}(k)\hat{\phi}'(k) \tag{5-210}$$

The expression for the interpolated force, Eq. (5-161), for momentum-conserving schemes can be written in integral form

$$F(x) = \frac{q}{H} \int W(x-x') E\dagger(x')\, dx' \tag{5-211}$$

by introducing the generalized function $E\dagger$ to represent the mesh-defined field values, where

$$E\dagger(x) = \text{Ш}\left(\frac{x}{H}\right) E'(x) \tag{5-212}$$

Hence

$$\hat{F}(k) = \frac{q}{H} \hat{W}(k) \hat{E}\dagger(k) \tag{5-213}$$

where

$$\hat{E}\dagger(k) = H\,\text{Ш}\left(\frac{k}{k_g}\right) * \hat{E}'(k) \tag{5-214}$$

$$= \sum_{n=-\infty}^{\infty} \hat{E}'(k - nk_g) \tag{5-215}$$

$$= -\hat{D}(k)\hat{\phi}(k) \tag{5-216}$$

The alias sum in Eq. (5-215) serves only to periodically repeat $\hat{E}'$ outside the interval $|k| < k_g/2$ since $E'(x)$ is, by construction, a band-limited function. This is reflected in Eq. (5-216) where the alias sum is replaced by the periodically repeating quantity $\hat{\phi}(k) \supset \{\phi(x_p)\}$ or $\phi\dagger(x)$.

For multidimensional calculations, Eq. (5-216) becomes

$$\hat{\mathbf{E}}\dagger(\mathbf{k}) = -\hat{\mathbf{D}}(\mathbf{k})\hat{\phi}(\mathbf{k}) \tag{5-217}$$

The exact gradient of a potential gives

$$\hat{\mathbf{E}}(\mathbf{k}) = -i\mathbf{k}\phi(\mathbf{k}) \tag{5-218}$$

Comparing Eqs. (5-217) and (5-218) we identify the effects of truncation errors in the harmonics $\hat{\mathbf{E}}\dagger$ of the field: Errors in magnitude cause $|\hat{\mathbf{D}}|$ to differ from $|\mathbf{k}|$, and errors in direction cause $\hat{\mathbf{D}} \times \mathbf{k}$ to be nonzero.

The expression for the force in energy-conserving schemes, Eq. (5-162), may be written as a convolution integral by defining the generalized potential function

$$\phi\dagger(\mathbf{x}) = \text{Ш}\left(\frac{x}{H}\right) \phi'(x) \tag{5-219}$$

giving (in one dimension)

$$F(x) = -\frac{q}{H}\int \frac{d}{dx} W(x-x')\phi\dagger(x')\,dx' \tag{5-220}$$

The transform immediately follows by applying the convolution theorem (Table A-3) and the derivative theorem (Table A-4):

$$\hat{F}(k) = -\frac{q}{H} ik\hat{W}(k)\hat{\phi}(k) \tag{5-221}$$

and for the multidimensional case

$$\hat{\mathbf{F}}(\mathbf{k}) = -\frac{q}{V_c} i\mathbf{k}\hat{W}(\mathbf{k})\hat{\phi}(\mathbf{k}) \tag{5-222}$$

Thus, although the energy-conserving schemes do not conserve momentum (cf. Sec. 5-5), their force interpolation schemes do avoid the errors in the direction of forces introduced by potential differencing in the momentum-conserving counterpart.

5-6-4 The Interparticle Force

The Fourier transform of the force on a charge q at position x_2 due to a charge $-q$ at position x_1 is obtained by collecting together the results of Secs. 5-6-1 to 5-6-3:

The force on the charge at x_2 is given by

$$F(x_2) = \int \frac{dk}{2\pi} \hat{F}(k)e^{ikx_2} \tag{5-223}$$

For momentum-conserving schemes the force transform is, from Eq. (5-213),

$$\begin{aligned}
\hat{F}(k) &= \frac{q}{H}\hat{W}\hat{E}\dagger \\
&= -\frac{q}{H}\hat{W}\hat{D}\hat{\phi} \qquad \text{[by Eq. (5-216)]} \\
&= -\frac{q}{H}\hat{W}\hat{D}\hat{G}\hat{\rho} \qquad \text{[by Eq. (5-192)]} \\
&= \frac{q^2}{H^2}\hat{W}\hat{D}\hat{G}\sum_n \hat{W}(k-nk_g)e^{-i(k-nk_g)x_1}
\end{aligned} \tag{5-224}$$

where in the final line of Eq. (5-224), Eq. (5-179), and Eq. (5-166) with $n(x) = \delta(x-x_1) \supset \hat{n}(k) = e^{-ikx_1}$ were used to eliminate $\hat{\rho}$. Repeating the same elimination procedure for energy-conserving schemes [Eq. (5-221)] gives

$$\hat{F}(k) = \frac{q^2}{H^2}\hat{W}ik\hat{G}\sum_n \hat{W}(k-nk_g)e^{-i(k-nk_g)x_1} \tag{5-225}$$

Combining Eqs. (5-223) and (5-224) gives the particle-mesh calculated interparticle force

$$F(x;x_1) = \int \frac{dk}{2\pi}\left(\frac{q^2}{H^2}\hat{W}\hat{D}\hat{G}\sum_n \hat{W}(k-nk_g)e^{-ink_g x_1}\right)e^{ikx} \tag{5-226}$$

where $x = x_2 - x_1$ is the particle separation. The alias sum term in Eq. (5-226), arising from the charge assignment term, is the sole cause of the loss of displacement invariance. The $n \neq 0$ terms in the sum introduce a periodicity of period length H in x space. The same is true for the interparticle force obtained by substituting Eq. (5-225) into Eq. (5-223).

Averaging F in Eq. (5-226) over all source charge positions gives

$$\langle F(x;x_1)\rangle = \frac{1}{H}\int_H F(x;x_1)\,dx_1 = \int \frac{dk}{2\pi}\frac{q^2}{H}(\hat{W}\hat{D}\hat{G}\hat{W})e^{ikx} \tag{5-227}$$

The terms in the bracket on the right-hand side of Eq. (5-227) in order of appearance respectively give the k space description of force interpolation, potential differencing, solving for the potential, and the average effect of charge assignment. Each has a shaping influence on the harmonics of the interparticle force, and their combined effect can be evaluated using expressions derived from Eqs. (5-223) to (5-227). Moreover, these expressions are an exact description, so conclusions drawn from them are not limited by smallness parameters as is usually the case in the x space analysis.

We defer the questions of generalization to higher dimensions and finite systems to Chapters 7 and 8, where we shall illustrate how the transform space analysis is used in interpreting the properties of particle models and in optimizing those properties within the economic constraints of computational cost.

CHAPTER

SIX

THE SOLUTION OF THE FIELD EQUATIONS

6-1 INTRODUCTION

An efficient method for the solution of the field equations is a necessary requirement for the practical implementation of the particle simulation algorithms that have been described. If we consider a typical computer experiment of 1,000 timesteps and an availability of one hour of computer time, it is clear that the solution of the field equations (which in a well-balanced calculation will be about half the cycle time) must not normally exceed a few seconds. For a machine with an average arithmetic-operation time of one microsecond, and a space mesh of about $N_g = 10{,}000$ points (say 20^3 or 100^2),* we require techniques that will obtain the solution in less than 100 arithmetic operations (or $\sim 100\,\mu s$) per point. In fact the best available transform methods, where applicable, can achieve solution times of about 20 operations per point.

It is important to realize that such a small operations count is an extremely stringent requirement, necessitating the use of special techniques. Indeed it may be necessary to alter the mathematical model to one on which fast special methods can be applied, in order to achieve it. For comparison, the straightforward solution of the 10,000 mesh point equations by Gauss elimination methods without sparse-matrix refinements ($\simeq N_g^3$ operations) would require $\sim 10^{12}$ operations or about one month of solid computing time per timestep!

The methods available for the solution of the partial-differential equations prescribing the fields depend very much on the details of the equations, and there is no single method that is universally the best. The issues that influence the choice of method are:

---

* N_g is the total number of mesh points, and in this chapter n is the number of mesh points in each dimension. Hence, for a uniform mesh in d dimensions we have $N_g = n^d$.

in the equations
(*a*) linearity,
(*b*) dimensionality,
(*c*) separability, or
(*d*) variation of coefficients
and
in the case of the boundary
whether the region is
(*a*) rectangular or arbitrary, has
(*b*) mixed or simple conditions, or is
(*c*) isolated

The available methods can be broadly classified as follows.

1. Mesh-relaxation methods In these, initially guessed mesh values are relaxed to the solution by systematically sweeping the mesh and adjusting the values. For example:

(*a*) Gauss–Seidel (GS)
(*b*) successive overrelaxation (SOR)
(*c*) successive line overrelaxation (SLOR)
(*d*) alternating direction implicit (ADI)

2. Matrix methods In these, the finite difference equations are treated as a matrix of linear equations and solved by variations of standard matrix methods. For example:

(*a*) Thomas tridiagonal algorithm
(*b*) sparse matrix methods (SM)
(*c*) conjugate gradient algorithm (CGA)
(*d*) Stone's strongly implicit procedure (SIP)
(*e*) incomplete Choleski-conjugate gradient (ICCG)

3. Rapid elliptic solvers In these, special techniques based on the fast Fourier transform (FFT) and cyclic reduction (CR) are used to solve certain classes of problems in at most $O(N_g \log_2 N_g)$ operations. For example:

(*a*) cyclic reduction or Buneman algorithm (CR)
(*b*) multiple Fourier transform (MFT)
(*c*) Fourier analysis/cyclic reduction (FACR)
(*d*) convolution methods
(*e*) Concus and Golub iteration

We will now discuss in more detail the interplay of the different aspects of the problem formulation on the choice of the method of solution.

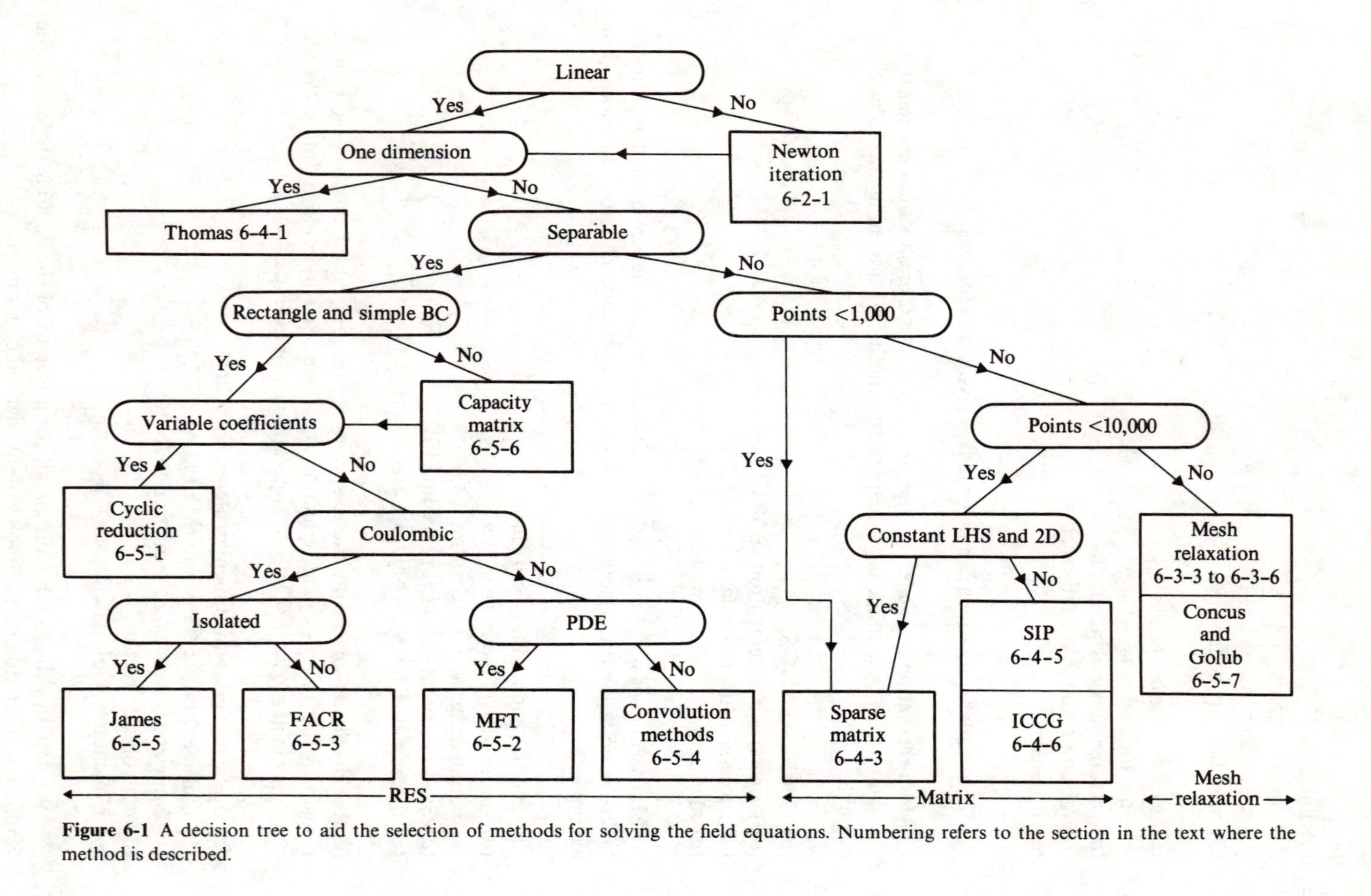

Figure 6-1 A decision tree to aid the selection of methods for solving the field equations. Numbering refers to the section in the text where the method is described.

6-1-1 Selection of Method

The selection of the most suitable method for the solution of the field equations will be assisted by Fig. 6-1 and Table 6-1. In the figure we give a decision tree to

Table 6-1 Operations counts and storage for all methods considered

Coefficient		Method	Ops/point		Storage/point		Ops/point‡	
			Decomp	Solve	Problem	Extra	Decomp	Solve
Special	RES	FACR(*) FACR(1)	$4.5 \log_2 (\log_2 N_g)$ $1.25 [\log_2 N_g + 5]$		1	0		17 25
Special	RES	MFT 3D (general)	$10 \log_2 N_g$		1	1		140
Special	RES	CR (Buneman)	$3 \log_2 N_g$		1	0		42
Special	RES	James	$5 \log_2 N_g$		1	0		70
Special	RES	CR general separable	$16 \log_2 N_g + 88$		1	$\frac{\log_2 N_g}{N_g^{1/2}}$	317	136
Arbitrary	Matrix	Thomas (1D)	3	5	4	1	3	5
Arbitrary	Matrix	Sparse 2D Matrix 3D	$10N_g^{1/2}$ $4N_g$	$12 \log_2 N_g$ $112N_g^{1/3}$	6	$3 \log_2 N_g$	1280 6×10^4	168 10^4
Arbitrary	Matrix	ICCG§	$27 \times 1.3N_g^{1/d}$		6	3		270†
Arbitrary	Matrix	Stones§ SIP	27×10		6	7		270†
Arbitrary	Relaxation	Concus & Golub§	$4.5 \log_2 (\log_2 N_g) \times 10$		6	2		170†
Arbitrary	Relaxation	SOR§ SLOR	$11 \times 2N_g^{1/d}$ $11 \times \sqrt{2}N_g^{1/d}$		6	1		2,816 2,000
Arbitrary	Relaxation	ADI§	$18 \times 12 \frac{\log_2 N_g}{d}$		6	1		1,500

† There is no strong evidence to discriminate the convergence of these methods. We assume a typical value of 10 iterations for all methods.

‡ On a two-timensional (128 × 128) mesh.

§ In form (ops per iteration) × (number of iterations for 10^{-6} error reduction). d dimensionality.

aid the selection, and in the table the required operations and storage per mesh point are summarized.

The first issue is linearity. If the equation is nonlinear, it will usually be solved by local linearization, followed by the solution of a succession of changing linear problems. This method of successive approximation is a generalization of Newton iteration for finding the roots of scalar equations. It is discussed in Sec. 6-2-1. The coefficients of the linear equations arising in the Newton iteration depend on the value of the current solution and thus cannot be assumed to have any special symmetry or values. We call this the arbitrary coefficient problem and with the exception of one-dimensional problems there is no generally accepted best method for solving it.

In one dimension the problem is that of solving a tridiagonal set of equations, and this may be efficiently done for arbitrary coefficients using a variation of Gauss elimination often called the Thomas algorithm (see Sec. 6-4-1).

In two or three dimensions one should look next to see if the equations can be expressed in the following separable form:

$$a(x)\frac{\partial^2 \phi}{\partial x^2} + b(x)\frac{\partial \phi}{\partial x} + c(x)\phi + \text{similar terms in } y \text{ and } z = g(x, y, z) \qquad (6\text{-}1)$$

in which the operator on the left-hand side is a sum of three terms, each depending only on one of the coordinates. This form includes the ∇^2 operator when expressed in all the common coordinate systems (cylindrical, spherical polar, etc.). If the equation is separable then special rapid elliptic solvers (RES) may be used and these are described in Secs. 6-5-1 to 6-5-7.

If the equation is nonseparable then either matrix or mesh relaxation methods may be used. The choice between them is liable to be determined by the number of mesh points involved. In small systems with $N_g < 1{,}000$, the arbitrary coefficient problem can be solved directly by sparse matrix methods (Sec. 6-4-3). Since the solution is then obtained with certainty in a known time, this method is recommended. If the system is larger, $N_g > 1{,}000$, then the time to decompose the matrix by sparse matrix methods is likely to become intolerably long for repeated use. If however the coefficients of the equations remain unchanged for a large number of right-hand sides, only one decomposition is required, and the time to solve for each right-hand side by SM methods will be acceptable for up to a few thousand equations. The operations per mesh point for the decompose phase and solve phase are given separately in Table 6-1.

If the arbitrary coefficients of the equation change for each right-hand side or the number of mesh points is $> 1{,}000$, then one is forced to use iterative methods of solution. For less than 10,000 points, the strongly implicit procedure of Stone (SIP) (see Sec. 6-4-5) or the incomplete Choleski/conjugate gradient method (see Sec. 6-4-6) are suitable. Three to seven extra meshes are required for intermediate vectors in the latter iterative methods and for meshes greater than 10,000 points this storage overhead is likely to make the methods impractical. Mesh relaxation methods (see Secs. 6-3-3 to 6-3-6) such as SLOR, or ADI, require only one extra mesh and may be used for such large problems. Alternatively, the iterative use of a

RES algorithm as proposed by Concus and Golub (see Sec. 6-5-7) is likely to give faster convergence, and is recommended. This method, however, requires two additional meshes.

We now return to the selection of methods when the equations are separable. An RES can be used directly on such problems if the region for solution is a square or rectangle and the boundary conditions are simple (for example, given value or gradient). If not, the capacity matrix procedure (see Sec. 6-5-6) can be used to hold the required boundary conditions on a region of arbitrary shape embedded in a rectangle. This requires the calculation of the capacity matrix, which is time-consuming, and subsequently two applications of an RES to obtain the solution. If the geometry is constant, the capacity matrix need be calculated only once and the solution time becomes that of the RES.

If the coefficients of the separable problem are variable, cyclic reduction may be used to solve the equations in $3\log_2 N_g$ operations per point (see Sec. 6-5-1). If the coefficients are constant, faster methods are available. If the boundary conditions are suitable, the general constant coefficient PDE can be solved by multiple Fourier transform (MFT) (see Sec. 6-5-2). If the potential is expressed as a sum of contributions from individual sources, then convolution methods using MFT are suitable. These cover the interaction of sources with an arbitrary-shaped Green's function, either in multiply-periodic geometry or as an assembly isolated in space (see Sec. 6-5-4). If the interaction is coulombic, faster methods are available for both the isolated problem (see James' algorithm, Sec. 6-5-5) and for Poisson's equation in the rectangle [see the optimal FACR algorithm, which has the smallest operation count of $4.5\log_2(\log_2 N_g)$, Sec. 6-5-3].

We now give a brief account of the different methods of solution, starting with methods for tackling the most general nonlinear problem and ending with special methods for the solution of Poisson's equation in the rectangle.

6-2 NONLINEAR PROBLEMS

6-2-1 Newton Iteration

If the field equation is nonlinear then the solution is most usually obtained by an iterative application of a method for the solution of a related linear problem.

For example, let

$$N(\phi) = 0 \tag{6-2}$$

be a nonlinear differential equation for the potential function $\phi(x, y, z)$ and let $\phi^{(t)}$ be the approximate solution to Eq. (6-2) at the tth iteration. If $\Delta\phi = \phi^{(t+1)} - \phi^{(t)}$ is the required change (assumed small) during the iteration, we may expand Eq. (6-2) about the present approximation, obtaining:

$$N(\phi^{(t)} + \Delta\phi) = N(\phi^{(t)}) + \left.\frac{\partial N}{\partial \phi}\right|^{(t)} \Delta\phi + \text{h.o.t.} \tag{6-3}$$

Both the equation N and its derivative on the right-hand side are evaluated using $\phi^{(t)}$ and are known operators.

If we ignore the higher-order terms (h.o.t.) in Eq. (6-3) and require $\phi^{(t+1)}$ to be a solution of Eq. (6-2), we obtain the first-order iterative scheme:

$$\left.\frac{\partial N}{\partial \phi}\right|^{(t)} \Delta\phi = -N(\phi^{(t)}) \tag{6-4a}$$

or

$$\left.\frac{\partial N}{\partial \phi}\right|^{(t)} \phi^{(t+1)} = -N(\phi^{(t)}) + \left.\frac{\partial N}{\partial \phi}\right|^{(t)} \phi^{(t)} \tag{6-4b}$$

In both the Eqs. (6-4) the left-hand side is a known linear operator, operating on the required solution; and the right-hand side is a known function of $\phi^{(t)}$ that is available from the last iteration. As shown in the example below, N will frequently depend on other given functions, which we designate here by $n(\mathbf{x})$.

In the form Eq. (6-4a) the residual $N(\phi)$ "drives" the change $\Delta\phi$ in the solution during one iteration. It is most suitable in computers with a short wordlength, in which the rounding error must be carefully controlled. Apart from the coefficient matrix which requires a storage of $4N_g$, four other meshes are required for n, $\phi^{(t)}$, $N(\phi)$, and $\Delta\phi$. In the form Eq. (6-4b), however, the new value $\phi^{(t+1)}$ is found directly and only three meshes, other than the coefficient matrix, are required: for n, $\phi^{(t)}$ and the right-hand side of Eq. (6-4b). The total storage required to solve the nonlinear problem by this method is $7N_g$ or $8N_g$ in addition to temporary storage required for the solution to the linear equation. This can be contrasted with the total storage of N_g required by a RES algorithm to solve particular constant-coefficient problems (see Secs. 6-5-1 to 6-5-7).

To see how this works out in practice, let us consider the nonlinear differential equation:

$$N(\phi) = \nabla^2\phi - (e^{\phi} - 1)n(\mathbf{x}) = 0 \tag{6-5}$$

which has a nonlinear dependence on the solution in the second term. In suitably normalized variables, this is the electrostatic potential in an electron plasma with a fixed positive-ion background $n(\mathbf{x})$ that is a function of the position vector $\mathbf{x}$. The exponential density is in thermal equilibrium according to the Boltzmann law. The equation could be used to calculate the potential in a hybrid model in which the slow-moving ions are advanced by the particle methods already described, and the light mobile electrons are assumed to adjust themselves instantaneously according to the Boltzmann law $n_e \propto \exp(e\phi/k_B T)$. In this case the ion background [represented by the function of position $n(\mathbf{x})$ in Eq. (6-5)] would be a known function.

Differentiating Eq. (6-5) with respect to ϕ we obtain

$$\frac{\partial N}{\partial \phi} = \nabla^2 - e^{\phi} n(\mathbf{x}) \tag{6-6}$$

Note that $n(\mathbf{x})$ is a function of position, but not of ϕ and hence differentiates to zero. Then, the iterative scheme in Eq. (6-4b) becomes

$$[\nabla^2 - e^{\phi^{(t)}} n(\mathbf{x})]\phi^{(t+1)} = [(1-\phi^{(t)})e^{\phi^{(t)}} - 1]n(\mathbf{x}) \tag{6-7}$$

If Eq. (6-7) is differenced in the normal way [see Chapter 2, Eq. (2-21)] the operator on the left-hand side is seen to be a three-, five-, or seven-point difference operator, depending on the number of dimensions. The coefficients of the operator, however, are variable because they depend on the last solution $\phi^{(t)}$ which is a function of position. This has the important consequence that the equations represented by (6-7) have different coefficients at each iteration, and the equations must be inverted afresh each time.

The iterative scheme in Eq. (6-4) is the generalization of the Newton–Raphson method for solving nonlinear algebraic equations and it can be shown that the convergence is very rapid—indeed it is quadratic—if the initial guess is sufficiently near the solution. This is the case in a time-stepping simulation if the guess for the solution at step n is taken as the solution at step $(n-1)$.

To find the convergence of Eq. (6-4), we expand about the exact solution ϕ^* and let $\varepsilon^{(t)}$ be the error at iteration t; then

$$\left.\begin{aligned} &\phi^{(t)} = \phi^* + \varepsilon^{(t)} \\ &N(\phi^{(t)}) = N(\phi^*) + \left.\frac{\partial N}{\partial \phi}\right|^* \varepsilon^{(t)} = \left.\frac{\partial N}{\partial \phi}\right|^* \varepsilon^{(t)} \\ &\left.\frac{\partial N}{\partial \phi}\right|^{(t)} = \left.\frac{\partial N}{\partial \phi}\right|^* + \left.\frac{\partial^2 N}{\partial \phi^2}\right|^* \varepsilon^{(t)} \end{aligned}\right\} \tag{6-8}$$

Subtracting $\partial N/\partial\phi|^{(t)}\phi^*$ from both sides of Eq. (6-4b) and substituting Eqs. (6-8), we obtain

$$\left\{\left.\frac{\partial N}{\partial \phi}\right|^* + \left.\frac{\partial^2 N}{\partial \phi^2}\right|^* \varepsilon^{(t)}\right\}\varepsilon^{(t+1)} = -\cancel{\left.\frac{\partial N}{\partial \phi}\right|^*} \varepsilon^{(t)} + \left\{\cancel{\left.\frac{\partial N}{\partial \phi}\right|^*} + \left.\frac{\partial^2 N}{\partial \phi^2}\right|^* \varepsilon^{(t)}\right\}\varepsilon^{(t)} \tag{6-9}$$

The first-order terms in ε on the right-hand side of Eq. (6-9) cancel, and, keeping only the most important lowest-order terms that remain on both sides, we obtain

$$\left.\frac{\partial N}{\partial \phi}\right|^* \varepsilon^{(t+1)} = \left.\frac{\partial^2 N}{\partial \phi^2}\right|^* (\varepsilon^{(t)})^2 \tag{6-10}$$

Equation (6-10) is a linear differential equation for the error at iteration $(t+1)$ in terms of the error at iteration (t). We note that the source term on the right-hand side is proportional to the square of the error function at the last iteration. Since the equation is linear in $\varepsilon^{(t+1)}$, the new error function is also proportional to the square of the last error function. This is called quadratic convergence and is extremely rapid. The typical magnitudes of errors at successive iterations are

$$\left.\begin{aligned} \varepsilon^{(0)} &\sim 10^{-1} \\ \varepsilon^{(1)} &\sim 10^{-2} \\ \varepsilon^{(2)} &\sim 10^{-4} \\ \varepsilon^{(3)} &\sim 10^{-8} \end{aligned}\right\} \tag{6-11}$$

and the solution is obtained to rounding error usually in four to five iterations.

We contrast this rate of convergence with the exponential convergence which is observed in most mesh relaxation schemes (see Secs. 6-3-1 to 6-3-6). In exponential convergence

$$\varepsilon^{(t+1)} = \lambda\varepsilon^{(t)} \tag{6-12}$$

where λ is a constant less than unity in modulus. In this case

$$\varepsilon^{(t+1)} \sim \lambda^{t}\varepsilon^{(0)} \tag{6-13}$$

and the number of iterations required to obtain an error reduction of 10^{-8} is $-8/\log_{10}\lambda$. For a typical value of $\lambda \sim 0.9$ about 100 iterations are required, instead of the 4 or 5 required if the convergence is quadratic.

The disadvantage with the Newton method is that the equations have variable coefficients and direct use cannot be made of rapid elliptic solvers (see Secs. 6-5-1 to 6-5-7). Sparse matrix methods are appropriate if there are only a few hundred mesh points (see Sec. 6-4-3). For larger numbers of mesh points some form of iteration is inevitable. Incomplete decomposition methods (Secs. 6-4-4 to 6-4-6) are suitable for a few thousand points and mesh relaxation or the iterative use of RES if the number of points exceeds about 10,000.

6-3 MESH RELAXATION

If the difference equations for each mesh point are written as a single matrix equation we have

$$A\boldsymbol{\phi} = \mathbf{q} \tag{6-14}$$

where A is an $(N_g \times N_g)$ matrix and $\boldsymbol{\phi}$ and $\mathbf{q}$ are, respectively, the vectors of N_g unknown values and the given N_g sources.

Mesh relaxation methods are defined by splitting the matrix A into two parts

$$A = B + R \tag{6-15}$$

where B is a matrix that is easily invertible and R is the remainder. Equation (6-14) can then be written

$$B\boldsymbol{\phi} = -R\boldsymbol{\phi} + \mathbf{q} \tag{6-16a}$$

or

$$= -(A-B)\boldsymbol{\phi} + \mathbf{q} \tag{6-16b}$$

Mesh relaxation proceeds by starting with an initial guess $\boldsymbol{\phi}^{(0)}$ and generating a series of iterates from

$$B\boldsymbol{\phi}^{(t+1)} = -(A-B)\boldsymbol{\phi}^{(t)} + \mathbf{q} \qquad t = 0, 1, \ldots \tag{6-17}$$

Since B has been chosen to be easily invertible Eq. (6-17) can readily be solved for the new iterate

$$\boldsymbol{\phi}^{(t+1)} = -B^{-1}(A-B)\boldsymbol{\phi}^{(t)} + B^{-1}\mathbf{q} \tag{6-18a}$$

or

$$\boldsymbol{\phi}^{(t+1)} = M\boldsymbol{\phi}^{(t)} + B^{-1}\mathbf{q} \tag{6-18b}$$

where $M = -B^{-1}(A-B)$ is called the iteration matrix.

If now we define the error vector at iteration t to be

$$\varepsilon^{(t)} = \phi^{(t)} - \phi \tag{6-19}$$

where ϕ is the exact solution, then by substitution in Eq. (6-18a) we obtain the equation for the error

$$\varepsilon^{(t+1)} = \boldsymbol{M}\varepsilon^{(t)} \tag{6-20a}$$

or

$$\varepsilon^{(t)} = \boldsymbol{M}^t\varepsilon^{(0)} \tag{6-20b}$$

where $\varepsilon^{(0)}$ is the initial error vector. Different iterative processes are defined by splitting the matrix A in different ways, that is to say, by defining different forms for the matrix $\boldsymbol{B}$.

We have assumed in the above discussion that the iteration matrix $\boldsymbol{M}$ remains the same at every iteration. This is not necessarily so, and one writes Eq. (6-20b) more generally as

$$\varepsilon^{(t)} = \boldsymbol{M}^{(t)}\varepsilon^{(0)} \tag{6-21}$$

where $\boldsymbol{M}^{(t)}$ may be a polynomial in $\boldsymbol{M}^{(1)}$.

The convergence of the process can be estimated by defining a single scalar measure, called a norm, of the error vector. We will use the L_2 norm, which is defined as

$$||\varepsilon|| = \sum_{i=1}^{N_g} \varepsilon_i^2 \tag{6-22}$$

where ε_i is the error at the ith mesh point.

Taking the norm of both sides of Eq. (6-21) one obtains

$$\frac{||\varepsilon^{(t)}||}{||\varepsilon^{(0)}||} \leqslant ||\boldsymbol{M}^{(t)}|| \tag{6-23}$$

That is to say the relative reduction in the size of the error after t iterations is guaranteed to be less than or equal to the norm of the matrix $||\boldsymbol{M}^{(t)}||$. Since the L_2 norm of a matrix is the modulus of its largest eigenvalue (called the spectral radius), convergence will take place provided, eventually, $||\boldsymbol{M}^{(t)}|| < 1$ for $t > t_0$. The convergence is clearly more rapid the smaller the value of $||\boldsymbol{M}^{(t)}||$. Hence the matrix $\boldsymbol{B}$, used in the splitting of A, must be chosen to minimize the norm of the iteration matrix, as well as being easily invertible. The problem with relaxation methods, in general, is that some eigenvectors of $\boldsymbol{M}^{(t)}$ have eigenvalues less than, but very close to, unity and the convergence of error vectors of this shape can be very slow indeed. This is seen in the Gauss–Seidel method below (Sec. 6-3-2).

In Fig. 6-2 we plot $||\boldsymbol{M}^{(t)}||$ for a variety of mesh relaxation methods. It is important to realize that Fig. 6-2 is not the graph of the error decay for any particular case. It is the envelope below which all error decay curves must lie. The equality in Eq. (6-23) guarantees that there is some initial error vector which will have a decay curve touching any selected point on the curves. In any particular case, all that is guaranteed is that the actual decay curve lies below the line for the process in Fig. 6-2, and actual examples may well not touch the curves.

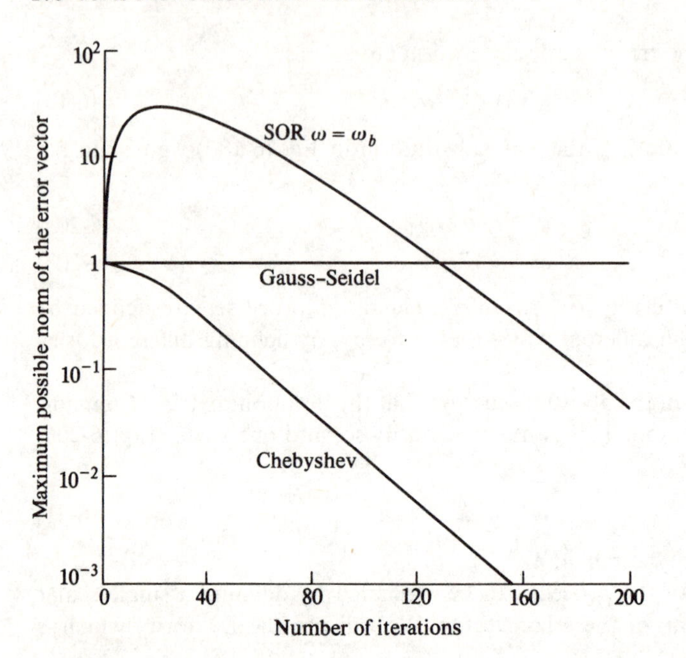

Figure 6-2 Theoretical error bounds for the variation of the maximum possible norm of the error vector (relative to its initial value) with the number of iterations, for the SOR, Gauss–Seidel, and Chebyshev methods, on a (128 × 128) mesh using odd/even ordering. (*From Hockney, 1970, courtesy of Methods in Computational Physics, © Academic Press Inc.*)

If it is required to obtain a solution to p decimal digits (error $< 10^{-p}$) then the number of iterations t^* will be given by

$$\|\boldsymbol{M}^{(t^*)}\| = 10^{-p} \tag{6-24a}$$

For simplicity, methods are usually compared on the basis of their asymptotic convergence for large t. If λ is the asymptotic convergence factor, then

$$\lambda^{t^*} = 10^{-p}$$

and

$$t^* = -p\frac{\ln 10}{\ln \lambda} \tag{6-24b}$$

Relaxation methods for the solution of PDEs are well described by Young (1962) and Forsythe and Wasow (1960). A more comprehensive discussion of the matrix theory and derivation of convergence rates is given by Varga (1962).

The matrix presentation of the relaxation procedure which has been given above is appropriate to the development of the theory of convergence. The matrices $\boldsymbol{A}$ and $\boldsymbol{B}$ which appear in the expressions are never, of course, explicitly stored or inverted. From the programmer's point of view, however, the relaxation schemes are best regarded in terms of the individual mesh point equations and the operations performed upon them.

The general five-point difference equation, corresponding to one row of the matrix A, may be written

$$a_{p,q}\phi_{p,q-1}+b_{p,q}\phi_{p,q+1}+c_{p,q}\phi_{p-1,q}+d_{p,q}\phi_{p+1,q}+e_{p,q}\phi_{p,q}=f_{p,q} \qquad p,q=1,\ldots,n \tag{6-25}$$

where the coefficients a, b, c, d, e in general vary from mesh point to mesh point and the subscripts p, q are the mesh point numbers in x and y, respectively. For the Poisson equation on a uniform mesh $a=b=c=d=1$ and $e=-4$, but this is a very special case and we would be quite wrong to use a general mesh-iterative method for this problem. Far quicker and more reliable methods are available and described in Secs. 6-5-1 to 6-5-7. These RES methods are usually five to ten times faster than the iterative methods now being described, but can be used only in such important special cases.

Equation (6-25) is converted to an explicit iterative method by solving for the value at the mesh point (p, q) in terms of its surrounding values as follows:

$$\phi^*_{p,q}=\frac{1}{e_{p,q}}(f_{p,q}-a_{p,q}\phi_{p,q-1}-b_{p,q}\phi_{p,q+1}-c_{p,q}\phi_{p-1,q}-d_{p,q}\phi_{p+1,q}) \tag{6-26a}$$

$$\phi^{\text{new}}_{p,q}=\omega\phi^*_{p,q}+(1-\omega)\phi^{\text{old}}_{p,q} \tag{6-26b}$$

where ω is a constant or variable relaxation factor chosen to optimize the convergence of the iteration.

In actual calculation, it is advantageous to express the Eqs. (6-26) in terms of the residual

$$r_{p,q}=a_{p,q}\phi_{p,q-1}+b_{p,q}\phi_{p,q+1}+c_{p,q}\phi_{p-1,q}+d_{p,q}\phi_{p+1,q}+e_{p,q}\phi_{p,q}-f_{p,q} \tag{6-27a}$$

The Eqs. (6-26) then become

$$\phi^{\text{new}}_{p,q}=\phi^{\text{old}}_{p,q}-\frac{r_{p,q}}{e_{p,q}}\omega \tag{6-27b}$$

In this formulation the norm of the residual vector may be conveniently accumulated and used as a criterion for the termination of the iteration (see Sec. 6-3-4).

The iterative procedures differ (1) in the selection of ω, (2) in the order in which the mesh points are processed, and (3) in the time, during the iteration, at which the new values replace the old values on the mesh.

Two orderings are important and are illustrated in Fig. 6-3. In typewriter ordering, the mesh points are processed in sequence and line by line, as in reading the words of a book. In odd-even ordering the mesh points are divided into two groups in the same manner as the black and white squares of a chessboard. The black, or odd, mesh points are traversed in the first half-sweep, and the white, or even, points in the second half-sweep. Since Eq. (6-26a) for an odd point refers only to values on even points, all the variables on the right-hand side are old ones from the last half-sweep. It follows that the order in which the odd (or even) points are

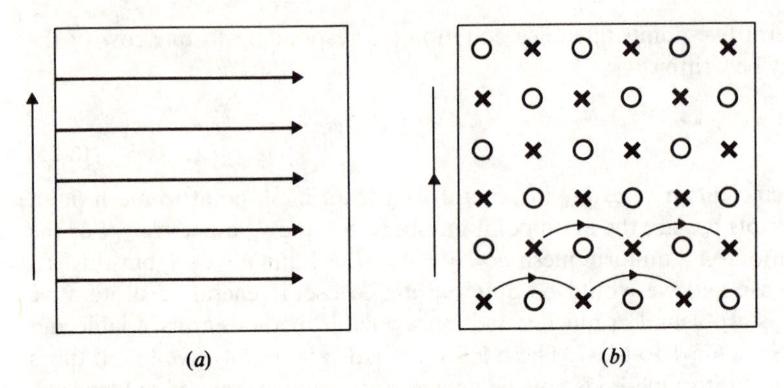

Figure 6-3 The orderings used in mesh relaxation schemes: (*a*) typewriter ordering; (*b*) odd/even or chequerboard ordering, odd points ×, even points O.

swept is immaterial and does not alter the algorithm. However, for convenience of programming, the points within each group are usually swept in typewriter order.

We will now describe the most important relaxation methods in terms of the ideas just presented.

6-3-1 Jacobi Method (J)

In the Jacobi method the matrix $\boldsymbol{B}$ is the diagonal matrix formed from the diagonal elements of $\boldsymbol{A}$. The solution for the next iteration is thus obtained by solving each mesh point equation for the value at that point, on the assumption that all the surrounding values are correct. The new values thus obtained do not replace the old values until all the mesh values have been recalculated. Thus, in terms of the mesh point formulation, all values of ϕ on the right-hand side of Eq. (6-26*a*) are old values from the last iteration and $\omega \equiv 1$. This method is only of theoretical interest because of its slow convergence. We mention it because the convergence rate of later methods are all expressed in terms of the spectral radius ρ of the Jacobi iteration matrix. We will take the solution of the five-point difference approximation to the Poisson's equation on a regular $(n \times n)$ mesh as a model problem for comparison purposes. A square region with $\phi \equiv 0$ on the boundary is assumed. For this model problem we have

$$\rho = \cos(\pi/n) \simeq 1 - \frac{1}{2}\frac{\pi^2}{n^2} \tag{6-28a}$$

and the asymptotic convergence factor is

$$\lambda_J = \rho \tag{6-28b}$$

6-3-2 Gauss–Seidel (GS)

If the matrix $\boldsymbol{B}$ is a lower triangular matrix formed from the lower triangular elements of $\boldsymbol{A}$, then we have the Gauss–Seidel iteration. The inversion of $\boldsymbol{B}$ is obtained by forward substitution. This method is computationally the same as the

Jacobi method, except that newly computed values for the next iterate replace old values from the last iteration as soon as they are calculated. In terms of the mesh point equations (6-26), the latest available values are taken on the right-hand side of Eq. (6-26*a*) and $\omega \equiv 1$.

The asymptotic convergence factor of the Gauss–Seidel method, using odd-even ordering, is

$$\lambda_{GS} = \rho^2 \tag{6-29}$$

where ρ is the spectral radius of the Jacobi method,

and
$$\lambda_{GS} = \cos^2(\pi/n) \simeq 1 - \frac{\pi^2}{n^2} \tag{6-30a}$$

Hence
$$t^* = \frac{\ln 10}{\pi^2}(pn^2) \simeq \frac{pn^2}{4} \simeq \frac{p(N_g)^{2/d}}{4} \tag{6-30b}$$

The number of operations per point is found from Eq. (6-26) to be 9, hence the number of operations required to solve the equations by Gauss–Seidel is

$$T_{GS} \simeq 2pn^4 = 2pN_g^2 \tag{6-31}$$

Since the number of mesh points N_g is likely to be at least 1,000, several thousand iterations will be required for reasonable convergence. For this reason the Gauss–Seidel method is utterly useless for computation purposes. We include it here as a warning, and as an aid to the description of the subsequent methods. This conclusion is confirmed in Fig. 6-2 which shows essentially no error reduction in a hundred iterations on the model problem with a mesh of (128 × 128).

6-3-3 Successive Overrelaxation (SOR)

Successive point overrelaxation recognizes that the convergence of Gauss–Seidel can be dramatically improved by taking a linear combination of the old value $\phi_{p,q}^{\text{old}}$ and the value given by Gauss–Seidel as is expressed in Eq. (6-26*b*). The relaxation factor ω lies in the range $1 \leqslant \omega \leqslant 2$ and corresponds to making an overcorrection at the point in order to anticipate future corrections. In some cases convergence can only be obtained with $\omega < 1$, in which case the process is known as underrelaxation.

If the spectral radius ρ of the Jacobi matrix is known, it can be shown that the best value to select for ω is

$$\omega_b = \frac{2}{1 + \sqrt{1 - \rho^2}} \tag{6-32}$$

in which case the asymptotic convergence factor is

$$\lambda_{\text{SOR}} = \omega_b - 1 \tag{6-33}$$

In the case of the two-dimensional Poisson equation in the square, using an $(n \times n)$ mesh, we have for large n

$$\rho = \cos(\pi/n) \simeq 1 - \frac{1}{2}\frac{\pi^2}{n^2} \tag{6-34a}$$

$$\omega_b \cong 2(1-\pi/n) \tag{6-34b}$$

$$\lambda_{\text{SOR}} = \omega_b - 1 = 1 - 2\pi/n \tag{6-34c}$$

The number of iterations t^* required asymptotically for a reduction of 10^{-p} in the error is given by

$$\lambda_{\text{SOR}}^t = (1-2\pi/n)^t = 10^{-p} \tag{6-35a}$$

or

$$t^* \approx \frac{\ln 10}{2\pi}(np) \simeq \frac{pn}{3} \tag{6-35b}$$

If we demand an accuracy of 10^{-3} ($p = 3$) then the number of iterations is approximately equal to the number of mesh points n along a side. For an accuracy of 10^{-6} we need twice that number of iterations. Thus, in a problem with a spatial resolution of about 1 percent ($n \simeq 100$), Eq. (6-35b) shows that any tolerably accurate solution will require a few hundred iterations. The number of operations per point in Eq. (6-26) is 11, hence we have the number of operations to solve Eq. (6-25) to an accuracy of 10^{-p}

$$\begin{aligned} T_{\text{SOR}} &= 11 N_g \frac{p}{3} (N_g)^{1/d} \\ &\cong 22 N_g (N_g)^{1/d} \quad \text{for } p = 6 \end{aligned} \tag{6-36}$$

where d is the dimensionality.

The important conclusion of this discussion is that the number of iterations required to solve a problem by SOR is not only proportional to the number of digits required to be accurate, namely p, but also to the number of mesh points along the side of the mesh—the time to solve being proportional to $N_g^{(1+1/d)}$ and not to N_g. This is a major disadvantage when the number of mesh points becomes large.

6-3-4 Chebyshev Acceleration

Figure 6-2 shows that, for the case of SOR, the asymptotic rate of convergence may not occur until after about n iterations. In fact, for the first n iterations the error vector may grow by a factor of about 20. This phenomenon of initial error growth is observed in practice and makes SOR a dangerous method to use, unless several times n iterations can be afforded. This is rarely the case.

A trivial alteration to SOR, known as Chebyshev acceleration, can be made which overcomes this problem. The Chebyshev method uses odd-even ordering and changes ω at each half-sweep, according to

$$\begin{aligned} &\omega^{(0)} = 1 \\ &\omega^{(1/2)} = 1/(1-\tfrac{1}{2}\rho^2) \\ &\omega^{(t+1/2)} = 1/(1-\tfrac{1}{4}\rho^2\omega^{(t)}) \qquad t = \tfrac{1}{2}, 1, \ldots \infty \\ &\omega^{(\infty)} \rightarrow \omega_b \end{aligned} \tag{6-37}$$

It can be shown that the maximum possible norm of the error vector always decreases as the number of iterations increases. The method is thus called "norm decreasing." The asymptotic convergence factor can be shown to be the same as SOR, but the guaranteed improved behavior for small numbers of iterations makes Chebyshev preferable to SOR. Indeed, since the operation count per iteration is the same for both cases, there would appear to be no valid reason for using SOR without Chebyshev acceleration.

In an obvious FORTRAN-like notation the code for the Chebyshev iteration is as follows:

$$
\begin{array}{l}
\qquad\qquad \omega = 1.0 \\
\qquad \text{DO 2} \quad t = 1, \text{MAXIT} \\
\qquad\qquad \text{Norm} = 0.0 \\
\qquad \text{DO 1} \quad p = 1, n \\
\qquad \text{DO 1} \quad q = 1, n \\
\text{IF (MOD}(p+q, 2)\text{ .NE. MOD}(t, 2))\text{ go to 1} \\
\text{Residual} = a_{p,q}\phi_{p,q-1} + b_{p,q}\phi_{p,q+1} + c_{p,q}\phi_{p-1,q} \\
\qquad\qquad + d_{p,q}\phi_{p+1,q} + e_{p,q}\phi_{p,q} - f_{p,q} \\
\text{Norm} = \text{Norm} + (\text{Residual})^2 \\
\phi_{p,q} = \phi_{p,q} - \omega \times \text{Residual}/e_{p,q} \qquad\qquad (6\text{-}38) \\
1 \quad \text{Continue} \\
{}^{*}\omega = 1.0/(1 - 0.25 \times \rho^2 \times \omega) \\
{}^{*}\text{IF } (t\text{ .EQ. }1) \quad \omega = 1.0/(1 - 0.5 \times \rho^2) \\
\text{IF (Norm .LT. EPS} \times \|\mathbf{f}\|)\text{ solution obtained} \\
2 \quad \text{Continue}
\end{array}
$$

In the above $\|\mathbf{f}\|$ is the precalculated norm of the right-hand side and EPS is a previously assigned accuracy limit (e.g., 10^{-6}). MAXIT is a preassigned maximum number of iterations, introduced to protect against infinite looping in cases of slow convergence. Typically MAXIT = 100 or 1,000.

The above code can be converted to straight Gauss–Seidel by omitting statements marked with an asterisk. If in addition the right-hand side of the first statement is replaced by ω_b we have the SOR iteration.

6-3-5 Block Methods

We have so far considered methods involving the pointwise adjustment of the approximate solution. However, our discussion of the one-dimensional solution of a tridiagonal matrix in Sec. 6-4-1 shows that we can equally easily solve directly for a line of unknown values, assuming that the solution values on the two neighboring lines are exact. A successive-line overrelaxation (SLOR) method may be defined in this way. The lines may be divided into odd and even groups and the process accelerated using the Chebyshev technique with Eqs. (6-37). For such one-line methods on the model problem, we have

$$\rho = \cos(\pi/n)/[2 - \cos(\pi/n)] \simeq 1 - \pi^2/n^2 \qquad (6\text{-}39a)$$

hence

$$\omega_b = 2/(1 + \sqrt{1-\rho^2}) \simeq 2 - 2\sqrt{2}\,\frac{\pi}{n} \qquad (6\text{-}39b)$$

and $$\lambda_{\text{SLOR}} = \omega_b - 1 \simeq 1 - 2\sqrt{2}\frac{\pi}{n} \tag{6-39c}$$

then $$t^*_{\text{SLOR}} = \frac{\ln 10}{2\sqrt{2}\pi}(np) \simeq \frac{pn}{4} \tag{6-39d}$$

$$T_{\text{SLOR}} = 11N_g\frac{pn}{4} \simeq 3pn^3 \tag{6-39e}$$

It is also possible to define a "two-line block SOR" method, which makes use of an efficient method for finding the values on two neighboring lines given the values on the next-adjacent lines. The convergence of the two-line method is $\sqrt{2}$ times better than single-line SOR.

The common feature of these methods is that the rate of convergence is improved as larger and larger blocks of the problem are solved implicitly. Other variations may be defined by alternatively solving for lines in the x, y, and z directions on successive sweeps of the mesh. Intuitively one may regard the error as flowing freely to the boundary in the direction of the line solution, but only slowly in the directions perpendicular to this. The object of the exercise is to drive the error to the boundary, where it is absorbed. It is clear that this is achieved more quickly if the direction of the line solution is changed cyclically.

The ultimate in block methods is described by Concus and Golub, who use a routine capable of solving Poisson's equation over the whole mesh as the basis of an iterative solution to a problem which resembles Poisson's equation. This may be regarded as a block relaxation and is described with rapid elliptic solvers in Sec. 6-5-7.

Block methods are less frequently used than point SOR principally because they are more complicated to program. However, their improved convergence rates show that they should be used if the effort or a library program is available.

6-3-6 Alternating Direction Implicit (ADI)

The alternating direction implicit (ADI) method is a block-iterative method developed originally for solving the time-dependent heat flow equation

$$\frac{\partial\phi}{\partial t} = \nabla^2\phi + q \tag{6-40}$$

It may be used to solve elliptic problems by observing that, when a steady-state solution to Eq. (6-40) is found with $\partial\phi/\partial t = 0$, ϕ is the solution of

$$-\nabla^2\phi = q \tag{6-41}$$

More generally one can apply the technique to the solution of the linear problem

$$L\phi = q \tag{6-42}$$

by solving the time-dependent problem

$$\frac{\partial \phi}{\partial t} = -L\phi + q \tag{6-43}$$

The operator L which may already be taken to be the desired finite-difference operator, is positive definite (like the operator $-\nabla^2$). We now split the operator L into two parts

$$L = L_x + L_y \tag{6-44}$$

where L_x represents the differencing in x and L_y that in y.

In the case of the five-point approximation to $-\nabla^2\phi$ on a uniform mesh, we would use

$$-\frac{\partial^2 \phi}{\partial x^2} \simeq L_x\phi = 2\phi_{p,q} - (\phi_{p-1,q} + \phi_{p+1,q}) \tag{6-45a}$$

$$-\frac{\partial^2 \phi}{\partial y^2} \simeq L_y\phi = 2\phi_{p,q} - (\phi_{p,q-1} + \phi_{p,q+1}) \tag{6-45b}$$

and

$$-\nabla^2\phi \simeq L\phi = [L_x + L_y]\phi \tag{6-45c}$$

More complicated operators may be similarly split, but it is wise to base the splitting on the natural splitting of the original physical differential equations, as was done above. We know that the physical problem has a steady-state solution, because the temperature ϕ is known to approach thermodynamic equilibrium after an initial transient. If the splitting and finite-differencing mimics this behavior we can have some assurance of the convergence of the method.

The time-dependent equation

$$\frac{\partial \phi}{\partial t} = -(L_x + L_y)\phi + q \tag{6-46}$$

is now differenced implicitly in two half-steps as follows:

$$\frac{\phi^{n+1/2} - \phi^n}{\Delta t/2} = -L_x\phi^{n+1/2} - L_y\phi^n + q \tag{6-47a}$$

$$\frac{\phi^{n+1} - \phi^{n+1/2}}{\Delta t/2} = -L_x\phi^{n+1/2} - L_y\phi^{n+1} + q \tag{6-47b}$$

Rearranging and writing in matrix vector notation, we have

$$[I + rL_x]\boldsymbol{\phi}^{n+1/2} = [I - rL_y]\boldsymbol{\phi}^n + r\mathbf{q} \tag{6-48a}$$

$$[I + rL_y]\boldsymbol{\phi}^{n+1} = [I - rL_x]\boldsymbol{\phi}^{n+1/2} + r\mathbf{q} \tag{6-48b}$$

where $r = \Delta t/2$.

The operators on the left-hand side of Eq. (6-48) are tridiagonal matrices that can be readily solved by applying the Thomas algorithm (Sec. 6-4-1) line by line. We note that the right-hand side of Eq. (6-48) must be evaluated first using always "old" values of $\boldsymbol{\phi}$ from the last sweep, before any of the tridiagonal equations are solved.

If we eliminate $\boldsymbol{\phi}^{n+1/2}$ we obtain the iteration matrix $\boldsymbol{M}$ for one double sweep

$$\boldsymbol{\phi}^{n+1} = \boldsymbol{M}\boldsymbol{\phi}^n + \boldsymbol{N} \tag{6-49a}$$

where

$$\boldsymbol{M} = (\boldsymbol{I}+r\boldsymbol{L}_y)^{-1}(\boldsymbol{I}+r\boldsymbol{L}_x)^{-1}(\boldsymbol{I}-r\boldsymbol{L}_x)(\boldsymbol{I}-r\boldsymbol{L}_y) \tag{6-49b}$$

and

$$\boldsymbol{N} = 2(\boldsymbol{I}+r\boldsymbol{L}_y)^{-1}(\boldsymbol{I}+r\boldsymbol{L}_x)^{-1}r\mathbf{q} \tag{6-49c}$$

If $\boldsymbol{L}_x$ and $\boldsymbol{L}_y$ commute and are symmetric they have the same eigenvectors. If the eigenvalues of $\boldsymbol{L}_x$ are λ_x and $\boldsymbol{L}_y$ are λ_y, then the eigenvalues of the iteration matrix $\boldsymbol{M}$ are

$$\lambda(\boldsymbol{M}) = \frac{(1-r\lambda_x)(1-r\lambda_y)}{(1+r\lambda_x)(1+r\lambda_y)} \tag{6-50}$$

It is clear from Eq. (6-50) that the eigenvalues of $\boldsymbol{M}$ are less than one for all positive r. This is reasonable because positive r corresponds to integrating the diffusion equation forward in time, which is physically a stable process. The error may be regarded as diffusing to the boundary where it is absorbed. The use of negative r, or timestep, would correspond to integrating backward in time. This negative diffusion results in the peaking or accumulation of errors and one would not expect it to be convergent.

The parameter r can be varied every double sweep in order to improve convergence. If

$$0 < \lambda_{\min} \leqslant \lambda_x, \lambda_y \leqslant \lambda_{\max} \tag{6-51}$$

then a series of r_k are selected by

$$r_k = 1/(\lambda_{\max} x^{k-1}) \qquad k = 1, 2, \dots, t \tag{6-52a}$$

where

$$x = \left(\frac{\lambda_{\min}}{\lambda_{\max}}\right)^{1/(t-1)} \tag{6-52b}$$

The length of the cycle t should be chosen to minimize

$$S_t = \left(\frac{1-\sqrt{x}}{1+\sqrt{x}}\, e^{-x^{3/2}/(1+x)}\right)^{4/t} \tag{6-53}$$

The parameters r_k span the interval from $r_1 = 1/\lambda_{\max}$ to $r_t = 1/\lambda_{\min}$ and for the model problem in the square we have

$$\lambda_{\min} = 4\sin^2(\pi/2n) \qquad \lambda_{\max} = 4\cos^2(\pi/2n) \qquad t = 9$$

For practical purposes, suitable values for $\lambda_{\min}$ and $\lambda_{\max}$ may be estimated by embedding the actual problem in a square and using the values just given.

The number of iterations required to obtain a specified accuracy with the ADI method is proportional to $\log N_g$ and not $N_g^{1/d}$ as in SOR or SLOR. However, the conditions under which ADI is known to converge are more restrictive than for overrelaxation, and it is not uncommon to find ADI divergent on problems which converge with SOR, although the reverse experience has also been reported. Hence it can be said that ADI should be tried on a problem in the first instance because, if successful, its convergence rate is likely to be superior to SOR. However, if it fails, one may usually fall back on successive overrelaxation.

6-4 MATRIX METHODS

In the following matrix methods the equations arising from finite differencing on a mesh are regarded simply as a large set of linear equations and various standard methods for solving general sets of equations are used. These include Gauss elimination, Choleski decomposition, and conjugate-gradient methods. Such methods have the advantage of applying to the most general linear problem, but equally have the disadvantage of making no use of the known structure of the equations.

6-4-1 Thomas Tridiagonal Algorithm

The application of finite differences with variable mesh spacing to a second-order ODE leads to a tridiagonal set of equations with arbitrary coefficients. Similar equations are obtained from the Newton method in one dimension (Sec. 6-2). Such equations may be efficiently solved in $5N_g$ operations for any coefficients, by standard Gauss elimination.

If the difference equations are

$$a_i\phi_{i-1}+b_i\phi_i+c_i\phi_{i+1} = d_i \qquad i = 1,\dots, N_g \tag{6-54}$$

the solution may be obtained with two FORTRAN DO-loops expressing the forward and backward recurrence relations.

Forward elimination:

$$\omega_1 = \frac{c_1}{b_1} \qquad \omega_i = \frac{c_i}{b_i-a_i\omega_{i-1}} \qquad i = 2, 3,\dots, N_g-1 \tag{6-55}$$

$$g_1 = \frac{d_1}{b_1} \qquad g_i = \frac{d_i-a_ig_{i-1}}{b_i-a_i\omega_{i-1}} \qquad i = 2, 3,\dots, N_g \tag{6-56}$$

Back substitution:

$$\phi_{N_g} = g_{N_g} \quad \phi_i = g_i-\omega_i\phi_{i+1} \qquad i = N_g-1, N_g-2,\dots, 1 \tag{6-57}$$

For the special periodic case with $a = c = 1$ and $b = -2$, an alternative algorithm is given in Chapter 2 [Eqs. (2-54) and (2-55)].

If the coefficients a_i, b_i, c_i change at each iteration, as they do in the Newton scheme, then the number of operations required is $8N_g$. If, however, the coefficients remain constant and only the source term d_i changes, then the intermediate vectors ω_i and $1/(b_i-a_i\omega_{i-1})$ can be precalculated and stored. In this case the solution is obtained in $5N_g$ operations. If, in addition $a_i = c_i = 1$, the operations can be further reduced to $4N_g$.

During the forward elimination g_i may overwrite the right-hand side d_i and in the backward substitution ϕ_i may overwrite g_i. Auxiliary storage is required, however, for the variables ω_i, giving a total storage requirement of $2N_g$ in addition to the coefficients.

6-4-2 Conjugate-Gradient Algorithm (CGA)

The conjugate-gradient algorithm is a general method for finding the minimum of a function. It may be used to solve a set of difference equations

$$A\boldsymbol{\phi} = \mathbf{q} \tag{6-58}$$

by defining the quadratic function

$$V(\boldsymbol{\phi}) = \tfrac{1}{2}\boldsymbol{\phi}^T A\boldsymbol{\phi} - \mathbf{q}^T\boldsymbol{\phi} \tag{6-59}$$

where $\mathbf{q}$ is the vector whose elements are the N_g charge values on a mesh, and $\boldsymbol{\phi}$ is the vector of unknown potentials. The $(N_g \times N_g)$ matrix A is the result of applying the difference equation (6-25) at each of the N_g points of the mesh.

The conjugate-gradient method applies if the matrix A is symmetric and positive definite. This can be made to be the case if A is the finite-difference representation of $-\nabla^2$. Differentiating Eq. (6-59) in this case, we obtain the gradient vector $\mathbf{r}$, which turns out to be the residual of the original equations

$$\frac{\partial V}{\partial \boldsymbol{\phi}} = A\boldsymbol{\phi} - \mathbf{q} = \mathbf{r} \tag{6-60}$$

and differentiating again we show that A is the matrix of second derivatives

$$A_{ij} = \frac{\partial^2 V}{\partial \phi_i \partial \phi_j} \tag{6-61}$$

Equations (6-60) and (6-61) show that all first derivatives are zero at a solution to the original equations (6-58), and if A is positive definite, this extremum is a minimum.

The CGA finds the minimum of a quadratic function of N_g variables, such as that defined by Eqs. (6-59), in a maximum of N_g iterations. In actual problems very much faster convergence is frequently reported, but it can be shown that, in the absence of rounding error, the exact solution of Eq. (6-58) is bound to be found after N_g iterations. The CGA successively finds minima in subspaces of the N_g variables, starting with a one-dimensional subspace. Having found the minimum in one subspace, it expands the subspace by one dimension and locates the new minimum. When the dimension of the subspace reaches N_g, the number of original equations, the solution is found.

In the two-dimensional case illustrated in Fig. 6-4, the process is as follows. The gradient $\mathbf{r}_0$ is evaluated at the position of the initial guess $\boldsymbol{\phi}_0$. The initial one-dimensional subspace is defined by the search direction $\mathbf{s}_0 = -\mathbf{r}_0$, which is in the steepest "downhill" direction at $\boldsymbol{\phi}_0$. The minimum of $V(\boldsymbol{\phi})$ is then computed along the direction $\mathbf{s}_0$. If ω is the distance along $\mathbf{s}_0$, $V(\boldsymbol{\phi}_0 + \omega\mathbf{s}_0)$ is a scalar quadratic

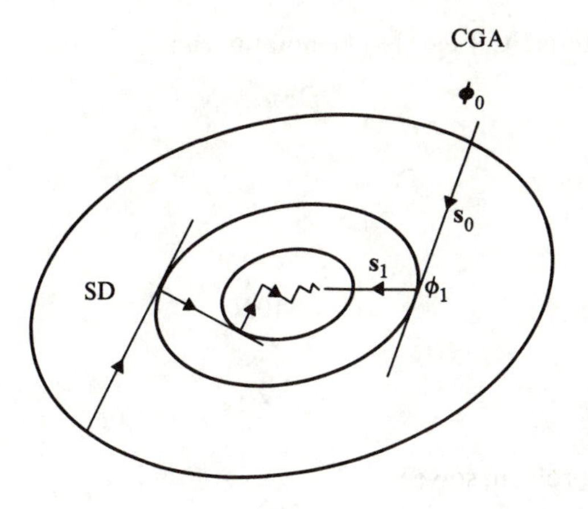

Figure 6-4 CGA: the conjugate gradient algorithm showing convergence in two steps. SD: the steepest descent algorithm shows slow convergence in a narrow valley with $\lambda_{max}/\lambda_{min} \gg 1$.

function of ω and the minimum at $\omega = \omega_0$ can be directly evaluated. This position $\boldsymbol{\phi}_1 = \boldsymbol{\phi}_0 + \omega_0 \mathbf{s}_0$ defines the next value in the iteration towards the solution. If we continue this process of searching along the line of steepest descent, we have described the method of "steepest descent." This method can be extremely slowly convergent if the ratio of the maximum to minimum eigenvalues of A (called the condition number) is large, that is to say the valley represented in Fig. 6-4 is very long and narrow. In this case successive iterates tend to bounce across the sides of the valley and only progress very slowly towards the minimum, as is illustrated at the left of Fig. 6-4. The CGA overcomes this problem by choosing the next search direction $\mathbf{s}_1$ as a linear combination of $\mathbf{s}_0$ and the new steepest-descent direction $-\mathbf{r}_1$ at $\boldsymbol{\phi}_1$, namely

$$\mathbf{s}_1 = -\mathbf{r}_1 + \beta \mathbf{s}_0 \tag{6-62a}$$

where

$$\beta = \frac{\mathbf{r}_1 \cdot \mathbf{r}_1}{\mathbf{r}_0 \cdot \mathbf{r}_0} \tag{6-62b}$$

It can be shown that $\mathbf{s}_1$ is the direction of the minimum in the two-dimensional subspace containing $\mathbf{r}_1$ and $\mathbf{s}_0$. In the two-dimensional example of Fig. 6-4, the minimum along $\mathbf{s}_1$ is the solution to the problem. It has been reached in two steps.

The two-dimensional case may be regarded as a slice through a three-dimensional problem, and the method extends naturally to higher dimensions. The successive residual vectors have the important property of being orthogonal to each other. Since there can be at most N_g different orthogonal vectors in an N_g-dimensional space, the N_g plus first residual vector must be the null vector and the solution has been found.

The orthogonality property is never exactly obeyed because of rounding error, and the CGA is not found to be useful for solving the finite-difference equations themselves. However, it is found to be particularly effective as an adjunct to methods in which A is approximately the identity matrix, as in the ICCG method (Sec. 6-4-6).

The general algorithm, in an obvious FORTRAN-like notation, is:

Initial guess: *Operations*

ϕ_0

EPS = required accuracy

→ DO 1 $j = 0, N_g - 1$

Residual:

$\mathbf{r}_j = A\boldsymbol{\phi}_j - \mathbf{q}$ $10N_g$

Error:

$\|\varepsilon_j\| = \mathbf{r}_j^T \mathbf{r}_j$ $2N_g$

Early finish:

if ($\|\varepsilon_j\| \leqslant$ EPS × $\|\mathbf{q}\|$) problem solved

Search direction:

$\mathbf{s}_0 = -\mathbf{r}_0$

$\left.\begin{array}{l}\beta_j = \|\varepsilon_j\|/\|\varepsilon_{j-1}\| \\ \mathbf{s}_j = -\mathbf{r}_j + \beta_j \mathbf{s}_{j-1}\end{array}\right\} j > 0$ $2N_g$

Distance to minimum:

$\omega_j = \dfrac{\|\varepsilon_j\|}{\mathbf{s}_j^T A \mathbf{s}_j}$ $11N_g$

Next iterate:

$\boldsymbol{\phi}_{j+1} = \boldsymbol{\phi}_j + \omega_j \mathbf{s}_j$ $2N_g$

— 1 Continue

Operations per iteration $27N_g$

The storage required for the solution of a problem by the CGA is $3N_g$ for storage of the mesh vectors $\boldsymbol{\phi}_j$, $\mathbf{r}_j$, and $\mathbf{s}_j$. This is in addition to the storage of $5N_g$ for the matrix A and N_g for the right-hand side $\mathbf{q}$. The operation count using three meshes is $27N_g$. This may be reduced to $17N_g$ if we introduced a fourth storage mesh for the vector $\mathbf{t}_j = A\mathbf{s}_j$ during the calculation of ω_j. The next residual $\mathbf{r}_{j+1}$ can then be computed from

$$\mathbf{r}_{j+1} = \mathbf{r}_j + \omega_j \mathbf{t}_j \tag{6-63}$$

without the need to evaluate $A\boldsymbol{\phi}_j$. However, for most PDE problems on substantial meshes, this extra storage will be unacceptable. The conjugate-gradient algorithm is extensively discussed by Faddeev and Faddeeva (1963) and Reid (1971). Concus, Golub and O'Leary (1976) consider the method to be most effective when used as an acceleration technique for existing iterations.

6-4-3 Sparse Matrix Methods (SM)

Variations of the Gauss elimination process for the direct solution of sets of equations have been extensively developed for application to matrices with a large number of zeros. Such sparse matrices arise in the finite-difference or finite-element descretizations of PDEs, and also in management science, power systems analysis, surveying, circuit theory, and structural analysis. An excellent survey has been given by Reid (1976). The application of the five-point finite difference approximation to Poisson's equations on a (128×128) mesh leads to $\sim 16{,}000$ equations each with only five variables. The complete matrix has 256 million entries, however only $\sim 80{,}000$ (or ~ 0.03 percent) of these are nonzero. This is the typical situation for which sparse matrix methods were developed. Such methods only store the nonzero elements and minimize the creation of further nonzeros during the elimination process. The method of storage usually involves some form of linked list, in which there is an address index associated with each nonzero element which points to the next nonzero element in the row or column. The insertion or deletion of nonzero elements can then be achieved by altering the values of these "links" without the need to move data about the store.

The general problem is to find the solution of the equation

$$A\mathbf{x} = \mathbf{b} \tag{6-64}$$

where A is the $(N \times N)$ matrix with elements $a_{i,j}$. The factorization of A by Gauss elimination proceeds in stages $k = 1, 2, \ldots, N$. At each stage a pivotal element is selected from the rows and columns of the matrix that have not yet been factorized (i.e. those for which $i, j \geqslant k$). These rows and columns are interchanged, although not necessarily moved in store, in order to bring the pivotal element to the position $a_{k,k}$. Multiples of the pivotal kth row are then added to all lower rows in order to eliminate the elements below the pivot, using the equations

$$a_{i,j}^{(k+1)} = a_{i,j}^{(k)} - l_{i,k} a_{k,j}^{(k)} \qquad i, j > k \tag{6-65}$$

$$b_i^{(k+1)} = b_i^{(k)} - l_{i,k} b_k^{(k)} \qquad i > k \tag{6-66}$$

where

$$l_{i,k} = [a_{k,k}^{(k)}]^{-1} a_{i,k}^{(k)}$$

for $k = 1, 2, \ldots N$ starting with $A^{(1)} = A$ and $\mathbf{b}^{(1)} = \mathbf{b}$. This eliminates the element $a_{i,k}$ in each row, but clearly may introduce in the new row other nonzero elements not previously present. This is called fill-in. The new values $a_{i,j}^{(k+1)}$ may overwrite $a_{i,j}^{(k)}$, and $l_{i,k}$ may overwrite $a_{i,k}^{(k)}$. Finally, at stage N, the triangular factorization is complete. The multipliers $l_{i,k} = a_{i,j}^{(N)} (i > j)$ are the elements of a lower triangular matrix $\boldsymbol{L}$ with unit diagonal elements that are not stored, and $a_{i,j}^{(N)} (i \leqslant j)$ are the elements of an upper triangular matrix $\boldsymbol{U}$ such that

$$A = LU \tag{6-67}$$

The solution of $A\mathbf{x} = \mathbf{b}$ is then obtained by forward substitution to obtain an auxiliary vector $\mathbf{y}$:

$$L\mathbf{y} = \mathbf{b} \tag{6-68}$$

followed by the backward substitution

$$U\mathbf{x} = \mathbf{y} \tag{6-69}$$

which obtains the solution $\mathbf{x}$.

In principle any nonzero element of the reduced matrix can be chosen as the next pivot. The choice is guided by three considerations:

1. numerical stability
2. minimization of fill-in
3. minimization of arithmetic

Numerical stability is guaranteed for a symmetric positive-definite matrix if we make symmetric interchanges of rows and columns during the pivotal elimination. For a nonsymmetric case it is important not to choose a pivot that is too small because then the factors $l_{i,k}$ become large. In this case precision is lost when the multiple of the pivotal row (now large numerically) is added to the nonpivotal row. A criterion that the pivotal element must be greater than some fraction (typically $\frac{1}{4}$) of the maximum element in its row is usually used to prevent the unreasonable growth in the size of matrix elements as elimination proceeds.

The minimization of the creation of new nonzero elements or fill-in of the matrix requires a calculation that is difficult to implement efficiently. However, a criterion due to Markowitz that minimizes the number of arithmetic operations is much less costly to implement and is found, empirically, to limit the fill-in almost as well as a minimum fill-in procedure (see Table 6-2). This table shows that minimum fill-in is up to six times more costly in computer time and is only marginally better in reducing fill-in than the Markowitz criterion. The Markowitz criterion selects as pivot that element which has the minimum product of other nonzeros in its row and other nonzeros in its column. That is to say, one pivots using the element with the minimum value of

$$\rho_{i,j} = (r_i - 1)(c_j - 1) \tag{6-70}$$

where r_i is the number of nonzeros in the ith row and c_j is the number of nonzeros in the jth column.

The organization of a sparse matrix program package, for example the Harwell MA18 (Curtis and Reid, 1971) and MA28 routines, is in three parts:

1. ANALYZE: the matrix structure is analyzed to determine an order of pivoting which will be stable numerically and reduce the creation of nonzero elements.

Table 6-2 The extent of fill-in during decomposition

Matrix order		54	57	199
Nonzeros in **A**		291	281	701
Nonzeros in **L/U**	Markowitz	381	315	1,387
	Min. of fill-in	361	291	1,372
ANALYZE time (ms on 370/165)	Markowitz	140	110	900
	Min. of fill-in	800	500	6,400

Source: Reid (1976); courtesy of Academic Press, © Institute of Mathematics and its Applications.

Table 6-3 Relative cost of analyze, factorize, and operate on the five-point difference approximation to Laplace's equation on a (31 × 31) mesh.

Matrix order			930
Number of nonzeros stored	Variable-band		20,245
	Markowitz		11,328
Time on IBM 370/165 ms	ANALYZE	(Markowitz)	12,200
	FACTOR	Variable-band	1,300
		Markowitz	600
	OPERATE	Variable-band	160
		Markowitz	120

Source: Reid (1976); courtesy of Academic Press, © Institute of Mathematics and its Applications.

Gauss elimination is applied to the matrix, thus permuted, to form the ***LU*** decomposition.

2. FACTORIZE: a new matrix with the same structure as one previously analyzed by (1) is factorized into ***L*** and ***U***.
3. OPERATE: Given the factorization obtained in (2) the solution for one right-hand side is obtained.

The routine ANALYZE must be used first on any completely new problem. If, now, another matrix is presented with the same structure (that is to say, the same positions of nonzero elements) as that previously ANALYZED, it may be decomposed by FACTORIZE without another call to ANALYZE. Similarly, if a variety of different right-hand sides are to be solved for a matrix that has already been FACTORIZED, then calls need only to be made to OPERATE. The importance of this subdivision is apparent from Table 6-3, which shows that ANALYZE is ten times more costly to use than FACTORIZE, and that FACTORIZE is ten times more costly to use than OPERATE.

The advantage of sparse matrix techniques is that they can, in principle, tackle the general nonseparable variable-coefficient PDE by direct methods, and thus get the solution to the completely general problem in a fixed and known time. Their practical application depends critically on the growth of nonzero elements and the consequential demands on storage. Although we know that one should always use an RES method for Poisson's equation in the square, this is a useful test problem with the same structure as the most general PDE with variable coefficients. As the time for a sparse matrix solution is dependent on the structure of the matrix and not on the numerical value of its coefficients, the computer times found will apply to the solution of the variable-coefficient problem to which RES methods cannot be applied (except in a Concus and Golub iteration, Sec. 6-5-7).

The best initial ordering for the finite-difference equations is that given by

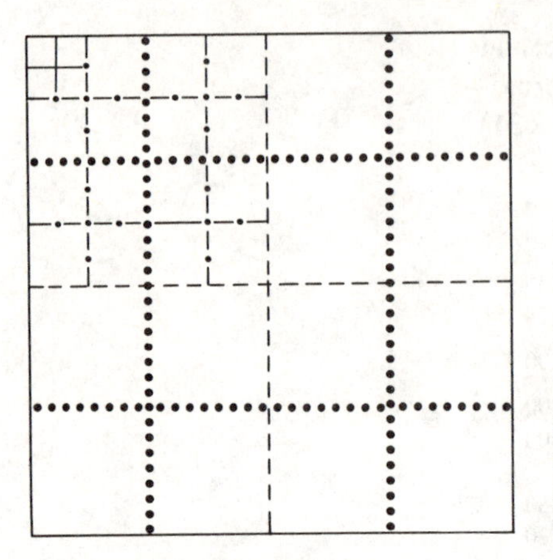

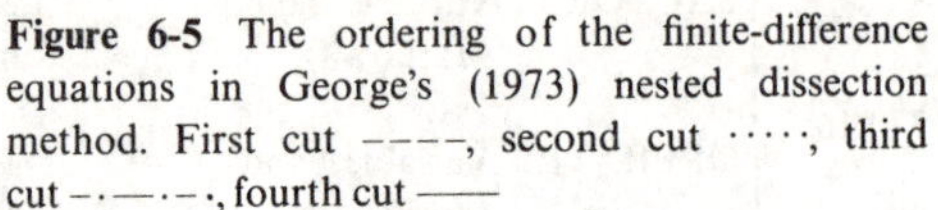

Figure 6-5 The ordering of the finite-difference equations in George's (1973) nested dissection method. First cut ----, second cut ····, third cut -·—·-·, fourth cut ——

George (1973) and called "nested dissection" (see Fig. 6-5). If the equations for all mesh points forming the central cross (called the first cut) are written last, the remaining equations form four unconnected sets for the interior points of each quarter. If all the equations for the top left quarter are written first, followed by those for the top right quarter, we obtain a block diagonal matrix for the remaining equations. The equations in each of these blocks may now be reordered, in exactly the same way as were the original equations, by introducing the second cut in the figure. The process repeats recursively until only the set of all odd points is left, and these become the first $N_g/2$ equations. The procedure has been generalized by Duff et al. (1976), without loss of effectiveness, to meshes with any number of mesh points. That is to say, the implied assumption, used in our example, that the number of mesh points is a power of two, is not necessary. Furthermore, it can be proved that the nested-dissection ordering gives minimum fill-in for the problem just discussed.

Duff et al. (1976) report the following results for George's initial ordering followed by the Markowitz criterion for subsequent elimination. For comparison purposes we show in square brackets the corresponding figures for an optimal RES algorithm [optimal FACR(*), Sec. 6-5-3].

Non-zero elements

$$\begin{aligned} &\text{in } \mathbf{A}: 5N_g && [5] \\ &\text{in } \mathbf{L} = \mathbf{U}^T: 3N_g \log_2 N_g && [0] \\ &\text{in } \mathbf{b}: N_g && [N_g] \end{aligned} \tag{6-71}$$

Arithmetic operations

$$\left.\begin{aligned} \text{FACTORIZE}: 10N_g^{3/2} \\ \text{OPERATE}: 12N_g \log_2 N_g \end{aligned}\right\} [4.5N_g \log_2 (\log_2 N_g)]$$

From these results, we see that the solution of a prefactorized matrix by sparse techniques is theoretically between eight and ten times slower than the optimal FACR algorithm for meshes in the range 32^2 to 128^2. In the case of the

Table 6-4 Time and storage to solve 5-Point Poisson difference equation

Time in ms for FORTRAN code scaled to equivalent IBM 360/168

Mesh	Sparse†		RES, POT1 = FACR(1)		
	Operate	Bytes‡	Solve		Bytes§
32 × 32	87	120 k	71§ 66††	71¶	4 k
64 × 64	(438)	576 k	192 140	223	16 k
128 × 128	(2028)	3 M	618 441	748	64 k
256 × 256	(9270)	12 M	[2,472] [1,764]	2491	256 k

† I. S. Duff, private communication.
‡ $k = 1024$, $M = k^2$.
§ 2.7 times measured time on CDC 7600
¶ 2.3 times measured time on IBM 360/195.
†† IBM 360/168 assembler.
( ) Extrapolated using $8.84 N_g \log_2 N_g$ ms.
[ ] Extrapolated.

nonoptimal FACR(1) algorithm [e.g., program POT1 (Hockney, 1970)], which has an operation count of $1.25N_g\,[\log_2 N_g + 4.8]$, the expected factor is 6.6. An actual comparison has been made by Hockney and Duff, between the FACR(1) program POT1 and the Harwell sparse matrix routine MA17 and is shown in Table 6-4. This shows that for a 32^2 mesh there is little difference between the methods, but that the sparse matrix is significantly slower than RES for meshes larger than 64^2. The speeds are approaching the expected ratio as the number of mesh points increases.

A major difference between the methods lies in the storage (see Table 6-4) required which is N_g for RES and $3N_g \log_2 N_g$ for sparse methods. If we are doing many solutions for different right-hand sides and wish to do an in-core calculation, as in a time-dependent physical simulation, this fact alone will limit the use of sparse techniques to relatively small meshes, e.g. (32 × 32), whereas RES is routinely applied to meshes of (256 × 256) with $N_g \cong 65{,}000$.

Since at least five iterations would be required for an RES algorithm to solve the variable coefficient problem by the Concus and Golub procedure, the sparse matrix method with the George nested-dissection ordering is preferred for the variable-coefficient problem, provided the left-hand side remains the same and sufficient storage is available. If the left-hand side alters then the factorize phase must also be used each time. Since this phase is between ten and twenty times slower than an RES iteration, the Concus and Golub iteration or the incomplete decomposition methods are to be preferred in this case.

For large meshes, $N_g \gg 1{,}000$, sparse matrix methods are likely to be impractical, because of the enormous storage required for $\boldsymbol{L}$ and $\boldsymbol{U}$. However, a solution by RES using the Concus and Golub iteration is quite feasible from both the storage ($2N_g$) and time required (~ 5 RES solutions).

In three dimensions the estimated operations and storage for the nested-dissection method are:

operations in factorize: $4N_g^2$
nonzero elements in $\boldsymbol{L}$: $28N_g^{4/3}$
operations in operate: $112N_g^{4/3}$

These figures are over 100 times larger than a RES or ICCG iteration and ten times larger than a mesh relaxation solution. It would appear, therefore, that sparse matrix methods are not competitive in three dimensions even with careful ordering.

6-4-4 Incomplete Decomposition

We can see from Eq. (6-71) that if we demand the exact (or complete) decomposition of a sparse matrix $\boldsymbol{A}$ into an $\boldsymbol{LU}$ product, the number of nonzero elements increases by a factor of about $\frac{3}{5}\log_2 N_g$. An alternative approach is to construct an approximate (or incomplete) decomposition $\boldsymbol{LU}$, which is as sparse as the original matrix $\boldsymbol{A}$ and then find the exact decomposition by iteration.

The approximate decomposition of $\boldsymbol{A}$ will be an exact decomposition of a matrix $\boldsymbol{P}$ which is close to $\boldsymbol{A}$

$$\boldsymbol{P} = \boldsymbol{A}+\boldsymbol{N} = \boldsymbol{LU} \tag{6-72}$$

where $\boldsymbol{N}$ is small if the method is to succeed. The iterative solution of the equation

$$\boldsymbol{A\phi} = \boldsymbol{\rho} \tag{6-73}$$

is defined by

$$(\boldsymbol{A}+\boldsymbol{N})\boldsymbol{\phi}^{(t+1)} = \boldsymbol{N\phi}^{(t)}+\boldsymbol{\rho}$$

or

$$\boldsymbol{LU\phi}^{(t+1)} = \boldsymbol{N\phi}^{(t)}+\boldsymbol{\rho} \tag{6-74a}$$

Since $\boldsymbol{LU}$ is as sparse as $\boldsymbol{A}$ the solution for the next iterate $\boldsymbol{\phi}^{(t+1)}$ is simple and rapid.

It is advantageous to express Eq. (6-74a) in terms of the residual vector

$$\mathbf{r} = \boldsymbol{A\phi}-\boldsymbol{\rho} \tag{6-74b}$$

and to solve for the change in the solution $\Delta\boldsymbol{\phi}$. The iteration then becomes

$$\boldsymbol{L}\mathbf{y} = -\mathbf{r} \tag{6-74c}$$

$$\boldsymbol{U}\Delta\boldsymbol{\phi} = \mathbf{y} \tag{6-74d}$$

and

$$\boldsymbol{\phi}^{(t+1)} = \boldsymbol{\phi}^{(t)}+\Delta\boldsymbol{\phi} \tag{6-74e}$$

Since $(\boldsymbol{LU})^{-1}$ is the approximate inverse for the solution over the whole region, we might anticipate much faster convergence than is obtained with SOR in which only a local point or linewise inverse is used.

We consider two methods based on this idea: the strongly implicit procedure due to Stone (1968) which uses the iteration just described, and the incomplete Choleski decomposition due to Meijerink and Vorst (1977), which uses the conjugate-gradient iteration. Another related algorithm used by Petravic and Kuo-Petravic (1979) combines incomplete $\boldsymbol{LU}$ decomposition with the conjugate gradient iteration.

6-4-5 Stone's Strongly Implicit Procedure (SIP)

Figure 6-6 shows the form of the matrices $\boldsymbol{A}$, $\boldsymbol{P}$, $\boldsymbol{L}$, and $\boldsymbol{U}$ in the SIP method for the case of a five-point difference scheme. $\boldsymbol{A}$ is five-diagonal with arbitrary elements arising from the general nonseparable PDE with variable coefficients, mesh intervals, and arbitrary-shaped boundaries. On an $(n \times n)$ mesh there are elements on the main diagonal and the lines immediately above and below the main diagonal. In addition there are two diagonals a distance n from the main diagonal. The lower and upper triangular matrices $\boldsymbol{L}$ and $\boldsymbol{U}$ are defined to have elements only in the positions where there are also elements in $\boldsymbol{A}$, so that both $\boldsymbol{L}$ and $\boldsymbol{U}$ have

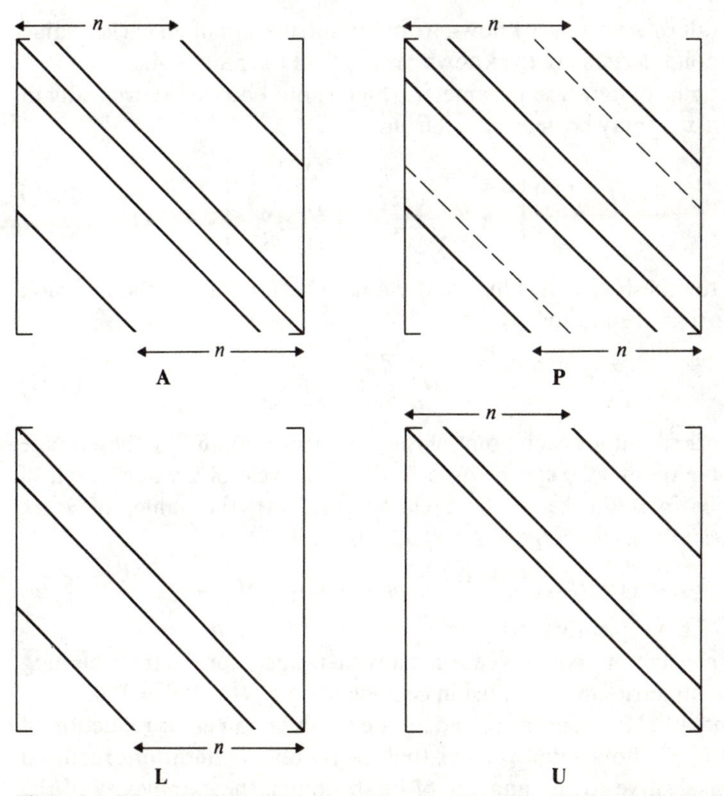

Figure 6-6 Structure of the matrix A, and its approximate L, U decomposition. P is the product of LU, showing dotted the unwanted elements that do not correspond to elements in A. A derives from the five-point difference of Poisson's equation on an $(n \times n)$ mesh. All elements not marked by lines are zero.

only $3N_g$ elements. The exact decomposition of $\boldsymbol{A}$ would, on the other hand have $\sim N_g^{3/2}$ elements because the matrix would be full, from the outer diagonal shown in Fig. 6-6 to the main diagonal. The product of an $\boldsymbol{L}$ and a $\boldsymbol{U}$ of this structure is the matrix $\boldsymbol{P}$ which in addition to elements corresponding to those of $\boldsymbol{A}$, has two extra diagonals which are shown dotted. These lie just inside the outer diagonals. These unwanted terms may be interpreted as arising from a nonexistent $\partial^2/(\partial x \partial y)$ terms in the original equations.

If the coefficients in the five diagonals of the matrix $\boldsymbol{A}$ (see Fig. 6-6) corresponding to the mesh point (p, q) are $(B_{p,q}, D_{p,q}, E_{p,q}, F_{p,q}, H_{p,q})$, the elements $(b_{p,q}, c_{p,q}, d_{p,q})$ of $\boldsymbol{L}$ and $(1, e_{p,q}, f_{p,q})$ of $\boldsymbol{U}$ are calculated explicitly as follows:

$$\begin{aligned}
b_{p,q} &= B_{p,q}/(1+\alpha e^*_{p,q-1}) \\
c_{p,q} &= D_{p,q}/(1+\alpha f^*_{p-1,q}) \\
d_{p,q} &= E_{p,q}+b_{p,q}(\alpha e^*_{p,q-1}-f^*_{p,q-1})+c_{p,q}(\alpha f^*_{p-1,q}-e^*_{p-1,q}) \\
e_{p,q} &= (F_{p,q}-\alpha b_{p,q}e^*_{p,q-1})/d_{p,q} \\
f_{p,q} &= (H_{p,q}-\alpha c_{p,q}f^*_{p-1,q})/d_{p,q}
\end{aligned} \tag{6-75}$$

where the starred values are already known from the calculation of an earlier value of p or q, and the capital letters are the known elements of the matrix $\boldsymbol{A}$.

The parameter α is an iteration parameter which should be varied cyclically in the range 0 to 1. An $\alpha_{\max}$ may be determined from

$$(1-\alpha_{\max}) = \min\left[\frac{2\Delta x^2}{1+\dfrac{DY\Delta x^2}{DX\Delta y^2}}, \frac{2\Delta y^2}{1+\dfrac{DX\Delta y^2}{DY\Delta x^2}}\right] \tag{6-76}$$

where Δx, Δy are the mesh spacings in x and y and DX and DY are the diffusion coefficients in the original equation

$$\frac{\partial}{\partial x}DX\frac{\partial}{\partial x}\phi+\frac{\partial}{\partial y}DY\frac{\partial}{\partial y}\phi = q \tag{6-77}$$

If an $\alpha_{\max}$ is worked out for each point of the mesh from Eq. (6-76), the average value over the whole mesh $\alpha^*_{\max}$ can be used to define a cycle of α values, each of which is used twice in sequence. If the cycle length is M, the values of α are calculated geometrically in the range $0 \leqslant \alpha \leqslant \alpha^*_{\max}$ from

$$1-\alpha_m = (1-\alpha^*_{\max})^{m/(M-1)} \qquad m = 0, 1, \ldots; M-1 \tag{6-78}$$

Values of $m = 4$ to 9 are typically used.

Stone reports satisfactory convergence for a wide range of problems including some with large asymmetries in the diffusion coefficients, e.g., $DX/DY = 100$.

Bewteen 20 and 30 SIP iterations are adequate to obtain an error reduction of between 10^{-5} and 10^{-6}. Stone (1968) claims that the number of iterations required for a solution is insensitive to the number of mesh points, the complexity of the boundary conditions and the variation of coefficients in the original equations. In contrast we note that the number of iterations required for a given accuracy with SOR is proportional to the number of mesh points along a side (see Sec. 6-3-3).

The coefficients of the $\boldsymbol{LU}$ decomposition may be obtained from Eq. (6-75) with $17N_g$ arithmetic operations and the forward and back substitution to obtain a new iterate from Eq. (6-74) in a further $19N_g$ operations. Since the same value of α and therefore the same $\boldsymbol{LU}$ is used for two iterations, an average total of $27N_g$ operations per iteration are required.

In addition to the storage required for the statement of the problem (A and ρ), $5N_g$ stores are required for the $\boldsymbol{LU}$ decomposition and $2N_g$ for $\boldsymbol{\phi}^{(t)}$ and $\Delta\boldsymbol{\phi}$.

A program for the Stone method comprises three DO loops as can be seen in the following outline:

```
      β = (1 − α*max)^(1/(M−1))
      DO 5  t = 1, MAXIT
c—adjust α, and iteration control
         m = MOD(t, M)
         α = 1 − β^m
      DO 1  q = 1, n
      DO 1  p = 1, n
c—perform incomplete LU decomposition
         Evaluate Eqs. (6-75)
         for b_{p,q}, c_{p,q}, d_{p,q}, e_{p,q} and f_{p,q}
1        Continue

      DO 4  I = 1, 2
c     repeat twice with same α
      DO 2  q = 1, n
      DO 2  p = 1, n
c     calculate residual
      r_{p,q} = B_{p,q}φ_{p,q−1} + H_{p,q}φ_{p,q+1} + D_{p,q}φ_{p−1,q} + F_{p,q}φ_{p+1,q} + E_{p,q}φ_{p,q} − ρ_{p,q}
c     forward elimination
      y_{p,q} = −(r_{p,q} + b_{p,q}y_{p−1,q} + c_{p,q}y_{p,q−1})/d_{p,q}
2        Continue
      DO 3  q = n, 1, −1
      DO 3  p = n, 1, −1
c—back substitution
c—calculate Δφ, the change in φ
      Δφ_{p,q} = y_{p,q} − e_{p,q}Δφ_{p+1,q} − f_{p,q}Δφ_{p,q+1}
c—update solution
      φ_{p,q} = φ_{p,q} + Δφ_{p,q}
3        Continue
      if (||r||.LT.EPS*||ρ||) solution obtained
4        Continue
5        Continue
```

A working program for the strongly implicit procedure has been published by Jesshope (1979). The program SIPSOL is available from the Belfast University Computer Physics Communications library and provides a suite of subprograms for the solution of the linear equations arising from partial differential equations.

6-4-6 Incomplete-Choleski–Conjugate-Gradient (ICCG)

If the matrix A is symmetric and positive definite an alternative incomplete decomposition has been given by Meijerink and Vorst (1977). This is based on the Choleski decomposition for a symmetric matrix followed by a conjugate-gradient iteration which is natural for the solution of a positive definite system.

The sparsity pattern defined for the matrices L and U are the same as those used in the SIP method. The calculation of the matrix elements is, however, different. Since A is symmetric, $U = L^T$ and we have only one matrix L to compute.

The Choleski decomposition of a symmetric matrix A can be written in a form that avoids a square root by introducing the diagonal matrix D, such that

$$A = LDL^T \tag{6-79}$$

The complete decomposition leads to a banded lower triangular matrix L with semibandwidth n and $\sim N_g^{3/2}$ elements. It would take $\sim N_g^2$ operations to compute. In the incomplete decomposition we force the same sparsity on L as exists in A, and only calculate $3N_g$ elements corresponding to the three diagonals in the lower part of A. In what follows L stands for this incomplete decomposition.

We denote the elements of A by a_i, b_i, and c_i, where a_i is the main diagonal, b_i the immediate upper diagonal, and c_i is the nth upper diagonal. The elements of L^T in corresponding positions to a_i, b_i, and c_i are denoted by $\tilde{a}_i$, $\tilde{b}_i$, and $\tilde{c}_i$ and the elements of D are denoted by $\tilde{d}_i$. Then the incomplete decomposition is calculated recursively from

$$\begin{aligned} &\tilde{b}_i = b_i, \qquad \tilde{c}_i = c_i \\ &\tilde{a}_i = a_i - \tilde{b}_{i-1}^2 \tilde{d}_{i-1} - \tilde{c}_{i-n}^2 \tilde{d}_{i-n} \qquad i = 1, 2, ,,,, N_g \\ &\tilde{d}_i = \tilde{a}_i^{-1} \end{aligned} \tag{6-80}$$

where elements, not defined, are replaced by zeros.

Equation (6-80) shows that, in contrast to Stone's method, the off-diagonal elements of L (and therefore L^T) are the same as the off-diagonal elements of A. The only extra storage required is that for the new diagonal elements $\tilde{d}_i$.

To get the exact solution we rewrite

$$A\phi = \rho \tag{6-81a}$$

as

$$\mathbf{B}\mathbf{y} = D^{-1}L^{-1}\rho \tag{6-81b}$$

and

$$L^T\phi = \mathbf{y} \tag{6-81c}$$

where

$$B = [D^{-1}L^{-1}A(L^T)^{-1}] \tag{6-81d}$$

The conjugate gradient algorithm is now used to solve Eq. (6-81*b*) (see Sec. 6-4-2). Since $\boldsymbol{A} \approx \boldsymbol{LDL}^T$, $\boldsymbol{B}$ is approximately the identity matrix. Hence the eigenvalues of $\boldsymbol{B}$ cluster around unity and the quadratic surface that is being minimized has almost spherical contours. In these circumstances all steepest descent directions point towards the minimum and the CGA converges very rapidly.

The number of operations that are required to compute the approximation is $6N_g$. This must be followed by $27N_g$ operations for every iteration required in the CGA. During the CGA iterations the matrix $\boldsymbol{B}$ remains unchanged so that the operation count is dominated by the CGA. The results given by Meijerink and Vorst (1977) show that the number of iterations required for a 10^{-6} error is approximately $1.3n$. This is the same dependence on mesh number as is found in optimized SOR.

The storage required during solution is $3N_g$ for the CGA vectors and N_g for the diagonal elements of the incomplete decomposition, giving a total of $4N_g$. This is $4N_g$ less than Stone's procedure and arises from the specialization that the matrix is symmetric, and the circumstance that the off-diagonal elements of $\boldsymbol{L}$ are the same as those of $\boldsymbol{A}$.

We now see that an important issue separating any of the matrix methods (sparse and incomplete decomposition) from the mesh-iterative methods such as SOR, ADI, and Concus and Golub is that of storage. The former require three to seven meshes for temporary storage during the calculation compared with only one extra mesh required for the latter methods (see Table 6-1). This means that, for a given amount of available store, fewer mesh points and hence less spatial resolution can be obtained with matrix methods than with mesh methods. For this reason matrix methods become increasingly less practical for $N_g > 1{,}000$.

6-5 RAPID ELLIPTIC SOLVERS (RES)

We use the term "rapid elliptic solver" to describe a class of direct methods that obtain the exact solution to a set of N_g difference equations in a number of operations proportional to $N_g \log_2 N_g$ (or better) and have a storage requirement of N_g. These methods are special to the extent that they cannot solve directly the general second-order PDE with variable coefficients and arbitrary shaped boundary. However, they can solve a sufficient variety of important special cases very much faster than more general methods, and therefore warrant separate treatment. Since these methods have no competitors for problems on which they can be applied, there is no excuse for using more general methods in such cases. The methods are based on cyclic reduction (CR) and the fast Fourier transform (FFT), or a combination of the two (the FACR algorithm). Comparative theoretical reviews of RES algorithms have been given by Dorr (1970), Swarztrauber (1977), and Hockney (1980), and the comparative performance of computer programs by Hockney (1978*a*), Temperton (1979*b*), and Schumann (1978). The latter reports on a competition between twenty programs when used to solve the Poisson equation on a (32×128) mesh with Dirichlet and Neumann boundary conditions. The programs were run under identical conditions on the IBM 370/168 using the same compiler. The competition was won by the optimized FACR(l) program, PSOLVE, of Temperton (1980).

The first RES algorithm was the FACR method of Hockney (1965) which predates even the fast Fourier transform (Cooley and Tukey, 1965). The method was generalized to an arbitrary number of levels of cyclic reduction and incorporated with the FFT by Hockney (1970). The existence of an optimum number of levels of reduction was noted empirically by Hockney (1970) and later shown analytically by Swarztrauber (1977). This optimized FACR algorithm has also been extensively studied in practice by Temperton (1980). The implementation of the FACR algorithm on very large 2D and 3D meshes, e.g., $(128 \times 128 \times 128)$, has also been considered by Temperton (1979*a*). The FACR algorithm is described in Sec. 6-5-3 and has been used since 1965 in the simulation of plasmas (see Chapter 9), electron devices (see Chapter 10) and the interaction of vortices (Christiansen, 1973*a*,*b*).

The second RES algorithm to appear was Buneman's (1969) stabilized cyclic reduction method, described in Sec. 6-5-1. This method was originally given for Poisson's equation with the number of mesh points equal to a power of two. It was extended to the general separable equation by Swarztrauber (1974) and to arbitrary numbers of mesh points by Sweet (1974, 1977). The theory and stability analysis has been given by Buzbee, Golub, and Nielson (1970).

Since the publication of the fast Fourier transform in 1965 many authors (e.g., Boris and Roberts, 1969; Le Bail, 1971) have used a multiple Fourier transform (Sec. 6-5-2) for the rapid solution of the Poisson and similar equations in rectangular regions, particularly if there is intrinsic interest in the amplitudes of the Fourier modes. The multiple Fourier transform also provides, via the convolution theorem, a method applicable to an arbitrary potential of interaction and for an isolated distribution of particles (Sec. 6-5-4). In the case of the coulombic interaction a more economic method for the isolated problem has been given by James (1977) and is described in Sec. 6-5-5.

Another class of RES algorithm uses procedures that are numerically unstable but that may nevertheless be used successfully on computers with very long word lengths (for example, CDC 7600 and CRAY-1). Examples are multiple marching procedures (Lorenz, 1976; Bank and Rose, 1977; Bank 1977; Ehrlich, 1979) and the method of point cyclic reduction (Detyna, 1979). These methods exchange unwanted arithmetic precision in the computer number representation for the speed of solution, and when applicable can be the fastest of all methods (Hockney, 1980). Interest in these methods has also been aroused because they can often tackle directly the difficult nonseparable problem. Nevertheless, they are not further described in this volume because of their limitation to use on certain computers.

RES algorithms were originally developed for simple geometries such as the square or rectangle. However a capacity matrix technique (Sec. 6-5-6) has been used by Hockney (1968, 1978*b*) for problems including arbitrarily shaped boundaries and further developed for larger numbers of boundary points by Widlund (1972) and Proskurowski and Widlund (1976). Martin (1974) has extended the method to more general boundary conditions and used it in the solution of aerodynamic flow problems. If the region of solution is the union of rectangles, methods are available for joining correctly solutions obtained by RES algorithms in the separate rectangles (Buzbee, Dorr, George, and Golub 1971).

6-5-1 Cyclic Reduction (CR)

Cyclic reduction, in the stable variant given by Buneman (1969), can be used to solve the general separable differential equation of the form.

$$a(x)\frac{\partial^2\phi}{\partial x^2}+b(x)\frac{\partial\phi}{\partial x}+c(x)\phi+d(y)\frac{\partial^2\phi}{\partial y^2}+e(y)\frac{\partial\phi}{\partial y}+f(y)\phi=g(x,y) \tag{6-82}$$

provided that the shape of the region and the boundary conditions are "simple." A set of FORTRAN routines for the solution of equations of this type has been published by Swarztrauber and Sweet (1975). The solution of a PDE can be expressed as the linear combination of eigenvalues of the left-hand side operator. In the case of the general separable form these eigenvectors are of the product form $U_n(x,y)=X_n(x)Y_n(y)$, as can be demonstrated by substitution, the functions $X_n(x)$, $Y_n(y)$ satisfying, for the eigenvalue n^2,

$$a(x)\frac{d^2X_n}{dx^2}+b(x)\frac{dX_n}{dx}+c(x)X_n=n^2X_n \tag{6-83a}$$

and

$$d(y)\frac{d^2Y_n}{dy^2}+e(y)\frac{dY_n}{dy}+f(y)Y_n=-n^2Y_n \tag{6-83b}$$

The general separable form includes, after minor manipulation, the Helmholtz equation in the common coordinate systems. For example:

Cartesian (x,y)

$$\frac{\partial^2\phi}{\partial x^2}+\frac{\partial^2\phi}{\partial y^2}+\lambda\phi=q \tag{6-84a}$$

Polar (r,θ)

$$\frac{1}{r}\frac{\partial}{\partial r}\left(r\frac{\partial\phi}{\partial r}\right)+\frac{1}{r^2}\frac{\partial^2\phi}{\partial\theta^2}+\lambda\phi=q \tag{6-84b}$$

Cylindrical (r,z)

$$\frac{1}{r}\frac{\partial}{\partial r}\left(r\frac{\partial\phi}{\partial r}\right)+\frac{\partial^2\phi}{\partial z^2}+\frac{\lambda}{r^2}\phi=q \tag{6-84c}$$

Axisymmetric spherical (r,θ)

$$\frac{1}{r^2}\frac{\partial}{\partial r}\left(r^2\frac{\partial\phi}{\partial r}\right)+\frac{1}{r^2\sin\theta}\frac{\partial}{\partial\theta}\left(\sin\theta\frac{\partial\phi}{\partial\theta}\right)+\frac{\lambda}{r^2\sin^2\theta}\phi=q \tag{6-84d}$$

Spherical surface (θ, ϕ)

$$\frac{1}{\sin\theta}\frac{\partial}{\partial\theta}\left(\sin\theta\frac{\partial u}{\partial\theta}\right)+\frac{1}{\sin^2\theta}\frac{\partial^2 u}{\partial\phi^2}+\lambda u = q \qquad (6\text{-}84e)$$

The general separable form includes a variable diffusion coefficient problem, of the form

$$\frac{\partial}{\partial x}D(x)\frac{\partial\phi}{\partial x}+\frac{\partial}{\partial y}E(y)\frac{\partial\phi}{\partial y}+\lambda\phi = g(x,y) \qquad (6\text{-}85a)$$

but not, unfortunately, the common and more general form,

$$\frac{\partial}{\partial x}D(x,y)\frac{\partial\phi}{\partial x}+\frac{\partial}{\partial y}E(x,y)\frac{\partial\phi}{\partial y}+\lambda\phi = q \qquad (6\text{-}85b)$$

We now explain what we mean by a 'simple' region and 'simple' boundary conditions. The region of the solution is the interior of the space defined by two pairs of values for the independent variables. For example a rectangle in cartesian coordinates defined by (x_1, y_1), (x_2, y_2), part of an anular ring defined by (r_1, θ_1), (r_2, θ_2) in polar coordinates. The boundary conditions allowed on these constant coordinate surfaces are any combination of

1. Given value of ϕ, (6-86a)
2. The mixed condition, for example

$$\alpha\phi(x,y)+\frac{\partial}{\partial x}\phi(x,y) = p(y) \qquad (6\text{-}86b)$$

along $x = a$, where $p(y)$ is given.
If $\alpha = 0$ we have the given gradient condition.

3. Periodicity. (6-86c)

If the general separable PDE is differenced, one obtains a set of algebraic equations of the form

$$l_j\boldsymbol{\phi}_{j-1}+\boldsymbol{\Lambda}_j\boldsymbol{\phi}_j+m_j\boldsymbol{\phi}_{j+1} = \mathbf{h}_j \qquad (6\text{-}87a)$$

where $\boldsymbol{\phi}_j$ and $\mathbf{h}_j$ are the the solution vector and right-hand side vector on the jth row of the mesh. The matrix $\boldsymbol{\Lambda}_j$ is tridiagonal with variable coefficients and of the form

$$\boldsymbol{\Lambda}_j = \boldsymbol{B}+p_j\boldsymbol{I} \qquad (6\text{-}87b)$$

Separability ensures that the tridiagonal matrix $\boldsymbol{B}$—which represents differencing in x along a row—is the same for all rows, and that the coefficients (l_j, p_j, m_j)—which represent the differencing in y between rows—are the same for all variables within a row. Consequently, the difference equations may be written in the form of the row equations (6-87a).

Swarztrauber (1974) has given the cyclic-reduction method for the general separable equation (6-87a). The details are too lengthy to give here, so we will illustrate the principle of the method by considering the simpler case when the coefficients of the equations are independent of y, then we may drop the subscript j on l, m, p, and therefore Λ. We note that with $l = m = 1$ and $p = -2$, we represent any PDE whose differential operator in one coordinate direction is the second derivative alone. This includes the Poisson and Helmholtz equations in cartesian, polar, cylindrical, and spherical surface coordinates [Eqs. (6-84*a*,*b*,*c*, and *e*)].

Taking three successive equations like Eq. (6-87*a*) we have

$$l\boldsymbol{\phi}_{j-2}+\Lambda\boldsymbol{\phi}_{j-1}+m\boldsymbol{\phi}_{j} = \mathbf{h}_{j-1} \tag{6-88a}$$

$$l\boldsymbol{\phi}_{j-1}+\Lambda\boldsymbol{\phi}_{j}+m\boldsymbol{\phi}_{j+1} = \mathbf{h}_{j} \tag{6-88b}$$

$$l\boldsymbol{\phi}_{j}+\Lambda\boldsymbol{\phi}_{j+1}+m\boldsymbol{\phi}_{j+2} = \mathbf{h}_{j+1} \tag{6-88c}$$

Multiplying the three equations respectively by l, $-\Lambda$, and m, and then adding, we obtain an equation referring only to $\boldsymbol{\phi}$ values on alternate rows of the mesh

$$l^{(1)}\boldsymbol{\phi}_{j-2}+\Lambda^{(1)}\boldsymbol{\phi}_{j}+m^{(1)}\boldsymbol{\phi}_{j+2} = \mathbf{h}_{j}^{(1)} \tag{6-89a}$$

where

$$\Lambda^{(1)} = 2mlI-\Lambda^{2} \tag{6-89b}$$

$$\mathbf{h}_{j}^{(1)} = l\mathbf{h}_{j-1}-\Lambda\mathbf{h}_{j}+m\mathbf{h}_{j+1} \tag{6-89c}$$

and

$$l^{(1)} = l^{2} \qquad m^{(1)} = m^{2} \tag{6-89d}$$

We have now completed one level of cyclic reduction, and thereby reduced the number of equations by a factor two. Further, the three equations (6-89) are of the same form as Eq. (6-87) and the whole process of reduction can be repeated until only a single equation is left for the central line of variables.

$$\Lambda^{(t)}\boldsymbol{\phi}_{n/2} = \mathbf{h}_{n/2}^{(t)}-l^{(t)}\boldsymbol{\phi}_{0}-m^{(t)}\boldsymbol{\phi}_{n} \tag{6-90}$$

In the case that the number of points in y is a power of two, $\boldsymbol{\phi}_0$ and $\boldsymbol{\phi}_n$ will be known boundary values, so that Eq. (6-90) can be solved for the central-line variables $\boldsymbol{\phi}_{n/2}$. The relationship between the known and calculated variables at each level of reduction is shown in Fig. 6-7.

The matrix $\Lambda^{(t)}$ can be expressed as a product of factors depending linearly only on the initial tridiagonal matrix $\Lambda^{(0)} = \Lambda$

$$\Lambda^{(t)} = -\prod_{k=1}^{2^t} (\Lambda-\beta_k I) \tag{6-91a}$$

For the case $l = m = 1$ the roots β_k are known analytically to be

$$\beta_k = 2\cos\left[\frac{2(k-1)\pi}{2^{t+1}}\right] \tag{6-91b}$$

In the general separable case the roots are not known analytically, but may be precalculated numerically. In either case Eqs. (6-90) can be solved by successive solutions of 2^t tridiagonal systems using the Thomas algorithm (see Sec. 6-4-1). The number of levels of reduction required will be $t = \log_2 n$ if the boundary conditions are periodic or $t = \log_2 n-1$ if the boundary values are given.

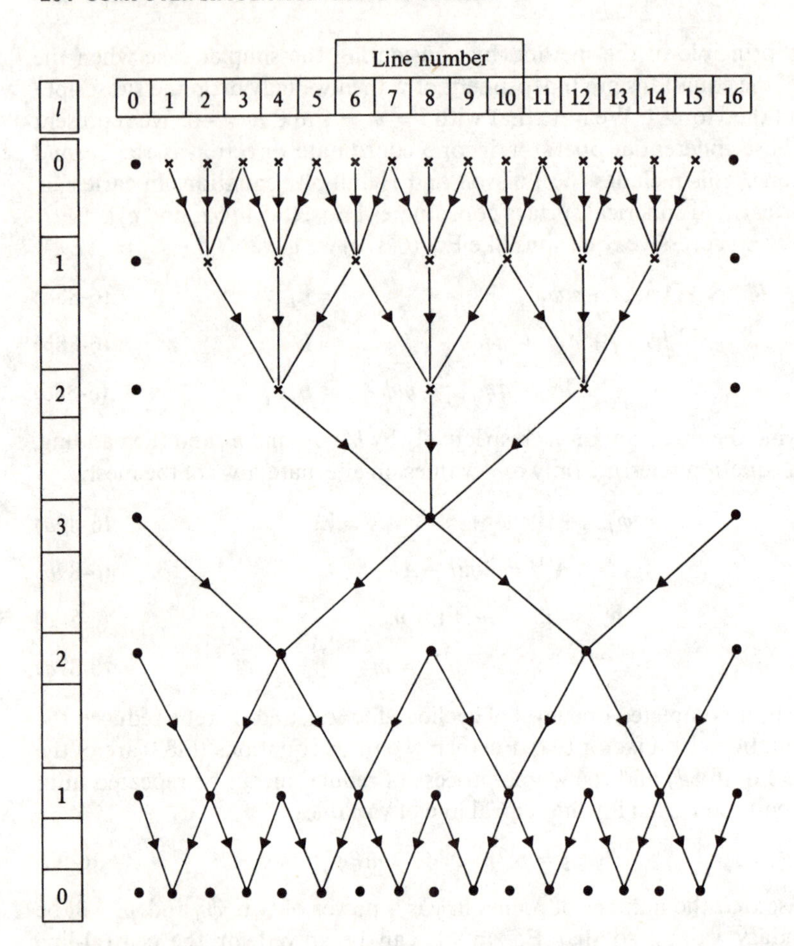

Figure 6-7 A diagram showing the lines of values computed at each level of cyclic reduction. Arrows indicate the data that contribute to a new value. Known boundary values and solution values are indicated by solid circles.

To obtain the unknown values on $\boldsymbol{\phi}_{n/4}$ and $\boldsymbol{\phi}_{3n/4}$ we use the equations we obtained at level $(t-1)$ where the value $\boldsymbol{\phi}_{n/2}$, just computed, can be put on the right-hand side.

$$\boldsymbol{\Lambda}^{(t-1)}\boldsymbol{\phi}_{n/4} = \mathbf{h}_{n/4}^{(t-1)} - l\boldsymbol{\phi}_0^{(t-1)} - m\boldsymbol{\phi}_{n/2}^{(t-1)} \tag{6-92}$$

and similarly for $\boldsymbol{\phi}_{3n/4}$. In this way the intermediate values of the solution can be successively filled in.

The calculation of the right-hand side by means of Eq. (6-89c) is numerically unstable and instead the scheme below, devised by Buneman (1969), must be used. The right-hand side at level r is expressed as

$$\mathbf{h}^{(r)} = \boldsymbol{\Lambda}^{(r)}\mathbf{p}_j^{(r)} + \mathbf{q}_j^{(r)} \tag{6-93}$$

where, for the case $l = m = 1$, the vectors $\mathbf{p}_j$ and $\mathbf{q}_j$ obey the recurrence relations

$$\mathbf{p}_j^{(r)} = \mathbf{p}_j^{(r-1)} + (A^{(r-1)})^{-1}(\mathbf{q}_j^{(r-1)} - \mathbf{p}_{j-2^{r-1}}^{(r-1)} - \mathbf{p}_{j+2^{r-1}}^{(r-1)})$$

and
$$\mathbf{q}_j^{(r)} = \mathbf{q}_{j-2^{r-1}}^{(r-1)} + \mathbf{q}_{j+2^{r-1}}^{(r-1)} - 2\mathbf{p}_j^{(r)} \tag{6-94}$$

with
$$\mathbf{p}_j^{(0)} = 0 \quad \text{and} \quad \mathbf{q}_j^{(0)} = \mathbf{h}_j$$

The asymptotic operation count for cyclic reduction with $l = m = 1$ is $\sim 6n^2 \log_2 n$ for an $(n \times n)$ mesh or $3N_g \log_2 N_g$. The storage required is N_g. This is increased to $16N_g \log_2 N_g$ for the general separable case which has variable coefficients. A time of 1.3 s on the CDC 7600 is quoted by Swarztrauber and Sweet (1975) for the solution of a variable-coefficient problem on a (127×127) mesh. This is to be compared with a time of 0.23 s for the constant coefficient Poisson problem using the FACR algorithm (see Sec. 6-5-3 and Hockney, 1978*a*). Both programs are in FORTRAN.

6-5-2 Multiple Fourier Transform (MFT)

The general second-order PDE with constant coefficients and periodic boundary conditions

$$a\frac{\partial^2 \phi}{\partial x^2} + b\frac{\partial \phi}{\partial x} + c\phi + d\frac{\partial^2 \phi}{\partial y^2} + e\frac{\partial \phi}{\partial y} + f\phi = g(x, y) \tag{6-95}$$

can be solved by using the periodic fast Fourier transform. If the mesh values of the unknown $\phi_{p,q}$ and right-hand side $g_{p,q}$ are expanded in a finite double Fourier series

$$\phi_{p,q} = \sum_{k,l} \hat{\phi}_{k,l} e^{2\pi i(pk+ql)/n}$$

and
$$g_{p,q} = \sum_{k,l} \hat{g}_{k,l} e^{2\pi i(pk+ql)/n} \tag{6-96}$$

and substituted into the usual central finite-difference form of Eq. (6-95), on a uniform mesh one obtains equations for the n^2 the harmonic amplitudes $\hat{\phi}_{k,l}$

$$\hat{\phi}_{k,l} = \hat{G}_{k,l}\hat{g}_{k,l} \qquad 0 < k, l < n-1 \tag{6-97a}$$

where for unit mesh separation

$$\hat{G}_{k,l} = \left[2a\left(\cos\frac{2\pi k}{n} - 1\right) + ib\sin\frac{2\pi k}{n} + c + 2d\left(\cos\frac{2\pi l}{n} - 1\right) + ie\sin\frac{2\pi l}{n} + d\right]^{-1} \tag{6-97b}$$

The procedure for solution is therefore:

1. Form the complex Fourier analysis of the right-hand side $\hat{g}_{k,l}$ using the FFT algorithm. $5N_g \log_2 N_g$ operations are required.
2. Multiply each harmonic by the known complex coefficient $\hat{G}_{k,l}$. This gives the Fourier transform of the solution $\hat{\phi}_{k,l}$. The $2N_g$ operations required is negligible compared with (1) and (3).

3. Fourier-synthesize the harmonics $\hat{\phi}_{k,l}$ to obtain the solution $\phi_{p,q}$, using the FFT. $5N_g \log_2 N_g$ operations are required.

The total operation count is $10N_g \log_2 N_g$ and the solution may overwrite the right-hand side. If the first derivative term is absent the transforms are all real and the operation count is halved. The method can be used for a constant-coefficient problem in three dimensions by performing the triple Fourier transform.

The multiple transform method is the simplest to program of all RES methods, if one has available a proven library program for the fast Fourier transform. However, it is an expensive method—compared with other RES methods—for solving the finite-difference form of PDEs, and is not therefore recommended for this purpose. The method, however, has attractions if the multiple Fourier transform is required in any case (as, for example, in the analysis of the power spectrum) or if force-shaping is required (for example in the P^3M algorithm or arbitrary force laws, see Sec. 6-5-4), or if an isolated system is being solved (for example in a galaxy of stars, see Sec. 6-5-5).

In the MFT method the Fourier modes in the expansion must be the eigenfunctions of the difference form of the left-hand side operator in Eq. (6-95), and must satisfy the boundary conditions of the problem. This ensures that the harmonic equations (6-97*a*) are independent of one another and may be solved separately by multiplying each harmonic with the numerical factor $\hat{G}_{k,l}$. The complex functions $\exp(2\pi ipk/n)$ in Eq. (6-96) clearly satisfy these conditions for the general constant coefficient problem, Eq. (6-95), with periodic boundary conditions. For zero value boundary conditions which are satisfied by a Fourier sine expansion, the sine functions are eigenfunctions of the operator only when the first derivative terms are absent ($b = e = o$). An analogous statement applies to the use of a Fourier cosine expansion for gradient boundary conditions.

Fortunately the first derivative terms can be removed from Eq. (6-95) by the substitution $\phi(x, y) = e^{-\alpha x} e^{-\beta y} \psi(x, y)$ with $\alpha = b/2a$ and $\beta = e/2d$. After multiplying through by $e^{(\alpha x + \beta y)}$ one obtains

$$a \frac{\partial^2 \psi}{\partial x^2} + (c - b^2/4a)\psi + d \frac{\partial^2 \psi}{\partial y^2} + (f - e^2/4d)\psi = e^{(\alpha x + \beta y)} g(x, y) \qquad (6\text{-}97c)$$

The solution $\phi(x, y)$ is obtained as follows:

1. Form the modified right-hand side $e^{(\alpha x + \beta y)} g(x, y)$.
2. Solve Eq. (6-97*c*) for ψ, using the technique below.
3. Form $\phi(x, y) = e^{-(\alpha x + \beta y)} \psi(x, y)$.

In the subsequent discussion we assume that the first derivative terms have been removed in this way. Alternatively, if the Eq. (6-95) is first differenced, the substitution $\psi_{i,j} = \alpha^i \beta^j \phi_{i,j}$ can be used to eliminate the effect of the first derivative (Le Bail, 1971, page 463), where now $\alpha = [(1 + bH/2a)/(1 - bH/2a)]^{1/2}$ and $\beta = [(1 + eH/2d)/(1 - eH/2d)]^{1/2}$. In this case the mesh separation H must be chosen such that $H < 2a/b$ and $< 2d/e$.

Separability allows the boundary conditions in x to be considered separately from those in y, and we discuss now the boundary conditions in x. An exactly analogous discussion applies to the y direction. Separability requires that the boundary condition in x must be independent of the y coordinate. Three obvious cases present themselves:

1. *Periodicity.* $\phi(x+L) = \phi(x)$, in which case the expansion (6-96) is appropriate with $L = nH$.
2. *Zero values* (i.e. Dirichlet conditions). $\phi(L) = \phi(0)$ in which case the finite Fourier sine transform is appropriate:

$$\phi_p = \sum_{k=1}^{n-1} \hat{\phi}_k \sin(\pi kp/n) \tag{6-98a}$$

and ignoring a constant factor which can be absorbed elsewhere in the calculation

$$\hat{\phi}_k = \sum_{p=1}^{n-1} \phi_p \sin(\pi kp/n) \tag{6-98b}$$

3. *Zero gradient* (i.e. Neumann conditions). $\left.\dfrac{\partial \phi}{\partial x}\right|_{x=0} = \left.\dfrac{\partial \phi}{\partial x}\right|_{x=L} = 0$

in which case the finite Fourier cosine transform is appropriate:

$$\phi_p = \sum_{k=0}^{n} E(k)\hat{\phi}_k \cos(\pi kp/n) \tag{6-99a}$$

and again ignoring a constant factor

$$\hat{\phi}_k = \sum_{p=0}^{n} E(p)\phi_p \cos(\pi kp/n) \tag{6-99b}$$

where

$$E(i) = \begin{cases} \frac{1}{2} & \text{if } i = 0 \text{ or } n \\ 1 & \text{otherwise} \end{cases} \tag{6-99c}$$

Cases (1) and (2) can be generalized to include given values (other than zero) by the method of equivalent charges:

4. *Given values.* The equation (6-95), when expressed in finite differences for the mesh point (p, q), is of the form

$$a_{p,q}\phi_{p,q-1} + b_{p,q}\phi_{p,q+1} + c_{p,q}\phi_{p-1,q} + d_{p,q}\phi_{p+1,q} + e_{p,q}\phi_{p,q} = f_{p,q} \tag{6-100}$$

where the mesh point coefficients $a_{p,q}, \ldots, f_{p,q}$ are functions of the constants $a, \ldots, f$ and $g(x, y)$ of the original equations. Since the first derivative in x is absent, $c_{p,q} = a_{p,q}$. Let us suppose that the left-hand x boundary is at $p = 0$, then the equation for a mesh point just to the right of the left-hand boundary $(p = 1)$ is

$$a_{1,q}\phi_{1,q-1} + b_{1,q}\phi_{1,q+1} + d_{1,q}\phi_{2,q} + e_{1,q}\phi_{1,q} = f_{1,q} - c_{1,q}\phi_{0,q} \tag{6-101}$$

where we have moved the known boundary value $\phi_{0,q}$ to the right-hand side of the equation. We note that Eq. (6-101) may be regarded as the equation for the problem with zero values given on the boundary, provided a layer of equivalent sources equal to $-c_{1,q}\phi_{0,q}$ is assigned to the column of mesh points just inside the boundary. The sine expansion of case 3 can then be used to solve the problem. Similarly, the given values on the right-hand boundary can be replaced by a layer of equivalent charge, $-d_{n-1,q}\phi_{n,q}$, on the column of mesh points just to the left of the right-hand boundary ($p = n-1$).

5. *Given gradient*. Let us consider as before the case of the left-hand boundary in x. A given gradient at point $(0, q)$ will be expressed in finite-difference form as

$$\frac{\phi_{1,q}-\phi_{-1,q}}{2H} \simeq \frac{\partial \phi}{\partial x} = h_q \tag{6-102a}$$

This equation must be satisfied in conjunction with the boundary equation at $(0, q)$

$$a_{0,q}\phi_{0,q-1}+b_{0,q}\phi_{0,q+1}+c_{0,q}\phi_{-1,q}+d_{0,q}\phi_{1,q}+e_{0,q}\phi_{0,q} = f_{0,q} \tag{6-102b}$$

where again $c_{0,q} = a_{0,q}$ because of the absence of the first derivative in x.

If we introduce the new variable

$$\phi^*_{-1,q} = \phi_{-1,q}+2Hh_q \tag{6-103}$$

the Eqs. (6-102) can be rewritten as

$$\phi_{1,q}-\phi^*_{-1,q} = 0 \tag{6-104a}$$

$$q_{0,q}\phi_{0,q-1}+b_{0,q}\phi_{0,q+1}+c_{0,q}\phi^*_{-1,q}+d_{0,q}\phi_{1,q}+e_{0,q}\phi_{0,q} = f_{0,q}+2Hh_q c_{0,q} \tag{6-104b}$$

The Eqs. (6-104) are the equations for zero gradient, providing a layer of equivalent charge equal to $2Hh_q c_{0,q}$ is assigned on the boundary points $(0, q)$. The equations may be solved by the finite Fourier cosine transformation, case (3) above.

6-5-3 FACR Method

The minimum operation count for the solution of the constant coefficient problem is obtained by a judicious combination of Fourier analysis and cyclic reduction in the FACR algorithm (Hockney, 1965, 1970). At the rth level of cyclic reduction one has the equations

$$l^{(r)}\phi_{j-2^r}+\Lambda^{(r)}\phi_j+m^{(r)}\phi_{j+2^r} = \mathbf{h}^{(r)} \tag{6-105}$$

In the FACR algorithm one notices that the reduction can be stopped at any level "r," and the reduced equations solved by taking the Fourier transform of Eq. (6-105). The level r at which reduction is stopped is now a parameter to be varied to minimize the number of arithmetic operations. The Fourier transform is taken in one dimension (the x direction) following the methods and conditions given in Sec. 6-5-2.

After Fourier transformation, Eqs. (6-105) become

$$l^{(r)}\hat{\phi}^k_{j-2^r}+\lambda^{(r)}_k\hat{\phi}^k_j+m^{(r)}\hat{\phi}^k_{j+2^r}=\hat{h}^k_j \tag{6-106}$$

where $\lambda^{(r)}_k$ is the eigenvalue of $\Lambda^{(r)}$ corresponding to the kth Fourier mode. This is one independent tridiagonal system in the y direction for each of the harmonic amplitudes. The tridiagonal systems are solved by the Thomas algorithm and Fourier synthesis gives the solution on the $n/2^r$ lines at the rth level. The fill-in of intermediate lines takes place as in the CR method.

In the simple case of Poisson's equation in the square, the number of operations per mesh point required by FACR, if reduction is stopped at the rth level, is (Hockney, 1970)

$$S(r)=3+4.5r+(5\log_2 n-4)2^{-r} \tag{6-107}$$

where the first two terms arise from the cyclic reduction and the last term from the Fourier analysis and synthesis on every 2^r line.

We usually call this an FACR(l) algorithm, where the maximum depth of reduction is given in parenthesis. However, here we use the symbol r instead of l, to avoid confusion with the coefficient in Eq. (6-105). The curve $S(r)$ is plotted in Fig. 6-8 for the case $n=128$ and shows a minimum at $r=2$. More generally, following Swarztrauber (1977), we can differentiate to find the analytic minimum

$$\frac{dS}{dr}=4.5-\ln 2\,(5\log_2 n-4)2^{-r}=0 \tag{6-108a}$$

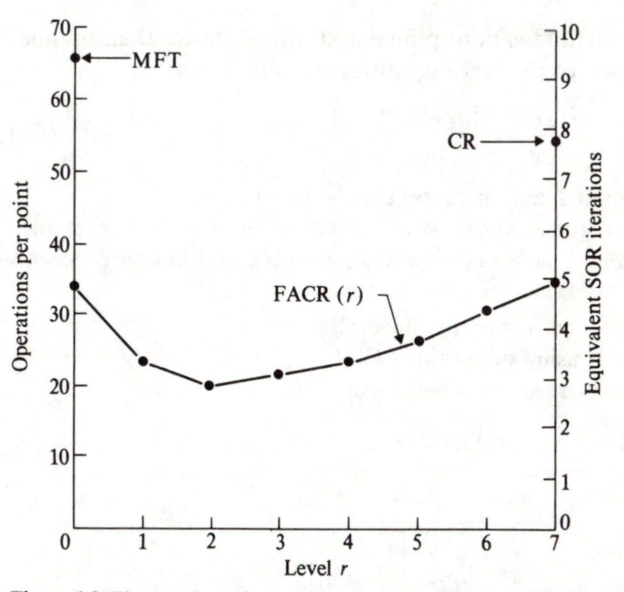

Figure 6-8 The number of operations per point for the FACR(r), MFT and CR algorithms on a (128 × 128) mesh. The equivalent number of SOR iterations is also given (assuming the most favorable value of seven operations per point per iteration). In all cases, the most favorable boundary conditions for each algorithm are chosen. Note the minimum number of operations at $r=2$. (*From Hockney, 1970, courtesy of Methods in Computational Physics*, © *Academic Press Inc.*)

hence at the minimum

$$2^r = \frac{\ln 2}{4.5}(5\log_2 n - 4) \tag{6-108b}$$

and
$$S_{\text{min}} = 3 + \frac{4.5}{\ln 2} + 4.5\log_2\left[\frac{\ln 2(5\log_2 n - 4)}{4.5}\right] \tag{6-108c}$$

Thus, in the limit of large n, the optimal FACR algorithm, which we designate as FACR(*), has the operation count of:

$$S_{\text{min}}(n \to \infty) = 4.5\log_2(\log_2 n) \tag{6-109a}$$

and
$$r(n \to \infty) = \log_2(\log_2 n) \tag{6-109b}$$

and a limiting total number of operations of

$$T_{\text{FACR}} = 4.5N_g\log_2(\log_2 N_g) \tag{6-109c}$$

For practical values of n the asymptotic formulae (6-109) are reasonably accurate. For $n = 128$ one obtains $r \sim 3$ and $S_{\text{min}} \sim 14$ compared with the values of $r \sim 2$ and $S_{\text{min}} \sim 20$ from the complete formulae (6-108). An extensive investigation of the minimum has been made by Temperton (1980) using a program with variable r.

The CR method (Sec. 6-5-1) uses $r = \log_2 n$ ($=7$ in the above example), and takes the reduction to completion. The number of operations in this case is

$$T_{\text{CR}} = 4.5N_g\log_2 N_g \tag{6-109d}$$

We thus find that the CR method is not optimal and will be slower than optimal FACR. The asymptotic times for the two algorithms are in the ratio

$$\frac{T_{\text{FACR}}}{T_{\text{CR}}} = \frac{\log_2(\log_2 N_g)}{\log_2 N_g} \tag{6-109e}$$

For $n \sim 128$ this ratio is about 2, and increases slowly for larger n.

The operation counts given above are for reduction by the "unstable" algorithm (6-89). On a machine such as the CDC 7600 with an arithmetic precision

Table 6-5 Computer time and error to solve 2D Poisson equation in the square using subroutine POT1

An FACR(1) algorithm carefully programmed in assembler code

Mesh	CDC 7600		IBM 360/195	
	ms	Error†	ms	Error†
32 × 32	9	1×10^{-12}	12	8×10^{-5}
64 × 64	36	5×10^{-12}	39	2×10^{-4}
128 × 128	145	2×10^{-11}	137	4×10^{-4}
256 × 256	—	—	559	1×10^{-3}

† Error in solution of the difference equation; *not* the truncation error.

Table 6-6 Computer time required to solve the 2D Poisson equation in the square by different methods on a (128 × 128) mesh

Time in milliseconds on CDC 7600. FORTRAN CODE.

Mesh	PWSCRT† CR	POT1‡ FACR(1)	POT3‡ MFT
32 × 32	49	26	34
64 × 64	211	71	142
128 × 128	926	229	604
256 × 256	—	[630]	—

† Swarztrauber and Sweet, 1975.
‡ FORTRAN FTN 4.6.
[ ] 0.6 times 360/195 FORTRAN, estimate for 7600.

of $\sim 10^{-14}$, it is quite permissible to lose several decimal digits of accuracy by the use of the "unstable" algorithm. The subroutine POT1 (Hockney, 1970) is an FACR(1) algorithm and therefore is not quite optimal. Its operation count is $1.25(\log_2 N_g + 4.8)$. Table 6-5 shows that the precision on this machine is $\sim 10^{-12}$ on a (128 × 128) mesh. On IBM machines in 32-bit precision ($\sim 10^{-6}$ rounding error) the accuracy is reduced to $\sim 10^{-4}$. Hence about two decimal digits are lost in one level of "unstable" reduction. In order to calculate with larger meshes on the IBM one can either go to 64-bit double-precision or perform the cyclic reduction steps by the "stable" Buneman algorithm (6-94). The disadvantage of using the Buneman algorithm is that it introduces further arithmetic operations and the number 4.5 in formulae (6-108) and (6-109) should be replaced by 6. However, the asymptotic ratio of optimal FACR to CR in Eq. (6-109*e*) remains unchanged. Table 6-6 compares the times for CR, FACR(1), and MFT on different meshes.

In the FACR algorithm Fourier analysis is applied only in the x direction. Consequently the algorithm may be applied to more general problems of the following form:

$$\nabla D(y) \nabla \phi(x, y) + k^2(y) \phi(x, y) = s(x, y) \tag{6-110}$$

in which the coefficients of the equation and the mesh spacing may vary in y but not in x. An example is the solution of Poisson's equation in cylindrical coordinates for which Hughes (1971) has published an FACR(0) program. FACR(1) programs for the same problem in cartesian coordinates have been published by Hockney (1970) and Christiansen and Hockney (1971).

6-5-4 Convolution Methods

It is often necessary to calculate the potential due to a distribution of sources when the potential of interaction cannot be represented by a partial-differential equation. An example of this is the Lennard-Jones interaction between atoms which has the variation $G(r) \propto \{r^{-12} - r^{-6}\}$, where $G(r)$ is the potential of two atoms separated by a distance r. If the interaction is pairwise additive, the potential at a mesh point (p, q) can always be written as the sum of contributions from all other source points (p', q')

$$\phi_{p,q} = \sum G_{p-p', q-q'} \rho_{p', q'} \tag{6-111}$$

Equation (6-111) expresses the potential as the convolution of the source distribution ρ with the Green's function of the interaction (or interaction potential) G. The convolution theorem (see Appendix A-3) states that the Fourier transform of a convolution is the product of the Fourier transforms of the quantities convolved. Hence Eq. (6-111) can be written as

$$\hat{\phi}_{k,l} = \hat{G}_{k,l} \hat{\rho}_{k,l} \tag{6-112}$$

where the circumflex denotes the Fourier transform and (k, l) the harmonic wave numbers.

The potential due to any periodic array of sources with an arbitrary potential of interaction can be found. First, the desired periodic interaction potential G is written on a mesh for a unit source at the origin. The Fourier transform $\hat{G}$ of this interaction potential is found and stored. This calculation need be done only once. During each timestep of a simulation, a source distribution ρ is obtained on a second mesh. The Fourier transform $\hat{\rho}$ of this distribution is found, and this may overwrite the original source distribution ρ. The Fourier transform of the source is multiplied by the Fourier transform of the interaction potential to give the Fourier transform of the potential distribution. The potential distribution due to the original source distribution is then obtained by a Fourier synthesis. All operations on the second mesh may overwrite each other and the total storage required is approximately $2N_g$.

A multiple real Fourier transform on N_g points requires $2\frac{1}{2}N_g \log_2 N_g$ operations (independent of the number of dimensions), hence the total number of operations is $5N_g \log_2 N_g$. The storage required for an arbitrary interaction is $2N_g$—one mesh for the source and the solution and a second mesh for the Green's function and its transform. However, if G is dependent only on the separation between particles and we are working in a cube, the symmetries can be used to reduce the storage of $\hat{G}$ to approximately $N_g/48$. In this case the total storage required is essentially N_g.

The convolution method described will solve a periodic system of sources with an arbitrary form of interaction. No conductors or boundaries are permitted in the system and, if these are present, the capacity matrix method (see Sec. 6-5-6) must be used. By an appropriate choice of Green's function the convolution method enables the potential to be found for an isolated source distribution. By isolated we mean a system, like a galaxy of stars, in which the only boundary condition is that the potential decays to zero correctly at infinity (i.e., $\phi \propto r^{-1}$).

The method, which we describe in the two dimensions, may be adapted to isolated systems if one is prepared to use only one-quarter of the available mesh points for the source distribution—say, the bottom left-hand corner, defined by $0 \leqslant p,\ q \leqslant n/2$. The source distribution over the remaining three-quarters of the system is made identically zero. Taking the interaction of point charges as an example, an interaction potential is constructed as follows:

$$\left.\begin{aligned} &G_{p,q} = (p^2+q^2)^{-1/2} \\ &G_{n-p,q} = G_{p,n-q} = G_{n-p,n-q} = G_{p,q} \end{aligned}\right\} \quad \begin{aligned} &0 \leqslant p,q \leqslant n/2 \\ &p+q \neq 0 \end{aligned} \qquad (6\text{-}113)$$
$$G_{0,0} = 1$$

When this potential is repeated periodically, one sees that the correct r^{-1} potential for a point charge at the origin is obtained within the region $-n/2 \leqslant p, q \leqslant n/2$. At the boundary of this region there is a cusp, and outside the region the potential is incorrect. However, if we use only the bottom left-hand corner for the charge distribution and use only the potential in this region, the correct potential for an isolated system is obtained. The potential outside the bottom left-hand corner is incorrect, containing as it does all the unphysical cusps of the interaction potential. But this does not matter, since this potential is never used. The use of the mesh for an isolated system is illustrated in Fig. 6-9.

If N_g is the total number of points in the physical region of calculation (the active mesh), the calculation must be performed on $2^d N_g$ points, where d is the dimensionality. This requires asymptotically $5(2^d N_g)\log_2 N_g$ operations. The period of the transforms is $n = 2N_g^{1/d}$.

The storage requirement in two-dimensions is apparently $4N_g$, however, this may be reduced to $2N_g$ as follows. We store only the bottom half of Fig. 6-9, and initially transform all the data in the x direction. This fills the lower half of Fig. 6-9 with numbers. We now take the data one line at a time in y and perform three operations. These are (1) Fourier-transform the line in y into a temporary vector of

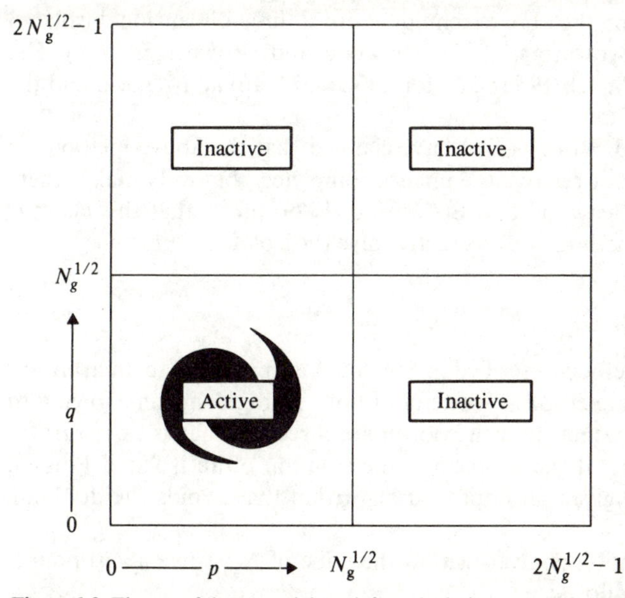

Figure 6-9 The use of the potential mesh for calculations on an isolated system. Only the active quarter of the mesh may be used for charges or masses. The remaining inactive regions are required during the calculation. (*From Hockney, 1970, courtesy of Methods in Computational Physics*, © *Academic Press Inc.*)

length $2N_g^{1/2}$, (2) multiply by $\hat{G}$, and (3) backtransform in y and store in the lower left quarter of Fig. 6-9. The data that might be stored in the upper left quarter will not influence the solution in the active region and is discarded. The total storage required is $2N_g+2N_g^{1/2} \approx 2N_g$. The number of computer operations required is $15N_g \log_2 N_g$ compared with $20N_g \log_2 N_g$ that would be required without the above reorganization.

Proceeding similarly in three-dimensions, one requires a storage for two active meshes ($2N_g$), plus a temporary plane of $2N_g^{2/3}$ points and a temporary vector of $2N_g^{1/3}$ points. One first Fourier-analyzes all the N_g initial data in the x direction. Then one takes a plane at a time and analyzes in the y direction. From this plane one takes a vector at a time, and analyzes in the z direction. Multiplication by the Green's function can now take place, followed by synthesis in z. This is repeated for each vector in z, until all the plane has been completed. Synthesis in y then takes place within this plane. When all planes in y have been completed the synthesis in x takes place, yielding the solution.

Generalized in this way, to d dimensions the number of arithmetic operations is

$$\frac{5 \times 2^d}{d} N_g \log_2 N_g$$

and the storage requirement is

$$2(N_g^{1/d}+N_g^{2/d}+\cdots+N_g) \approx 2N_g$$

for the calculation. The additional storage required for the Green's function can be reduced to N_g/d by using the symmetries in G. A program POT5A has been written by Brownrigg using the above techniques, for a three-dimensional isolated distribution of masses (Brownrigg, 1975; Eastwood and Brownrigg, 1979). The execution time on the IBM 360/195 is 4.6 s for a $(32 \times 32 \times 16)$ active mesh and the storage requirement is $\sim 2.5N_g$.

Maruhn, Welton, and Wong (1976) have claimed that the above method for the isolated system does not remove the images completely, but only makes them more distant. However Eastwood and Brownrigg (1979) prove that this claim is false and that the above method does correctly solve the isolated problem.

6-5-5 James' Algorithm

The MFT convolution method described in Sec. 6-5-4 is applicable to an arbitrary interaction potential. The method is wasteful of both storage and time, owing to the doubling up of the system in each coordinate direction that is necessary to eliminate "image" systems. If the problem is the solution of an isolated Poisson system James (1977) has given an improved algorithm that avoids the doubling up.

If the source distribution is given within the cube of $N_g = (n \times n \times n)$ points, the solution proceeds as follows:

1. Calculate the potential ϕ in the interior of the cube by the MFT method. The potential is arbitrarily assigned the value zero on the boundary.

2. Taking the potential $\phi = 0$ outside the cube, apply the Poisson operator to the boundary points to determine what implied "screening" charge q is present on the boundary, from

$$q = -\nabla^2 \phi \tag{6-114}$$

3. Calculate the correction ψ due to minus these screening charges. The doubling up procedure (Sec. 6-5-4) is used to calculate the isolated potential due to the screening charges, but the doubling up is only of the boundary values and *not* the whole mesh.

The exact solution is the sum of the original ϕ and the correction ψ due to the fictitious boundary charges. Further, the correction for the boundary charges can be applied in transform space halfway through the MFT solution in step (1), in a manner similar to the introduction of boundary electrodes in the capacity matrix methods (see Sec. 6-5-6).

The James' algorithm is roughly three times faster than MFT convolution and uses approximately half the storage. James reports the following performance, in FORTRAN (FTN 4.4) using the CDC 7600, on a $(33 \times 33 \times 33)$ mesh:

1. FFT on entire mesh.	1.0 s
($15n^3 \log_2 n$ ops)	
2. Calculate screening charge.	0.3 s
($6n^3$ ops)	
3. Calculate boundary potential.	0.6 s
($18n^3 + 108n^2 \log_2 n$ ops)	
4. Organization overhead.	0.1 s
Total time	2.0 s

Since the calculation only takes place over the active mesh of N_g points, only slightly more than a storage of N_g is required. The asymptotic operation count is $5N_g \log N_g$.

6-5-6 Capacity Matrix Method

The direct methods just described for the solution of Poisson's equation may appear to have rather limited application—particularly to electron device simulation (Chapter 10, especially Sec. 10-4-7)—since electrodes are not permitted in the interior of the region, although of course they may form the boundaries to the problem. One way around this restriction is to precalculate a capacity matrix **C**, say, which relates the potential and charge on a number of points in the interior. These points are to be electrodes held at given potentials and the purpose is to calculate the surface charge induced on the electrode by the surrounding space charge. The solution is then obtained by solving Poisson's equation twice, as follows. First, Poisson's equation is solved with zero charge on the electrode points and the error in the potential at the electrode points from the desired values is recorded. This error, when multiplied by the capacity matrix and negated, gives the desired surface charge on each electrode point. Poisson's equation is solved again with the surface charge present and the solution is now correct everywhere in the region including the electrode points.

The straightforward way to find the capacity matrix is to place unit charge on each electrode point in turn (with zero charge on the other electrode points) and solve for the potential. The potential values on the electrode points form the elements of one column of the inverse capacity matrix $\boldsymbol{A} = \boldsymbol{C}^{-1}$. Repeating this process for each of the l electrode points fills in the l columns of $\boldsymbol{A}$. The capacity matrix is then found by inversion.

If the boundary conditions of the region are periodic, the potential due to a unit charge depends only on the vector distance between the potential and charge points. In this case the inverse capacity matrix can be built up from a single solution for the potential obtained from a unit charge placed at any point by appropriate selection of the data (Widlund, 1972). Cases of given values on opposite surfaces or given gradient can be reduced to the periodic case over an appropriately doubled-up region and similarly treated. These techniques reduce the number of potential solutions required to find the elements of $\boldsymbol{A}$ to at most four but they do not avoid the more time-consuming task of inverting $\boldsymbol{A}$ to obtain $\boldsymbol{C}$.

The need to store and invert $\boldsymbol{A}$ limits the application of the capacity matrix method as described above to 100 to 200 points. The capacity matrix $\boldsymbol{C}$, which may overwrite $\boldsymbol{A}$ in store, is symmetric so that the storage required is $l(l+1)/2$ for l electrode points. Since the capacity matrix depends only on the geometry of the problem and not on the space charge distribution, it may be calculated once and for all at the beginning and stored. The time to multiply the error by the capacity matrix is negligible, hence the computer time required for determining the potential is doubled if electrodes are included in this fashion. We shall call this method the direct capacity matrix (DCM) method.

Martin (1974) has shown that the method may be extended to include a wider range of conditions than the specification of potential. The error at an "electrode" point can be calculated more generally as

$$\varepsilon = a\phi + b\frac{\partial \phi}{\partial n} - c \tag{6-115}$$

Since the method makes $\varepsilon = 0$, the first and last terms give a fixed potential, and the second term allows for discontinuities of diffusion coefficient, dielectric constant, or conductivity. The capacity matrix is calculated as before, except that the elements of the inverse capacity matrix are the values of ε from Eq. (6-115), arising from unit charges at the electrode points.

The program POT4 (Hockney, 1978*b*) makes use of the FACR(1) algorithm and the direct capacity matrix method to solve a wide class of problems arising in semiconductor geometries. Figure 6-10 shows the conditions which can be solved

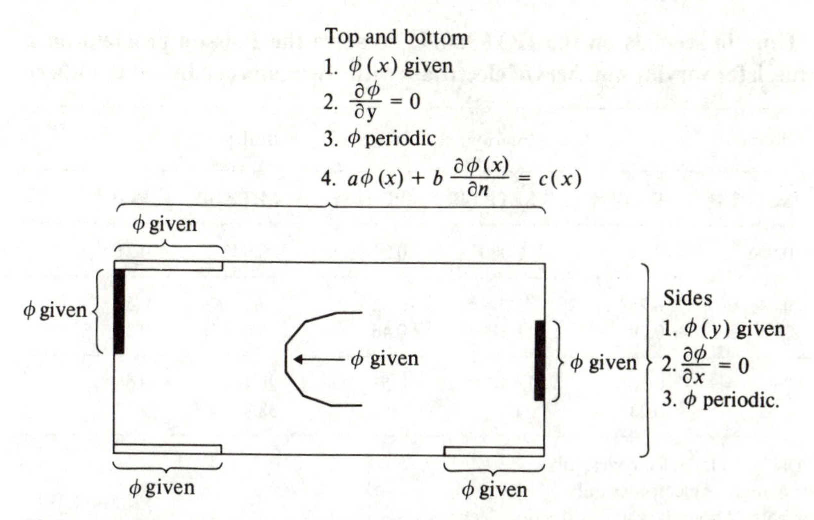

Figure 6-10 Examples of the type of Poisson problem solved by POT4. The open lines are boundary electrodes and the solid lines are interior electrodes. (*From Hockney, 1978b, courtesy of Advance Publications.*)

by this program. A rectangular geometry is a good representation of an epitaxial semiconductor. If electrodes are specified on the top and bottom of the rectangle, the capacity matrix correction can be applied halfway through the FACR algorithm and the complete solution is obtained in one application of FACR. If, however, any electrode exists in the interior or on the left or right sides, two applications of the FACR algorithm are required with the capacity correction applied between the two applications. Tables 6-7 and 6-8 give the time for this

Table 6-7 Time in seconds on the IBM 360/91 to solve a test problem with 100 boundary and 100 internal electrodes with Neumann conditions elsewhere using POT4 on a (NX × NY) mesh

NX	NY	Error	SETPT4†	POT4‡	No. electrodes§
128	32	8.4×10^{-4}	38.8	0.73	0.23
256	64	5.1×10^{-4}	99.2	1.98	0.79
512	128	1.3×10^{-3}	360	7.0	3.20
1024	256	—¶	1440	28.0	13.4

† Time to compute the capacity matrices.

‡ Time to solve the potential problem when the capacity matrices are known.

§ Time to solve the potential problem without the electrodes.

¶ There is insufficient storage available to keep the exact solution, thus no error can be computed.

Source: Hockney (1978*b*); courtesy Advance Publications.

Table 6-8 Time in seconds on the IBM 360/91 to solve the Poisson problem on a (128 × 32) mesh for varying numbers of electrodes with Neumann conditions elsewhere

	Boundary†		Interior‡		Both§	
N	SETPT4¶	POT4††	SETPT4¶	POT4††	SETPT4¶	POT4††
0	0.006	0.21	0.006	0.21	0.006	0.21
25	0.94	0.25	5.5	0.43	7.5	0.51
50	2.16	0.29	11.2	0.46	15.6	0.56
75	3.94	0.30	17.5	0.51	26.4	0.65
100	6.52	0.33	24.9	0.58	38.8	0.73

† Time for *N* boundary electrodes only.
‡ Time for *N* interior electrodes only.
§ Time for both *N* boundary and *N* interior electrodes.
¶ Time to execute subroutine SETPT4. This is primarily the time to compute the capacity matrix.
†† Time to solve the Poisson problem when the capacity matrix is given. When interior electrodes are present this involves two calls of the subroutine POT4.
Source: Hockney, 1978*b*; courtesy Advance Publications.

algorithm for a variety of meshes and numbers of electrodes. The preparation of the capacity matrix (columns headed SEPT4) is an overhead that is required only if the geometry of the electrodes changes. The capacity matrix does not change if the charge distribution or applied potentials change. Hence, in a time-dependent simulation in constant geometry, the capacity matrix is calculated only once. The times in columns headed POT4 are the relevant cost per timestep in this case.

If the number of electrode points is too large to permit the calculation and storage of a capacity matrix, the surface charge can be adjusted iteratively in response to the local error at the electrode point, with the solution of Poisson's equation still being obtained by a direct method. Alternatively, direct methods may be abandoned entirely, and any of the methods described for the solution of the general nonseparable equation can be used. Depending on the number of mesh points, sparse matrix (see Sec. 6-4-3), incomplete decomposition (Secs. 6-4-4 to 6-4-6) or mesh relaxation (Secs. 6-3-3 to 6-3-6 and 6-5-7) might be suitable.

In the iterative capacity matrix (ICM) method devised by Proskurowski and Widlund (1976), a set of correction dipoles $\boldsymbol{\mu}$ on the boundary points is determined by solving $\boldsymbol{A\mu} = -\varepsilon$, where ε is the error in the boundary conditions. The matrix $\boldsymbol{A}$ (confusingly and, in the terminology of electrostatics, incorrectly called a capacity matrix by Proskurowski) is the dipole analogue of the inverse capacity matrix. The solution is found iteratively by the conjugate gradient method, and the use of correction dipoles, rather than charges, ensures that the eigenvalues of $\boldsymbol{A}$ are near unity and that the convergence is rapid (see Sec. 6-4-2). Since the CG method requires only the calculation of the residual $\mathbf{r} = \boldsymbol{A\mu} + \varepsilon$, which may be found by one application of an RES, there is no need to find or store the elements of $\boldsymbol{A}$. The total storage is thereby reduced to about $30l$. Actually two applications of an RES per CG iteration are required because A is unsymmetric and one must work with the symmetric matrix AA^T.

The Proskurowski and Widlund method is suitable for regions either completely exterior or completely interior to a closed boundary. Since a line of dipoles causes a discontinuity of potential across the boundary, the correction is not analogous to the action of a physical electrode in which charges, not dipoles, are induced. This may mean that the method is unsuitable for boundaries that do not form a closed curve, such as the interior electrodes in Fig. 6-10. Using his program HLMHLZ, Proskurowski (1977) has solved the Poisson equation with Dirichlet boundary conditions in the interior of a circle with 528 boundary points on a (256×256) mesh. Sixteen CG iterations (32 RES solutions) are required to reduce the error to 10^{-6} and took 24 s on the CDC 7600. The method has been extended to three-dimensions by O'Leary and Widlund (1979).

The choice between the direct and iterative capacity matrix methods will be influenced primarily by the number of electrode points l. Given the capacity matrix, the DCM method can obtain the solution in the time of two RES solutions or one ICM iteration. The direct method is therefore preferable if many solutions are to be found in a fixed geometry, in which case the cost of precalculating the capacity matrix can be ignored. If, on the other hand, one is considering a one-off solution then one must compare the time to invert an $(l \times l)$ matrix ($\sim 2l^3$ ops for DCM using Widlund's periodic method for obtaining A) with the time for ~ 32 RES solutions [$\sim 80n^2 \log_2 n$ for ICM on an $(n \times n)$ mesh]. If we take $l = 2n$ and $n = 128$ we find that DCM is faster than ICM if $l \leqslant 140$. This is close to the practical limit for matrix inversion so that the DCM method is likely to be favored whenever it can practicably be applied. The ICM method is most suitable if the number of electrode points is several hundreds or thousands when DCM obviously cannot be used. This would be the case in the accurate description of any reasonably complex geometry such as might occur in an electron gun.

The most general arbitrary-coefficient field equation will consist of regions with a relative smooth variation of coefficients, separated by boundaries at which the coefficients have discontinuities of value or gradient. Any iterative method (e.g., Concus and Golub iteration, Sec. 6-5-7) that assumes a smooth variation of coefficients throughout the problem will show very slow or no convergence on such problems, because of the discontinuities. In these cases, it is recommended that the discontinuities be included by the capacity matrix adjustment, thus leaving a problem with smooth variation of coefficients for solution by the iterative method.

6-5-7 Concus and Golub Iteration

The most general equation that can be solved directly by RES methods is the general separable equation Eq. (6-82). Concus and Golub (1973) have shown that more general equations can be solved by the iterative use of an RES algorithm. They tackle the general nonseparable variable-coefficient problem in the rectangle

$$-\nabla[D(x,y)\nabla\phi] = f(x,y) \tag{6-116}$$

This equation represents diffusion with a variable diffusion coefficient or the electrostatic potential problem with variable dielectric constant. In the case that D is constant we have the model Poisson equation on which all RES methods may be directly applied. If $D(x,y) = X(x)Y(y)$ is a product form then Eq. (6-116) becomes a separable equation and CR can be used directly (Sec. 6-5-1). Such a variation of D is highly artificial and does not correspond to any obvious physical situation. We seek therefore the solution of Eq. (6-116) with an arbitrary variation of $D(x,y)$.

If $D(x,y)$ has a second derivative, we can introduce a change of variable

$$W(x,y) = [D(x,y)]^{1/2}\phi(x,y) \tag{6-117}$$

Then Eq. (6-116) becomes

$$-\nabla^2 W + p(x,y)W = q(x,y) \tag{6-118a}$$

where
$$p(x,y) = D^{-1/2}\nabla^2(D^{1/2}) \tag{6-118b}$$

and
$$q(x,y) = D^{-1/2}f \tag{6-118c}$$

The change of variable has thus transformed the differential part of the operator to that most suited for solution by RES methods.

An iterative procedure is defined as follows

$$(-\nabla^2 + K)W^{(n+1)} = [K - p(x,y)]W^{(n)} + q(x,y) \tag{6-119a}$$

where n is the iteration number and K is an iteration parameter that may be varied to improve convergence. The choice of K that minimizes the maximum eigenvalue of the iteration matrix is

$$K = \tfrac{1}{2}(\text{min of } D + \text{max of } D) \tag{6-119b}$$

The solution of the discretized form of Eq. (6-119a) is carried out by an RES algorithm since the left-hand side is a constant coefficient problem. If p is in fact constant then the solution is obtained in one iteration. This occurs, for example, if $D(x,y) = \sin^2(x+y)$.

The discretized form of Eq. (6-119a) is

$$(-\nabla_h^2 + KI)\mathbf{W}^{(n+1)} = (KI - P)\mathbf{W}^{(n)} + \mathbf{Q} \tag{6-120}$$

where ∇_h^2 and P, are matrices obtained from the finite difference of ∇^2 and $p(x,y)$, and $\mathbf{Q}$ is a vector of right-hand side values $q(x,y)$.

The error obeys the equation

$$\varepsilon^{(n+1)} = M\varepsilon^{(n)} \tag{6-121a}$$

where the iteration matrix is

$$M = (-\nabla_h^2 + KI)^{-1}(KI - P). \tag{6-121b}$$

If the eigenvalues of M are in the range $[-\rho, \rho]$ Chebyshev acceleration can be applied to successive iterates

$$\mathbf{W}^{*(n+1)} = \omega_{n+1}(\mathbf{W}^{(n+1)} - \mathbf{W}^{*(n-1)}) + \mathbf{W}^{*(n-1)} \tag{6-122}$$

where
$$\omega_0 = 1,$$
$$\omega_1 = 2/(2-\rho^2) \tag{6-123}$$
and
$$\omega_{n+1} = 1/(1-\rho^2\omega_n/4) \qquad n = 1, 2, \ldots$$

$\mathbf{W}^{*(n+1)}$ is the improved value of $\mathbf{W}^{(n+1)}$, and $\mathbf{W}^{(n+1)}$ in Eq. (6-122) is the solution of Eq. (6-120) with $\mathbf{W}^{*(n)}$ replacing $\mathbf{W}^n$ on the right-hand side. The spectral radius ρ can be estimated from the ratio

$$\rho \simeq \|\mathbf{W}^{(n)} - \mathbf{W}^{(n-1)}\|/\|\mathbf{W}^{(n-1)} - \mathbf{W}^{(n-2)}\| \tag{6-124}$$

Concus and Golub observe a convergence to an error of $\sim 10^{-6}$ in 4 iterations if the variation of $D(x, y)$ is relatively smooth. For examples with a discontinuity, errors of $\sim 10^{-2}$ remain after 10 iterations. Matrix methods (see Secs. 6-4-1 to 6-4-6) are recommended for such problems. An important conclusion is that, unlike the case with SOR iterative techniques, the number of iterations required for a solution is essentially independent of the number of mesh points.

Apart from the storage of $p(x, y)$ and $q(x, y)$ necessary for the statement of the problem, the unaccelerated C and G scheme requires a mesh of N_g points for the solution vector $\mathbf{W}^{(n)}$. If Chebyshev acceleration is used a further mesh of N_g is required for the storage of $\mathbf{W}^{(n-1)}$.

Further refinements to the Concus and Golub iteration are possible. Since an optimal FACR algorithm can solve a problem whose coefficients depend only on one of the two dependent variables (see Sec. 6-5-3), we can propose the following iteration:

$$[-\nabla^2 + X(x)]W^{(n+1/2)} = [X(x) - p(x, y)]W^{(n)} + q \tag{6-125a}$$
$$[-\nabla^2 + Y(y)]W^{(n+1)} = [Y(y) - p(x, y)]W^{(n+1/2)} + q \tag{6-125b}$$

where
$$\left.\begin{aligned} X(x) &= \frac{1}{L}\int p(x, y)\,dy \qquad 0 \leqslant y \leqslant L \\ Y(y) &= \frac{1}{L}\int p(x, y)\,dx \qquad 0 \leqslant x \leqslant L \end{aligned}\right\} \tag{6-125c}$$
and

Equation (6-125a) would be solved by FACR with Fourier analysis in the y-direction, and Eq. (6-125b) with Fourier analysis in the x-direction.

6-6 CONCLUDING REMARKS

In this chapter we have tried to give a survey of the issues involved in selecting a method for solving the field equations and a brief account of the main features of the most important methods. We hope this will be adequate to indicate the type of method likely to be the most useful in any particular case, and avoid the choice of grossly inefficient methods. Reference should then be made to the original papers or to the description of appropriate library programs for many of the details of the methods we could not include here.

CHAPTER

SEVEN

COLLISIONLESS PARTICLE MODELS

7-1 INTRODUCTION

The physical systems simulated by particle models can be mathematically described by a hierarchy of distribution functions (cf. Chapter 1). The mathematical analysis of the systems is performed by truncating the hierarchy by means of some closure assumption, then finding approximate solution to the resulting set of nonlinear equations.

The lowest-order closure is established for dilute systems, where expansion in terms of the smallness of potential energy with respect to the kinetic energy leads to the collisionless (Vlasov) equations. Analytic treatment of the Vlasov equations (cf. Chapter 2) is pursued for short timescales by linearization and Fourier–Laplace transformation of the linearized equations. The result of this approach for plasmas is the plasma wave theory (Stix, 1962), and for gravitating systems the tight-spiral approximation density wave theory (Lin and Lau, 1979). In the latter cases (and in the former when magnetic fields are present) the analysis becomes somewhat involved because of the necessity of integrating along unperturbed orbits to obtain the linearized response. The linear response theory is extended to slowly changing systems by allowing the zeroth order distribution to vary slowly with time (quasilinear theory, see for example Bernstein and Engelmann, 1966).

The next-highest-order closure, again for dilute systems, retains equations both for the one-particle distribution function and for the pair correlation function. Methods of analysis of this system of equations range from a simple treatment of the graininess by a sequence of impulsive binary collisions to the comprehensive analysis leading to the Lenard–Balescu–Guernsey equation

(Lenard, 1960; Balescu, 1960; Guernsey, 1962). The result of all of these methods of analysis is to lead to Fockker–Plank type diffusion equations for the one-particle distribution function

$$\left(\frac{\partial f}{\partial t}\right)_c = \frac{\partial}{\partial v_i} D_{ij} \frac{\partial f}{\partial v_j} + \frac{\partial}{\partial v_i} A_i f \tag{7-1}$$

where the forms of the diffusion coefficients D_{ij} and drag coefficients A_i are determined by the approximations used. In the simplest formulation diffusion is characterized by a collision time τ_c and drag is characterized by a slow-down time τ_s (cf. Chapter 9).

The theory for dense systems follows the same pattern as for dilute systems, but employs more sophisticated expansion methods to obtain the required closure equations (Hansen and McDonald, 1976; Rice and Gray, 1965). The outcome of such expansions is a variety of closures which have a range of degrees of success depending on the circumstances in which they are used.

To encompass the gamut of statistical theory of many body systems is far beyond the scope of this book. However, such theory is intimately linked with the particle approach to simulation, being the motivating object for many of the computations which have been undertaken. In this chapter we will illustrate for collisionless systems (using an electrostatic plasma as our model) the reverse link: Namely, how physical theory may be applied to the analysis of particle models. The link, when used in both directions, creates a bootstrapping process where the theoretical analysis leads to improved simulation models which in turn provide more precise information required for improving the approximations of the theory.

In the next section we shall obtain the statistical equations for particle models. This is followed in Sec. 7-3 by a brief illustration of linearized mode analysis of collisionless plasma simulation models and in Sec. 7-5 by some salient results concerning the collision terms in the kinetic equations. Sections 7-6, 7-7, and 7-8 show how the results of the analysis are fed back into improvements in the design of the simulation models. Although the results we obtain are for the simple case of an electrostatic collisionless plasma, the methods of analysis can be extended to other particle–mesh simulation models.

7-2 THE KINETIC EQUATIONS

The starting point for the derivation of the kinetic equations for the physical system of N_p particles is either Liouville's equation in $6N_p$ dimensional Γ space, Eq. (1-9), or the equivalent equation in 6-dimensional μ space

$$\frac{\partial F}{\partial t} + \mathbf{v} \cdot \frac{\partial F}{\partial \mathbf{x}} + \mathbf{a} \cdot \frac{\partial F}{\partial \mathbf{v}} = 0 \tag{7-2}$$

where F is the exact one-particle distribution function defined by Eq. (1-10).

7-2-1 Small-Timestep Limit

The spatial mesh used in particle–mesh calculations is constrained by limitations of computer storage to be relatively coarse. The choice of timestep is less restricted, and in practical computations it is chosen to be much less than the stability and accuracy limits set by the difference approximations to time derivatives in the equations of motion. Consequently, it is of value to investigate first the modifications of the continuum mathematical description which incorporate the effects of the spatial mesh while omitting the effects of finite timestep DT.

In the limit $DT \to 0$, the only difference between the equations of motion for the exact case and the mesh-approximated case is in the force. Hence the conservation arguments used in the continuum case can be carried over, changed only in the definition of the force (or equivalently acceleration) appearing in the various equations. Liouville's equation for probability density ρ in Γ space becomes

$$\frac{\partial \rho}{\partial t} + \sum_{i=1}^{N_p} \left\{ \mathbf{v}_i \cdot \frac{\partial \rho}{\partial \mathbf{x}_i} + \frac{\mathbf{F}_i}{m} \cdot \frac{\partial \rho}{\partial \mathbf{v}_i} \right\} = 0 \tag{7-3}$$

where the forces $\{\mathbf{F}_i\}$ are given by the PM force calculation. The hierarchy of equations for the s particle distribution function is, as outlined in Chapter 1, obtained by integrating Eq. (7-3) over the remaining $N_p - s$ coordinates. For s equal to one, this yields the equation for the ensemble-averaged one-particle distribution function

$$\frac{\partial f_1}{\partial t} + \mathbf{v} \cdot \frac{\partial f_1}{\partial \mathbf{x}} + \frac{\mathbf{F}^{\text{ext}}}{m} \cdot \frac{\partial f_1}{\partial \mathbf{v}} + \int \frac{\mathbf{F}(\mathbf{x};\mathbf{x}')}{m} \cdot \frac{\partial f_2}{\partial \mathbf{v}} \, d\mathbf{x}' \, d\mathbf{v}' \tag{7-4}$$

where $\mathbf{F}^{\text{ext}}$ are external forces acting on the system, $\mathbf{F}(\mathbf{x};\mathbf{x}')$ is the force on a particle at $\mathbf{x}$ due to another at $\mathbf{x}'$, and $f_2(\mathbf{x}, \mathbf{v}, \mathbf{x}', \mathbf{v}', t)$ is the ensemble-averaged two-particle distribution function.

The Vlasov (or collisionless) equation for the particle-mesh model is obtained by setting $f_2 = f_1 f_1$ in Eq. (7-4). The equations which form the starting point for the derivation of the Fockker–Planck type equation of Lenard (1960) are obtained by setting $f_2 = f_1 f_1 + g$ in Eq. (7-4) and $f_2 = f_1 f_1 + g$, $f_3 = f_1 f_1 f_1 + \Sigma f_1 g$ in the corresponding equation for f_2, and then using the timescale ordering approximation that g relaxes to its asymptotic value on a timescale much less than that of the variation of f_1. This, together with the assumption of spatial homogeneity, reduces the equation for g to an expression where g is only a functional of f_1 and the equation for f_1 to

$$\frac{\partial f_1}{\partial t} = -\int \frac{\mathbf{F}(\mathbf{x};\mathbf{x}')}{m} \cdot \frac{\partial}{\partial \mathbf{v}} g(\mathbf{x}, \mathbf{x}', \mathbf{v}, \mathbf{v}') \, d\mathbf{x}' \, d\mathbf{v}' \tag{7-5}$$

Therefore, by replacing the exact interparticle force by the mesh approximation and repeating the steps of the analysis of the physical system (using the transform space analysis methods described in Chapter 5) one can obtain kinetic equations in the limit $DT \to 0$ for particle models.

7-2-2 Finite Timestep

If a finite timestep is retained in the analysis, a direct copy of the differential analysis for the continuum system is no longer possible. Instead of the equations of motion providing the characteristics for a differential equation for probability density, they now provide transition rules for successive states of the system. Thus, if the state of a leapfrog system at timestep n is described by the coordinates $\{\mathbf{x}_i = \mathbf{x}_i^n; \bar{\mathbf{v}}_i = (\mathbf{v}_i^{n+1/2} + \mathbf{v}_i^{n-1/2})/2; i \in [1, N_p]\}$, then the conservation law in Γ space may be expressed as

$$\rho(\{\mathbf{x}_i\}, \{\bar{\mathbf{v}}_i\}, t^n) = \rho(\{\mathbf{x}'_i\}, \{\bar{\mathbf{v}}'_i\}, t^{n+1}) \tag{7-6}$$

In the collisionless (Vlasov) approximation, the particles are uncorrelated, so the probability density ρ may be written as a product of one-particle distribution functions

$$\rho(\{\mathbf{x}_i\}, \{\bar{\mathbf{v}}_i\}, t^n) = \prod_{i=1}^{N_p} f_1(\mathbf{x}_i, \bar{\mathbf{v}}_i, t^n) \tag{7-7}$$

in which case Eq. (7-6) may be reduced to the Vlasov equation for particle models with finite timestep:

$$f_1(\mathbf{x}, \bar{\mathbf{v}}, t^n) = f_1(\mathbf{x}', \bar{\mathbf{v}}', t^{n+1}) \tag{7-8}$$

where $\mathbf{x} = \mathbf{x}^n$, $\bar{\mathbf{v}} = (\mathbf{v}^{n+1/2} + \mathbf{v}^{n-1/2})/2$, $\mathbf{x}' = \mathbf{x}^{n+1}$, $\bar{\mathbf{v}}' = (\mathbf{v}^{n+3/2} + \mathbf{v}^{n+1/2})/2$ are related by the leapfrog equations of motion. The next-highest-order closure is obtained by setting

$$f_2(\mathbf{x}, \bar{\mathbf{v}}, \mathbf{x}', \bar{\mathbf{v}}', t^n) = f_1(\mathbf{x}, \bar{\mathbf{v}}, t^n) f_1(\mathbf{x}', \bar{\mathbf{v}}', t^n) + g(\mathbf{x}, \mathbf{v}, \mathbf{x}', \mathbf{v}', t^n) \tag{7-9}$$

and

$$\rho = \prod_{i=1}^{N_p} f_1 + \prod_{i=1}^{j-1} g \prod_{k \neq i, j}^{N_p} f_1 \tag{7-10}$$

Integration of Eq. (7-6) over all but one and two of the coordinates can then be performed to obtain the kinetic equations for weakly correlated systems.

7-3 THE DISPERSION RELATION

The dispersion relation $\omega = \omega(\mathbf{k})$ is the consistency condition for small amplitude waves. It provides information on the characteristic time and scale lengths, wave phase and group velocities, and on the stability of a given configuration. In the continuum case, the range of wavenumbers is unrestricted. The introduction of a discrete mesh of spacing H reduces the range of possible wavenumbers to the principal zone (cf. Sec. 5-6), $|k| < \pi/H$. Any harmonic with wavenumber k beyond $|k| = \pi/H$ (i.e., of wavelength less than $2H$) appears to the mesh as a contribution to the harmonic in the principal zone with wavenumber $k' = k \bmod k_g$. Discretization in time has a similar effect. The introduction of a finite timestep DT limits frequencies ω to the range $|\omega| < \pi/DT$. However, since the stability of the time integration scheme requires $|\omega| DT < 2$ (for leapfrog—cf. Chapter 4), a timestep too large to represent the frequencies of the system will give computational results which are obviously incorrect.

7-3-1 Small-Timestep Limit

Setting f_2 equal to the product of one-particle distribution functions in Eq. (7-4) gives the Vlasov equation for particle models in the limit of zero timestep:

$$\frac{\partial f}{\partial t} + \mathbf{v} \cdot \frac{\partial f}{\partial \mathbf{x}} + \left\{ \frac{\mathbf{F}^{\text{ext}}_{(\mathbf{x})}}{m} + \int \frac{\mathbf{F}(\mathbf{x}; \mathbf{x}')}{m} f(\mathbf{x}', \mathbf{v}', t)\, d\mathbf{x}'\, d\mathbf{v}' \right\} \cdot \frac{\partial f}{\partial \mathbf{v}} = 0 \tag{7-11}$$

In Eq. (7-11) and in the following material, the subscript one is dropped from the one-particle distribution function.

For simplicity, let us take the one-dimensional case of Eq. (7-11) and assume a mesh spacing H in an infinite system. Equation (7-11) becomes

$$\frac{\partial f}{\partial t} + v \frac{\partial f}{\partial x} + \left\{ \frac{F^{\text{ext}}}{m} + \int \frac{F(x, x')}{m} n(x')\, dx' \right\} \frac{\partial f}{\partial v} = 0 \tag{7-12}$$

where $n(x)$ is the number density

$$n(x) = \int f(x', v')\, dv' \tag{7-13}$$

$F(x, x')$ is the force at position x due to a charge at x' as calculated by the particle mesh scheme. The compact notation introduced in Sec. 5-6 enables the steps of the force calculation cycle to be written as follows.

Charge assignment

$$\rho\dagger(x) = \text{Ш}\left(\frac{x}{H}\right) \frac{q}{H} W(x) * n(x) \tag{7-14}$$

Potential solution

$$\phi'(x) = G'(x) * \rho\dagger(x) \tag{7-15}$$

Force interpolation (momentum-conserving scheme)

$$E'(x) = -D(x) * \phi'(x) \tag{7-16}$$

$$E\dagger(x) = \text{Ш}\left(\frac{x}{H}\right) E'(x) \tag{7-17}$$

$$F(x) = \frac{q}{H} W(x) * E\dagger(x) \tag{7-18}$$

where the primes denote continuous band-limited functions constructed from discrete sets of values and the daggers indicate functions comprising sets of impulses, where the impulse strengths give mesh values at the locations of the impulses. In particular, for a single-source charge at $x = x'$, Eq. (7-14) becomes

$$\rho\dagger(x) = \text{Ш}\left(\frac{x}{H}\right) \frac{q}{H} W(x) * \delta(x - x') \tag{7-19}$$

Combining Eqs. (7-15) to (7-19) gives

$$F(x, x') = -\frac{q^2}{H^2} W(x) * \text{Ш}\left(\frac{x}{H}\right) \times \left[D(x) * G'(x) * \text{Ш}\left(\frac{x}{H}\right) (W(x) * \delta(x - x')) \right] \tag{7-20}$$

The only dependence of $F(x, x')$ on x' comes through the $\delta(x-x')$ source charge term, so we can symbolically represent Eq. (7-20) as

$$F(x, x') = h(x) * \delta(x - x') \tag{7-21}$$

Therefore

$$\int F(x, x') n(x') \, dx' = h(x) * \int \delta(x - x') n(x') \, dx'$$

$$= h(x) * n(x)$$

$$= -\frac{q^2}{H^2} W(x) * \text{Ш}\left(\frac{x}{H}\right) \left[D(x) * G'(x) * \text{Ш}\left(\frac{x}{H}\right) (W(x) * n(x)) \right] \tag{7-22}$$

Equation (7-22) is the expression for the force at position x, calculated by the particle mesh scheme and arising from a density $n(x)$. The total force at position x is obtained by adding the external force term to Eq. (7-22), where the external force term includes contributions from fixed neutralizing background charge. Thus Eq. (7-12) may be rewritten, in the same manner as for the continuum case, as

$$\frac{\partial f}{\partial t} + v \frac{\partial f}{\partial x} + \frac{F}{m} \frac{\partial f}{\partial v} = 0 \tag{7-23}$$

where now F describes the total force at position x due to the distribution f and background charge. The same holds for the energy-conserving variants of the particle mesh calculation. In both cases the results generalize to multispecies systems by adding extra variables for extra species and sums over species where appropriate.

Equation (7-23) is of the same form as the exact μ space description of the N_p body system, Eq. (7-2). It differs from the exact form in two respects: The exact one-particle distribution function is replaced by the smooth ensemble-averaged distribution function f and the acceleration term is that arising from the ensemble-averaged distribution rather than the exact distribution of particle coordinates.

The simplest ensemble-averaged (collisionless) case to analyze is the homogeneous charge-neutral system. This yields the equilibrium

$$f = f(v) \qquad F = 0 \tag{7-24}$$

Small departures from equilibrium are described by setting $f = f^0(v) + f^1(x, v, t)$, $F = F^1$, etc., in Eq. (7-23) and retaining only first-order terms:

$$\frac{\partial f^1}{\partial t} + v \frac{\partial f^1}{\partial x} = -\frac{F^1}{m} \frac{\partial f^0}{\partial v} \tag{7-25}$$

Fourier-transforming in space and time gives the steady-state response equation

$$-i(\omega-kv)\hat{f}^1(k,v,\omega) = -\frac{\hat{F}(k,\omega)}{m}\frac{\partial f^0}{\partial v} \tag{7-26}$$

[Strictly, a Laplace transform in time should be used, and Eq. (7-26) extracted from the asymptotic behaviour of the initial value problem; see for example Stix, 1962.] Rearranging Eq. (7-26) and integrating over v gives the first-order density harmonics

$$\hat{n}(k) = \int \hat{f}^1\,dv = -i\frac{\hat{F}(k)}{m}\int \frac{1}{(\omega-kv)}\frac{\partial f^0}{\partial v}\,dv \tag{7-27}$$

Fourier-transforming Eqs. (7-14) to (7-18) gives

$$\begin{aligned}\hat{\rho}(k) &= \frac{q}{H} H\,\text{Ш}\left(\frac{k}{k_g}\right) * \hat{W}(k)\hat{n}(k)\\ &= \frac{q}{H}\sum_{n=-\infty}^{\infty} \hat{W}(k-nk_g)\hat{n}(k-nk_g)\end{aligned} \tag{7-28}$$

$$\hat{\phi}'(k) = \hat{G}'(k)\hat{\rho}(k) \tag{7-29}$$

$$\hat{E}'(k) = -\hat{D}(k)\hat{\phi}'(k) \tag{7-30}$$

$$\begin{aligned}\hat{E}(k) &= H\,\text{Ш}\left(\frac{k}{k_g}\right) * \hat{E}'(k)\\ &= -\hat{D}(k)\hat{\phi}(k)\end{aligned} \tag{7-31}$$

$$\hat{F}(k) = \frac{q}{H}\hat{W}(k)\hat{E}(k) \tag{7-32}$$

where the daggers have been omitted since $\hat{\rho}\dagger = \hat{\rho}$ and $\hat{E}\dagger = \hat{E}$. In Eq. (7-31) the periodicity of $\hat{D}$ and the band-limited property of the primed quantities is used to write $H\,\text{Ш}(k/k_g) * \hat{\phi}'(k) = \hat{\phi}(k)$ (see Sec. 5-6 for further details). Multiplying Eq. (7-27) by $q\hat{W}/H$ and forming the sum appearing on the right-hand side of Eq. (7-28) gives

$$\hat{\rho}(k) = -i\sum_n \frac{q\hat{W}(k_n)}{H}\frac{\hat{F}(k_n)}{m}\int \frac{1}{(\omega-k_n v)}\frac{\partial f^0}{\partial v}\,dv \tag{7-33}$$

Using Eqs. (7-29) to (7-32) to write $\hat{F}$ in terms of $\hat{\rho}$, using the periodicity of $\hat{\rho}$, $\hat{G}$, and $\hat{D}$ to extract them from the sum in the right-hand side of Eq. (7-33), and finally dividing through by $\hat{\rho}$, gives the dispersion relation for momentum-conserving (collisionless) particle-mesh plasma model

$$1 = -(-i\hat{D})(\epsilon_0\hat{G})\frac{\omega_p^2}{n_0}\sum_{n=-\infty}^{\infty}\left(\frac{\hat{W}(k_n)}{H}\right)^2\int \frac{1}{(\omega-k_n v)}\frac{\partial f^0}{\partial v}\,dv \tag{7-34}$$

In Eqs. (7-33) and (7-34), the abbreviated form k_n is used to denote $k-nk_g$, and the integrals over v are taken along the Landau contour in the complex v plane. These integrals may be rewritten in terms of the principal parts plus contributions from the singularities:

$$1 = -(-i\hat{D})(\epsilon_0 \hat{G}) \frac{\omega_p^2}{n_0} \sum_{n=-\infty}^{\infty} \left(\frac{\hat{W}(k_n)}{H} \right)^2 \times \left(P \int_{-\infty}^{\infty} \frac{\frac{\partial f^0}{\partial v} dv}{\omega - k_n v} - \frac{i\pi}{|k_n|} \frac{\partial f^0}{\partial v} \bigg|_{\omega/k_n} \right) \tag{7-35}$$

A similar dispersion relation emerges for the energy-conserving schemes. Following the procedure outlined above, but replacing Eqs. (7-16) to (7-18) by their energy-conserving counterparts:

$$\phi\dagger(x) = \text{III}\left(\frac{x}{H}\right)\phi'(x) \tag{7-36}$$

$$F(x) = -\frac{q}{H}\left(\frac{d}{dx} W(x)\right) * \phi\dagger(x) \tag{7-37}$$

where

$$F(x) \supset \hat{F}(k) = -\frac{q}{H} ik\hat{W}(k)\hat{\phi}(k) \tag{7-38}$$

yields

$$1 = -(\epsilon_0 \hat{G}) \frac{\omega_p^2}{n_0} \sum_{n=-\infty}^{\infty} k_n \left(\frac{\hat{W}(k_n)}{H} \right)^2 \times \left(P \int_{-\infty}^{\infty} \frac{\frac{\partial f^0}{\partial v} dv}{(\omega - k_n v)} - \frac{i\pi}{|k_n|} \frac{\partial f^0}{\partial v} \bigg|_{\omega/k_n} \right) \tag{7-39}$$

7-3-2 Finite Timestep

If finite DT is retained in the dispersion analysis, the appropriate form of the Vlasov equation to use is Eq. (7-8). Using the leapfrog equations [Eqs. (4-3) and (4-4)], Eq. (7-8) may be expressed for a one-dimensional system as

$$f(x, \bar{v}, t) = f\left(x + \bar{v}DT + a(x, t^n) \frac{DT^2}{2}, \bar{v} + [a(x, t^n) + a(x', t^{n+1})] \frac{DT}{2}, t + DT \right) \tag{7-40}$$

where $a(x, t^n)$ is the acceleration at position x at time t^n. The corresponding expression for multidimensional systems is obtained by replacing x, $\bar{v}$, and a by their vector counterparts.

Taylor-expanding Eq. (7-40) gives

$$f(x, \bar{v}, t) = f(x + \bar{v}DT, \bar{v}, t + DT) + a(x, t^n) \frac{DT^2}{2} \frac{\partial f}{\partial x} + [a(x, t^n) + a(x', t^{n+1})] \frac{DT}{2} \frac{\partial f}{\partial v} + \cdots \tag{7-41}$$

Eq. (7-41) is linearized by setting

$$f = f^0 + f^1 \qquad a = a^0 + a^1 \tag{7-42}$$

and retaining only first-order terms. In particular, for a homogeneous plasma

$$f^0 = f^0(v) \qquad a^0 = 0 \tag{7-43}$$

yielding

$$f^1(x, v, t) = f^1(x+vDT, v, t+DT) + [a^1(x, t^n) + a^1(x+vDT, t+DT)]\frac{DT}{2}\frac{\partial f^0}{\partial v} \tag{7-44}$$

Transforming Eq. (7-44) in space and time [to give harmonics $\sim \exp i(kx-\omega t)$], using the shift theorem (Table A-4) on arguments $x+vDT$ and $t+DT$ yields

$$\hat{f}^1(k, v, \omega) = \hat{f}^1(k, v, \omega)e^{ikvDT}e^{-i\omega DT} + [\hat{a}^1(k, \omega) + \hat{a}^1(k, \omega)e^{ikvDT}e^{-i\omega DT}]\frac{DT}{2}\frac{\partial f^0}{\partial v} \tag{7-45}$$

which upon rearrangement gives

$$\hat{f}^1(k, v, \omega) = -i\frac{DT}{2}\frac{\hat{F}}{m}\cot\left((\omega - kv)\frac{DT}{2}\right)\frac{\partial f^0}{\partial v} \tag{7-46}$$

where $\hat{F}/m$ has been substituted for $\hat{a}$. The steps summarized by Eqs. (7-27) to (7-34) can now be repeated, using Eq. (7-46) rather than Eq. (7-26) to yield the dispersion relation for momentum-conserving particle mesh plasma models:

$$1 = -(-i\hat{D})(\epsilon_0\hat{G})\frac{\omega_p^2}{n_0}\sum_{n=-\infty}^{\infty}\left(\frac{\hat{W}(k_n)}{H}\right)^2\int\frac{DT}{2}\cot\left((\omega - k_n v)\frac{DT}{2}\right)\frac{\partial f^0}{\partial v}\,dv \tag{7-47}$$

Similarly, for energy-conserving models, we find

$$1 = -(\epsilon_0\hat{G})\frac{\omega_p^2}{n_0}\sum_{n=-\infty}^{\infty}k_n\left(\frac{\hat{W}(k_n)}{H}\right)^2\int\frac{DT}{2}\cot\left((\omega - k_n v)\frac{DT}{2}\right)\frac{\partial f^0}{\partial v}\,dv \tag{7-48}$$

The integrals in Eqs. (7-47) and (7-48) can be written in the same form as is found in the continuum case by using the identity

$$\cot x = \sum_{n=-\infty}^{\infty}\frac{1}{(x-\pi n)} \tag{7-49}$$

i.e.,

$$\int\frac{DT}{2}\cot\left((\omega - k_n v)\frac{DT}{2}\right)\frac{\partial f^0}{\partial v}\,dv = \sum_{m=-\infty}^{\infty}\int\frac{\dfrac{\partial f^0}{\partial v}}{(\omega_m - k_n v)}\,dv \tag{7-50}$$

where

$$\omega_m = \omega - \frac{2\pi}{DT}m \tag{7-51}$$

The result of the finite timestep is to introduce a periodicity in the frequency domain with period $2\pi/DT$. The frequencies which can be represented by the particle mesh models are restricted to the principal zone $|\omega| \leqslant \pi/DT$. The frequencies ω_m appear in the finite timestepping computations as the principal frequency ω.

7-3-3 The Warm-Plasma Approximation

The consequences of the discrete spatial mesh, and of the forms of influence function, $\hat{G}$ and force interpolation, on the dispersive properties of the simulation plasma can be investigated by comparing the dispersion relations obtained from Eq. (7-35) or (7-39) with that found for the continuum case, Eq. (2-7). The effects of finite timestep can be included by using Eqs. (7-47) and (7-48). If magnetic fields are present, dispersion relations for the discrete models are obtained by expressing f^1 in terms of an integral along unperturbed orbits, following the procedure used for the continuum system (Stix, 1962). However, even in the simplest approximations, the presence of alias sums in the equations for the particle–mesh system generally preclude the explicit analytic solution to obtain $\omega = \omega(k)$, although closed forms involving trignometric function can be obtained in certain instances.

In the warm-plasma approximation, the maxwellian distribution function

$$f^0(v) = \frac{n_0}{v_T\sqrt{2\pi}} \exp\left(-\frac{v^2}{2v_T^2}\right) \tag{7-52}$$

is approximated by its first two moments in evaluating the velocity space integrals in the dispersion relations. Formally, this procedure may be undertaken by replacing Eq. (7-52) by

$$f^0(v) = \frac{n_0}{2\sqrt{3}v_T} \Pi\left(\frac{v}{2\sqrt{3}v_T}\right) \tag{7-53}$$

in the velocity space integrands. This yields

$$P\int_{-\infty}^{\infty} \frac{\partial f^0/\partial v}{\omega - k_n v}\, dv \simeq \frac{-n_0 k_n}{\omega^2 - 3k_n^2 v_T^2} \tag{7-54}$$

Similarly, for a drifting maxwellian distribution, with drift velocity v_0

$$P\int_{-\infty}^{\infty} \frac{\partial f^0/\partial v}{\omega - k_n v}\, dv \simeq \frac{-n_0 k_n}{(\omega - k_n v_0)^2 - 3k_n^2 v_T^2} \tag{7-55}$$

Substituting Eq. (7-54) into Eq. (7-34) gives the warm-plasma approximation dispersion relation for a momentum-conserving simulation plasma:

$$1 = (-i\hat{D})(\epsilon_0 \hat{G})\omega_p^2 \sum_n \left(\frac{\hat{W}}{H}\right)^2 \frac{k_n}{(\omega^2 - 3k_n^2 v_T^2)} \tag{7-56}$$

In particular, for the one-dimensional plasma model described in Chapter 2, the two-point finite-difference approximation (FDA) for the derivative, the three point

FDA laplacian and the NGP assignment/interpolation function, respectively, give

$$(-i\hat{D}) = \frac{\sin kH}{H} = \frac{\sin\dfrac{kH}{2}\cos\dfrac{kH}{2}}{(H/2)} \tag{7-57}$$

$$(\epsilon_0 \hat{G}) = \frac{(H/2)^2}{\sin^2\dfrac{kH}{2}} \tag{7-58}$$

$$\frac{\hat{W}}{H} = \frac{\sin\dfrac{kH}{2}}{\dfrac{kH}{2}} \tag{7-59}$$

Therefore, Eq. (7-56) becomes

$$1 = \omega_p^2 \cot\frac{kH}{2} \sum_n \frac{\sin^2\dfrac{k_n H}{2}}{\dfrac{k_n H}{2}} \frac{1}{\omega^2 - 3k_n^2 v_T^2} \tag{7-60}$$

The sine term in the sum can be removed by using the result that

$$\sin^2\frac{k_n H}{2} = \sin^2\left(\frac{kH}{2} - n\pi\right) = \sin^2\frac{kH}{2} \tag{7-61}$$

giving

$$1 = \omega_p^2 \cos\frac{kH}{2} \sin\frac{kH}{2} \sum_{n=-\infty}^{\infty} \frac{1}{\dfrac{k_n H}{2}(\omega^2 - 3k_n^2 v_T^2)} \tag{7-62}$$

In the "cold" plasma limit $v_T \to 0$, the identity, Eq. (7-49) may be used on Eq. (7-62) to yield

$$\omega^2 = \omega_p^2 \cos^2\frac{kH}{2} \tag{7-63}$$

The hierarchy of assignment functions NGP ($p = 1$), CIC ($p = 2$), TSC ($p = 3$), etc.

$$\frac{\hat{W}(k)}{H} = \left(\frac{\sin\dfrac{kH}{2}}{\dfrac{kH}{2}}\right)^p \tag{7-64}$$

give the warm-plasma dispersion relation

$$1 = \omega_p^2(-i\hat{D})(\epsilon_0\hat{G})\left(\sin\frac{kH}{2}\right)^{2p} \frac{2}{H} \sum_n \frac{1}{\left(\dfrac{k_n H}{2}\right)^{2p-1}} \frac{1}{(\omega^2 - 3k_n^2 v_T^2)} \tag{7-65}$$

The identity
$$\frac{(-1)^s}{s!}\frac{d^s}{dx^s}\cot x = \sum_{n=-\infty}^{\infty}\frac{1}{(x-\pi n)^{s+1}} \tag{7-66}$$

enables Eq. (7-65) to be written in the cold-plasma limit as

$$\omega^2 = \omega_p^2(-i\hat{D})(\epsilon_0\hat{G})\frac{2}{H}\times\begin{cases}\cos\dfrac{kH}{2}\sin\dfrac{kH}{2} & \text{NGP}\\[2ex] \cos\dfrac{kH}{2}\sin\dfrac{kH}{2} & \text{CIC}\\[2ex] \cos\dfrac{kH}{2}\dfrac{\left(2+\cos^2\dfrac{kH}{2}\right)\sin\dfrac{kH}{2}}{3} & \text{TSC}\end{cases} \tag{7-67}$$

If a finite timestep DT is retained in the analysis, the warm-plasma dispersion relation, Eq. (7-65), is replaced by

$$1 = \omega_p^2(-i\hat{D})(\epsilon_0\hat{G})\left[\sin\frac{kH}{2}\right]^{2p}\left(\frac{2}{H}\right)\sum_n\sum_m\frac{1}{\left(\dfrac{k_nH}{2}\right)^{2p-1}}\frac{1}{(\omega_m^2-3k_n^2v_T^2)} \tag{7-68}$$

In the cold-plasma limit, the application of Eq. (7-66) to Eq. (7-68) leads to dispersion relations differing from Eq. (7-67) only in the replacement of ω^2 by $\sin^2(\omega DT/2)/(DT/2)^2$.

Closed trigonometric forms for the warm-plasma dispersion relation can be obtained by using the method of partial fractions on the expressions appearing in the alias sum of the dispersion relation, and applying Eq. (7-66) to evaluate the sums. For example, the dispersion relation for a warm plasma, with thermal velocity v_T, in the one-dimensional NGP model described in Chapter 2 is, for $DT = 0$,

$$1 = \omega_p^2\cos\frac{kH}{2}\sin\frac{kH}{2}\sum_n\frac{1}{\dfrac{k_nH}{2}[\omega^2-3k_n^2v_T^2]} \tag{7-69}$$

The term in the alias sum may be written as

$$\frac{y^2}{\omega^2}\frac{1}{x_n(y-x_n)(y+x_n)} = \frac{1}{\omega^2}\left[\frac{1}{x_n}+\tfrac{1}{2}\left(\frac{1}{x_n-y}+\frac{1}{x_n+y}\right)\right] \tag{7-70}$$

where
$$y = \frac{\omega H}{2\sqrt{3}v_T} \qquad x_n = \frac{k_nH}{2} \tag{7-71}$$

giving after some rearrangement

$$\omega^2 = \omega_p^2\left[\cos^2\frac{kH}{2}+\tfrac{1}{2}\left(\cot\frac{\omega H}{\sqrt{3}v_T}-\cot kH\right)^{-1}\right] \tag{7-72}$$

Summaries of the electrostatic dispersion relations for momentum-conserving and energy-conserving schemes are given in Tables 7-1 and 7-2, respectively. The

Table 7-1 Dispersion relations for momentum-conserving particle-mesh plasma models

$DT \neq 0$

$$1 = -(-i\hat{D})(\epsilon_0 \hat{G}) \frac{\omega_p^2}{n_0} \sum_{n=-\infty}^{\infty} \sum_{m=-\infty}^{\infty} \left(\frac{\hat{W}(k_n)}{H} \right)^2 \int_C \frac{\dfrac{\partial f^0}{\partial v}}{\omega_m - k_n v} \, dv$$

$DT = 0$

$$1 = -(-i\hat{D})(\epsilon_0 \hat{G}) \frac{\omega_p^2}{n_0} \sum_{n=-\infty}^{\infty} \left(\frac{\hat{W}(k_n)}{H} \right)^2 \int_C \frac{\dfrac{\partial f^0}{\partial v}}{\omega - k_n v} \, dv$$

$DT = 0$, *averaged force*

$$1 = -(-i\hat{D})(\epsilon_0 \hat{G}) \frac{\omega_p^2}{n_0} \left(\frac{\hat{W}(k)}{H} \right)^2 \int_C \frac{\dfrac{\partial f^0}{\partial v}}{\omega - k v} \, dv$$

Warm plasma, $DT \neq 0$

$$1 = (-i\hat{D})(\epsilon_0 \hat{G}) \omega_p^2 \sum_n \sum_m \left(\frac{\hat{W}(k_n)}{H} \right)^2 \frac{k_n}{(\omega_m - k_n v_0)^2 - 3k_n^2 v_T^2}$$

$$\int_C \frac{\dfrac{\partial f^0}{\partial v}}{\omega_m - k_n v} \, dv = P \int_{-\infty}^{\infty} \frac{\dfrac{\partial f^0}{\partial v}}{\omega_m - k_n v} - \frac{i\pi}{|k_n|} \frac{\partial f^0}{\partial v} \bigg|_{\omega_m/k_n}$$

$$k_n = k - n \frac{2\pi}{H}$$

$$\omega_m = \omega - m \frac{2\pi}{DT}$$

v_0 = drift velocity

v_T = thermal velocity

averaged-force dispersion relation is that which obtains when the displacement averaged mesh force is used, the averaging causing all the $n \neq 0$ terms in the alias sum to disappear (cf. Sec. 5-6-4).

The averaged-force dispersion relations can be written as

$$1 = -\frac{\omega_p^2 k}{n_0 k^2} \hat{S}_e^2(k) \int \frac{\dfrac{\partial f^0}{\partial v}}{\omega - kv} \, dv \tag{7-73}$$

where for momentum-conserving schemes

$$\hat{S}_e^2(k) = \left(-\frac{i\hat{D}}{k} \right) (\epsilon_0 \hat{G} k^2) \left(\frac{\hat{W}}{H} \right)^2 \tag{7-74}$$

Table 7-2 Dispersion relations for energy-conserving particle-mesh plasma models

DT $\neq 0$

$$1 = -(\epsilon_0 \hat{G}) \frac{\omega_p^2}{n_0} \sum_{n=-\infty}^{\infty} \sum_{m=-\infty}^{\infty} \left(\frac{\hat{W}(k_n)}{H} \right)^2 k_n \int_C \frac{\frac{\partial f^0}{\partial v}}{\omega_m - k_n v} dv$$

DT $= 0$

$$1 = -(\epsilon_0 \hat{G}) \frac{\omega_p^2}{n_0} \sum_{n=-\infty}^{\infty} \left(\frac{\hat{W}(k_n)}{H} \right)^2 k_n \int_C \frac{\frac{\partial f^0}{\partial v}}{\omega - k_n v} dv$$

DT $= 0$, *averaged force*

$$1 = -(\epsilon_0 \hat{G}) \frac{\omega_p^2}{n_0} \left(\frac{\hat{W}(k_n)}{H} \right)^2 k \int_C \frac{\frac{\partial f^0}{\partial v}}{\omega - k v} dv$$

Warm plasma, DT $\neq 0$

$$1 = (\epsilon_o \hat{G}) \omega_p^2 \sum_n \sum_m \left(\frac{\hat{W}(k_n)}{H} \right)^2 \frac{k_n^2}{(\omega_m - k_n v_0)^2 - 3k_n v_T^2}$$

and for energy-conserving schemes

$$\hat{S}_e^2(k) = (\epsilon_0 \hat{G} k^2) \left(\frac{\hat{W}}{H} \right)^2 \tag{7-75}$$

$S(x) \supset \hat{S}(k)$ can be interpreted as the effective finite-sized particle shape of the particles in the simulation models. As the work of Birdsall et al. (1970), Langdon and Birdsall (1970), and Okuda and Birdsall (1970) has shown, if the width associated with $S(x)$ is of the same order as the Debye length, then the properties of the finite-size particle model correspond closely to those of the exact collisionless system.

The dispersion relation for the warm streaming plasma Eq. (7-56), with ω replaced by $(\omega - k_n v_0)$, where v_0 is the streaming velocity, has been numerically evaluated and confirmed by results of computer simulation by Birdsall and Maron (1980). They show that nonphysical instabilities present in a cold streaming plasma (Langdon, 1970*a*) saturate when the ratio of the Debye length to the mesh spacing, λ_D/H, becomes sufficiently large, and do not appear if the initial thermal spread of velocities is sufficiently large. Figure 7-1 summarizes their results. If a simulation is started initially in the region marked "unstable," nonphysical growth in the energy leads to a temperature increase until the system has moved to the region marked "stable." If the initial conditions of a simulation are above the marginal curve, then the instability is absent. Unlike instabilities arising from

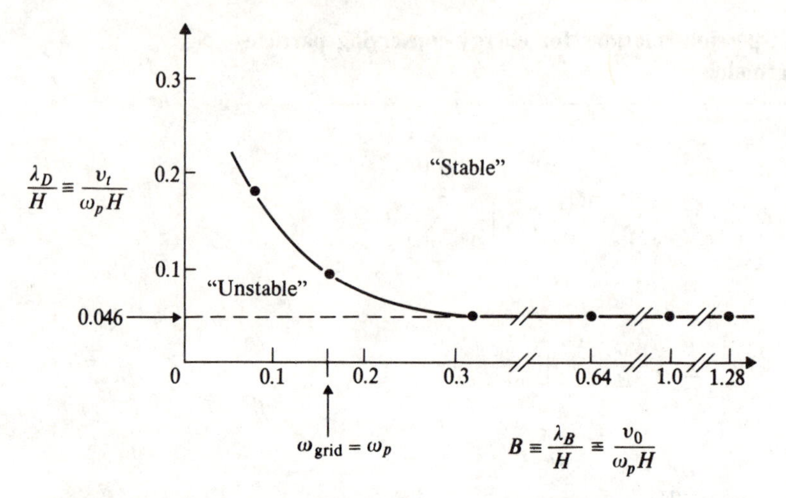

Figure 7-1 Experimental determination of the thermal spread needed for stability of a beam in a momentum-converving plasma simulation program. (*After Birdsall and Maron, 1980, Fig. 10, courtesy of Journal of Computational Physics.*)

taking too large a timestep ($DT > 2/\omega_p$, cf. Chapter 4), this instability, arising from taking $H \gg \lambda_D$, is benign in that its effect is to move the initial state towards one which can be represented adequately by a discrete mesh of values.

7-3-4 Mode Coupling

Langdon (1970*b*) numerically solved Eq. (7-35) and the corresponding averaged-force dispersion relation (Table 7-1) for f^0 Maxwellian and $\hat{D}$, $\hat{G}$, and $\hat{W}$ given by Eqs. (7-57)–(7-59). The results of his computations are summarized in Figs. 7-2 and 7-3. Figure 7-2 shows the real and imaginary parts of the frequency as functions of wavenumber k in the principal zone for $\lambda_D = H$. In this instance, the inclusion of the $n \neq 0$ terms in the alias sum has negligible impact except for very short wavelengths. The situation is drastically different in Fig. 7-3 where the cell width H is ten times the Debye length. Both the averaged-force and the exact real part of the frequency show the same qualitative features, giving a group velocity $\partial\omega/\partial k$ which is of opposite sign to that obtained for a Vlasov plasma ($\partial\omega/\partial k \simeq 6k\lambda_D^2$). More striking is the imaginary part of the frequency, where weak Landau damping at relatively long wavelengths is changed by the alias contributions to instability. Thus, if $\lambda_D \ll H$, the behavior of the simulation plasma is qualitatively as well as quantitatively incorrect, since physically a maxwellian plasma is stable. The presence of this instability and its growth rate have been verified in simulations by Okuda (1970, 1972*a*).

The nonlinear outcome of the instability is similar to that found for the cold beam. The instability causes a nonphysical heating of the plasma, increasing λ_D to a point where the aliases are relatively unimportant. Both examples should be regarded as cautionary: Although the nonphysical instabilities are not cata-

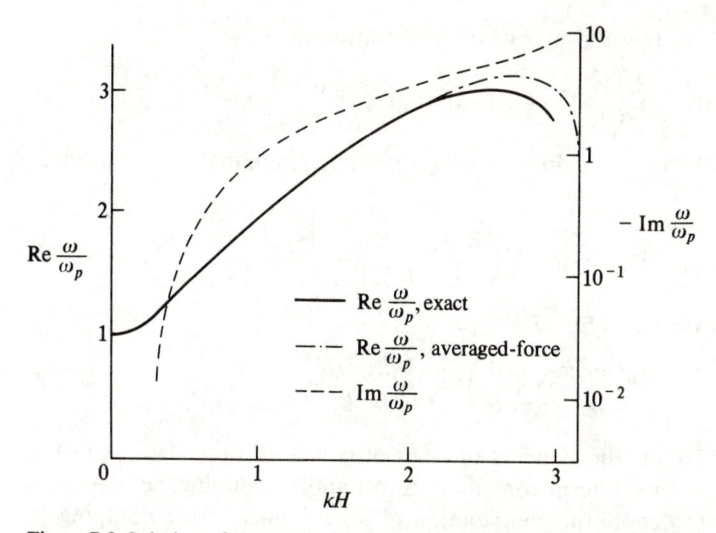

Figure 7-2 Solution of the exact and the averaged-force dispersion relations for a Maxwell velocity distribution $\lambda_D = H$ in a one-dimensional NGP simulation model. (*After Langdon, 1970b, Fig. 5, courtesy of Journal of Computational Physics.*)

strophic, they can be misleading since they manifest themselves in a form very similar to physical instabilities which may be the object of study.

The cause of the nonphysical spatial mesh instabilities is, as mentioned above, the nonphysical coupling of modes which arises from the undersampling of the density distribution by the mesh. If we approximate the principal part of the

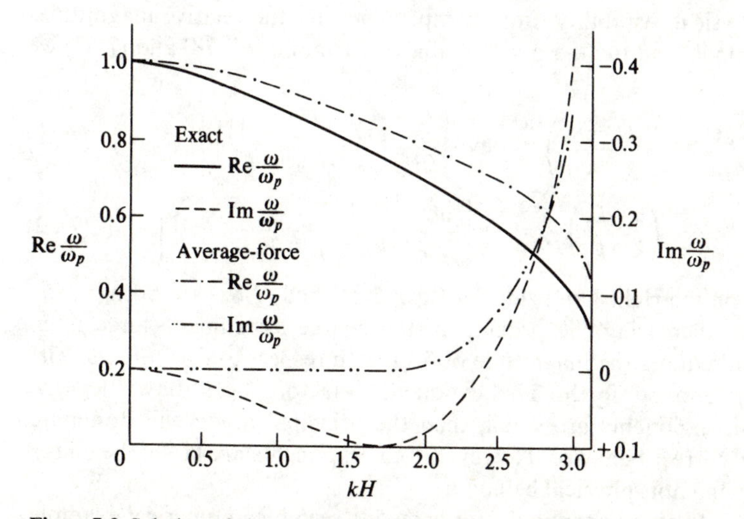

Figure 7-3 Solution of the exact and averaged-force dispersion relations with $10\lambda_D = H$ in a one-dimensional NGP model. (*After Langdon, 1970b, Fig. 6, courtesy of Journal of Computational Physics.*)

integral in Eq. (7-35) by the cold-plasma approximation then

$$1 = (-i\hat{D})(\epsilon_0 \hat{G})\omega_p^2 \sum_n \left(\frac{\hat{W}}{H}\right)^2 \frac{k_n}{\omega^2} + (-i\hat{D})(\epsilon_0 \hat{G}) \frac{\omega_p^2}{n_0} \sum_n \left(\frac{\hat{W}}{H}\right)^2 \frac{i\pi}{|k_n|} \frac{\partial f^0}{\partial v}\bigg|_{\omega/k_n} \tag{7-76}$$

Multiplying Eq. (7-76) by ω^2, writing $\omega = \omega_r + i\omega_i$, and assuming $|\omega_i| \ll |\omega_r|$ gives the imaginary part

$$\omega_i = (-i\hat{D})(\epsilon_0 \hat{G})\omega_p^2 \sum_n \left(\frac{\hat{W}(k_n)}{H}\right)^2 \frac{\pi}{2} \frac{\omega_r}{|k_n|} \frac{1}{n_0} \frac{\partial f^0}{\partial v}\bigg|_{\omega_p/k_n} \tag{7-77}$$

For a stationary maxwellian [Eq. (7-52)],

$$\omega_i = -\frac{\epsilon_0 \hat{G} \omega_p^2 \pi}{\sqrt{2\pi v_T^3 2}} \sum_n \left(\frac{\hat{W}}{H}\right)^2 \frac{-i\hat{D}(k)}{k_n} \frac{\omega_r}{|k_n|} \exp \frac{-\omega_r^2}{2k_n^2 v_T^2} \tag{7-78}$$

Since the sign of $(-i\hat{D})$ is the same as the sign of k, and all other factors in Eq. (7-78) are positive, the $n = 0$ term contribution to Landau damping, i.e., $\delta_0\omega_i < 0$ where $\delta_n\omega_i$ is used to denote the contribution of the nth alias to the damping or growth rate (proportional to $e^{\omega_i t}$). Similarly, since

$$\operatorname{sgn}(-i\hat{D}) = \operatorname{sgn}(k + |n|k_g) \tag{7-79}$$

all alias terms with $n < 0$ contribute to damping. However, for all positive aliases the sign reverses

$$\operatorname{sgn}(-i\hat{D}) = -\operatorname{sgn}(k - nk_g) \tag{7-80}$$

since $|k| < k_g/2$ and $n > 0$. Therefore all positive aliases give a positive contribution to ω_i.

If the sum of the positive aliases is greater in magnitude than the sum of the contributions of the principal term and negative aliases, then ω_i will be positive, indicating nonphysical instability. For example, consider the relative magnitudes of the principal mode, and the $+1$ and -1 aliases. From Eqs. (7-78) and (7-64) we find

$$\frac{\delta_0\omega_i}{\delta_1\omega_i} = -\left(\frac{k-k_g}{k}\right)^{2p+2} \exp\left[\frac{\omega_r^2}{2v_T^2}\left(\frac{1}{(k-k_g)^2} - \frac{1}{k^2}\right)\right] \tag{7-81}$$

$$\frac{\delta_1\omega_i}{\delta_{-1}\omega_i} = -\left(\frac{k+k_g}{k-k_g}\right)^{2p+2} \exp\left[\frac{\omega_r^2}{2v_T^2}\left(\frac{1}{(k+k_g)^2} - \frac{1}{(k-k_g)^2}\right)\right] \tag{7-82}$$

The first factors on the right-hand sides of Eqs. (7-81) and (7-82) are both greater than one. Hence, increasing the order p of the charge assignment–force interpolation function reduces the magnitude of $\delta_1\omega_i$ with respect to $\delta_0\omega_i$, $\delta_{-1}\omega_i$ with respect to $\delta_1\omega_i$, and so forth. The exponential factors both have negative arguments: If the exponents are small, then the principal mode will dominate, giving physically correct behavior. However, if the exponents are large, the aliases will dominate, giving non-physical behavior.

The influence of the exponential factor is readily seen by a numerical example. If we take an NGP scheme ($p = 1$), approximate ω_r by the plasma frequency ω_p, and take a wavenumber in the middle of the principal zone $k = k_g/4 = \pi/2H$, then Eqs. (7-81) and (7-82) become

$$\frac{\delta_0 \omega_i}{\delta_1 \omega_i} = -(3)^4 \exp\left[-\frac{16}{9\pi^2}\left(\frac{H}{\lambda_D}\right)^2\right] \tag{7-83}$$

and

$$\frac{\delta_1 \omega_i}{\delta_{-1} \omega_i} = -\left(\frac{5}{3}\right)^4 \exp\left[-\frac{32}{225\pi^2}\left(\frac{H}{\lambda_D}\right)^2\right] \tag{7-84}$$

If $\lambda_D \gtrsim H$, the exponentials in Eqs. (7-83) and (7-84) are approximately equal to one, and so the relative strengths of the principal and alias modes are determined

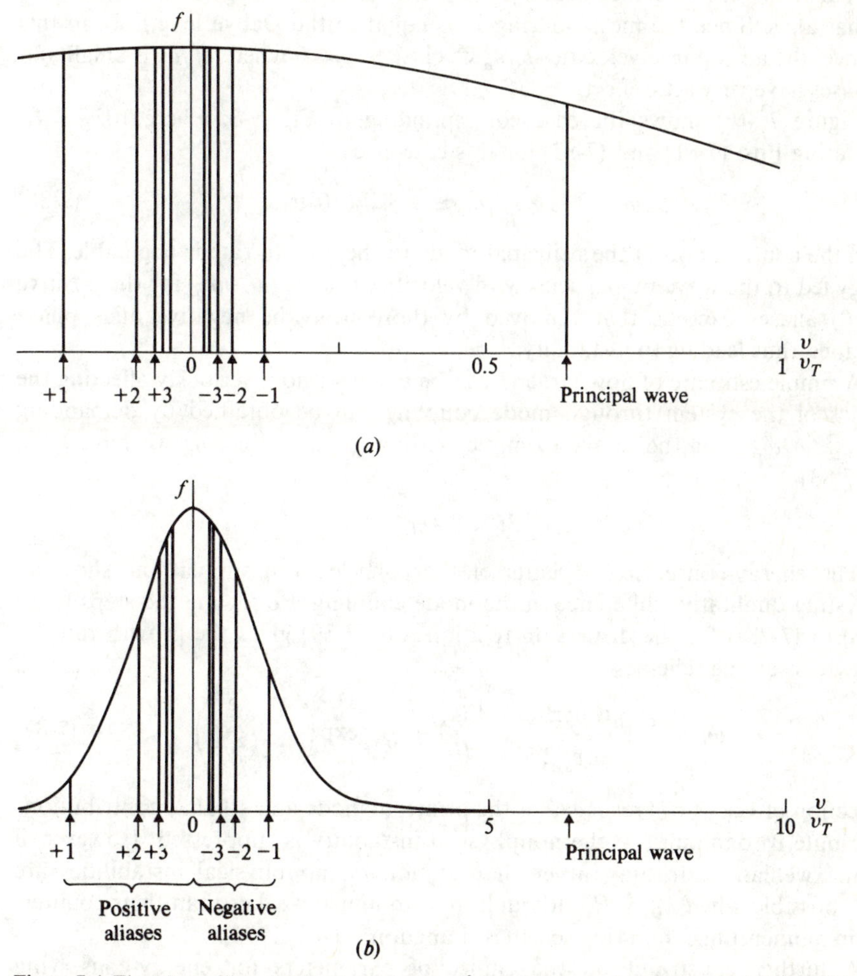

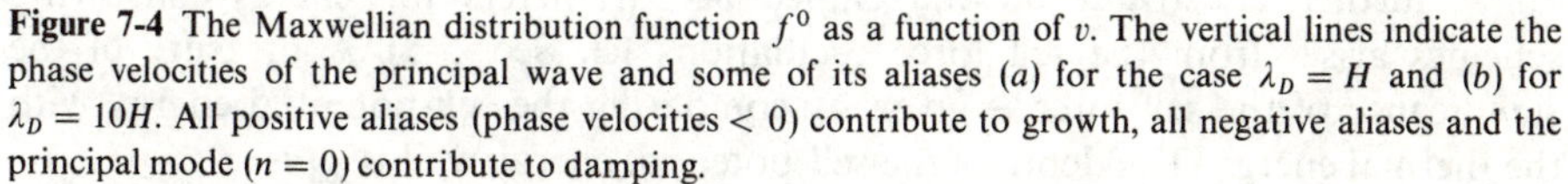

Figure 7-4 The Maxwellian distribution function f^0 as a function of v. The vertical lines indicate the phase velocities of the principal wave and some of its aliases (a) for the case $\lambda_D = H$ and (b) for $\lambda_D = 10H$. All positive aliases (phase velocities < 0) contribute to growth, all negative aliases and the principal mode ($n = 0$) contribute to damping.

almost solely by the charge assignment function: The closer $\hat{W}$ is to being band-limited, the weaker the aliases become.

For $H = \lambda_D$, corresponding to the case shown in Fig. 7-2,

$$\delta_0 \omega_i = -68\delta_1 \omega_i = 515\delta_{-1}\omega_i \tag{7-85}$$

A physical interpretation of the mode coupling for this case can be obtained from Fig. 7-4(*a*). The principal mode with phase velocity ω/k is Landau-damped, as is its physical counterpart. The mesh sees the wavelengths $k_n = k - nk_g$ as k (cf. Sec. 5-6-1), so particles with phase velocities $\omega/(k - nk_g)$ are also in resonance with the wave of phase velocity ω/k, and if the sign of $(\partial f/\partial v)|_{\omega/k_n}$ differs from that of ω/k, then particles with velocities near ω/k_n will lead to a net energy flow to the wave for that alias. Since the mesh spacing H is equal to the Debye length λ_D in this instance, the alias phase velocities ω/k_n lie close to $v = 0$ where $\partial f/\partial v$ is small and so aliases have very little effect.

Figure 7-4(*b*) shows the case corresponding to Fig. 7-3, where $10\lambda_D = H$. Evaluating Eqs. (7-81) and (7-82) for this case gives

$$\delta_1 \omega_i = -1.8\delta_{-1}\omega_i \simeq -8.2 \times 10^5 \delta_0 \omega_i \tag{7-86}$$

Now, the contribution of the principal mode to the growth rate is negligible. The energy fed to the wave by particles with velocities near $\omega/(k - nk_g)$ for the positive $(n > 0)$ aliases exceeds that removed by those near the negative alias phase velocities, thus leading to instability.

A simple estimate of how large H can be taken without seriously affecting the physics of the system through mode coupling can be obtained by demanding $|\delta_0 \omega_i| \gg |\delta_1 \omega_i|$. For the above example, setting $|\delta_0 \omega_i| > 10|\delta_1 \omega_i|$ we have from Eq. (7-83)

$$H < 3.4\lambda_D$$

The energy-conserving variants of the particle-mesh simulations show an interesting qualitative difference in the mode coupling. Following the steps [Eqs. (7-76) to (7-78)] for the dispersion relation, Eq. (7-39) gives the growth rate for energy-conserving schemes

$$\omega_i = -\frac{\epsilon_0 \hat{G}\omega_p^2 \pi}{\sqrt{2\pi v_T^3 2}} \sum_n \left(\frac{\hat{W}(k_n)}{H}\right)^2 \frac{\omega_r}{|k_n|} \exp\left(-\frac{\omega_r^2}{2k_n^2 v_T^2}\right) \tag{7-87}$$

Inspection of Eq. (7-87) reveals that the principal mode and all alias contributions contribute to damping, so the nonphysical instability is suppressed. However, if the maxwellian is drifting rather than stationary, nonphysical instabilities are again possible when $\lambda_D \ll H$, although they are much weaker than their counterpart in momentum conserving schemes (Langdon, 1970*b*, 1973).

A further constraint on the choice of parameters for energy-conserving schemes arises from the self force oscillations (cf. Sec. 5-5). A measure of the significance of the self force is given by comparing the self-potential energy with the thermal energy. The depth of the self-potential energy well W_{self} is

$$W_{\text{self}} = \frac{m}{2}\omega_{\text{self}}^2\left(\frac{H}{2}\right)^2 = \frac{m}{2}\frac{\omega_p^2}{n_0 H}\left(\frac{H}{2}\right)^2 \tag{7-88}$$

Hence
$$\frac{W_{\text{self}}}{\frac{1}{2}mv_T^2} = \frac{1}{4}\left(\frac{H}{\lambda_D}\right)^2 \frac{1}{(n_0 H)} \tag{7-89}$$

Thus, if $H \gg \lambda_D$ is used in energy-conserving schemes, the number of particles per cell $n_0 H$ must also be large if the self-potential energy is to be insignificant. Therefore, we conclude that for both energy- and momentum-conserving schemes we require the cell width H to be less than or equal to the characteristic length λ_D for the dispersive properties of the collisionless simulations to adequately model the dispersive properties of the physical system.

7-4 FINITE MULTIDIMENSIONAL SYSTEMS

The discussion of dispersive properties in the previous section and of the interparticle force in Sec. 5-6 have been presented for a one-dimensional infinite system. The results we have obtained apply, with the addition of a restriction on the allowable values of wavenumber, to finite periodic systems. This restriction is that the wavenumbers k are limited to a discrete set of equispaced values $k = lk_0$, where l is integer, $k_0 = 2\pi/L$, and L is the periodic box length. Furthermore, if we take periodic systems of length $2L$ and restrict functions to be odd (even) in that period then the results can be applied systems of length L with homogeneous Dirichlet (Neumann) boundary conditions.

Generalization of the finite one-dimensional results to two and three dimensions is also straightforward. The only significant change introduced by extra dimensions is that of directional dependence of the force, dispersive properties, etc.

7-4-1 Periodicity

One method of showing that the results for periodic systems differ from those of an infinite system only in the restriction on the allowable wavenumbers is to repeat the analysis from scratch, using $FS(i)$ in place of FT and FFT in place of $FS(ii)$ (see Table A-1). Symbolically, the change can be represented by

$$\int \frac{dk}{2\pi} \rightarrow \frac{1}{L}\sum_l = \sum_l \frac{k_0}{2\pi} \tag{7-90}$$

where l is the integer denoting the value of wavenumber, and takes values from $-N/2$ to $N/2-1$ in the principal zone. N is the number of mesh points used in the period length L in x space.

The correspondence between results for periodic and infinite systems can be demonstrated by using the sampling function Ш. A periodic system can be regarded as the superposition of a set of infinite systems, each one shifted by the periodic length L with respect to the previous one: One example we have already referred to is the method of images in electrostatics. The process of shifting and replicating, as we showed in Chapter 5, can be described by convolution with Ш.

If we let $f(x) \supset \hat{f}(k)$ be the function in the infinite system, then its periodic counterpart is $g(x)$ given by

$$g(x) = \frac{1}{L} Ш\left(\frac{x}{L}\right) * f(x) \tag{7-91}$$

$$= \frac{1}{L} \int \sum_n \delta\left(\frac{x' - nL}{L}\right) f(x - x')\, dx'$$

$$= \sum_{n=-\infty}^{\infty} f(x - nL) \tag{7-92}$$

which is clearly periodic with period L and is given by summing displaced "images" of $f(x)$. Now by Eq. (A-29) and the similarity theorem (Table A-4) we have

$$\frac{1}{L} Ш\left(\frac{x}{L}\right) \supset Ш\left(\frac{k}{k_0}\right) \tag{7-93}$$

Therefore,
$$\hat{g}(k) = Ш\left(\frac{k}{k_0}\right) \hat{f}(k) \tag{7-94}$$

To reduce $g(x)$ to a single period length, we multiply it by $\Pi(x/L)$, i.e.,

$$g'(x) = \Pi\left(\frac{x}{L}\right) g(x) \tag{7-95}$$

where $g'(x) \supset \hat{g}'(k) = L \operatorname{sinc}\left(\frac{k}{k_0}\right) * Ш\left(\frac{k}{k_0}\right) \hat{f}(k) = \sum_l \operatorname{sinc}\left(\frac{k - lk_0}{k_0}\right) \hat{f}(lk_0)$ (7-96)

Comparing Eq. (7-96) with Eq. (5-183), we recognize the form of Eq. (7-96) as that of a continuous function $\hat{g}(k)$, which is equivalent to the set of values $\{\hat{f}(lk_0)\}$, i.e., given some function $f \supset \hat{f}$ for the infinite system, the transform of its periodic counterpart is given by $\hat{f}(lk_0)$; l integer. The relationships between FT, FS, and FFT are summarized in Figs. A-2 and A-3.

7-4-2 Two and Three Dimensions

The Fourier transforms along the x and y directions in two dimensions and the x, y, and z directions in three dimensions are independent of each other. Consequently, the results for two and three dimensions can be obtained by repeating the steps of the one-dimensional analysis, replacing scalars by vectors and sums by double or triple sums, as appropriate. The rules for the changes are summarized in Appendix A, Sec. A-6.

To illustrate the manipulation of multidimensional transforms, let us now repeat the derivation of the electrostatic plasma dispersion relation, Eq. (7-34), for two- and three-dimensional systems. We again employ the compact notation introduced in Sec. 5-6; primes denote continuous functions corresponding to mesh values and daggers denote generalized functions representing the sets of mesh values. Apart from the charge density ρ', all primed quantities are constructed from mesh values in such a manner as to have band-limited transforms (cf. Sec. 5-6-1). The equations appearing in the derivation apply equally well to periodic or to infinite systems, the only difference in the two instances being the definitions of (1) the transform pairs (see Table A-1), (2) the convolution integral (see Table A-3), and (3) the sampling function (see Sec. A-4-5).

Charge density

$$\rho'(\mathbf{x}) = \frac{q}{V_c} W(\mathbf{x}) * n(\mathbf{x}) \supset \hat{\rho}'(\mathbf{k}) = \frac{q}{V_c} \hat{W}(\mathbf{k})\hat{n}(\mathbf{k}) \tag{7-97}$$

$$\begin{aligned} \rho\dagger(\mathbf{x}) = \text{Ш}(\mathbf{x};\mathbf{H})\rho'(\mathbf{x}) \supset \hat{\rho}(\mathbf{k}) &= V_c \text{Ш}(\mathbf{k};\mathbf{k}_g) * \hat{\rho}'(\mathbf{k}) \\ &= \frac{q}{V_c}\sum_{\mathbf{n}} \hat{W}(\mathbf{k_n})\hat{n}(\mathbf{k_n}) \end{aligned} \tag{7-98}$$

Potential

$$\phi'(\mathbf{x}) = G'(\mathbf{x}) * \rho\dagger(\mathbf{x}) \supset \hat{\phi}'(\mathbf{k}) = \hat{G}'(\mathbf{k})\hat{\rho}(\mathbf{k}) \tag{7-99}$$

$$\phi\dagger(\mathbf{x}) = \text{Ш}(\mathbf{x};\mathbf{H})\phi'(\mathbf{x}) \supset \hat{\phi}(\mathbf{k}) = \hat{G}(\mathbf{k})\hat{\rho}(\mathbf{k}) \tag{7-100}$$

Electric field

$$\mathbf{E}'(\mathbf{x}) = -\mathbf{D}(\mathbf{x}) * \phi'(\mathbf{x}) \supset \hat{\mathbf{E}}'(\mathbf{k}) = -\hat{\mathbf{D}}(\mathbf{k})\hat{\phi}'(\mathbf{k}) \tag{7-101}$$

$$\mathbf{E}\dagger(\mathbf{x}) = \text{Ш}(\mathbf{x};\mathbf{H})\mathbf{E}'(\mathbf{x}) \supset \hat{\mathbf{E}}(\mathbf{k}) = -\hat{\mathbf{D}}(\mathbf{k})\hat{\phi}(\mathbf{k}) \tag{7-102}$$

Force

$$\mathbf{F}(\mathbf{x}) = \frac{q}{V_c} W(\mathbf{x}) * \mathbf{E}\dagger(\mathbf{x}) \supset \hat{\mathbf{F}}(\mathbf{k}) = \frac{q}{V_c} \hat{W}(\mathbf{k})\hat{\mathbf{E}}\dagger(\mathbf{k}) \tag{7-103}$$

In three dimensions

$$V_c = H_1 H_2 H_3 = \text{cell volume} \tag{7-104}$$

$$\mathbf{H} = (H_1, H_2, H_3) = \text{cell vector} \tag{7-105}$$

$$\text{Ш}(\mathbf{x};\mathbf{H}) = \text{Ш}\left(\frac{x}{H_1}\right)\text{Ш}\left(\frac{y}{H_2}\right)\text{Ш}\left(\frac{z}{H_3}\right) \tag{7-106}$$

$$\mathbf{n} = (n_1, n_2, n_3) = \text{alias number} \qquad n_i \text{ integer} \tag{7-107}$$

$$\mathbf{k}_g = 2\pi\left(\frac{1}{H_1}, \frac{1}{H_2}, \frac{1}{H_3}\right) = \text{grid wavenumber} \tag{7-108}$$

$$\mathbf{k_n} = \mathbf{k} - 2\pi\left(\frac{n_1}{H_1}, \frac{n_2}{H_2}, \frac{n_3}{H_3}\right) \tag{7-109}$$

The one-dimensional sampling functions on the right-hand side of Eq. (7-106) are given by Eq. (A-28) for infinite systems or Eq. (A-31) for periodic systems. If, as is almost invariably the case, $\hat{W}(\mathbf{k})$ is a product function then

$$\hat{W}(\mathbf{k}) = \hat{w}(k_1)\hat{w}(k_2)\hat{w}(k_3) \tag{7-110}$$

where $\hat{w}(k_i)$ is the transform of the one-dimensional assignment function for component i. Two-dimensional expressions corresponding to Eqs. (7-104) to (7-110) are obtained by dropping the z (or 3-) components, i.e., $V_c = H_1 H_2$, etc.

The vector equivalent of Eq. (7-27) is

$$\hat{n}(\mathbf{k}) = -\frac{i}{m}\int \frac{\hat{\mathbf{F}}(\mathbf{k})\cdot\dfrac{\partial f^0}{\partial \mathbf{v}}}{(\omega - \mathbf{k}\cdot\mathbf{v})}\,d\mathbf{v} \tag{7-111}$$

which, upon elimination of $\hat{n}$, $\hat{\mathbf{F}}$, $\hat{\rho}$, $\hat{\phi}$, and $\hat{\mathbf{E}}$ using Eqs. (7-97) to (7-103), yields the result

$$1 = -(\epsilon_0 \hat{G})\frac{\omega_p^2}{n_0}\sum_{\mathbf{n}}\left(\frac{\hat{W}(\mathbf{k_n})}{V_c}\right)^2 \int_L \frac{-i\hat{\mathbf{D}}(\mathbf{k})\cdot\dfrac{\partial f^0}{\partial \mathbf{v}}}{(\omega - \mathbf{k_n}\cdot\mathbf{v})}\,d\mathbf{v} \tag{7-112}$$

7-5 COLLISIONS

In Sec. 7-3 we saw that if the cell size is of the same order as the Debye length, then the dispersive properties of computer simulation model closely follow those of a finite-sized particle plasma, where the transforms of the effective particle shape $\hat{S}_e$ are given by Eq. (7-74) or (7-75). Thus, provided one works within the range of parameters where the averaged-force dispersion relation approximation is good, one would expect the averaged-force description of collision processes also to apply: This premise is, in fact, borne out by experience.

Collision processes in the (averaged-force) finite-size particle approximation have been studied extensively by Okuda and Birdsall (1970), Langdon and Birdsall (1970), Birdsall et al. (1970). In those papers, the authors show that the kinetic equation for a point particle plasma (Lenard, 1960):

$$\left(\frac{\partial f}{\partial t}\right)_c = \frac{\partial}{\partial \mathbf{v}}\int d^3v' \frac{1}{8\pi^2}\int d^3k\,\delta(\mathbf{k}\cdot\mathbf{v} - \mathbf{k}\cdot\mathbf{v}')\frac{\mathbf{k}\mathbf{k}\hat{\Phi}(\mathbf{k})}{|\epsilon(\mathbf{k}\cdot\mathbf{v},\mathbf{k})|^2}\cdot \left(\frac{\partial f}{\partial \mathbf{v}}(\mathbf{v})f(\mathbf{v}') - \frac{\partial f(\mathbf{v}')}{\partial \mathbf{v}}f(\mathbf{v})\right) \tag{7-113}$$

is modified only in the replacement of the transform of the pair potential energy $\hat{\Phi}(\mathbf{k})$ and the plasma dielectric function $\epsilon(\omega = \mathbf{k}\cdot\mathbf{v}, \mathbf{k})$ (the zeros of which give the dispersion relation) by their finite-sized particle counterparts:

$$\hat{\Phi}(\mathbf{k}) = \frac{q^2\hat{S}_e^2(\mathbf{k})}{\epsilon_0 k^2} \tag{7-114}$$

$$\epsilon(\omega, \mathbf{k}) = 1 + \frac{\omega_p^2}{n_0 k^2} \hat{S}_e^2(\mathbf{k}) \int \mathbf{k} \cdot \frac{\dfrac{\partial f}{\partial \mathbf{v}}}{\omega - \mathbf{k} \cdot \mathbf{v}} \, d\mathbf{v} \tag{7-115}$$

Equation (7-113) can be written in the form of Eq. (7-1) by separating the first- and second-derivative terms. The resulting diffusion and drag coefficients for a maxwellian distribution $f(\mathbf{v}')$ and gaussian-shaped clouds $[\hat{S}_e(\mathbf{k}) = \exp(-2k^2R^2)]$ have been numerically evaluated by Okuda and Birdsall (1970) for both two- and three-dimensional systems: They show that cloud width R of H to $2H$ leads to reduction of the magnitude of these coefficients by one to two orders of magnitude below corresponding values for point particles, with further reductions ensuing from increased R.

Mathematically, the reduction in the diffusion and drag coefficients can be identified as arising from the factor $\hat{S}_e^4(\mathbf{k})$ which is introduced into Eq. (7-113) by the finite particle width. Indeed, this factor eliminates the divergence problems at large k encountered in the point particle calculation. It follows that if $\hat{S}_e(\mathbf{k})$ is made to decay more rapidly (corresponding to large values of R), then the value of the diffusion coefficients will decrease.

A simple physical picture of the effect of finite particle size can be obtained by considering the motion of a test particle through a plasma which has a relatively small number of particles in a Debye sphere. The orbit of the test particle in a "smeared" out (i.e., ensemble-averaged) collisionless plasma is a straight line. In the point particle plasma, the test particle sees a strongly fluctuating field from the small number of particles in the Debye sphere around it, and so will suffer many strong perturbations, causing it to deviate rapidly (correspondingly to a large velocity diffusion) from the collisionless straight-line orbit. Adding a finite size eliminates the strong scattering of short-range encounters. If the size and number of particles is sufficiently large for many of them to overlap the test particle, then the test particle will "see" the finite-size particle background as a uniform medium, i.e., a collisionless plasma.

A limit to the maximum acceptable radius R of the finite-size particles is set by the collective properties. When R is increased much beyond the Debye length it takes over the role of the Debye length, causing collective properties to become significantly different from those of the physical system: This change is dramatically illustrated by the dispersion curves shown in Figs. 7-2 and 7-3.

For the purposes of simulating collisionless systems, be they plasmas or gravitating systems, it is sufficient to know whether or not the simulation model will behave as a collisionless system for the timespan of the calculation. Even in systems where collisional processes are added on to the collisionless motions, as is done in the semiconductor simulations (cf. Chapter 10), it is important that the physical collision processes dominate the numerical ones. A simple estimate of the time for which a simulation can be run before collisional effects become important is given by the collision time τ_c. Experimental determination of τ_c (see Sec. 9-2-3) shows that, in the range of particle sizes used in plasma simulation models, the collision time in two-dimensional models closely follows the empirical law

$$\frac{\tau_c}{\tau_p} = N_D\left[1+\left(\frac{W}{\lambda_D}\right)^2\right] = \frac{\omega_p/2\pi}{\nu_c} \tag{7-116}$$

where $\omega_p = 2\pi/\tau_p$ = plasma frequency
ν_c = collision frequency
N_D = number of particles in a Debye square $(= n\lambda_D^2)$
$W(=2R)$ = particle width

The empirical relationship (Hockney, 1971, Hockney et al. 1974) summarized by Eq. (7-116) shows good agreement with theoretical collision times obtained by Okuda and Birdsall (1970) for the finite-sized plasma model. Figure 7-5 shows the computed values of cross-sections found for uniform-density finite-sized particles. Since the collision frequency is approximately given by $\nu \sim nv_T\sigma$, Fig. 7-5 may be used to estimate the changes in collision time brought about by varying the particle width. Similar collision frequency estimates by Langdon and Birdsall (1970) for effective particle shapes $S_e(\mathbf{x}) \supset \hat{S}_e(k) = \exp[-k^2R^2/2]$ and

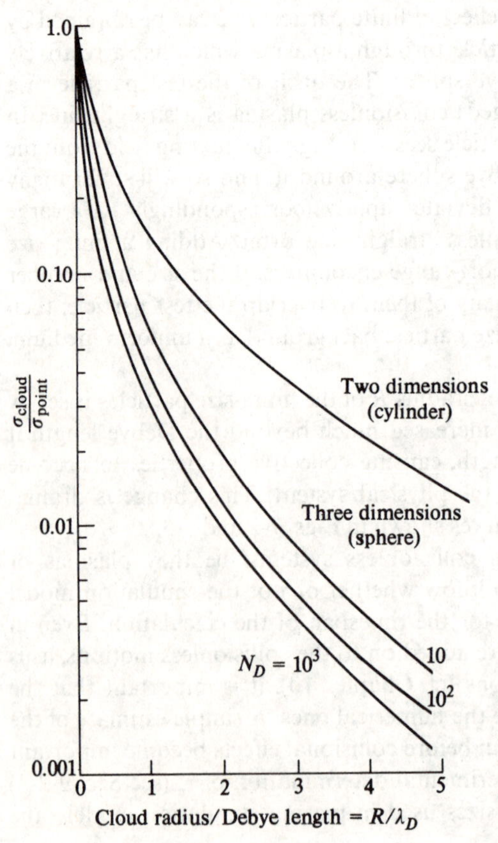

Figure 7-5 Cross sections for two- and three-dimensional uniform-density finite-size particles normalized to the point particle cross section. (*After Okuda and Birdsall, 1970, Fig. 7, courtesy of Physics of Fluids.*)

$S_e \supset \hat{S}_e(k) = \Pi(2k/R)$ show that the collision frequency is insensitive to particle shape in the range of particle radii R of practical importance around $R = \lambda_D$.

Further experimental results on collisional effects in finite-sized particle plasmas (Orens et al., 1970; Okuda, 1972*b*; Matsuda and Okuda, 1975) bear out the results of the averaged-force finite-sized particle kinetic theory. A more general kinetic theory of particle-mesh simulation models which includes both the aliasing effects of the finite spatial mesh and finite timestep has been developed by Langdon (1970*a*, 1979*a,b*). We refer readers to those articles for further details of the general kinetic theory of collisionless plasma simulation models.

7-6 CONSERVATION LAWS

The majority of particle simulation models correspond to conservative physical systems, i.e., to physical systems which conserve energy, momentum, and angular momentum. The presence of discretization errors in the simulation models causes some or all of the conservation laws to become only approximate. In practical computations, the physically conserved quantities are usually monitored continuously to provide global checks of the quality of the simulation. Similar tests can often be devised for nonconservative systems, since the computer experimenter generally has available information on the inputs to and losses from the system which enables him to construct total energy and momentum balance equations.

Many factors contribute to the breakdown of the conservation laws; aliasing, finite-difference errors in computing fields, finite-timestep errors, roundoff errors, and errors in the measurement of the physical quantities. The last item can be treated separately, since its effect is to introduce a quantifiable experimental error in the observed quantities without affecting the dynamics. Roundoff error can be ignored since its effect is small compared to those of truncation and undersampling, provided that the scheme is numerically stable (cf. Chapter 4). To comprehensively treat the effects of the remaining errors still presents a formidable analytic task. However, as we shall show below, a few simplifying assumptions allow us to quantify the dominant factors.

7-6-1 Energy

The total energy E in a conservative system is the sum of the kinetic and potential energies. The kinetic energy $\mathcal{T}$ is given by the sum of the particle kinetic energies

$$\mathcal{T} = \sum_{i=1}^{N_p} \tfrac{1}{2} m v_i^2 \tag{7-117}$$

and the potential energy V is given by the sum of the pair potential energies $\Phi_{ij} = \Phi(x_i; x_j)$

$$V = \tfrac{1}{2} \sum_{\substack{i=1 \\ i \neq j}}^{N_p} \sum_{j=1}^{N_p} \Phi_{ij} \tag{7-118}$$

The measurement of $\mathcal{T}$ in particle-mesh simulation models presents no problems. To measure V we require the potential at particle positions: This we can do by interpolating the value of the potential at particle positions from known mesh values in the same manner as used for interpolating force values from the mesh-defined field (Eq. 5-161). If a particle at position x_i with charge q leads to a set of mesh-defined potential values $\{\phi(x_p)\}$, or, equivalently, to the generalized potential function $\phi\dagger$, then its pair potential energy with a charge q at position x_j is given by

$$\Phi(x_i, x_j) = q \sum_p W(x_j - x_p)\phi(x_p) = \frac{q}{H} \int W(x_j x')\phi\dagger(x')\, dx' \qquad \text{(7-119)}$$

Eliminating $\phi(x_p)$ using Eqs. (5-158) and (5-159) gives

$$\Phi(x_i; x_j) = q^2 \sum_p \sum_{p'} W(x_j - x_p) G(x_p - x_{p'}) W(x_i - x_{p'}) \qquad \text{(7-120)}$$

Substituting Eq. (7-120) into Eq. (7-118) (ignoring the restriction $i \neq j$) and rearranging yields the approximate potential energy

$$V = \frac{H}{2} \sum_p \left(\frac{q}{H} \sum_j W(x_j - x_p) \right) \left[H \sum_{p'} G(x_p - x_{p'}) \left(\frac{q}{H} \sum_i W(x_i - x_{p'}) \right) \right] \qquad \text{(7-121)}$$

$$= \frac{H}{2} \sum_p \rho(x_p)\phi(x_p) = \tfrac{1}{2} \int \rho' \phi\dagger \, dx = \tfrac{1}{2} \int \rho\dagger \phi' \, dx \qquad \text{(7-122)}$$

Dropping the $i \neq j$ restriction causes the self-potential energy term to be included in Eq. (7-121). Since this term is finite for particle-mesh models its inclusion causes no material change to the definition of the energy (note, however, that the self-potential energy is not exactly constant, owing to aliasing errors). Identifying the terms in round brackets in Eq. (7-121) as the mesh-defined charge density, and the term in the square brackets as potential, enables the potential energy to be written in terms of known mesh quantities or their equivalent generalized functions [Eq. (7-122)].

In any practical calculation, Eq. (7-122) would be used in preference to Eq. (7-118) in computing the potential energy because of its much smaller operations count. However, for two and three dimensions it is common practice to overwrite charge density values with potential values at the potential solver step to reduce storage requirement, in which case $\{\rho(x_p)\}$ and $\{\phi(x_p)\}$ are not simultaneously available. One alternative is to accumulate V at the force interpolation step of the calculation, since the contribution of particle i to the total potential energy is, from Eq. (7-121),

$$\delta V_i = q \sum_p W(x_i - x_p)\phi(x_p) \qquad \text{(7-123)}$$

If s mesh points are involved in force interpolation to the ith particle, the extra overhead cost in computing V in using Eq. (7-123) is sN_p additions and sN_p multiplications, compared with N_g additions and N_g multiplications using Eq. (7-122). As the number of particles N_p is generally much greater than the number of mesh points N_g, the latter approach is computationally cheaper.

A second alternative method of computing V is possible when fast Fourier transforms are used to solve the field equation (see Chapter 6). Using the power theorem (Table A-4), we can rewrite Eq. (7-122) as

$$V = \frac{1}{2} \sum_{l=0}^{N_g - 1} \left(\frac{k_0}{2\pi} \right) \hat{\rho}^*(k) \hat{\phi}(k) \tag{7-124}$$

for a periodic system of period length L. k_0 is the k space "cell" width, $k_0 = 2\pi/L$. $\hat{\rho}^*$ is the complex conjugate of $\hat{\rho}$ [$\hat{\rho}^*(k) = \hat{\rho}(-k)$ since ρ real $\supset \hat{\rho}$ hermitian]. If V is accumulated in the calculation loop where $\hat{\phi} = \hat{G}\hat{\rho}$ is computed, then an operations count $\sim 2N_g$ is again recovered. Similar results hold for calculations using sine or cosine transforms and for two- and three-dimensional calculations.

The pair-potential energy given by Eq. (7-119) shows the same mesh dependence (arising from aliasing errors in charge assignment) as does the interparticle force. Following the steps used for the interparticle force (Sec. 5-6-4) we can write the pair-potential energy of two particles of charge q and separation x as

$$\phi(x; x_1) = \int \frac{dk}{2\pi} e^{ikx} \left(\frac{q^2}{H^2} \hat{W}(k) \hat{G}(k) \sum_n \hat{W}(k_n) e^{-ink_g x_1} \right) \tag{7-125}$$

giving a displacement-averaged potential

$$\langle \phi(x; x_1) \rangle = \int \frac{dk}{2\pi} \left(q^2 \frac{\hat{W}^2}{H^2} \hat{G} \right) e^{ikx} \tag{7-126}$$

Expressing the averaged-pair potential harmonic in Eq. (7-126) in the form of Eq. (7-114) reveals the effective particle shape

$$\hat{S}_e^2(k) = \left(\frac{\hat{W}}{H} \right)^2 (\epsilon_0 \hat{G} k^2) \tag{7-127}$$

Comparing Eq. (7-127) with Eqs. (7-74) and (7-75), we see that the averaged-force and averaged-pair-energy effective particle shapes are identical for energy-conserving schemes, but differ for momentum-conserving schemes. In the case of momentum-conserving schemes, a better approximation to the potential energy is obtained by replacing Eq. (7-124) by

$$V = \frac{1}{2} \sum_l \frac{k_0}{2\pi} \hat{\rho}^*(k) \hat{\phi}(k) \left(\frac{-i\hat{D}(k)}{k} \right) \tag{7-128}$$

The factor $(-i\hat{D}/k)$ causes the (displacement-averaged) effective particle shape to be the same both for the force and for the potential-energy calculation. The result is that if $\hat{W}$ and/or $\hat{n}$ decay sufficiently rapidly with k for $\hat{\rho} = q\hat{W}\hat{n}/H$ to be a good approximation (i.e., aliasing is negligible), then forces and energies both describe the same meshless finite-sized-particle system.

The energy errors discussed so far are those of measurement. More significant are errors introduced by inaccuracies of the dynamics. In the limit of zero timestep, the energy-conserving variants of the particle-mesh schemes do have an energy constant by virtue of the fact that the force, although only approximate, can be written as the gradient of a scalar (cf. Sec. 5-5). The momentum conservation symmetry requirements (Sec. 5-3-3) prevent the force in momentum-conserving schemes being written as gradient of the scalar, and it is the nonconservative part of this force which causes the growth of total energy. If we write the interparticle force in momentum-conserving schemes as

$$\mathbf{F}(x; x_1) = -\nabla\Phi + \mathbf{f} \tag{7-129}$$

then we can identify $-\nabla\Phi$ as the displacement-averaged part of the force, and $\mathbf{f}$ as the (fluctuating) nonconservative remainder.

Consider a homogeneous, collisionless particle-mesh plasma where the force is defined by a momentum-conserving scheme. If the density of particles is sufficiently large for the system to be collisionless, then the energy of a test particle is given by its kinetic energy

$$E(t) = \frac{mv^2}{2} \tag{7-130}$$

The energy one timestep later is given by

$$E(t+DT) = \frac{m}{2}(\mathbf{v}+\Delta\mathbf{v})^2 \tag{7-131}$$

where

$$\Delta\mathbf{v} = \frac{DT}{m}\sum_{i=1}^{N_p}\mathbf{f}_i \tag{7-132}$$

In Eq. (7-132) we have made the simplifying assumption that the conservative part of the force sums to zero. The force $\mathbf{f}_i$ is the nonconservative part of the force on the test particle due to particle i, where the position $\mathbf{x}_i$ of particle i is located randomly in the computational box volume V_b. Treating $E(t+DT)$ as a functional of $E(t)$ and the set of particle coordinates $\{x_i; i = 1, N_p\}$ enables the expectation value of the energy $E(t+DT)$ to be written as

$$\langle E(t+DT)\rangle = \int EQ(E)\,dE = \int E\prod_{i=1}^{N_p}P(\mathbf{x}_i)\,d\mathbf{x}_i \tag{7-133}$$

where $Q(E)$ is the probability distribution of energies and $P(\mathbf{x}_i)(=V_b^{-1})$ is the probability distribution of positions of particle i. In obtaining Eq. (7-133), it is assumed that particles positions are uncorrelated. Substituting Eq. (7-131) into Eq. (7-133) gives

$$\langle E(t+DT)\rangle = \langle E(t)\rangle + m\mathbf{v}\cdot\int\frac{DT}{m}\sum_j\mathbf{f}_j\prod_i P(\mathbf{x}_i)\,d\mathbf{x}_i + \frac{DT^2}{2m}\int\sum_j\mathbf{f}_j\cdot\sum_{j'}\mathbf{f}_{j'}\prod_i P(\mathbf{x}_i)\,d\mathbf{x}_i \tag{7-134}$$

Setting $P = V_b^{-1}$ and using the result $\int\mathbf{f}_j\,d\mathbf{x}_j = 0$ allows Eq. (7-134) to be integrated to give

$$R = \frac{\langle E(t+DT)\rangle - \langle E(t)\rangle}{DT} = \frac{DT n_0}{2m}\int|\mathbf{f}(\mathbf{x};\mathbf{x}')|^2\,d\mathbf{x} \tag{7-135}$$

Treating each particle in turn as the test particle gives the ensemble-averaged specific heating rate $\langle r \rangle$:

$$\langle r \rangle = \frac{\langle R \rangle}{\frac{1}{2} m v_T^2} = \frac{DTn_0}{m^2 v_T^2} \frac{1}{V_c} \int_{V_c} d\mathbf{x}' \int d\mathbf{x} |\mathbf{f}(\mathbf{x};\mathbf{x}')|^2 \tag{7-136}$$

In Eq. (7-136), periodicity of **f** has been used to reduce the averaging integral to a single cell. The further assumption of no time correlation allows Eq. (7-136) to be applied over successive timesteps, giving the linear energy growth characteristic of stochastic processes (cf. Sec. 9-2-2).

Applying the power theorem (Table A-4) to the integral over x in Eq. (7-136), and using Eqs. (5-223) and (5-224) for the transform $\hat{f}$ of f, gives (for a one-dimensional system)

$$\frac{1}{H} \int dx' \int dx |f(x,x')|^2 = \int_{-\infty}^{\infty} \frac{dk}{2\pi} \frac{q^4}{H^4} \hat{W}^2 |\hat{D}|^2 \hat{G}^2 \left[\sum_n \hat{W}^2(k_n) - \hat{W}^2(k) \right] \tag{7-137}$$

$$= \int_{k_g} \frac{dk}{2\pi} q^4 |\hat{D}|^2 \hat{G}^2 \left[\left(\sum_n \frac{\hat{W}^2(k_n)}{H^2} \right)^2 - \sum_n \frac{\hat{W}^4(k_n)}{H^4} \right] \tag{7-138}$$

For a finite system of period length L, $\int_{k_g} dk/2\pi$ is replaced by $\Sigma_{l=0}^{N_g-1} k_0/2\pi = \Sigma_l 1/L$, and for multidimensional systems k is replaced by $\mathbf{k}$, D by $\mathbf{D}$, etc., as described in Sec. 7-4-2. It follows from Eq. (7-138) that if $\hat{W}$ were band-limited, then the fluctuations would disappear and (in the limit $DT \to 0$) energy would be conserved. For practical schemes (NGP, CIC, TSC, etc.), the term in the square brackets in Eq. (7-138) is largest near the principal zone boundaries $|k| = \pi/H$, since the alias contributions are largest there, although the factor $|\hat{D}|^2 \hat{G}^2 \sim 1/k^2$ may cause the largest contribution to the total force fluctuations to occur at longer wavelengths. For example, using Eqs. (7-59) and (7-66) we find that for the NGP scheme

$$\left(\sum_n \frac{\hat{W}^2(k_n)}{H^2} \right)^2 - \sum_n \frac{\hat{W}^4(k_n)}{H^4} = \frac{2}{3} \sin^2 \frac{kH}{2} \tag{7-139}$$

Giving

$$\langle f^2 \rangle = \frac{1}{H} \int_H dx' \int dx |f(x,x')|^2 = \int_{k_g} \frac{dk}{2\pi} \frac{q^4}{\epsilon_0^2} |\hat{D}|^2 (\epsilon_0 \hat{G})^2 \left(\frac{2}{3} \sin^2 \frac{kH}{2} \right) \tag{7-140}$$

It is clear from Eq. (7-140) that taking a wider finite-sized particle width (causing $\hat{G}$ to decay more rapidly with k) will greatly reduce $\langle f^2 \rangle$ and thus improve energy conservation; cf. Chapter 9 and Orens et al. (1970).

Equation (7-140) may be integrated for the NGP model described in Chapter 2. Using Eqs. (7-57) and (7-58) and introducing a cutoff in $\hat{G}$ at $|k| = (\alpha k_g)/2$; $0 < \alpha \leqslant 1$ to represent a particle-widening factor reduces Eq. (7-140) to

$$\langle f^2 \rangle = \frac{q^4}{\epsilon_0^2} \frac{H^2}{3} \int_0^{(\alpha k_g)/2} \frac{dk}{2\pi} \cos^2 \frac{kH}{2} = \frac{q^4}{\epsilon_0^2} H \frac{\alpha}{12} [1 + \operatorname{sinc} \alpha] \tag{7-141}$$

and gives a specific heating rate Eq. (7-136)

$$\langle r \rangle_{\mathrm{NGP}} = \omega_p(\omega_p DT)\frac{H}{\lambda_D}\frac{1}{n_0\lambda_D}\left(\frac{\alpha}{12}(1+\operatorname{sinc}\alpha)\right) \tag{7-142}$$

Similarly, for the CIC scheme

$$\langle f^2 \rangle = \int_{k_g} \frac{dk}{2\pi}\frac{q^4}{\epsilon_0^2}|\hat{D}|^2(\epsilon_0\hat{G})^2\left[\frac{2}{45}\sin^4\frac{kH}{2}\left(1+\frac{2}{7}\sin^2\frac{kH}{2}\right)\right] \tag{7-143}$$

Substituting for $\hat{D}$ and $\hat{G}$ in Eq. (7-143) using Eqs. (7-57) and (7-58) gives the heating rate for the CIC scheme of

$$\langle r \rangle_{\mathrm{CIC}} = \frac{\omega_p^2 H\, DT}{n_o\lambda_D^2}\left[\frac{\alpha}{315}\left(1-\frac{\operatorname{sinc}2\alpha}{2}-\operatorname{sinc}\alpha\frac{\sin^2\pi\alpha}{24}\right)\right] \tag{7-144}$$

If all the harmonics are retained ($\alpha = 1$) the reduction in heating rate brought about by replacing the NGP assignment function by the CIC function is

$$\frac{\langle r \rangle_{\mathrm{NGP}}}{\langle r \rangle_{\mathrm{CIC}}} = \frac{\frac{1}{12}}{\frac{1}{315}} = 26\tfrac{1}{4}$$

i.e., the time for stochastic heating to significantly alter the total energy in the one-dimension CIC scheme is approximately twenty-five times greater than that for NGP. Moreover, if a wavenumber cutoff is introduced ($\alpha < 1$), then the heating rate is reduced in both schemes, although CIC shows a greater reduction.

The results of the simple analysis of the energy growth in simulation plasmas presented here shows quite good agreement with measured results (Peiravi and Birdsall, 1978), although its applicability is limited to small $\omega_p DT$, $H \gtrsim \lambda_D$, and $(n_0\lambda_D) \gg 1$. If these restrictions do not apply, then time integration errors, nonphysical mode coupling, and correlation (i.e., kinetic) effects complicate the picture. Despite these provisos, results such as Eqs. (7-142) and (7-144) give a useful quantitative guide to the changes in energy conservation brought about by changing characteristic parameters such as $(\omega_p DT)$, (H/λ_D), or $(n\lambda_D)$, or by altering the form of the force calculation algorithm (implying changes in $\hat{D}$, $\hat{G}$, and $\hat{W}$).

7-6-2 Momentum

The lack of energy conservation in momentum-conserving schemes is due principally to the nonconservative part of the force introduced by undersampling. The loss of momentum conservation in energy-conserving schemes is another manifestation of the same errors. Changing the form of the mesh-approximated force to enable the force to be written as the gradient of a scalar potential is incompatible with the symmetry required for momentum conservation.

The symmetry-breaking part of the force in energy-conserving schemes is the self force term (cf. Chapter 5). Thus, the change of total momentum Δp in a single timestep is

$$\Delta p = \sum_{i=1}^{N_p} DT\, F^{\text{self}} \tag{7-145}$$

Following the same random-walk arguments as used in the previous section for the energy, we find the expectation values for the rate of change of total momentum and for the rate of change of the total squared momentum to be respectively

$$\frac{\langle \Delta p \rangle}{DT} = 0 \qquad \frac{\langle \Delta p^2 \rangle}{DT} = DT \frac{N_p}{H} \int_{-H/2}^{H/2} (F^{\text{self}})^2 \, dx \tag{7-146}$$

Eqs. (7-146) state that, on average, we would expect the energy-conserving schemes to conserve momentum, but that deviations from conservation may increase (in the positive or negative sense) linearly with time.

To ensure that the drift in total momentum remains small, we require that the specific increase of the momentum uncertainty per timestep be small, i.e.,

$$\frac{\langle \Delta p^2 \rangle}{N_p m^2 v_T^2} = \frac{DT^2}{m^2 v_T^2 H} \int (F^{\text{self}})^2 \, dx \ll 1 \tag{7-147}$$

For the one-dimensional energy-conserving scheme using piecewise linear-basis functions (Sec. 5-5) the self force is given by (cf. Secs. 5-2-4 and 5-5)

$$F^{\text{self}} = -\frac{q^2}{\epsilon_0 H} x \qquad -\frac{H}{2} < x \leqslant \frac{H}{2} \tag{7-148}$$

giving

$$\frac{\langle \Delta p^2 \rangle}{N_p m^2 v_T^2} = \frac{\omega_p^2 DT^2}{12(n_0 \lambda_D)^2} = \frac{\omega_{\text{self}}^2 DT^2}{12(n_0 \lambda_D)} \left(\frac{H}{\lambda_D} \right) \tag{7-149}$$

Thus, if $\omega_p DT \ll 1$ and there are a large number of particles per Debye length ($n_0 \lambda_D \gg 1$), secular drift of the total momentum will not be a serious problem in energy conserving schemes. Experimental measurements of total momentum drift are presented by Lewis et al. (1972).

7-6-3 Angular Momentum

Angular momentum is an important conserved quantity in galaxies (cf. Chapter 11). The rate of change of the total angular momentum of a system comprising N_p particles is equal to the torque:

$$\frac{d\mathbf{L}}{dt} = \sum_{i=1}^{N_p} \mathbf{x}_i \times \mathbf{F}_i = \sum_{i=1}^{N_p - 1} \sum_{j=i+1}^{N_p} (\mathbf{x}_i - \mathbf{x}_j) \times \mathbf{f}_{ij} \tag{7-150}$$

where $\mathbf{F}_i$ is the total force on particle i and $\mathbf{f}_{ij}$ is the force on particle i due to particle j. For particles with symmetric charge (or mass for gravitating systems) distributions, i.e., distributions which are only functions of distance from the centre of the particles, $\mathbf{f}_{ij}$ is parallel to $(\mathbf{x}_i - \mathbf{x}_j)$. In this case the torque is zero and $\mathbf{L}$ is conserved. However, images destroy conservation in periodic systems as in Chapter 12.

For a system of particles moving according to the leapfrog equations of motion, the torque is given by $\sum_i \mathbf{x}_i^n \times \mathbf{F}_i^n$ where n denotes the timelevel. If we define the angular momentum at timelevel $n+\frac{1}{2}$ as

$$\mathbf{L}^{n+1/2} = m\sum_i \mathbf{x}_i^n \times \mathbf{v}_i^{n+1/2} = \frac{m}{DT}\sum_i \mathbf{x}_i^n \times \mathbf{x}_i^{n+1} \tag{7-151}$$

then, by the leapfrog equations of motion

$$\frac{\mathbf{L}^{n+1/2} - \mathbf{L}^{n-1/2}}{DT} = \sum_i \mathbf{x}_i^n \times \mathbf{F}_i^n \tag{7-152}$$

Thus, if $\mathbf{f}_{ij}^n$ is parallel to $(\mathbf{x}_i^n - \mathbf{x}_j^n)$, angular momentum is also conserved in a system moving according to the leapfrog equations. However, as we saw in Chapter 5, the mesh-approximated interparticle force is generally not parallel to the line joining the two particles, with the error being due to assignment/interpolation errors, anisotropies in the potential, and directional errors introduced by the difference approximation to the gradient (for momentum-conserving schemes). However, by using the Q-minimizing method described below and in Chapter 8, the errors in the directional dependence of the force can be made very small.

7-7 OPTIMIZATION

In setting up a particle mesh code to simulate a collisionless system, we are confronted with a wide range of possible charge assignment, potential-solving, differencing, and force interpolation schemes. For two- and three-dimensional simulations, the limited size and speed of computers makes inevitable a compromise between accuracy of representation of the physical system and computational cost.

The immediate impact of limited computer resources is that the computer experimenter must manage with relatively few particles and a coarse mesh. If the number of particles in a Debye length $N_D = n_0\lambda_D$ (or $n_0\lambda_D^2$ and $n_0\lambda_D^3$ in two and three dimensions) becomes small, particle correlations can significantly alter the collective behavior of the model plasma. A measure of the time for such correlations to dominate is given by the collision time τ_c. As is shown in Secs. 7-5 and 9-2-3, τ_c can be made acceptably large for relatively small N_D ($\sim$5–10) by increasing the effective finite-sized particle width. The sparseness of the mesh has a dual effect. It limits collisional effects by restricting the effective particle width to be greater than the cell width H, and it restricts the range of collective modes to those with wavenumbers in the principal zone $|k| < \pi/H$.

The effect of finite particle widths on collective properties can be assessed by the averaged-force dispersion relation. It is clear from the warm-plasma approximation (Sec. 7-3-3) that the slope of the dispersion curve $\omega = \omega(k)$ is reduced by increasing the effective particle width, and for certain particle shapes which have widths much greater than the Debye length the group velocity $\partial\omega/\partial k$ may have a sign opposite to that in the physical system. In addition, mode coupling introduced by undersampling the charge density (Sec. 7-3-4) affects growth rates and if the cell width H is greater than one or two Debye lengths it can cause nonphysical instabilities. However, provided that $H \gtrsim \lambda_D$, the consequences of aliases are mainly limited to a slow nonphysical increase in the total energy of the system (cf. Secs. 7-6-1 and 9-2-3).

The objective in optimizing collisionless simulation models is to get the best representation of the collective properties for a given computational cost. Aliasing restricts the maximum practicable cell width to approximately the Debye length. The length L of the computational box must be much greater than the wavelengths of importance for the density of permitted wavenumbers (spacing $k_0 = 2\pi/L$) to be large enough to describe physical nonlinear mode coupling, i.e., $H \leqslant \lambda_D \ll L$. Fixing H and L fixes the number of mesh points $N_g = L/H$. The number of particles is then constrained principally by the need for the collision time to be greater than the time of the computation, i.e., $n_{\max} DT \ll \tau_c$.

7-7-1 The Interparticle Force

The interparticle force is the important quantity in determining the behavior of the computer simulation model. The deviations of the properties of the simulation model from the physical system are direct consequences of finite timestep, limited particle number, and the difference of the interparticle forces in the two instances. Given the constraints on the mesh size and particle number introduced by limited computer resources, the question of optimization reduces to one of finding the optimal combination of the elements of the force calculation cycle. At best, the particle mesh force calculation can only represent some band-limited reference force $\mathbf{R}(\mathbf{x}) \supset \hat{\mathbf{R}}(\mathbf{k})$, where $\mathbf{R}(\mathbf{x})$ is chosen (using the finite-sized particle theory) to give the desired collective properties.

A quantitative measure Q of the quality of a particular scheme is given by the squared deviation of the particle-mesh calculated force $\mathbf{F}(\mathbf{x}, \bar{\mathbf{x}})$ from the reference force $\mathbf{R}(\mathbf{x})$:

$$Q = \frac{1}{V_c} \int_{V_c} d\bar{\mathbf{x}} \int d\mathbf{x}\, (\mathbf{F}(\mathbf{x}, \bar{\mathbf{x}}) - \mathbf{R}(\mathbf{x}))^2 \tag{7-153}$$

where the averaging over positions with respect to the mesh $\bar{\mathbf{x}}$ is reduced to a single mesh cell by the periodicity of $\mathbf{F}$ with respect to that argument. Using the power theorem (Table A-4) enables Q to be written in terms of the harmonics $\hat{\mathbf{R}}(\mathbf{k}) \supset \mathbf{R}(\mathbf{x})$ and $\hat{\mathbf{F}}(\mathbf{k}, \bar{\mathbf{x}}) \supset \mathbf{F}(\mathbf{x}, \bar{\mathbf{x}})$

$$Q = \frac{1}{V_c} \int_{V_c} d\bar{\mathbf{x}} \int \frac{d\mathbf{k}}{(2\pi)^d} \{|\hat{\mathbf{F}}(\mathbf{k}, \bar{\mathbf{x}})|^2 + |\hat{\mathbf{R}}(\mathbf{k})|^2 - 2\hat{\mathbf{F}}(\mathbf{k}, \bar{\mathbf{x}}) \cdot \hat{\mathbf{R}}^*(\mathbf{k})\} \tag{7-154}$$

where d (= 1, 2 or 3) is the dimensionality and $\hat{\mathbf{R}}^*$ is the complex conjugate of $\hat{\mathbf{R}}$. Expressing $\hat{\mathbf{F}}$ in terms of the transforms of the elements of the force calculation cycle (see Secs. 5-6-4 and 7-4) and reducing the integrals over $\mathbf{k}$ to the principal zone gives, for unit charges in a momentum-conserving scheme,

$$Q = \int_{\mathbf{k}_g} \frac{d\mathbf{k}}{(2\pi)^d} \left\{ \hat{G}^2 |\hat{\mathbf{D}}|^2 \left[\sum_{\mathbf{n}} \hat{U}^2 \right]^2 - 2\hat{G}\hat{U}^2 \hat{\mathbf{D}} \cdot \hat{\mathbf{R}}^* + |\hat{\mathbf{R}}|^2 \right\} \tag{7-155}$$

where $\hat{U} = \hat{W}/H$. In periodic systems, the integral $\int_{\mathbf{k}_g} d\mathbf{k}/(2\pi)^d$ is replaced by the sum over harmonics $\Sigma_{\mathbf{l}} V_b^{-1}$ (cf. Sec. 7-4).

If finite Fourier transforms (cf. Chap. 6) are used to solve for the potential, we are free to choose $\hat{G}$ to be any real even function in the principal zone. Clearly, the best choice is that which minimizes Q, which from Eq. (7-155) gives the optimal influence function

$$\hat{G}(\mathbf{k}) = \frac{\hat{\mathbf{D}}(\mathbf{k}) \cdot \hat{\mathbf{R}}^*(\mathbf{k}) \hat{U}^2(\mathbf{k})}{|\hat{\mathbf{D}}(\mathbf{k})|^2 \left[\sum_{\mathbf{n}} \hat{U}^2(\mathbf{k_n}) \right]^2} \tag{7-156}$$

and the minimum Q

$$Q = \int_{\mathbf{k}_g} \frac{d\mathbf{k}}{(2\pi)^d} |\hat{\mathbf{R}}|^2 \left\{ 1 - \frac{|\hat{\mathbf{D}} \cdot \mathbf{k}|^2 \hat{U}^4}{|\hat{\mathbf{D}}|^2 |\mathbf{k}|^2 \left[\sum_{\mathbf{n}} \hat{U}^2 \right]^2} \right\} \tag{7-157}$$

Equation (7-156) states that a more accurate force calculation results if compensating errors are introduced into the potential solver to offset the average charge-sharing–force-interpolation errors and the smoothing effects of potential differencing. The residual errors [Eq. (7-157)] are caused by the difference gradient not being parallel to the true gradient (i.e., $|\hat{\mathbf{D}} \cdot \mathbf{k}| \neq |\hat{\mathbf{D}}|\,|\hat{\mathbf{k}}|$) and by the assignment function not being band-limited [i.e., $\Sigma_{\mathbf{n}} \hat{W}(\mathbf{k_n}) \neq \hat{W}(\mathbf{k})$].

The improvements obtained by altering the form of $\hat{G}$ make no difference to cost of solving for the potential if fast Fourier transform methods are used. More accurate but computationally more expensive schemes are obtained by employing higher-order gradient operators $\mathbf{D}$ (to reduce the directional errors in the gradient) and/or higher-order assignment functions to obtain a more rapid decay of $\hat{W}$ with increasing $\mathbf{k}$. An alternative to higher-order assignment functions, interlacing, is described in Sec. 7-8. The method of matching gradient operators and assignment schemes is pursued further in the next chapter.

7-7-2 One-Dimensional Schemes

The effects of potential differencing in one-dimensional schemes can be completely eliminated by a compensating factor in $\hat{G}$, so, for a one-dimensional periodic system of length L, Q becomes

$$Q = \frac{1}{L} \sum_{l=0}^{N_g - 1} |\hat{R}|^2 \left\{ 1 - \frac{\hat{W}^4}{\left[\sum_n \hat{W}^2 \right]^2} \right\} = \frac{1}{L} \sum_l Q(k) \tag{7-158}$$

The root-mean-squared percentage error in the lth harmonic ($k = 2\pi l/H$) is given by

$$\sqrt{Q\dagger} = 100 \frac{Q(k)}{|\hat{R}(k)|^2} \tag{7-159}$$

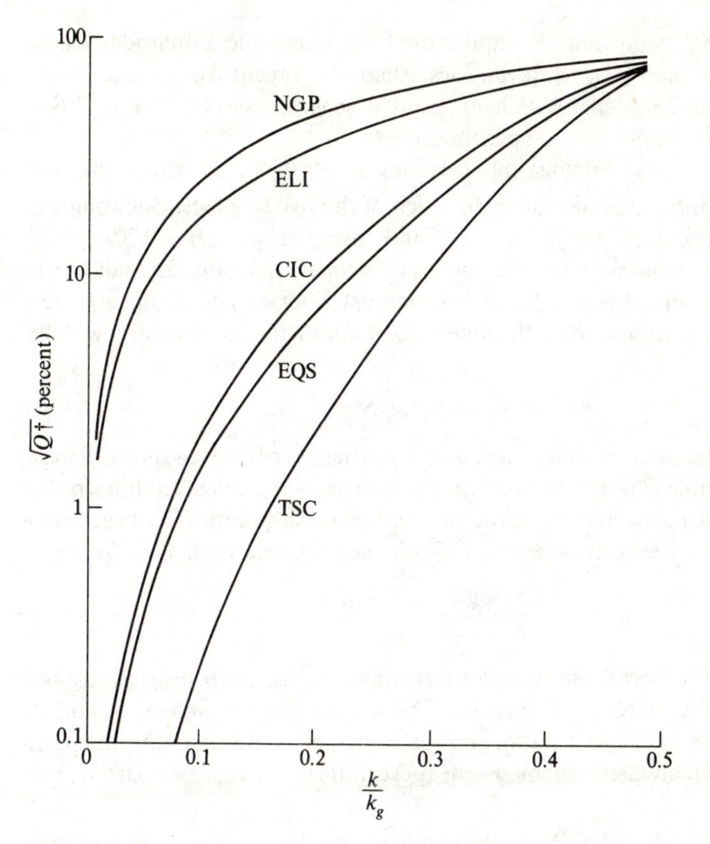

Figure 7-6 The root-mean-squared percentage deviation of the mesh-calculated harmonics from the reference force harmonic amplitudes as a function of wavenumber for an arbitrary band-limited force. The labeling of the curves corresponds to the assignment/interpolation functions (see text). (*After Eastwood, 1975, courtesy of Journal of Computational Physics.*)

Figure 7-6 shows plots of $\sqrt{Q\dagger}$ for the NGP, CIC, and TSC momentum-conserving schemes and for energy-conserving schemes using linear interpolation (ELI) and quadratic splines (EQS). ELI is the CIC/NGP scheme and EQS is the corresponding TSC/CIC scheme. There are two contributory factors to $\sqrt{Q\dagger}$; a fluctuating part arising from undersampling in charge assignment, and a non-fluctuating part arising from the mesh force not being band-limited (cf. Eastwood, 1975). The fluctuating part gives a measure of the strength of mode coupling (cf. Sec. 7-3-4) and the contribution to stochastic heating (cf. Sec. 7-6-1) as a function of wavenumber, and the nonfluctuating part gives a measure of the deviation of the real part of the dispersion relation from that given by the reference force. Thus, $\sqrt{Q\dagger}/k$ gives a simple measure of the quality of the representation of the collective properties.

The curves shown in Fig. 7-6 enable the most cost-effective scheme for a particular problem to be chosen. For example, assume that we want to simulate a system of length $100\lambda_D$ and that we require the force-harmonic amplitudes for all wavelengths greater than $2\lambda_D$ to deviate less than 10 percent from those of the point particle physical system. Which of the five schemes shown in Fig. 7-6 is cheapest in terms of cost per particle per timestep?

We obtain from Fig. 7-6 maximum values μ_i of (k/k_g) at which the line $\sqrt{Q\dagger} = 10$ percent intersects the curves for each of the five schemes. Equating the wavelength corresponding to μ_i to $2\lambda_D$ and using $L = N_g H = 100\lambda_D$ and $k_g = 2\pi/H$ gives the minimum number of mesh points $N_i = 50\mu_i$ for which the criterion can be met for scheme i. Assuming that fast Fourier transforms are used to solve the field equations, then the operations count T_i for scheme i will be approximately

$$T_i = \alpha_i N_p + \beta N_i \log N_i \tag{7-160}$$

Here, α_i measures the number of operations per particle per timestep in the charge sharing, force interpolation and time integration parts of the calculation loop, N_p is the number of particles, and N_i is the number of mesh points. The breakeven number of particles m_{ij} between scheme i and scheme j is given when $T_i = T_j$:

$$m_{ij} = \frac{-\beta(N_i \log N_i - N_j \log N_j)}{(\alpha_i - \alpha_j)}$$

Using real arithmetic operations count to estimate α_i, we have respectively for TSC–EQS, EQS–CIC, CIC–ELI, and ELI–NGP, $\Delta\alpha = \alpha_i - \alpha_j = 5, 6, 3,$ and 4. Using $\beta = 5$ (see Sec. 6-5-2) and appropriate values of N_i we find the range of values of N_p over which each scheme is cheapest in terms of cost per particle per timestep:

NGP: $N_p > 8{,}000$
CIC: $325 < N_p < 8{,}000$
TSC: $N_p < 325$

Nowhere do the energy-conserving schemes prove most cost-effective.

To obtain harmonics and hence a dispersion relation which correspond to those of the point particle physical system, $\hat{R}(k)$ is set equal to the point particle value for wavenumbers less than $k_{\max} = 2\pi/(2\lambda_D)$. Wavenumbers greater than $k_{\max}$ are suppressed (by setting $\hat{G} = 0$ for $k > k_{\max}$) to improve energy conservation and to reduce the number of particles required to obtain a sufficiently large collision time.

The collision time constraint $N_D \gg 1$ precludes the regime where TSC is most economical. In the range of N_D ($\sim 10-20$) which gives satisfactory collision times, CIC emerges as the best scheme (with $N_g = 256$) for this example. Given this choice, a-priori evaluation of dispersive properties (cf. Sec. 7-3) collision times (Sec. 7-5) and heating times (Sec. 7-6-1), together with experimental checks with the program, can be employed to ensure that the results of simulation experiments will reflect physical rather than numerical phenomena.

The advantage of optimized schemes over other plausible schemes is illustrated by Fig. 7-7. If we require the dispersive properties of the particle-mesh

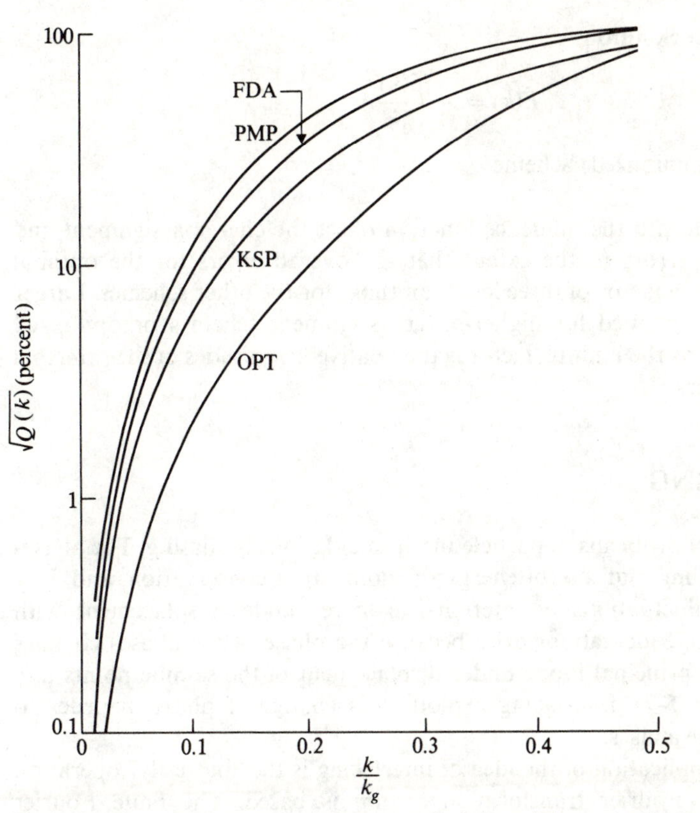

Figure 7-7 The root-mean-squared percentage deviation of the mesh-calculated harmonics from the harmonic amplitudes of a Coulomb force law for four CIC schemes. (*After Eastwood, 1975, courtesy of Journal of Computational Physics.*)

models to be given by the point particle result for the wavelengths retained, then a comparison of the different algorithms is provided by the relative deviation of the harmonic amplitudes from the reference force

$$\hat{R}(k) = -\frac{ik}{\epsilon_0 k^2}$$

Figure 7-7 shows such a comparison for four CIC schemes, where the labeling of the curves indicates the method of determining the field from the charge density:

PMP: The "Poor Man's Poisson Solver" (Boris and Roberts, 1969):

$$E_p = \frac{\phi_{p-1} - \phi_{p+1}}{2H} \qquad \hat{\phi}(k) = \frac{\hat{\rho}(k)}{\epsilon_0 k^2}$$

FDA: Finite-Difference Approximation

$$E_p = \frac{\phi_{p-1} - \phi_{p+1}}{2H} \qquad \frac{\phi_{p-1} - 2\phi_p + \phi_{p+1}}{2H} = -\frac{\rho_p}{\epsilon_0}$$

KSP: Direct k Space Solution

$$\hat{E}(k) = -\frac{i\hat{\rho}(k)}{\epsilon_0 k}$$

OPT: Optimal (Q-minimized) Scheme

The errors built into the influence function offset the charge assignment and force interpolation errors to the extent that the overall errors for the optimal scheme are typically a factor of three less than those for the other schemes. Larger improvements are obtained for higher-order assignment schemes or optimized interlaced schemes, as the limiting factor is the relative magnitudes of the principal modes to their aliases.

7-8 INTERLACING

The major source of problems in particle mesh calculations is aliasing. The aliases lead to mode coupling and loss of energy or momentum conservation, and, in a simpler guise, to fluctuations of interparticle forces under displacement with respect to the mesh. Fluctuations arise because the phase of the aliases changes with respect to the principal mode under displacement of the sample points (see Sec. 5-6-1 and Fig. 5-7). Interlacing exploits this change of phase in order to suppress some of the aliases.

The simplest application of the idea of interlacing is the "butterfly" operation on which the fast Fourier transform algorithm is based. The finite Fourier transform $\{D_l; l \in [0, 2N-1]\}$ of a set of $2N$ values $\{D_p\}$ is defined by

$$\hat{D}_l = \sum_{p=0}^{2N-1} D_p e^{-i2\pi lp/2N} \tag{7-161}$$

Splitting the sum on the right-hand side of Eq. (7-161) into terms with p even and terms with p odd, and writing $D_{2p} = A_q$, $D_{2p+1} = B_q$ gives

$$\hat{D}_l = \sum_{q=0}^{N-1} A_q e^{-i2\pi lq/N} + e^{-i\pi l/N} \sum_{q=0}^{N-1} B_q e^{-i2\pi ql/N} \tag{7-162}$$

Comparing the sums in Eq. (7-162) with the transform definition, Eq. (7-161), we see that the sums define finite transforms over half the number of points, i.e.,

$$\left.\begin{aligned} \hat{D}_l &= \hat{A}_l + e^{-i\pi l/N}\hat{B}_l \\ \hat{D}_{l+N} &= \hat{A}_l - e^{-i\pi l/N}\hat{B}_l \end{aligned}\right\} \quad 0 \leqslant l \leqslant N-1 \tag{7-163}$$

Equations (7-163) define the butterfly operation, so called because of the form of the data flow diagram describing the computation of values of $\hat{D}$ from $\hat{A}$ and $\hat{B}$. The important point to note from Eqs. (7-163) is that if we know separately the transform of the p-even and p-odd values, then by the "butterfly" sum and difference operations the transform of all p values can be found.

To apply the ideas of odd and even samples to continuous quantities, we employ the sampling function Ш. If we have some function $A(x) \supset \hat{A}(k)$, then the "even" sample is given by $C_1^\dagger = \text{Ш}(x/H)A(x)$, and the "odd" sample is given by

$$C_2^\dagger = \text{Ш}\left(\frac{x}{H}+\tfrac{1}{2}\right)A(x).$$

Using the convolution theorem, the shift theorem, and the definition of Ш, we find

$$C_1^\dagger = \text{Ш}\left(\frac{x}{H}\right)A(x) \supset \hat{C}_1 = H\text{Ш}\left(\frac{k}{k_g}\right) * \hat{A}(k) = \sum_n \hat{A}(k-nk_g)$$

$$C_2^\dagger = \text{Ш}\left(\frac{x}{H}+\tfrac{1}{2}\right)A(x) \supset \hat{C}_2 = H\text{Ш}\left(\frac{k}{k_g}\right)e^{+ikH/2} * \hat{A}(k) \tag{7-164}$$

$$= \sum_n (-1)^n \hat{A}(k-nk_g)$$

Hence

$$\hat{C} = \tfrac{1}{2}(\hat{C}_1+\hat{C}_2) = \sum_{\substack{n=-\infty \\ n\,\text{even}}}^{\infty} \hat{A}(k-nk_g) \tag{7-165}$$

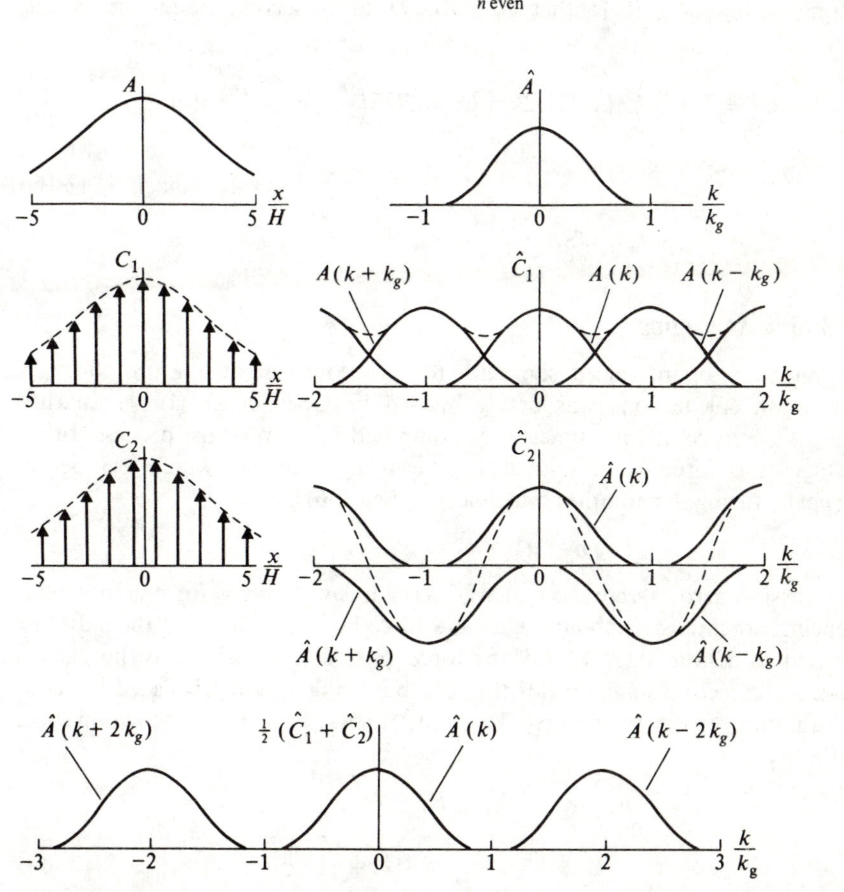

Figure 7-8 A pictorial representation of equations (7-164) and (7-165). Averaging the transforms of the "even" sample $C_1^\dagger \supset \hat{C}_1$ and the "odd" sample $C_2^\dagger \supset \hat{C}_2$ leads to the cancellation of all odd aliases of A in $\hat{C} = (\hat{C}_1+\hat{C}_2)/2$.

A pictorial interpretation of these results is given in Fig. 7-8. The alternate aliases in the second (odd) sample are reversed in sign by the factor $e^{-ink_gH/2} = (-1)^n$.

Corresponding but slightly more complicated results may be obtained for two interlaced samples which are not midpoint-centered. Three interlaced samples may be used to eliminate all but the aliases where n is a multiple of 3, four samples may be combined to leave only every fourth alias, etc. The three interlaced samples correspond to a radix 3 fast Fourier transform algorithm, four to a radix 4 algorithm, and so forth.

The phase (or "twiddle") factor $e^{-i\pi l/N}$ appearing in Eqs. (7-163) is also present in Eq. (7-165) but does not appear explicitly because the definition of the transform

$$\hat{C}_2 = \int dx\, \text{Ш}\left(\frac{x}{H} - \frac{1}{2}\right) A(x) e^{-ikx}$$

absorbs it. The factor can be made explicit by performing the equivalent operation of shifting the function $A(x)$ rather than $\text{Ш}(x/H)$ in the second sample. In this case we find

$$C_2^\dagger = \text{Ш}\left(\frac{x}{H}\right) A(x - H/2) \supset \hat{C}_2 = H\text{Ш}\left(\frac{k}{k_g}\right) * e^{-ikH/2}\hat{A}(k)$$

$$= e^{-ikH/2} \sum_n (-1)^n \hat{A}(k - nk_g) \qquad \text{(7-166)}$$

$$\Rightarrow \hat{C} = \tfrac{1}{2}(\hat{C}_1 + e^{-ikH/2}\hat{C}_2)$$

7-8-1 Force Averaging

The application of interlaced sampling to the reduction of aliasing effects in particle mesh calculations was first proposed by Chen et al. (1974). In their algorithm, the force at each timestep is computed in a two-stage process. In the first stage, the force is computed in the usual manner, which we denote symbolically (using the notation introduced in Sec. 5-6) as

$$n \rightarrow \rho_1^\dagger \rightarrow \phi_1' \rightarrow E_1^\dagger \rightarrow F_1$$

The successive arrows represent charge assignment, solving for the potential, differencing potentials to obtain fields and force interpolation, and the subscript "1" is used to denote stage 1. Half the force, $F_1/2$, is then added to the particle velocities. The second stage repeats the first, but using a mesh displaced by half a cell width with respect to the first. Thus, the timestep cycle may be summarized symbolically as

$$\begin{array}{ccc} & \rho_1^\dagger \rightarrow \phi_1' \rightarrow E_1^\dagger \rightarrow F_1 & \\ \nearrow & & \searrow \\ n & & F = \dfrac{(F_1 + F_2)}{2} \\ \searrow & & \nearrow \\ & \rho_2^\dagger \rightarrow \phi_2' \rightarrow E_2^\dagger \rightarrow F_2 & \end{array}$$

The Fourier transform of the force F_1 for the first stage of the timestep loop is identical to that of the noninterlaced scheme:

$$\hat{F}_1 = q^2 \frac{\hat{W}}{H} \hat{D}\hat{G} \sum_{n=-\infty}^{\infty} \frac{\hat{W}(k_n)}{H} \hat{n}(k_n) \tag{7-167}$$

The Fourier transform of the steps of the second stage require only the additional use of the shift theorem (Table A-4) in their derivation:

$$\rho_2^\dagger = \text{III}\left(\frac{x}{H}+\tfrac{1}{2}\right)\rho(x) \supset \hat{\rho}_2(k) = H\,\text{III}\left(\frac{k}{k_g}\right)e^{ikH/2} * \hat{\rho}' = \frac{q}{H}\sum_n (-1)^n \hat{W}(k_n)\hat{n}(k_n) \tag{7-168}$$

$$\phi_2' = G' * \rho_2^\dagger \supset \hat{\phi}_2' = \hat{G}'\hat{\rho}_2 \tag{7-169}$$

$$\begin{aligned} E_2^\dagger = \text{III}\left(\frac{x}{H}+\tfrac{1}{2}\right)D * \phi_2' \supset \hat{E}_2 &= H\,\text{III}\left(\frac{k}{k_g}\right)e^{ikH/2} * \hat{D}\hat{\phi}_2' \\ &= \hat{D}\sum_{n'} (-1)^{n'} \hat{G}'(k-n'k_g)\hat{\rho}_2(k-n'k_g) \end{aligned} \tag{7-170}$$

$$F_2 = \frac{q}{H} W * E_2^\dagger \supset \hat{F}_2 = \frac{q}{H}\hat{W}\hat{E}_2 \tag{7-171}$$

Collecting together Eqs. (7-168) to (7-171) yields, after a little rearrangement,

$$\hat{F}_2(k) = q^2 \frac{\hat{W}}{H}\hat{D}\hat{G} \sum_{n=-\infty}^{\infty} (-1)^n \frac{\hat{W}(k_n)}{H}\hat{n}(k_n) \tag{7-172}$$

where, as before, $\hat{G}(k) = \sum_n \hat{G}'(k_n)$. Therefore

$$F = \frac{(F_1+F_2)}{2} \supset \hat{F} = q^2 \frac{\hat{W}}{H}\hat{D}\hat{G} \sum_{\substack{n=-\infty \\ n\ \text{even}}}^{\infty} \frac{\hat{W}(k_n)}{H}\hat{n}(k_n) \tag{7-173}$$

The analysis of dispersive properties, etc., carries over to the interlaced schemes in a similar manner. Wherever a charge assignment alias sum appears it is replaced by a sum with the restriction of n even. Similar results hold for energy-conserving schemes. In both variants of the basic force calculation, the method extends to i interlaces to reduce n to multiples of i.

The two-stage interlaced scheme is almost equivalent to a conventional single-stage force calculation scheme using half the mesh spacing $H/2$. Its results differ in that $\hat{G}$ and $\hat{D}$ have a periodicity with period length $k_g = 2\pi/H$, whereas in conventional schemes with mesh spacing $H/2$ they have periodicity $4\pi/H$. However, by restricting choices of $\hat{G}$, $\hat{D}$, and $\hat{W}$ in the conventional scheme, the result [Eq. (7-173)] can be recovered.

Purely from the point of view of an operations count, the force-averaging interlaced schemes are nonstarters. They double the cycle time for what amounts to half a single stage of "butterfly" operations in the solution of Poisson's equation using fast Fourier transforms. However, in two and three dimensional schemes where limits on fast random access memory become an important factor, interlacing does become a more viable proposition.

7-8-2 Harmonic Averaging

A computationally cheaper method of implementing interlacing is to combine the harmonics of the interlaced charge density samples before computing potentials, etc. (Eastwood, 1976*b*). Symbolically, the algorithm is represented by

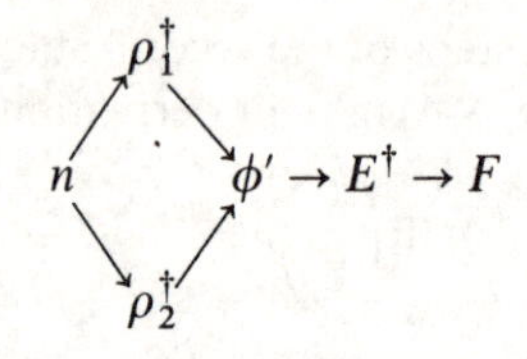

The algorithm is more costly than that of Chen et al. (1974) in terms of storage, since it requires two meshes rather than one to store the charge density values. It can be generalized to i interlaced charge density samples, where the scheme employing i interlaces still requires only two charge density meshes.

The two interlaced-mesh variant of the harmonic-averaging scheme comprises the following steps:

1. Assign charge to mesh 1.
2. Displace the mesh (or equivalently, the particle coordinates) by $H/2$ and assign charge to mesh 2.
3. Fourier-transform charges on both mesh 1 and mesh 2.
4. Combine the harmonics

$$\hat{\rho}(k) = (\hat{\rho}_1(k) + e^{+ikH/2}\hat{\rho}_2(k))/2 \tag{7-174}$$

5. Solve for potential, field, and force on mesh 1 using the conventional particle mesh method.

In practice, steps 1 and 2 are combined in a single pass through particle coordinates in order to exploit the simple relationship between the fractions of charge (i.e., assignment function values) assigned to meshes 1 and 2 by each particle. Similarly, the factor of 2 appearing in step 4 is absorbed into the stored values of the influence function.

The Fourier analysis of the harmonic-averaged interlaced scheme follows that outlined for the force-averaged scheme. However, in this case it is more instructive to regard the particle coordinates, rather than the mesh, being displaced in order to make explicit the "twiddle" factors required in the practical implementation of the algorithm:

1. Assign charge to mesh 1.

$$\rho_1' = \frac{q}{H} W * n \supset \hat{\rho}_1' = \frac{q}{H} \hat{W}\hat{n} \tag{7-175}$$

$$\rho_1^\dagger = \text{Ш}\left(\frac{x}{H}\right)\rho_1' \supset \hat{\rho}_1 = H\text{Ш}\left(\frac{k}{k_g}\right)\hat{\rho}_1' = \frac{q}{H}\sum_n \hat{W}(k_n)\hat{n}(k_n) \tag{7-176}$$

2. Shift coordinate in $+x$ direction by $H/2$ and assign charge to mesh 2.

$$\rho_2' = \frac{q}{H} W(x) * n(x - H/2) \supset \hat{\rho}_2' = \frac{q}{H} \hat{W}\hat{n}e^{-ikH/2} \tag{7-177}$$

$$\rho_2^{\dagger} = \text{Ш}\left(\frac{x}{H}\right)\rho_2' \supset \hat{\rho}_2 = e^{-ikH/2}\frac{q}{H}\sum_n (-1)^n \hat{W}(k_n)\hat{n}(k_n) \tag{7-178}$$

3. Harmonic average.

$$\hat{\rho} = \tfrac{1}{2}(\hat{\rho}_1 + e^{+ikH/2}\hat{\rho}_2) = \frac{q}{H} \sum_{\substack{n \\ n \text{ even}}} \hat{W}\hat{n} \tag{7-179}$$

4. Solve for potential, field and force.

$$\phi' = G' * \rho \supset \hat{\phi}' = \hat{G}'\hat{\rho}' \tag{7-180}$$

$$E^{\dagger} = \text{Ш}\left(\frac{x}{H}\right) D * \phi' \supset \hat{E} = H\text{Ш}\left(\frac{k}{k_g}\right) * \hat{D}\hat{\phi}' = \hat{D}\hat{\phi} \tag{7-181}$$

$$F = \frac{q}{H} W * E^{\dagger} \supset \hat{F} = \frac{q}{H} \hat{W}\hat{E} \tag{7-182}$$

Combining Eqs. (7-179) to (7-182) again recovers the result Eq. (7-173).

The cost of the harmonic-averaging interlaced scheme is only marginally greater than that of the conventional scheme using twice the number of mesh points in one dimension. For example, using the CIC scheme to assign charge in the interlaced scheme involves an extra two arithmetic operations per particle, compared with a twenty-five percent reduction in time for the potential solutions. Under conditions where the discarded harmonics ($\pi/H \leqslant |k| \leqslant 2\pi/H$) are unimportant, the large number of particles per mesh point will make the conventional scheme the most cost-effective. However, as for the force-averaging scheme, this conclusion must be reexamined when limitations of computer storage are encountered in two and three dimensions.

7-8-3 Multidimensional Schemes

The two- and three-dimensional analogues of the two interlaced one-dimensional schemes average forces or harmonics from four and eight interlaced meshes, respectively. Repeating the analysis of the force field gives the vector analog of Eq. (7-173)

$$\mathbf{F} \supset \hat{\mathbf{F}} = q^2 \frac{\hat{W}}{V_c} \hat{\mathbf{D}}\hat{G} \sum_{\substack{\mathbf{n} \\ \mathbf{n} \text{ even}}} \frac{\hat{W}(\mathbf{k_n})}{V_c} \hat{n}(\mathbf{k_n}) \tag{7-183}$$

where the restriction $\mathbf{n}$ even on the alias sum indicates that the component values n_1, n_2, and (in three dimensions) n_3 are even integers.

The dominant alias in the alias sums are those closest to the principal zone, i.e., those with $\mathbf{n} = (10)$ or (01) or in three dimensions $\mathbf{n} = (100)$, (010), (001). Intermediate interlaced schemes which remove these dominant terms are obtained by using only two interlaced meshes (cf. the intermediate charge assignment which became possible in more than one dimension). The most convenient positioning of mesh points of the second mesh with respect to the first in the two-interlaced-mesh multidimensional schemes are at the cell corners of the first mesh. Analysis of these schemes yields the force field given by Eq. (7-183), but now with the restriction that the sum of the components of $\mathbf{n}$ (n_1+n_2 or $n_1+n_2+n_3$) are even. If one pictures k space as a checkerboard, with the principal zone on a white square, then the two-interlaced scheme removes all the aliases centered on the black squares.

Along the principal directions (x = constant, etc., in x space or k_1 = constant, etc., in k space), the alias contributions to the force dispersion relation, etc., are the same as would be obtained using a conventional scheme with a mesh of spacing $H/2$, and are everywhere smaller than for a conventional scheme of mesh spacing H. Thus, if as must be the case for realistic simulation of collisionless systems, wavelengths $|\mathbf{k}| > \pi/H$ are unimportant, then the two interlaced-mesh algorithms with mesh spacing H give a quality of model comparable to that using a single mesh of spacing $H/2$. If the harmonic averaging scheme is used, the three-dimensional, two-interlaced-mesh potential solver uses approximately one-quarter of the CPU time and storage required for the finer-meshed conventional scheme, and therefore may be competitive in terms of computer time as well as storage.

A further application of the ideas of interlacing is to the removal of periodic x space images in potential solvers used for isolated gravitational systems (cf. Chapter 11). For further details on this application we refer readers to Eastwood and Brownrigg (1979).

CHAPTER

EIGHT

PARTICLE-PARTICLE–PARTICLE-MESH (P^3M) ALGORITHMS

8-1 INTRODUCTION

Particle-Particle–Particle-Mesh (P^3M) algorithms are a class of hybrid algorithms developed by the authors (Hockney et al., 1973; Eastwood and Hockney, 1974; Eastwood, 1975, 1976*b*; Eastwood et al. 1980). These algorithms enable correlated systems with long-range (for example, coulombic) forces to be simulated for much larger ensembles of particles than was hitherto possible. The essence of the method is to express the interparticle force as the sum of two component parts; the short-range part $\mathbf{f}^{sr}$, which is nonzero only for particle separations less than some cutoff radius r_e, and the smoothly varying part $\mathbf{R}$, which has a transform which is approximately band-limited (that is to say, is approximately nonzero for only a limited range of $\mathbf{k}$). The total short-range force on a particle $\mathbf{F}^{sr}$ is computed by direct particle-particle (PP) pair force summation and the smoothly varying part is approximated by the particle-mesh (PM) force calculation. Although we restrict discussions in this chapter to long-range forces which are purely coulombic, the results we obtain may be generalized to arbitrary force laws by replacing charge by another suitable measure of the strengths of force centers and by replacing the coulombic reference force $\mathbf{R}$ by the appropriate alternative.

Two meshes are employed in P^3M algorithms: the charge-potential mesh and a coarser mesh, the chaining mesh. The charge-potential mesh is used at different stages of the PM calculation to store, in turn, charge density values, charge harmonics, potential harmonics, and potential values. The chaining mesh is a regular array of cells whose sides have lengths greater than or equal to the cutoff radius r_e of the short-range force. Associated with each cell of this mesh is an entry in the head-of-chain array: This addressing array is used in conjunction with an extra particle coordinate, the linked-list coordinate, to locate pairs of neighboring particles in the short-range calculation. In practice, the head-of-chain array and linked-list coordinates use the same storage space as the charge-potential mesh (cf. Sec. 8-7) and so introduce no extra storage overhead.

Particle orbits are integrated forwards in time using the leapfrog scheme

$$\mathbf{x}_i^{n+1} = \mathbf{x}_i^n + \frac{\mathbf{P}^{n+1/2}}{m_i} DT \tag{8-1}$$

$$\mathbf{P}_i^{n+1/2} = \mathbf{P}_i^{n-1/2} + (\mathbf{F}_i + \mathbf{F}_i^{\mathrm{sr}})DT \tag{8-2}$$

Positions $\{\mathbf{x}_i\}$ are defined at integral timelevels and momenta $\{\mathbf{P}_i\}$ are defined at half-integral timelevels. In Eq. (8-2), $\mathbf{F}_i$ is the mesh-calculated approximation to

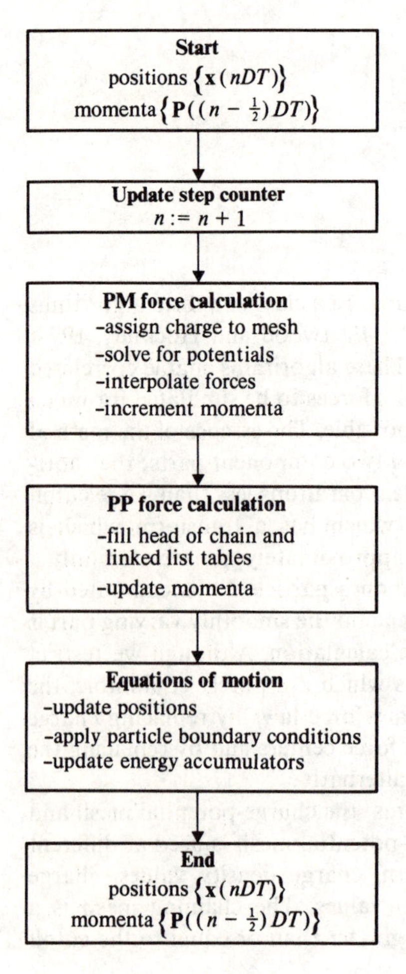

Figure 8-1 The timestep loop of the P³M algorithm.

the total smoothly varying part of the force on particle i and $\mathbf{F}_i^{sr}$ is the total short-range force. Momenta $\{\mathbf{P}_i\}$ are used rather than velocities in P³M computations for reasons of computational economy.

The timestep loop of the P³M algorithm is shown in block form in Fig. 8-1. At the start of the loop the set of particle coordinates are $\{\mathbf{x}^n, \mathbf{P}^{n-1/2}\}$ and at the end of the loop they are $\{\mathbf{x}^{n+1}, \mathbf{P}^{n+1/2}\}$. The PM part of the force calculation is an accurate version of the PM force calculation used for collisionless systems. The way in which the accuracy is obtained is discussed further in Secs. 8-3 and 8-6. The PP part (see Sec. 8-4) is completely separate from the PM force calculation, and could be used by itself to compute interparticle forces in the absence of long-range forces.

The object of the elaborate procedure for calculating the force is computational speed. P³M algorithms, as is shown in Sec. 8-5, have cycle times which scale approximately linearly with particle numbers, whereas conventional schemes with long-range forces have cycle times which increase with the square of particle number. The algorithms are such that thermodynamic quantities, the radial distribution function for small r, and the transform of the pair correlation function are available at each timestep with minimal overhead computational cost. In addition, the computationally costly problem of including periodic images that is present in direct summation methods disappears when P³M is used. If fast Fourier transforms are used to solve for the potential, then periodicity is automatically introduced into the mesh part of the force. Periodic images in the short-range part of the force are needed only for those particles within a distance r_e of the boundary of the computational box.

Despite its obvious speed advantage, the P³M algorithm may seem unattractive because of the difficulties in setting up the force calculation. It will be shown in Secs. 8-3 and 8-6 that such difficulties are illusory. Implementing the Q-minimizing method (cf. Sec. 7-7) in conjunction with the program-timing equations enables optimal combinations of mesh size, force splitting, cutoff radius, charge assignment scheme, influence function, and difference operator to be found automatically for a given choice of force accuracy and particle number. All that is left for a potential user to do is to obtain a P³M program (Eastwood et al., 1980) select the force law he requires and decide what compromise between force accuracy, particle number, and cycle time is appropriate for the configuration he wishes to study.

8-2 FORCE SPLITTING

The change in momentum of particle i at each timestep is determined by the total force on that particle. Thus we are free to choose how we apportion the total force between the short range and the smoothly varying part, or, equivalently, how to apportion the interparticle force between $\mathbf{f}^{sr}$ and $\mathbf{R}$.

The reference force $\mathbf{R}$ is the interparticle force we wish the mesh calculation to represent. For reasons of economy, we require the cutoff radius of $\mathbf{f}^{sr}$ to be as small as possible, and therefore **R** to be equal to the total interparticle force down to as small a particle separation as possible. Unfortunately, we are unable to make the reference and total interparticle forces identical down to arbitrarily small separations, since the mesh used in the PM part of the calculation cannot properly represent any harmonic content of $\hat{\mathbf{R}} \supset \mathbf{R}$ beyond the principal zone boundaries (i.e., for wavenumbers **k** whose components k_i are such that $|k_i| > \pi/H_i$). For example, if we wished to represent a point particle coulombic force, there would be little point in setting $\mathbf{R} \propto \mathbf{x}/|\mathbf{x}|^3$ as this would give $|\hat{\mathbf{R}}| \propto k^{-1}$.

The harmonic content of the reference force is reduced by smoothing. A suitable form of reference force for a coulombic long-range force is one which follows the point particle force law beyond the cutoff radius r_e, and goes smoothly to zero within that radius (see, for example, Fig. 12-7). The smoother the decay of $\mathbf{R}(\mathbf{x})$ as $\mathbf{x} \to 0$ and the larger r_e becomes, the more rapidly the harmonics $\hat{\mathbf{R}}(\mathbf{k})$ decay with increasing **k**. In practice, the errors due to band-limiting of the reference force can be made negligible in comparison to other errors by making $\mathbf{R}(\mathbf{x})$ continuous in one or more derivative and using a cutoff radius greater than two potential mesh cell widths.

Smoothing the coulombic reference interparticle force is equivalent to ascribing a finite size to the charged particle (cf. Sec. 1-5-3). Consequently, a straightforward method of including smoothing is to ascribe some simple density profile $S(\mathbf{x})$ to the reference interparticle force. Examples of shapes which have been found in practice to give comparable total force accuracies are the uniformly charged sphere (shape $S1$), the sphere with uniformly decreasing density (shape $S2$)

$$S2\colon S(r) = \begin{cases} \dfrac{48}{\pi a^4}\left(\dfrac{a}{2} - r\right) & r < a/2 \\ 0 & \text{otherwise} \end{cases} \tag{8-3}$$

and the gaussian distribution of density (shape $S3$). We will henceforth restrict consideration to the $S2$ shape as it gave marginally the better accuracies in three-dimensional schemes for given effective cutoff radii r_e. Reference force harmonics (for unit charges) are related to the shape profiles $S \supset \hat{S}$ by

$$\hat{R}(\mathbf{k}) = -\frac{i\mathbf{k}\hat{S}^2}{k^2} \tag{8-4}$$

The cutoff radius of the short-range force implied by Eq. (8-3) is a rather than r_e. In practice one may take r_e significantly smaller than a because continuity of derivatives at $r = a$ causes the reference force to closely follow the point particle force for radii somewhat less than a. It has been found empirically that a good measure of the lower bound on r_e is given by the cube root of the autocorrelation volume of the charge shapes, i.e.,

$$r_e^3 \geqslant \frac{\int d\mathbf{x} \int d\mathbf{x}'\, S(\mathbf{x})S(\mathbf{x}+\mathbf{x}')}{\int S^2(\mathbf{x})\, d\mathbf{x}} \tag{8-5}$$

giving for the shape $S2$

$$r_e \geqslant \left(\frac{5\pi}{48}\right)^{1/3} a \simeq 0.7\text{a} \tag{8-6}$$

Once the reference interparticle force $\mathbf{R}$ for the PM part of the calculation is chosen, the short-range part $\mathbf{f}^{\text{sr}}$ can be found by subtracting $\mathbf{R}$ from the total interparticle force $\mathbf{f}^{\text{tot}}$:

$$\mathbf{f}^{\text{sr}} = \mathbf{f}^{\text{tot}} - \mathbf{R} \tag{8-7}$$

Figure 12-7 shows a typical splitting of interparticle forces for ionic liquid simulations: The mesh approximation to $\mathbf{R}$ is a purely coulombic force for $S2$ shaped charges. The short-range force $\mathbf{f}^{\text{sr}}$ is the difference between the $S2$ shaped charge coulombic force law and the point charge coulombic force, plus a rapidly varying short-ranged force which represents the repulsive ionic cores. $\mathbf{f}^{\text{sr}}$ is set to zero beyond $r_e = 0.735a$ for the example shown in Fig. 12-7.

8-3 THE MESH FORCE

The total "mesh" part of the force on particle i,

$$\mathbf{F}_i\left(\simeq \sum_{j=1}^{N_p} \mathbf{R}(\mathbf{x}_i - \mathbf{x}_j)\right),$$

is computed using the PM method, i.e.,

1. Assign charge to the charge-potential mesh.
2. Solve for the potential.
3. Difference potentials to find mesh-defined fields.
4. Interpolate to find forces $\mathbf{F}_i$ on particles $i \in [1, N_p]$.

Conceptually it is useful to regard the steps in the PM part of the force calculation solely as artefacts for getting the best approximation $\mathbf{F}_i$ to $\Sigma_j \mathbf{R}(\mathbf{x}_i - \mathbf{x}_j)$. It is the accuracy of the forces which determines the accuracy of the modeling of the dynamics. Errors introduced in intermediate steps are of little consequence if they are self-canceling. The optimization of the PM force calculation leads to (intentional) inaccuracies in the potentials in order to cancel unavoidable errors introduced by assignment, interpolation, and potential differencing (cf. Sec. 7-7 for similar results for collisionless systems).

8-3-1 Charge Assignment

In Chapter 5 we showed that necessary properties of a charge assignment function are that assignment should be over a small number of mesh points, fields should tend towards the correct fields far from the source charge (long range constraint), and that assignment should vary smoothly with the location of the particle (smoothness constraint). The choice of charge assignment scheme for the accurate PM calculation required by P^3M is the result of the compromise between the number of mesh points and smoothness. The number of mesh points determines the computation cost. Smoothness determines the harmonic content of the mesh-calculated force beyond the principal zone. If that harmonic content is large, then aliasing (Sec. 5-6) will cause the computed forces to fluctuate, give erroneous correlations, and lead to poor energy conservation.

Implications of smoothness are best understood in terms of the transform space description of charge assignment. If the assignment function W is smooth, then its convolution ρ' with the density n has a transform $\hat{\rho}'$ which decays rapidly with increasing k. A function W which is continuous in n derivatives has a transform which decays as $k^{-(n+1)}$, so $\hat{\rho}' \propto \hat{W}\hat{n}$ decay much more rapidly than $\hat{n}$. Sampling ρ' leads to periodicity in k space, with the overlap of the periodic repeats with the principal zone giving rise to aliasing.

The ideal assignment function is a low-pass filter. It has a transform $\hat{W}$ which filters out all harmonics of n for $|k_i| > \pi/H$. Such a function is impracticable as it would require assignment from each particle to all mesh points in the computation box. A sequence of increasingly better (and increasingly costly in terms of numbers of arithmetic operations per particle per timestep) approximations to low-pass filters are the hierarchy of assignment functions which satisfy progressively higher-order long-range and smoothness constraints; i.e., NGP (0th order), CIC (1st order), TSC (2nd order), etc. In three dimensions,

$$W(\mathbf{x}) = \frac{1}{V_c} w\left(\frac{x}{H_1}\right) w\left(\frac{y}{H_2}\right) w\left(\frac{z}{H_3}\right) \tag{8-8}$$

where

$$w(x) \supset \hat{w}(k) = \left(\frac{\sin k/2}{k/2}\right)^{p+1} \tag{8-9}$$

In addition to reducing the amplitude of alias harmonics with respect to the principal harmonics, these polynomial assignment functions also change the amplitude of the harmonics within the principal zone. One of the compensating errors included in the potential is to offset the amplitude change of the principal harmonics. For a more detailed discussion of the hierarchy of assignment scheme, the transform space interpretation, sampling, and aliasing we refer readers back to Chapters 5 and 7.

8-3-2 The Force Calculation

The momentum-conserving variant of the PM force calculation is used in P^3M algorithms in order to get the best possible force accuracy for a given order of charge assignment function. Thus, by the momentum conservation constraint (Sec. 5-3-3), the force interpolation function is set equal to the charge assignment function.

If no interlacing is employed, then the steps of the force calculation may be formally expressed as:

Charge assignment:

$$\rho^\dagger(\mathbf{x}) = \frac{q}{V_c} \text{Ш}(\mathbf{x};\mathbf{H}) W(\mathbf{x}) * n(\mathbf{x}) \tag{8-10}$$

Potential solution:

$$\phi'(\mathbf{x}) = G'(\mathbf{x}) * \rho^\dagger(\mathbf{x}) \tag{8-11}$$

Potential differencing:

$$\mathbf{E}^\dagger(\mathbf{x}) = -\text{Ш}(\mathbf{x};\mathbf{H})\mathbf{D}(\mathbf{x}) * \phi'(\mathbf{x}) \tag{8-12}$$

Force interpolation:

$$\mathbf{F}(\mathbf{x}) = \frac{q}{V_c} W(\mathbf{x}) * \mathbf{E}^\dagger(\mathbf{x}) \tag{8-13}$$

Using the results of Secs. 5-6-4 and 7-4, Eqs. (8-10) to (8-13) may be Fourier-transformed and combined to give the mesh-calculated force on a unit charge at position $\mathbf{x}_2$ due to a negative unit charge at $\mathbf{x}_1$:

$$\mathbf{F}(\mathbf{x}_2;\mathbf{x}_1) = \frac{1}{V_b} \sum_{\mathbf{l}} \frac{\hat{W}}{V_c} \hat{\mathbf{D}}\hat{G} \sum_{\mathbf{n}} \frac{\hat{W}(\mathbf{k}_\mathbf{n})}{V_c} \exp(-i\mathbf{k}_\mathbf{n} \cdot \mathbf{x}_1) \exp(i\mathbf{k} \cdot \mathbf{x}_2) \tag{8-14}$$

In Eq. (8-14), we assume a periodic system, so the integral over wavenumbers becomes $V_b^{-1}\Sigma_\mathbf{l}$, where $\mathbf{l}$ is the integer triplet (in three dimensions) labeling harmonics (cf. Sec. 7-4).

The economics of simulations using the P^3M algorithm are such that the optimal operation point has less than one particle per cell. Under these conditions, the harmonic-averaging interlaced-mesh algorithm (Sec. 7-8-2) becomes a competitive alternative to the conventional PM algorithm. (The opposite is generally true for collisionless simulations where a large number of particles per cell is required if $H \sim \lambda_D$.) For example, if the charge distribution is sampled on two meshes, with the second being displaced by $\mathbf{H}/2$ with respect to the first, then the alias sum over $\mathbf{n}$ in Eq. (8-14) is reduced to the sum of values of $\mathbf{n} = (n_1, n_2, n_3)$ where $n_1 + n_2 + n_3$ are even.

8-3-3 Errors in the Force

A measure of the quality of the mesh-calculated approximation $\mathbf{F}(\mathbf{x} = \mathbf{x}_2 - \mathbf{x}_1; \mathbf{x}_1)$ to the reference interparticle force $\mathbf{R}(\mathbf{x})$ is given by the displacement-averaged total squared deviation of $\mathbf{F}$ from $\mathbf{R}$:

$$Q = \frac{1}{V_c} \int_{V_c} d\mathbf{x}_1 \int_{V_b} d\mathbf{x} |\mathbf{F}(\mathbf{x};\mathbf{x}_1) - \mathbf{R}(\mathbf{x})|^2 \tag{8-15}$$

Given a particular choice W of charge-assignment–force-interpolation function and a particular difference operator $\mathbf{D}$, then, by Eqs. (8-10) to (8-13) and Eq. (8-15), Q may be regarded as a quadratic functional of the set of discrete Green's function values $\{G_{\mathbf{p}} = G(\mathbf{x_p});\ p_i \in [0, N_i/2]\}$. The components p_i of $\mathbf{p} = (p_1, p_2, p_3)$ are restricted to half the number of mesh points $N_i/2$ along the i axis of the periodic computational box since G must be an even function of its arguments. In principle, we can apply variational methods to Q to obtain discrete sets of equations for the optimal set of values $\{G_q^{\text{opt}};\ q_i \in [0, N_i/2]\}$, i.e.,

$$\frac{\partial Q}{\partial G_{\mathbf{q}}} = \frac{1}{V_c}\int d\mathbf{x}_1 \int_{V_b} d\mathbf{x}\, 2(\mathbf{F}-\mathbf{R})\cdot\frac{\partial \mathbf{F}}{\partial G_{\mathbf{q}}} = 0 \tag{8-16}$$

Equation (8-16) gives an algebraic equation for $G_{\mathbf{q}}$ and hence defines the optimal (Q-minimizing) form of the algebraic equations relating mesh potential values to charge density values [cf. Eq. (5-153) for energy conserving schemes].

Optimization can be extended to a limited set of difference gradient operators by expressing $\mathbf{D}$ as a linear combination of gradient operations $\mathbf{D}_m$ whose components i are two point-centered difference derivatives taken over points of separation $2mH_i$, $m = 1, 2$, etc. and then minimizing Eq. (8-15) with respect to the linear combination factors. For instance, in the case of a four-point difference derivative [Eq. (5-137)]

$$\mathbf{D} = \alpha\mathbf{D}_1 + (1-\alpha)\mathbf{D}_2 \tag{8-17}$$

and Q_{opt} is given by minimizing Q with respect to α and $\{G_{\mathbf{p}}\}$.

In practice, the x space integrals in Eq. (8-16) prove difficult to evaluate. One approach is to write the Green's function as the sum of a Green's function satisfying a finite-difference approximation to Poisson's equation plus correction factors with undetermined multipliers, then numerically minimize Q with respect to the undetermined factors (Eastwood and Hockney, 1974). A more powerful approach is to express Eq. (8-15) in terms of Fourier components and work in terms of the influence function values $\{\hat{G}(\mathbf{k})\}$ (Eastwood, 1975, 1976*b*).

Applying the power theorem to Eq. (8-15) yields

$$Q = \frac{1}{V_c}\int_{V_c} d\mathbf{x}_1 \frac{1}{V_b}\sum_{\mathbf{l}} [|\hat{\mathbf{F}}(\mathbf{k};\mathbf{x}_1)|^2 - 2\hat{\mathbf{R}}^*(\mathbf{k})\cdot\hat{\mathbf{F}}(\mathbf{k};\mathbf{x}_1) + |\hat{\mathbf{R}}(\mathbf{k})|^2] \tag{8-18}$$

From Eq. (8-14),

$$\hat{\mathbf{F}}(\mathbf{k};\mathbf{x}_1) = \frac{\hat{W}}{V_c}\hat{\mathbf{D}}\hat{G}\sum_{\mathbf{n}}\frac{\hat{W}(\mathbf{k_n})}{V_c}\exp(i(\mathbf{k}-\mathbf{k}_n)\cdot\mathbf{x}_1) \tag{8-19}$$

Therefore

$$Q = \frac{1}{V_b}\sum_{\mathbf{l}}\left\{\hat{U}^2|\hat{\mathbf{D}}|^2\hat{G}^2\sum_{\mathbf{n}}\hat{U}^2(\mathbf{k_n}) - 2\hat{\mathbf{R}}^*\cdot\hat{\mathbf{D}}\hat{G}\hat{U}^2 + |\hat{\mathbf{R}}|^2\right\} \tag{8-20}$$

where $\hat{U} = \hat{W}/V_c$. The periodicity of $\hat{\mathbf{D}}$ and $\hat{G}$, i.e., $\hat{\mathbf{D}}(\mathbf{k_n}) = \hat{\mathbf{D}}(\mathbf{k})$, $\hat{G}(\mathbf{k_n}) = \hat{G}(\mathbf{k})$, can be used to rewrite Eq. (8-20) in terms of the harmonics in a single $\mathbf{k}$ space period interval

$$Q = \frac{1}{V_b}\sum_{\mathbf{l}=0}^{\mathbf{N}-1}\left\{|\hat{\mathbf{D}}|^2\hat{G}\left[\sum_{\mathbf{n}}\hat{U}^2\right]^2 - 2\hat{G}\hat{\mathbf{D}}\cdot\sum_{\mathbf{n}}\hat{\mathbf{R}}^*\hat{U}^2 + \sum_{\mathbf{n}}|\hat{\mathbf{R}}|^2\right\} \tag{8-21}$$

Minimizing Q with respect to the set of influence function values $\{\hat{G}(\mathbf{k})\}$ gives the optimal influence function

$$\hat{G}(\mathbf{k}) = \frac{\hat{\mathbf{D}}(\mathbf{k}) \cdot \sum_{\mathbf{n}} \hat{U}^2(\mathbf{k_n})\hat{\mathbf{R}}(\mathbf{k_n})}{|\hat{\mathbf{D}}(\mathbf{k})|^2 \left[\sum_n \hat{U}^2(\mathbf{k_n}) \right]^2} \tag{8-22}$$

and the minimized $Q = Q_{\text{opt}}$

$$Q_{\text{opt}} = \frac{1}{V_b} \sum_{\mathbf{l}=0}^{\mathbf{N}-1} \left\{ \sum_{\mathbf{n}} |\hat{\mathbf{R}}|^2 - \frac{\left| \hat{\mathbf{D}} \cdot \sum_{\mathbf{n}} \hat{U}^2 \hat{\mathbf{R}}^* \right|^2}{|\hat{\mathbf{D}}|^2 \left[\sum_{\mathbf{n}} \hat{U}^2 \right]^2} \right\} \tag{8-23}$$

Optimization is extended to a limited set of operators **D** by expressing $\hat{\mathbf{D}}$ as a linear combination of transforms of two point operators [cf. Eq. (8-17)] and minimizing Eq. (8-23) with respect to the linear combination factors.

The residual errors making up $Q_{\text{opt}} = \min_{\{\alpha, \hat{G}\}} [Q]$ arise because charge-sampling aliases are not completely suppressed, the difference gradient is not always parallel to the true gradient, and the reference force **R** and mesh-calculated force **F** do not have band-limited transforms. The problem of determining the relative importance of the contributory factors to the overall errors in the mesh force is resolved by evaluating three subsidiary quantities E_{ref}, P, and Z.

Z is the total squared deviation of the displacement-averaged mesh force from reference force, and so gives a measure of deviation of the nonfluctuating part of **F** from **R**. From Eq. (8-19) we have

$$\frac{1}{V_c} \int_{V_c} d\mathbf{x}_1\, \hat{\mathbf{F}}(\mathbf{k}; \mathbf{x}_1) = \langle \hat{\mathbf{F}}(\mathbf{k}; \mathbf{x}_1) \rangle = \hat{U}\hat{\mathbf{D}}\hat{G}\hat{U}$$

hence

$$\begin{aligned} Z &= \int d\mathbf{x} |\langle \mathbf{F}(\mathbf{x}; \mathbf{x}_1) \rangle - \mathbf{R}(\mathbf{x})|^2 \\ &= \frac{1}{V_b} \sum_{\mathbf{l}} |\langle \hat{\mathbf{F}}(\mathbf{k}; \mathbf{x}_1) \rangle - \hat{\mathbf{R}}(\mathbf{k})|^2 \\ &= \frac{1}{V_b} \sum_{\mathbf{l}} \{ |\hat{\mathbf{D}}|^2 \hat{G}^2 \hat{U}^4 - 2\hat{G}\hat{\mathbf{D}} \cdot \hat{\mathbf{R}}^* \hat{U}^2 + |\hat{\mathbf{R}}|^2 \} \\ &= \frac{1}{V_b} \sum_{\mathbf{l}=0}^{\mathbf{N}-1} \left\{ |\hat{\mathbf{D}}|^2 \hat{G}^2 \left[\sum_{\mathbf{n}} \hat{U}^4 \right] - 2\hat{G}\hat{\mathbf{D}} \cdot \left[\sum_{\mathbf{n}} \hat{\mathbf{R}}^* \hat{U}^2 \right] + \sum_{\mathbf{n}} |\hat{\mathbf{R}}|^2 \right\} \end{aligned} \tag{8-24}$$

P is the total squared deviation of the mesh force from its displacement average:

$$\begin{aligned} P &= \frac{1}{V_c} \int_{V_c} d\mathbf{x}_1 \int_{V_b} d\mathbf{x}\, |\mathbf{F}(\mathbf{x}; \mathbf{x}_1) - \langle \mathbf{F}(\mathbf{x}; \mathbf{x}_1) \rangle|^2 \\ &= \frac{1}{V_b} \sum_{\mathbf{l}} |\hat{\mathbf{D}}|^2 \hat{G}^2 \hat{U}^2 \left\{ \left[\sum_{\mathbf{n}} \hat{U}^2 \right] - \hat{U}^2 \right\} \\ &= \frac{1}{V_b} \sum_{\mathbf{l}=0}^{\mathbf{N}-1} |\hat{\mathbf{D}}|^2 \hat{G}^2 \left\{ \left[\sum_n \hat{U}^2 \right]^2 - \sum_n \hat{U}^4 \right\} \end{aligned} \tag{8-25}$$

P measures the contribution of aliasing to Q. If the charge assignment function were band-limited, rather than only approximately band-limited, then P would be zero. E_{ref} gives the smallest residual deviation of the mesh force from the reference force. If both the assignment function were band-limited and the directional errors in the gradient operator were absent, then the optimum $\hat{G}$ would lead to total cancellation of errors in the principal zone ($\mathbf{l} \in [-\mathbf{N}/2, \mathbf{N}/2-1]$ or equivalently $\mathbf{l} \in [0, \mathbf{N}-1]$) and to a residual error

$$E_{ref} = \frac{1}{V_b} \sum_{\mathbf{l}=0}^{\mathbf{N}-1} \left\{ \left[\sum_{\mathbf{n}} |\hat{\mathbf{R}}|^2 \right] - |\hat{\mathbf{R}}|^2 \right\} \tag{8-26}$$

It follows from the definitions of Q, P, and Z that

$$Q = P + Z \tag{8-27}$$

Dominance of P indicates that charge undersampling is the main source of errors. Significant improvements in this case may be realized by using interlacing or a high-order assignment function. If Z dominates, then directional errors in the gradient plus a too-large harmonic content of the reference force beyond the principal zone are the controlling factor. The value of Z relative to that of E_{ref} helps determine whether directional errors or band limiting are the main factors in Z. If Z is much larger than E_{ref}, significant improvements are possible by using a higher-order gradient operator, otherwise a different reference force or larger cutoff radius must be used to obtain reduction in Z and thence in Q. The practical application of these quantities is discussed further in Sec. 8-6.

The expressions Eqs. (8-20) to (8-25) are for the conventional PM calculation. If interlacing is employed, Eq. (8-21) becomes

$$Q = \frac{1}{V_b} \sum_{\mathbf{l}=0}^{\mathbf{N}-1} \left\{ |\hat{\mathbf{D}}|^2 \hat{G}^2 \left[\sum_{\mathbf{n}} \hat{U}^2 \right] \left[\sum_{\mathbf{n}'}{}' \hat{U}^2 \right] - 2\hat{G}\hat{\mathbf{D}} \cdot \sum_{\mathbf{n}} \hat{\mathbf{R}}^* \hat{U}^2 + \sum_{\mathbf{n}} |\hat{\mathbf{R}}|^2 \right\} \tag{8-28}$$

where the prime on the sum over $\mathbf{n}'$ indicates that the sum is over values $\mathbf{n}'$ left after interlacing. For two interlaced meshes, the prime represents the sum over $\mathbf{n}'$ values where $n_1' + n_2' + n_3'$ is even. All other quantities and sums in Eq. (8-28) are the same as for the conventional schemes. The optimal influence function becomes

$$\hat{G}(\mathbf{k}) = \frac{\hat{\mathbf{D}} \cdot \sum_{\mathbf{n}} \hat{\mathbf{R}}^* \hat{U}^2}{|\hat{\mathbf{D}}|^2 \left[\sum_{\mathbf{n}} \hat{U}^2 \right] \left[\sum_{\mathbf{n}'}{}' \hat{U}^2 \right]} \tag{8-29}$$

The expression for Z [Eq. (8-24)] is unchanged and Eq. (8-25) becomes

$$P = \frac{1}{V_b} \sum_{\mathbf{l}=0}^{\mathbf{N}-1} |\hat{\mathbf{D}}|^2 \hat{G}^2 \left\{ \left[\sum_{\mathbf{n}} \hat{U}^2 \right] \left[\sum_{\mathbf{n}'}{}' \hat{U}^2 \right] - \sum_{n} \hat{U}^4 \right\} \tag{8-30}$$

8-4 THE SHORT-RANGE FORCE

The total short-range part of the force on a particle i at position $\mathbf{x}_i$ is given by the sum of the interparticle short-range forces

$$\mathbf{F}_i^{\text{sr}} = \sum_{j=1}^{N_p} \mathbf{f}_{ij}^{\text{sr}} \tag{8-31}$$

The elementary method of evaluating $\mathbf{F}_i^{\text{sr}}$ is to sweep through all particles $j = 1, \ldots, N_p$, test whether the separation $r_{ij} = |\mathbf{x}_i - \mathbf{x}_j|$ is less than r_e, and, if so, compute $\mathbf{f}_{ij}^{\text{sr}}$ and add it to $\mathbf{F}_i^{\text{sr}}$. Such an approach is clearly impractical, since for each of the N_p values of i one would have to test $N_p - 1$ separations r_{ij} giving an operations count scaling as N_p^2.

8-4-1 The Chaining Mesh

The computational cost of locating those particles j which contribute to the short-range force on particle i is greatly reduced if the particle coordinates are ordered such that the tests for locating particles j such that $r_{ij} \leqslant r_e$ need only be performed over a small subset N_n of the total number of particles N_p. It is for this reason that the chaining mesh is introduced. The chaining mesh (in three dimensions) is a regular lattice of $(M_1 \times M_2 \times M_3)$ cells, covering the computational box (of side $L_1 \times L_2 \times L_3$) in much the same manner as the $(N_1 \times N_2 \times N_3)$ cells of the much

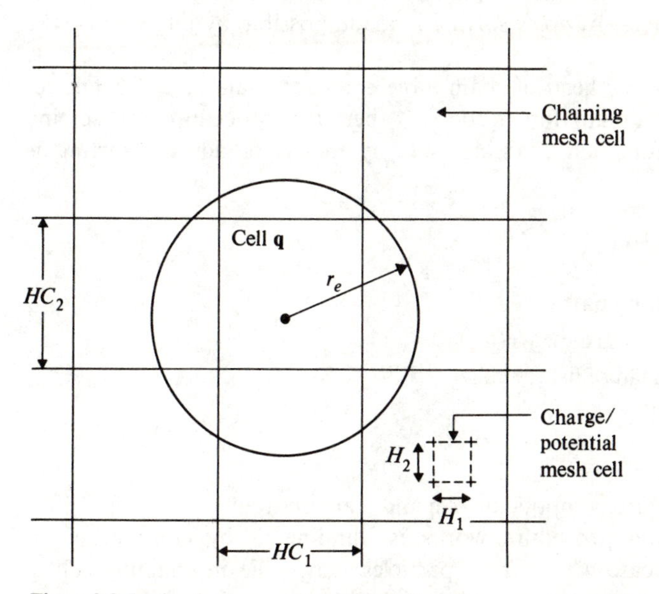

Figure 8-2 In the short-range force calculation, the computational box is divided into chaining cells. Contributions to the force $\mathbf{F}_i^{\text{sr}}$ on particle i in cell $\mathbf{q}$ are nonzero only from particles j in cell $\mathbf{q}$ and the neighboring cells.

finer charge-potential mesh. The number of cells M_s along the s direction is given by the largest integer less than or equal to L_s/r_e. Consequently, the lengths of the sides of the cells of the chaining mesh are always greater than or equal to the cutoff radius r_e.

Figure 8-2 depicts a chaining mesh in two dimensions. Typically, the side lengths of the chaining mesh cells HC_s are between three and four times greater than the side lengths H_s of the cells of the charge-potential mesh. The circle of radius r_e centered on particle i in chaining cell $\mathbf{q}$ delineates the area in which particles j must lie if they are to have a nonzero contribution to $\mathbf{F}_i^{sr}$. Since $HC_s \geqslant r_e$ for all s, it follows that those particles j which have nonzero contributions to $\mathbf{F}_i^{sr}$ must either lie in the same cell $\mathbf{q}$ as particle i or in one of the eight neighboring cells. If the particle coordinates are sorted into lists for each chaining cell, then to find the force $\mathbf{F}_i^{sr}$ on particle i involves approximately $9N_C$ tests, where $N_C(=N_p/M_1M_2)$ is the average number of particles per chaining cell. Therefore, if Newton's third law is used, the total number of tests in finding all the short-range forces is approximately $N_nN_p \simeq 4.5N_CN_p$ as compared with N_p^2 for the elementary approach. Similarly, in three dimensions, sorting coordinates into chaining cells gives the number of tests $N_nN_p \sim 13N_CN_p$.

8-4-2 The Linked Lists

For serial computers (but not necessarily for vector or array processor machines) it is computationally more efficient to sort the coordinate addresses rather than the coordinates themselves. Address sorting is made possible by introducing the linked-list array LL.

If we let HOC($\mathbf{q}$) be the head-of-chain table entry for chaining cell $\mathbf{q}$, and let LL(i) be the link coordinate for particle i, then the procedure for sorting coordinates into lists for each chaining cells by means of address sorting is summarized as follows:

1. set HOC($\mathbf{q}$) = 0 for all $\mathbf{q}$.
2. do for all particles i.
   (a) locate cell containing particle
       $\mathbf{q}\!:= \text{int}\,(x_1/HC_1, x_2/HC_2, x_3/HC_3)$
   (b) add particle i to head of list for cell $\mathbf{q}$
       LL(i): = HOC($\mathbf{q}$)
       HOC($\mathbf{q}$): = i

In two dimensions, the third components of $\mathbf{q}$ and $\mathbf{x}$ are omitted.

The way the sorting procedure works is illuminated by considering an example. Consider the case where three particles i_1, i_2, i_3 lie in chaining cell $\mathbf{q}$, where $i_1 < i_2 < i_3$. We represent the coordinates by a three-partition box.

i	$\mathbf{X}_i$	LL_i

where i is the address (or array element in FORTRAN), $\mathbf{X}_i$ are the physical particle coordinates $(\mathbf{x}_i, \mathbf{P}_i)$, and LL_i is the linked-list coordinate. If particle coordinates are swept through in increasing i values then the linked list for cell $\mathbf{q}$ develops as follows:

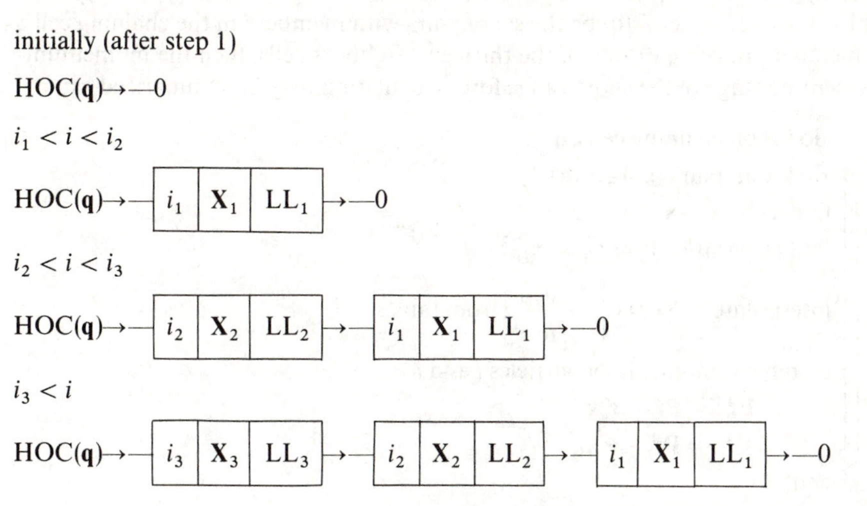

The speed and simplicity of creating the linked lists from scratch make it pointless saving and updating them timestep by timestep. The whole sorting process requires only three real arithmetic operations per particle in three dimensions or two in two dimensions.

Once the HOC and LL tables have been filled, a zero entry in $\mathrm{HOC}(\mathbf{q})$ indicates that there are no particles in chaining cell $\mathbf{q}$. A nonzero entry gives the address of the coordinates of the first particle in the list. The link coordinate of a particle either gives the address of the coordinates of the next particle in the list, or is zero to indicate the end of the list. Therefore, given HOC and LL, coordinates in each cell can be looked up without any searching.

8-4-3 The Momentum Change

The computation of the short-range force contribution to the momentum change at each timestep is a two-stage process. The first stage is to fill the HOC and LL arrays as described above, and the second is to accumulate the changes in momentum.

In three dimensions, particles which have a nonzero contribution to $\mathbf{F}_i^{\mathrm{sr}}$ must lie either in the same chaining cell as particle i or in one of the twenty-six neighboring chaining cells. To eliminate the unnecessary double computation of values $\mathbf{f}_{ij}^{\mathrm{sr}}$ used in incrementing the momenta of both particles i and j, we want to process each particle pair (i,j), $i \neq j$ only once. One method of achieving this is to sweep through each chaining cell, which we will call the current cell. For the current-chaining cell, $\mathbf{f}_{ij}^{\mathrm{sr}}$ is obtained for all pairs (i,j) with both members in the current cell, and for all pairs with one member i in the current cell and one member in one of thirteen of the neighboring cells. The same thirteen neighbors are used for each current cell. For example, one may choose the neighbors of chaining cell $\mathbf{q} = (q_1, q_2, q_3)$ to be $\{(q_1+s,\ q_2-1,\ q_3+t),\ (q_1+s,\ q_2,\ q_3-1),\ (q_1-1,\ q_2,\ q_3);\ s, t \in [-1, 1]\}$. If we let $\mathscr{P}(\mathbf{q})$ be the set of pairs with member i in the chaining cell $\mathbf{q}$ and member j in cell $\mathbf{q}$ or one of the thirteen neighbors cells, then the momentum-incrementing stage of the short-range force calculation may be summarized as:

```
┌─do for all chaining cells, q
│┌─do for all pairs (i,j) ∈ 𝒫(q)
││ find x_ij = x_i − x_j
││ find (separation)² = r²_ij = |x_ij|²
││ interpolate 𝓕 = DT[f^sr(r_ij)/r_ij] from tables
││ increment momenta of particles i and j
││       P*_i := P*_i + 𝓕 x_ij
││       P*_j := P*_j − 𝓕 x_ij
└└─end
```

Considerable computational savings are made in the momentum-incrementing loop by tabulating the values of $\mathscr{F}$ at uniform intervals of r^2, thus avoiding the computation of square roots in the pairwise summation loop. For example, if linear interpolation on the table is used, then, given r^2, $\mathscr{F}$ can be obtained from tabular values τ_t using only four real arithmetic operations:

$$\xi = \frac{r^2}{\Delta r^2} \qquad \text{(1 op)}$$

$$t = \operatorname{int}(\xi)$$

$$\mathscr{F} = \tau_t + (\xi - t)(\tau_{t+1} - \tau_t) \qquad \text{(3 ops)}$$

Further computational savings in the momentum-incrementing stage of the short-range force calculation can be made by refining the linked-list address sorting. For example, the linked lists for each chaining cell can be accumulated so that particles are ordered in the list in increasing y values. This enables the pair separation tests for each current-cell–neighboring-cell pair to be terminated on the first encounter of $\Delta y = y_i - y_j > r_e$. For the nine neighboring cells $\{(q_1+s, q_2, q_3+t);\ (s,t) \in [-1, 1]\}$ of chaining cell $\mathbf{q}$ this will lead, on average, to a 50 percent reduction of the number of pair separations to be tested. In general, the saving in time in the momentum-incrementing stage more than offsets the increased cost in creating the linked lists. The change in the linked-list creation algorithm consists of testing the y value y_i of the new entry i against successive existing y values in the list, and inserting entry i in the list before the first existing entry j encountered with $y_j > y_i$:

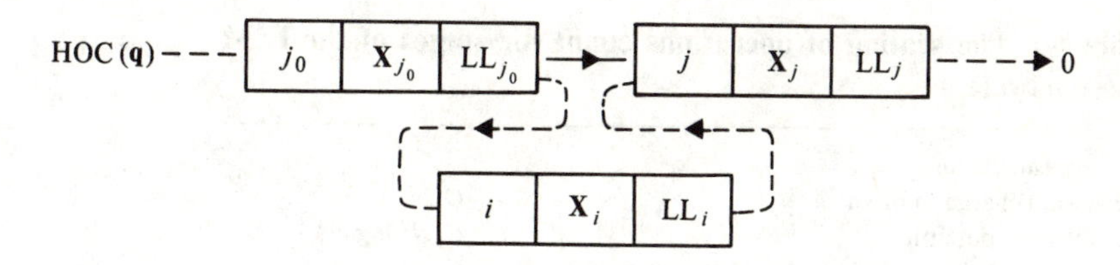

The operations involved in replacing the old link by the new link (broken lines) in this example are

$$\mathrm{LL}(i) := j$$
$$\mathrm{LL}(j_0) := i$$

8-5 THE TIMING EQUATION

The Q-minimizing method outlined in Sec. 8-3-3 gives optimal combinations of assignment function, influence function, and difference operator. Many optimal combinations can be found by choosing different assignment functions, using interlacing, varying the restricted class of difference operator, or simply by changing the reference force. Each makes different demands on computer time and storage. The value of Q provides a measure of the quality for all such optimal combinations. The timing equation provides the extra information required to determine which of the many ways of achieving a given value of Q is best, the optimal scheme being that which gives the desired force accuracy for least computational cost.

A further factor entering the determination of computational costs is storage limitations. Although partitioning the particle and/or mesh data and using serial access backing store presents no fundamental problems, practical experience shows that it is best avoided where possible. This eliminates the costly step of physically sorting particle coordinates and prevents calculations becoming dominated by disk transfer times. Here, we assume storage limitations are introduced in the simplest manner. If a given scheme needs more random-access memory than is available, then it is removed from the list of alternatives. Cycle times of the remaining possibilities are then compared to find the optimal variant.

Table 8-1 lists the stages of the timestep cycle, and gives the dependence of the number of operations at each stage on the number of particles N_p, on the number of mesh points in the charge potential mesh N^3, and on the cutoff radius r_e of the short-range force. For clarity, we restrict consideration to a cubic charge-potential mesh of side N. The quantities C_i; $i = 1, \ldots, 11$ are constants, the values of which depend on the detailed form of the algorithm used for each stage. To first order, $\{C_i\}$ are proportional to the number of real arithmetic operations. The functional dependence on N_p, N, etc., is established by investigating the scaling of operations count. The $N^3 \log N^3$ scaling of the fast Fourier transform solver was discussed in Chapter 6. The quantity $(r_e^3 N_p / H^3 N^3)$ in the short-range pair force calculation is, to within a constant, the average number of neighbors within the sphere of radius

Table 8-1 The scaling of operations count for stages of the P^{3}M timestep cycle

PM force calculation	
1. Assign charge to mesh	$C_1 N_p$
2. Solve for potential	$C_2 N^3 \log N^3$
3. Interpolate forces and accelerate particles	$C_3 N_p$
PP force calculation	
1. Fill HOC and LL arrays	$C_4 N_p$
2. Interpolate forces and accelerate particles	$C_5\left(\frac{r_e^3 N_p}{H^3 N^3}\right) N_p$
Equations of motion	
Move particles	$C_6 N_p$
Diagnostic overheads	
Internal energy	$C_7 N^3 + C_8 N_t$
Kinetic pressure	$C_9 N_p$
Interaction pressure	$C_9 N_t$
Radial distribution function (for $r < r_e$)	$C_{10}\left(\frac{r_e^3 N_p}{H^3 N^3}\right) N_p$
Structure factor (for $kH < \pi$)	$C_{11} N^3$

r_e around each particle. Further consideration of the operations count and scaling for some of the force calculation steps is given in Sec. 8-7. Establishing the operations count and scaling for the remaining steps we leave as an exercise for the reader.

The final section of Table 8-1, labeled "Diagnostic overheads," has been included to emphasize that measurement of physical quantities in a P^{3}M program is a small overhead. The measurement of the radial distribution function g for $r \leqslant r_e$ requires one extra operation per particle pair within the short-range force calculation loop if it is binned using an NGP scheme, or three if linear inverse interpolation is employed. Once g is determined, the short-range parts of the interaction pressure and internal energy can be obtained at the cost of two operations per tabular point of g. N_t is the total number of tabular points used in accumulating g. Terms scaling as N_t take a negligible portion of the total cycle time. The kinetic pressure is obtained from the particle kinetic energy. The long-range part of the internal energy and the structure factor (i.e., the transform of the density autocorrelation function) for small k are obtained directly from the density harmonics. Further details on physical measurement are given in Sec. 12-3-4.

The total cycle time of the P^{3}M timestep cycle T is obtained by summing the contributions of all the stages. This gives

$$T = \alpha N_p + \beta N^3 \log N^3 + \gamma \left(\frac{r_e}{H}\right)^3 \frac{N_p^2}{N^3} \tag{8-32}$$

where coefficients α, β, and γ depend on the particular form of the various stages of the calculation cycle. Note that in Table 8-1 and Eq. (8-32), only the dominant factors are retained—factors such as surface terms in the potential solution (scaling like N^2) are ignored.

The cost per particle per timestep Γ is given by

$$\Gamma = \frac{T}{N_p} = \alpha + \frac{\beta N^3 \log N^3}{N_p} + \gamma\left(\frac{r_e}{H}\right)^3 \frac{N_p}{N^3} \tag{8-33}$$

If N_p is small, then the dominant factor is the time used in solving for the potential [the $N^3 \log N^3$ term in Eq. (8-33)]. As N_p is increased, the cost per particle of the potential solution declines and the dominant term becomes time for computing the pair forces in the PP force calculation. For a given number of mesh points N^3 the optimal operating point is achieved when

$$\frac{\partial \Gamma}{\partial N_p} = -\frac{\beta N^3 \log N^3}{N_p^2} + \gamma\left(\frac{r_e}{H}\right)^3 \frac{1}{N^3} = 0 \tag{8-34}$$

i.e., when

$$\beta N^3 \log N^3 = \gamma\left(\frac{r_e}{H}\right)^3 \frac{N_p^2}{N^3} \tag{8-35}$$

Comparison of Eq. (8-35) with Eq. (8-32) shows that the most cost-effective operating point for a P³M program is achieved when the computer time for solving for the potential is equal to that used in evaluating the short-range part of the forces.

Rearranging Eq. (8-35) gives the optimal number of particles per cell

$$\left(\frac{N_p}{N^3}\right)_{\text{opt}} = \left(\frac{\beta \log N^3}{\gamma (r_e/H)^3}\right)^{1/2} \tag{8-36}$$

If one works at the optimal point, then the timing equation Eq. (8-32) reduces to

$$T = \left[\alpha + 2\left(3\gamma\beta\left(\frac{r_e}{H}\right)^3 \log N\right)^{1/2}\right] N_p \tag{8-37}$$

Thus, apart from the weak nonlinear dependence on N_p through the $\sqrt{\log N}$ term, P³M schemes have a cycle time which scales linearly with particle number. The consequence of the scaling law is that P³M algorithms have made possible the simulations of correlated systems of previously unattainable size.

8-6 OPTIMIZATION

The results of Secs. 8-3-3 and 8-5 may be combined to give a quantitative method for designing P³M programs and for selecting the numerical parameters for use in such programs. To illustrate this method, we shall consider its application to the design of the program P3M3DP (Eastwood, 1976*b*; Eastwood et al., 1980). Examples of computations performed using P3M3DP are described in Chapters 11 and 12.

8-6-1 Calculation of Force Accuracy

The first stage in the design optimization is to identify the combination of elements of the force calculation which give acceptable force accuracies. The expressions for the errors in the interparticle forces obtained in Sec. 8-3-3 enable this step to be performed for a wide range of alternatives without undertaking the onerous task of coding (and debugging!) each variant. All variants are parameterized by three transform space quantities $\hat{\mathbf{D}}$, $\hat{U}$, and $\hat{\mathbf{R}}$ plus restrictions on the values of $\mathbf{n}$ in the alias sums for interlaced schemes. Obviously, there is little point in using any choice of $\hat{G}$ apart from the optimal ones [Eq. (8-22) or Eq. (8-29)] which are fully determined by $\hat{\mathbf{D}}$, $\hat{U}$, $\hat{\mathbf{R}}$ and the restrictions on $\mathbf{n}$. Given these quantities, the accuracy of the interparticle forces can be measured by evaluating Q, P, Z, and E_{ref} using Eqs. (8-23), (8-24), (8-25), and (8-26) or the corresponding equations for interlaced schemes.

Although the expressions for Q, etc., contain triple infinite sums (over $\mathbf{n}$) they prove simple to evaluate. The sums are such that they either converge very rapidly or can be written in closed functional form. Consider, for example, Eq. (8-23). Using Eq. (8-4) the first infinite sum may be written

$$\sum_{\mathbf{n}} |\hat{\mathbf{R}}(\mathbf{k}_{\mathbf{n}})|^2 = \sum_{\mathbf{n}} \frac{\hat{S}^4(\mathbf{k}_{\mathbf{n}})}{|\mathbf{k}_{\mathbf{n}}|^2} \tag{8-38}$$

We restrict consideration to a cubic mesh ($H = H_1 = H_2 = H_3$), so $\mathbf{k}_{\mathbf{n}} = \mathbf{k} + \mathbf{n}k_g$ where $k_g = 2\pi/H$. Practical choices of $S \supset \hat{S}$ are restricted to spherically symmetric functions $S(\mathbf{x}) = S(r)$, $r = |\mathbf{x}|$. Other choices would necessitate the use of computationally expensive multidimensional force correction tables in the short-range force calculation. For spherically symmetric shape factors S, the three-dimensional Fourier transform reduces to

$$\hat{S}(k) = 4\pi \int_0^\infty r^2\, dr\, S(r) \frac{\sin kr}{kr} \tag{8-39}$$

where
$$k = |\mathbf{k}| = (k_1^2 + k_2^2 + k_3^2)^{1/2} \tag{8-40}$$

Note that in Eq. (8-39) we have used the infinite-system Fourier transform. We can likewise use the infinite-system transform for the force, since sampling the infinite-system harmonics gives the harmonics of the periodic system. This simple relationship between harmonics is in marked contrast with the complications introduced by periodicity in x space. If we wanted to evaluate Q using Eq. (8-15), then the integrals would involve elliptic functions (in two dimensions) or expansions in spherical harmonics (in three dimensions). Evaluating Eq. (8-39) for the $S2$ particle-shaped [Eq. (8-3)] gives

$$\hat{S}(k) = \frac{12}{(ka/2)^4}\left(2 - 2\cos\frac{ka}{2} - \frac{ka}{2}\sin\frac{ka}{2}\right) \tag{8-41}$$

Thus, for an $S2$ particle shape, S^4 decays as $(ka/2)^{-16}$. Provided that $a/H \gtrsim 1$, terms where the components of $\mathbf{n}$ have magnitudes greater than one or two are negligible.

Terms in the second sum in Eq. (8-23) which involves harmonics of the reference force also exhibit a rapid decrease in magnitude as components of $\mathbf{n}$ move further from zero. $\hat{\mathbf{R}}^*$ decays as k^{-9} and $\hat{U}^2$ decays as $k^{-2(p+1)}$ where p is the order of the assignment function. In particular, for the NGP–CIC–TSC–··· hierarchy, $\hat{U}(\mathbf{k}) = \hat{U}(k_1, k_2, k_3)$ is given by

$$\hat{U}(\mathbf{k}) = \left(\frac{\sin \dfrac{k_1 H}{2} \sin \dfrac{k_2 H}{2} \sin \dfrac{k_3 H}{2}}{\dfrac{k_1 H}{2} \quad \dfrac{k_2 H}{2} \quad \dfrac{k_3 H}{2}} \right)^{p+1} \tag{8-42}$$

The remaining alias sum in Eq. (8-23) can be written in closed form using the trigonometric function identities introduced in the previous chapter. For instance, the TSC scheme ($p = 2$) gives the sum

$$\sum_{\mathbf{n}} \hat{U}^2 = \prod_{i=1}^{3} \left[\sin^6 \frac{k_i H}{2} \sum_{n_i=-\infty}^{\infty} \frac{1}{\left(\dfrac{k_i H}{2} - \pi n_i \right)^6} \right] \tag{8-43}$$

The identity

$$\sin^{-6} x \left(1 - \sin^2 x + \frac{2}{15} \sin^4 x \right) = \sum_{n=-\infty}^{\infty} \frac{1}{(x - \pi n)^6} \tag{8-44}$$

reduces Eq. (8-43) to

$$\sum_{\mathbf{n}} \hat{U}^2 = \prod_{i=1}^{3} \left(1 - \sin^2 \frac{k_i H}{2} + \frac{2}{15} \sin^4 \frac{k_i H}{2} \right) \tag{8-45}$$

The Fourier transform of the gradient difference operator $\hat{\mathbf{D}}$ is obtained by transforming the equation relating mesh potentials to mesh fields. For two-point centered differences over points of separation $2mH$, $m = 1, 2, \ldots$, this gives for component j ($j = 1, 2, 3$)

$$\hat{D}_j = \frac{i \sin m k_j H}{mH} \tag{8-46}$$

Thus for four-point finite-difference approximations (Eq. 8-17)

$$\hat{D}_j = i\alpha \frac{\sin k_j H}{H} + i(1-\alpha) \frac{\sin 2k_j H}{2H} \tag{8-47}$$

Combining the results summarized by Eqs. (8-38) to (8-47) with Eq. (8-23) gives a relatively simple expression to be evaluated to find the force quality measure Q. A practical point concerning the computation of Q is that Eq. (8-23) is the sum over quantities $\hat{Q}(k) = \hat{Q}(l_1, l_2, l_3)$ which are even and periodic with respect to their arguments:

$$Q = \frac{1}{V_b} \sum_{l_1=0}^{N_1-1} \sum_{l_2=0}^{N_2-1} \sum_{l_3=0}^{N_3-1} \hat{Q}(l_1, l_2, l_3) \tag{8-48}$$

Therefore an eightfold reduction in the amount of computation is obtained by combining like terms:

$$Q = \frac{1}{V_b} \sum_{l_1=0}^{N_1/2} \sum_{l_2=0}^{N_2/2} \sum_{l_3=0}^{N_3/2} \alpha(l_1)\alpha(l_2)\alpha(l_3)\hat{Q}(l_1, l_2, l_3) \tag{8-49}$$

where

$$\alpha(l_i) = \begin{cases} 1 & \text{if } l_i = 0 \text{ or } N_i/2 \\ 2 & \text{otherwise} \end{cases} \tag{8-50}$$

Moreover, if the computational box is a cube ($N = N_1 = N_2 = N_3$), then the full 48-fold symmetry of a cubic lattice can be exploited to obtain a further factor of six reduction in the number of terms, i.e.,

$$Q = \frac{1}{V_b} \sum_{l_1=0}^{N/2} \sum_{l_2=0}^{l_1} \sum_{l_3=0}^{l_2} \alpha(l_1)\alpha(l_2)\alpha(l_3)\gamma(l_1, l_2, l_3)\hat{Q}(l_1, l_2, l_3) \tag{8-51}$$

where

$$\gamma(l_1, l_2, l_3) = \begin{cases} 1 & \text{if } l_1 = l_2 = l_3 \\ 3 & \text{if } l_1 = l_2 \neq l_3,\ l_3 = l_1 \neq l_2 \text{ or } l_2 = l_3 \neq l_1 \\ 6 & \text{otherwise} \end{cases} \tag{8-52}$$

Values of $\hat{G}$, P, Z, and E_{ref} are found in a similar fashion.

8-6-2 Comparison of Schemes

Examples of the numerical evaluation of Q, P, Z, and E_{ref} are shown in Figs. 8-3 to 8-5 for schemes using TSC charge-assignment–force-interpolation. The large assignment and interpolation errors of the lower-order NGP or CIC schemes make them unsuitable for accurate force modeling. Higher-order schemes than TSC can be generally discounted on grounds of computational cost. The next higher-order scheme than TSC increases the number of assignment points from twenty-seven to sixty-four, leading to approximately a factor of three increase in the magnitude of α in the timing equation. Unless one is working in the computationally expensive regime where r_e/H is large, little improvement in force accuracy is achieved for the extra computations, since accuracy is limited by the number of harmonics the mesh can handle.

The functional dependence of the error quantities shows only weak dependence on the cloud shape (cf. Sec. 8-2). Thus, although results are shown only for the $S2$ cloud shape, similar evaluation for the uniform charge distribution (shape $S1$) and the gaussian distribution (shape $S3$) show the same general behavior. To a good approximation, the curves shown in Figs. 8-3 to 8-5 may be applied to any smoothly varying spherically symmetric reference force particle shape, where r_e is the characteristic width associated with the distribution [cf. Eq. (8-5)], i.e., we can regard Q, etc., as a function of r_e rather than a functional of the shape $S1$, $S2$, $S3$, etc.

A second factor which has very little influence on Q, P, Z, and E_{ref} is the size of the computational box. If the cell size H is held constant, and the number of mesh points is increased, then changes in Q, etc., are negligible except for cases where

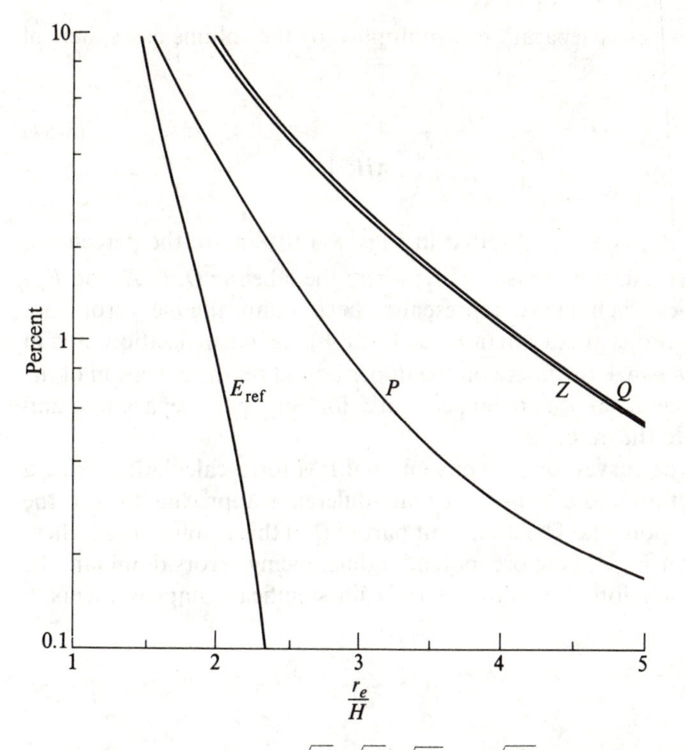

Figure 8-3 The dependence of $\sqrt{Q^\dagger}$, $\sqrt{P^\dagger}$, $\sqrt{Z^\dagger}$, and $\sqrt{E^\dagger_{\text{ref}}}$ on the effective cutoff radius r_e of the short-range force for an $S2$ reference force shape and a conventional PM force calculation using a TSC assignment function and two-point finite-difference approximations for components of the potential gradient. See text for details of normalization. (*After Eastwood, 1976b, courtesy of Advance Publications Ltd.*)

$r_e \simeq L/2$. (L is the side length of the periodic box.) In x space this property may be seen to be due to the spatial localization of errors in the interparticle force. For particle separations beyond a few cell widths, the deviation of the PM force from the reference force is negligible. In k space, the sequence of expressions for Q obtained by increasing the number of mesh points can be interpreted as a sequence of trapezium rule approximations on increasingly fine k space meshes to the integrals obtained in the limit as $N \to \infty$, i.e.,

$$Q_\infty = \int \frac{d^3\mathbf{k}}{(2\pi)^3}\hat{Q}(\mathbf{k}) = \lim_{\substack{\mathbf{N}\to\infty \\ \mathbf{H}=\text{const.}}} \left(\sum_{\mathbf{l}=-\mathbf{N}/2}^{\mathbf{N}/2-1} \frac{k_0^3}{(2\pi)^3}\hat{Q}(\mathbf{k}) \right) \tag{8-53}$$

where $k_0 = 2\pi/L = 2\pi/NH$. The integral in Eq. (8-53) is over the principal zone of harmonics (components k_j of $\mathbf{k}$ lie in the range $|k_j| \leqslant \pi/H$).

The normalization of Q, etc. in Figs. 8-3 to 8-5 is chosen to reflect the spatial localization of errors. Q, P, Z, and E_{ref} have dimensions of (force)^2 × volume. They are reduced to dimensionless quantities by dividing by the square of the coulombic force between unit charges of separation H multiplied by the volume of a sphere of radius H. Thus

$$Q^{\dagger} = \frac{Q}{\left(\dfrac{1}{4\pi\epsilon_0 H^2}\right)^2 \left(\dfrac{4}{3}\pi H^3\right)} \tag{8-54}$$

and similarly for $P^{\dagger}$, $Z^{\dagger}$, and $E^{\dagger}_{\text{ref}}$. Plotted in Figs. 8-3 to 8-5 are the percentage quantities $100\sqrt{Q^{\dagger}}$, etc., as functions of r_e/H, where the labeling Q, P, Z, and E_{ref} indicate the quantities each curve represents. Since nonnegligible errors are encountered for separations greater than H (cf. Fig. 8-8), the normalization used in Figs. 8-3 to 8-5 gives a (weak) estimate on the upper bound on the deviation of the PM interparticle force from the reference force for any pair separation and location with respect to the mesh.

Figure 8-3 gives the curves for the conventional PM force calculation using a TSC assignment function and a two-point finite-difference approximation to the potential gradient components. The dominant part of Q in this instance is Z, where Z is much greater than E_{ref}. Therefore, potential differencing errors dominate the error in the interparticle force for all r_e. To obtain significant improvements, a

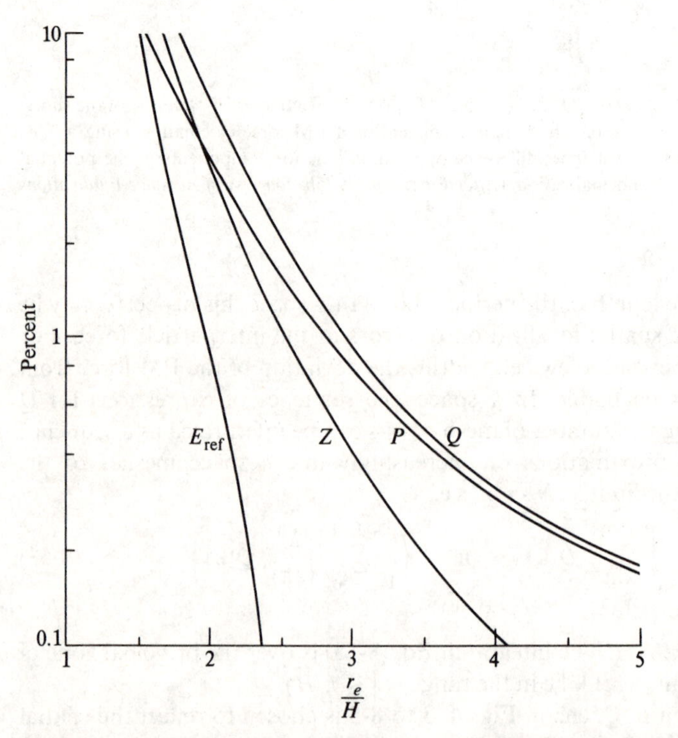

Figure 8-4 As Fig. 8-3, but with the two-point finite-difference approximations replaced by optimized four-point differences. (*After Eastwood, 1976b, courtesy of Advance Publications Ltd.*)

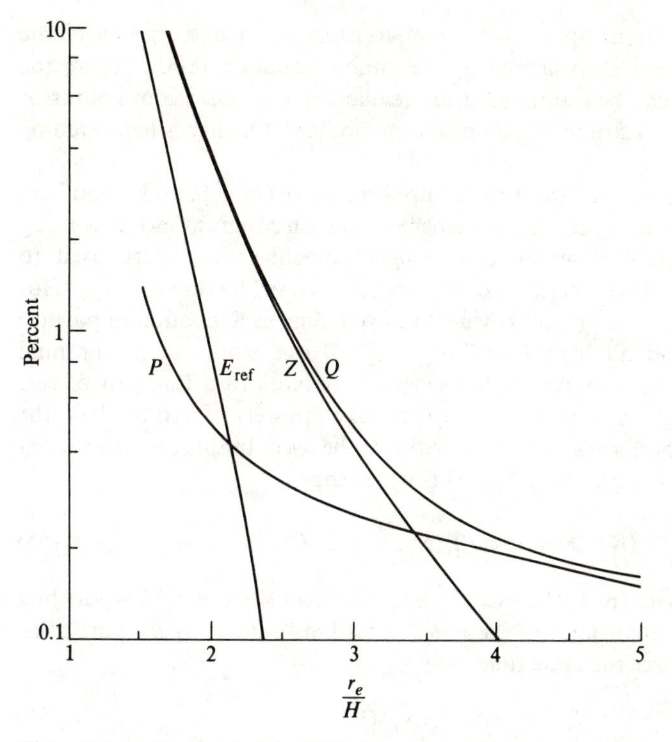

Figure 8-5 As Fig. 8-4, but for TSC charge assignment on two interlaced meshes. (*After Eastwood, 1976b, courtesy of Advance Publications Ltd.*)

higher-order potential differencing is needed. A higher-order assignment function would have an imperceptible effect.

Figure 8-4 shows the dramatic improvement resulting from replacing the two-point difference approximation by a four-point approximation where the linear combination coefficient α ($\simeq 4/3$) has been chosen to minimize Q. In contrast with Fig. 8-3, the fluctuating part P is now the dominant contribution to Q, except for small r_e. At small r_e, the limited range of wavenumbers which can be handled by the mesh cause Z to become the dominant factor. The most effective way of decreasing the overall error for a given r_e is now to use a higher-order charge assignment function or to introduce interlacing. Figure 8-5 shows the result of introducing one interlace (i.e., charge assignment to two meshes). For small r_e, Z again dominates, indicating that to reduce Q a higher-order differencing scheme is needed, while at large r_e fluctuations dominate, showing that higher-order charge sharing or more interlacing would be the most effective way of decreasing Q.

8-6-3 The Cost-Quality Relationship

The question as to whether it is best to increase r_e or introduce higher-order elements into the PM force calculation in order to achieve the desired Q is resolved by considering the relationship between computational cost and quality of the various possibilities. Cost is measured by the timing equation (8-32), where the coefficients α, β, and γ can be estimated using real arithmetic operations counts or by performing timing measurements on program modules. Quality is measured by Q.

To obtain quantitative results, we shall use here values of α, β, and γ obtained by averaging a number of timing measurements made on program modules using the FORTRAN H compiler on an IBM 360/195 machine. The data used to construct Figs. 8-3 to 8-5 is inverted to construct the inverse function $r_e = r_e(Q)$ for each variant. Equation (8-32) then provides the cycle time as a function of particle number, mesh size, and quality; $T = T(N_p, N, Q)$. If one works at the optimal operating point, the timing equation for scheme s reduces to a function of two variables, $T_s = T_s(N_p, Q)$. In practice, N is restricted to powers of two to allow the efficient radix-2 fast Fourier transform algorithm to be used. In this case the timing equation is reduced to a function of N_p and Q by setting

$$T_s(N_p, Q) = \min_r \{T_s(N_p, N = 2^r, Q)\} \tag{8-55}$$

Equation (8-55) gives the cycle time of a single variant s of the P^{3}M algorithm. The optimal among a set variants of the P^{3}M algorithms is given by the particular scheme s which minimizes the cycle time

$$T_{\text{opt}} = \min_s \{T_s(N_p, Q)\} \tag{8-56}$$

Figure 8-6 shows plots of contours of the optimal cycle times in the $\sqrt{Q^\dagger}/N_p$

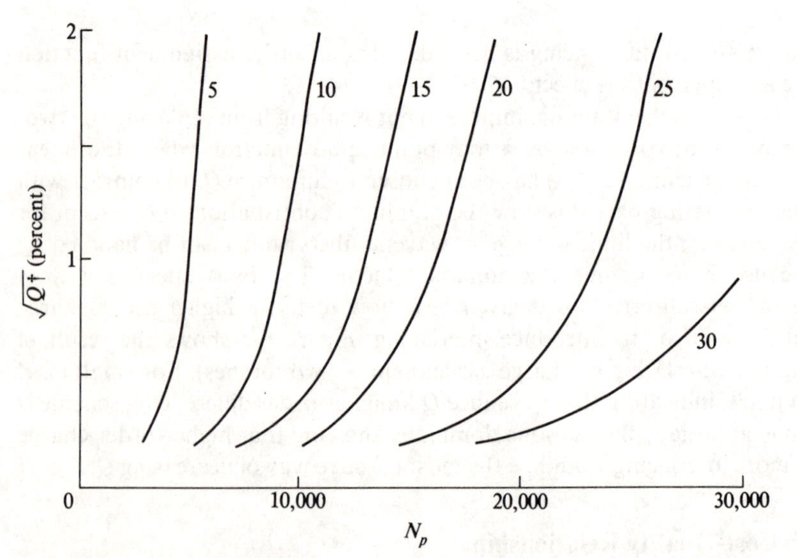

Figure 8-6 Contours of constant cycle time for the basic P^{3}M timestep cycle in the quality/particle number plane. Numbers beside curves are cycle times in seconds as measured on an IBM 360/195. (*After Eastwood, 1976b, courtesy of Advance Publications Ltd.*)

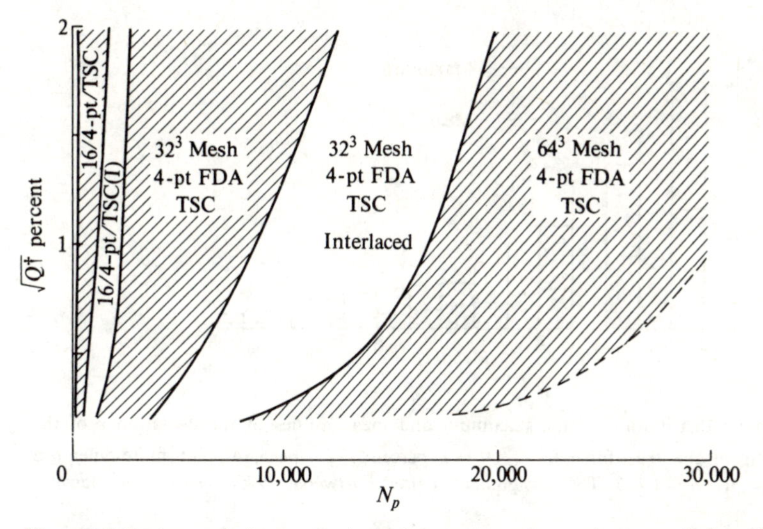

Figure 8-7 Regions of the quality/particle number plane where particular P^3M variants are computationally cheapest. The narrow region at the extreme left of the plot is where the direct summation (PP) method is most economical. The broken curve at the right-hand side is the 30-second cycle time contour. (*After Eastwood, 1976b, courtesy of Advance Publications Ltd.*)

plane. Contours in Fig. 8-6 are nonuniformly spaced (uniformity implying $T_{\text{opt}} \propto N_p$) because the choice $N = 2^r$ prevents the optimal matching [Eq. (8-35)] of N_p and N over continuous ranges of $\sqrt{Q^\dagger}$ and N_p. The cycle times labeling the curves are for the algorithm programed in FORTRAN with the exception of the Fourier transform routines, which were hand-coded in assembler language. The timings refer to a uniform distribution of particles filling the whole computational box. In certain computations, such as the clustering of galaxies, somewhat longer (by approximately fifty percent) cycle times may be encountered because the short-range force calculation times are sensitive to nonuniformity of the particle distribution.

Figure 8-7 shows the regions of $\sqrt{Q^\dagger}/N_p$ parameter space where each variant proved most cost effective. This diagram, together with the results shown in Figs. 8-3 to 8-6, reduces the problem of setting up a P^3M simulation to an automatic process. All the potential user of a P^3M program has to do is select the compromise between force accuracy, cycle time, and particle number (Fig. 8-6). The corresponding point on Fig. 8-7 then specifies the form of the PM part of the force calculation and the number of mesh points. Figure 8-3, 8-4, or 8-5, as appropriate, gives the cutoff radius r_e of the short-range force corresponding to the chosen value of $\sqrt{Q^\dagger}$. Equation (8-22), or, for interlaced schemes, Eq. (8-29), specifies the influence function values.

Practical experience has shown that a value of $\sqrt{Q^\dagger}$ of approximately one percent provides sufficient accuracy for most purposes. Selecting $\sqrt{Q^\dagger} = 1$ percent, $N_p = 5{,}000$ gives, from Fig. 8-6, $T \sim 5$ seconds. Figure 8-7 shows that this

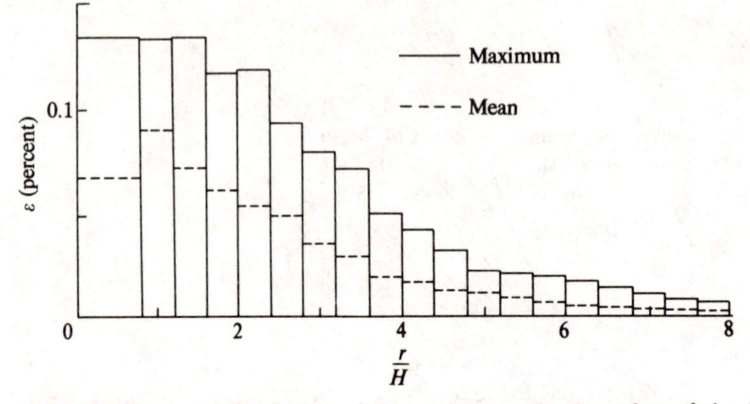

Figure 8-8 The radial distribution of the maximum and mean values of the deviation ϵ of the interparticle force from the true force for $\sqrt{Q^{\dagger}} = 1$ percent, 32^3 mesh, 4-point finite-difference approximation, $r_e = 3H$, $\alpha = 1.435$, TSC assignment. (*After Eastwood, 1976b, courtesy of Advance Publications Ltd.*)

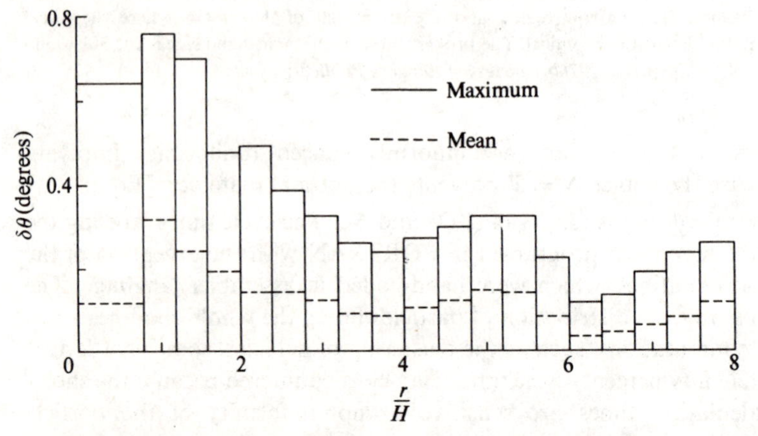

Figure 8-9 The distribution of errors in the direction of the interparticle forces for the case described in Fig. 8-8. (*After Eastwood, 1976b, courtesy of Advance Publications Ltd.*)

choice requires a 32^3 mesh, a four-point finite-difference derivative operator, TSC assignment function, and no interlacing. Figure 8-4 gives the appropriate value of r_e/H ($\simeq 3$). Inputting these quantities into the program P3M3DP provided interparticle forces whose errors in magnitude and direction are summarized by Figs. 8-8 and 8-9. The errors in the force magnitude shown in Fig. 8-8 are expressed as a percentage of the coulombic interparticle force at separation $r = H$.

8-7 PRACTICAL CONSIDERATIONS

The value of a numerical algorithm lies solely in its practical application. Often, the step from numerical algorithm to simulation program is the most time-consuming part of the development of a computer experiment. Despite this, practical problems of program development generally receive little attention. In this section, we will describe some of the practical questions encountered in the development of the P^3M program, P3M3DP. For a more detailed discussion, we refer readers to Eastwood et al. (1980) and to the program P3M3DP, which is available from the Computer Physics Communication program library.

8-7-1 The Program

The OLYMPUS programing methodology (Chapter 3) defines conventions for notation, layout, and documentation. The skeleton program CRONUS (Fig. 3-4) provides the canonical program architecture and a "test bed" in which the program modules can be developed. Flexibility is introduced into the program control structure by inclusion of calls to ⟨0.4⟩ EXPERT and of logical switches, NLOMT1, etc., at strategic points throughout the code. The overall grouping of data is prescribed by the OLYMPUS groups (Table 3-2).

The principal modules of the timestep loop (Fig. 8-1) are (1) the PM force calculation, (2) the PP force calculation and (3) integration of equations of motion. These three modules are, to a large extent, independent. They are linked only by principal quantities, where principal quantities are the minimal set of parameters which define the physical, numerical, and housekeeping state of the program. Secondary quantities are defined in separate common blocks for intercommunication between subprograms of a given module.

The main calculations performed by each of the modules is controlled by the timestep loop control subprogram ⟨2.1⟩ STEPON. Associated with each of the modules called in the timestep loop are initialization routines controlled by ⟨1.5⟩ AUXVAL or ⟨1.6⟩ INITAL, as appropriate. The initialization stage of the PM force calculation module generates values of $\hat{G}$, α, and various addressing pointers from an input of N and r_e. The PP force calculation initialization loads the short-range tables, given values of r_e, N_t and the functional form of the short-range force. Equations of motion are initialized by specifying initial particle positions and momenta.

Each of the principal modules may in turn be divided into modules. For example, the PM force calculation module consists of a charge assignment, potential solution, and force interpolation section. The potential solution section is further subdivided, calling upon a fast Fourier transform module for the major part of the computations. Similar hierarchical subdivision of tasks is performed throughout the program.

8-7-2 Data Organization

The principal storage requirement of a P^3M program is space for influence function values, charge density values and harmonics, potential values and harmonics, particle coordinates, short-range force (and energy) correction tables, and for the head-of-chain and linked-list coordinates. Influence function values and short-range correction table values are calculated only once in the program initialization and retained throughout a simulation run. New particle coordinates $\{\mathbf{x}_i, \mathbf{P}_i;\ i \in [1, N_p]\}$ overwrite old values as they are computed. Remaining quantities, which are needed only at particular stages of the calculation, can be arranged to occupy the same physical storage area in the computer, thereby reducing the total storage requirement of the program.

One way the same physical storage area can be used by different variables is by employing the FORTRAN EQUIVALENCE statement. In practice, it is found that it is easier to maintain simple, flexible, and efficient data management by reducing the use of the EQUIVALENCE statement to a minimum. Instead, the large multidimensional arrays needed for mesh data are explicitly mapped onto one-dimensional arrays. A full description of the mapping for the program P3M3DP is given by Eastwood et al. (1980). We restrict ourselves here to an illustration of the mapping for the charge-potential mesh and for the head-of-chain and linked-list coordinates.

Let the workspace array for the charge-potential mesh values be called CORE. If the charge-potential mesh comprises a cubic lattice of $(N \times N \times N)$ points labeled by $\mathbf{p} = (p_1, p_2, p_3)$, where $0 \leqslant p_i < N-1$, then values of charge density $\rho(\mathbf{p})$ and potential $\phi(\mathbf{p})$ are stored in element

$$L_0 + p_1 + p_2 L_2 + p_3 L_3$$

of array CORE. L_0 is the location of element $(0, 0, 0)$. L_2 and L_3 are the address increments between successive data values encountered along lines of mesh points parallel to the y and z axes, respectively. L_2 and L_3 are related to the mesh side N by

$$L_2 = N + 2L_B \qquad L_3 = (L_2)^2 \tag{8-57}$$

where L_B ($\geqslant 0$) is the depth of the bordering layer. If there is sufficient computer storage available, then the bordering layer can be used to eliminate tests for boundaries during charge assignment and force interpolation. In charge assignment, charge is assigned to neighboring mesh points from each particle, irrespective of the particle's position with respect to the boundary. Periodic boundary conditions are then imposed by adding charge densities in the bordering layer to the corresponding charge densities in the periodic length covered by the ranges $0 \leqslant p_i \leqslant N-1$. Similar savings are effected in force interpolation by periodically extending potential values before interpolating forces to the particles. If L_B is nonzero then components p_i of $\mathbf{p}$ take values in the range $-L_B \leqslant p_i \leqslant N-1+L_B$ and L_0 must be greater than $L_B(L_2 + L_3)$.

The Fourier transforms of ρ and ϕ are hermitian and periodic, and therefore can be stored in the same space as the values themselves. The storage arrangement for transform values and the overwriting sequence used is discussed further in Sec. 8-7-4.

Once the PM force calculation is complete, potential values in the array CORE can be discarded. Thus the integer workspace array ICORE for the head-of-chain tables and linked lists may be equivalenced to the array CORE to give

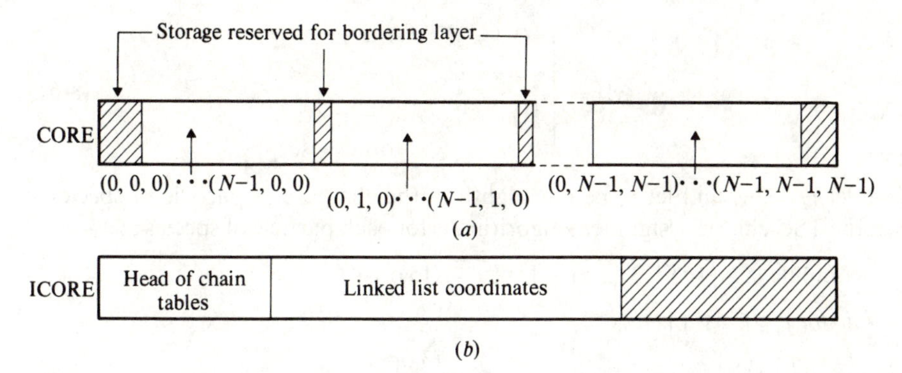

Figure 8-10 The storage of (*a*) charge and potential values and (*b*) head-of-chain and linked-list coordinate values in the workspace array CORE.

further storage savings. Figure 8-10(*b*) schematically shows the storage of the head-of-chain table and linked lists. The head-of-chain table for the three-dimensional (M × M × M) chaining mesh is mapped in exactly the same manner as the charge-potential values. In the absence of bordering, the value HOC(**q**), where $\mathbf{q} = (q_1, q_2, q_3)$ and $0 \leqslant q_i \leqslant M-1$, is stored in element

$$L_C = M_0 + q_1 + q_2 M + q_3 M^2$$

of the array ICORE, where $L_C > 0$, and M_0 is the location of element (0, 0, 0).

The linked-list coordinates $\{\mathrm{LL}(i);\ i \in [1, N_p]\}$ are stored in ICORE immediately after the M^3 head-of-chain values. The linked-list coordinates are in the same order as the position and momentum coordinates, so particle i has position $\mathbf{x}_i$, momentum $\mathbf{P}_i$, and link coordinate LL(i).

8-7-3 Assignment and Interpolation

In all particle simulation programs it is good practice to use computationally convenient dimensionless units in the main calculation loop (cf. Chapter 2). Using dimensionless units where $H = 1$ reduces the assignment function [Eq. (8-8)] to

$$W(\mathbf{x}) = w(x)w(y)w(z) \tag{8-58}$$

where, for the TSC scheme

$$w(x) = \Pi(x) * \wedge (x) \tag{8-59}$$

For practical purposes, it is more useful to express Eqs. (8-58) and (8-59) in terms of the fraction of charge assigned to each mesh point. If a particle lies in cell **p** and its position relative to the mesh point at the centre of the cell is $\mathbf{x}' = \mathbf{x} - \mathbf{x}_\mathbf{p}$, then the fraction of the charge assigned to mesh point $\mathbf{p}+\mathbf{t}$ is given by

$$W_{\mathbf{p}+\mathbf{t}} = w_{t_1}(x')w_{t_2}(y')w_{t_3}(z') \tag{8-60}$$

where, from Eqs. (5-86) and (5-87),

$$w_t(x') = \begin{cases} \frac{1}{2}(\frac{1}{2}+x')^2 & t = 1 \\ \frac{3}{4}-x'^2 & t = 0 \\ \frac{1}{2}(\frac{1}{2}-x)^2 & t = -1 \\ 0 & \text{otherwise} \end{cases} \tag{8-61}$$

If we let $v_t = 2w_t$ and let q_s be one-eighth of the charge of a particle of species s, then the TSC charge assignment algorithm is for each particle of species s:

compute v_{t_i}; $\quad i \in [1, 3], t \in [-1, 1]$ (15 ops)

do for $t_1 \in [-1, 1]$
$T_1 = q_s v_{t_1}(x')$ (3 ops)
do for $t_2 \in [-1, 1]$
$T_2 = T_1 v_{t_2}(y')$ (9 ops)
do for $t_3 \in [-1, 1]$
$T_3 = T_2 v_{t_3}(z')$ (27 ops)
$\rho(\mathbf{p}+\mathbf{t}) = \rho(\mathbf{p}+\mathbf{t}) + T_3$ (27 ops)

The quantities in the brackets refer to the number of real arithmetic operations per particle. Summing these quantities gives a real arithmetic operations count per particle per timestep for charge assignment of $3 \times 27 = 81$.

The force interpolation algorithm differs from the assignment algorithm in that charge assignment in the inner loop is replaced by interpolation, i.e., the line

$$\rho(\mathbf{p}+\mathbf{t}) = \rho(\mathbf{p}+\mathbf{t}) + T_3 \tag{8-62}$$

is replaced by
$$\mathbf{P}_{i,s} = \mathbf{P}_{i,s} + T_3 \mathbf{E}(\mathbf{p}+\mathbf{t}) \tag{8-63}$$

where $\mathbf{P}_{i,s}$ is the momentum of particle i of species s and $\mathbf{E}(\mathbf{p}+\mathbf{t})$ is the electric field (constructed by differencing potential values) at mesh point $\mathbf{p}+\mathbf{t}$.

8-7-4 The Potential Solver

The Fourier transform method for solving for the potential (Sec. 6-5-2) consists of the three steps

1. Transform $\rho \supset \hat{\rho}$.
2. Compute $\hat{\phi} = \hat{G}\hat{\rho}$.
3. Transform $\hat{\phi} \supset \phi$.

The three-dimensional Fourier transform of the charge density is, for $H = 1$,

$$\hat{\rho}(\mathbf{k}) = \sum_{\mathbf{p}=0}^{N-1} \rho(\mathbf{p}) e^{-i\mathbf{k}\cdot\mathbf{p}} \tag{8-64}$$

Eq. (8-64) may be written explicitly in terms of the integers $\mathbf{l}(\mathbf{k} = 2\pi\mathbf{l}/N)$ labeling the harmonics:

$$\hat{\rho}(l_1, l_2, l_3) = \sum_{p_3=0}^{N-1} e^{-i2\pi l_3 p_3/N} \times \left[\sum_{p_2=0}^{N-1} e^{-i2\pi l_2 p_2/N} \left(\sum_{p_1=0}^{N-1} e^{-i2\pi l_1 p_1/N} \rho(p_1, p_2, p_3) \right) \right] \quad (8\text{-}65)$$

The term in the braces in Eq. (8-65) describes one-dimensional FFTs for lines parallel to the x axis of values of charge density, i.e.,

$$\tilde{\rho}(l_1, p_2, p_3) = \sum_{p_1=0}^{N-1} e^{-i2\pi l_1 p_1/N} \rho(p_1, p_2, p_3) \quad (8\text{-}66)$$

The term in the square brackets describes a set of one-dimensional transforms along lines parallel to the y axis:

$$\tilde{\rho}(l_1, l_2, p_3) = \sum_{p_2=0}^{N-1} e^{-i2\pi l_2 p_2/N} \tilde{\rho}(l_1, p_2, p_3) \quad (8\text{-}67)$$

The outer sum in Eq. (8-65) gives transforms along lines parallel to the z axis:

$$\hat{\rho}(l_1, l_2, l_3) = \sum_{p_3=0}^{N-1} e^{-i2\pi l_3 p_3/N} \tilde{\rho}(l_1, l_2, p_3) \quad (8\text{-}68)$$

The algorithmic description of Eq. (8-66) is

do for $p_3 = 1, N-1$
do for $p_2 = 1, N-1$
copy $\{\rho(p_1, p_2, p_3); p_1 = 1, N-1\}$ to FFT input buffer
Fourier-transform values
copy FFT output buffer to storage locations occupied by $\{\rho(p_1, p_2, p_3); p_1 = 1, N-1\}$

Figure 8-11 shows diagramatically the transform for a single line. Values are copied from the workspace array CORE to the FFT input buffer AX. The FFT package generates from the contents of AX the transformed quantities in the output buffer AY. The first $N/2+1$ values in AY are the real (even) part of the harmonics for $l_1 \in [0, N/2]$ and the subsequent $N/2-1$ values are the imaginary (odd) part of the harmonics for $l_1 \in [1, N/2-1]$. Since the imaginary part is odd, the imaginary parts of the harmonics $\tilde{\rho}(0, p_2, p_3)$ and $\tilde{\rho}(N/2, p_2, p_3)$ are zero, so are not stored.

The procedures for Eqs. (8-67) and (8-68) are the same as for Eq. (8-66). For Eq. (8-67) data values along lines parallel to the y axis of the charge-potential mesh are copied to the FFT buffer, the transform is performed, and the FFT output overwrites those data values. These operations are repeated for lines parallel to the z axis. The final result is harmonics $\{\hat{\rho}\}$ occupying the same storage area as the density values. Each harmonic is stored in eight parts:

$$\hat{\rho}(l_1, l_2, l_3) = (\hat{\rho}_{eee} + \hat{\rho}_{eoo} + \hat{\rho}_{oeo} + \hat{\rho}_{ooe}) + i(\hat{\rho}_{eeo} + \hat{\rho}_{eoe} + \hat{\rho}_{oee} + \hat{\rho}_{ooo}) \quad (8\text{-}69)$$

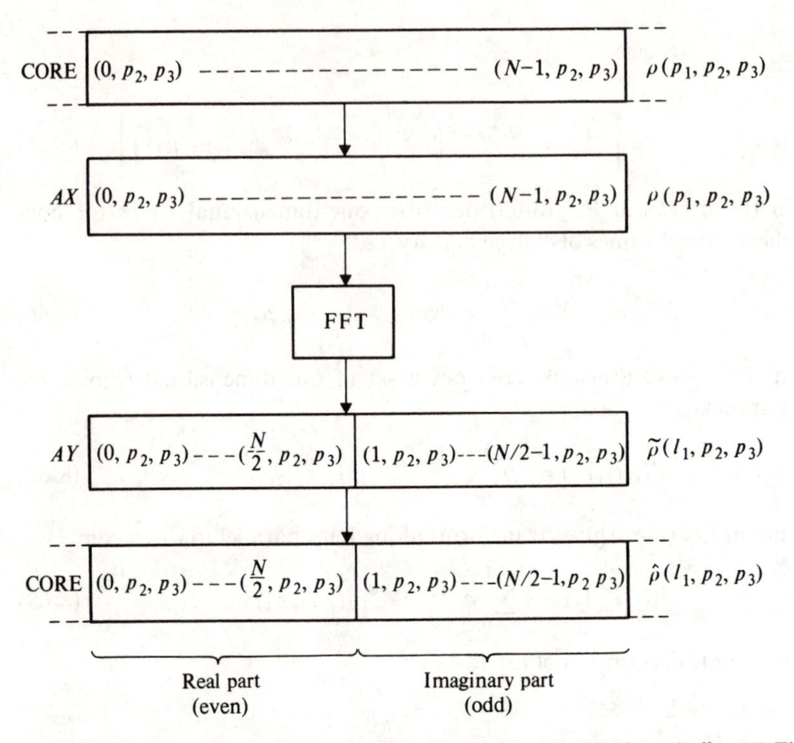

Figure 8-11 A line of charge density data values is copied to the FFT input buffer AX. The result of the transform in buffer AY is copied back to the workspace array CORE, with the transform values occupying the same storage locations as the input values.

The subscript "e" and "o" denote parts of $\hat{\rho}$ which are even or odd in indices l_1, l_2, l_3. For example, $\hat{\rho}_{oeo}$ is odd in indices l_1 and l_3 and even in index l_2. Harmonics outside the range $0 \leqslant l_1 \leqslant N/2$ can be obtained by using Eq. (8-69) plus the symmetry and periodicity of each component.

The storage location of the parts of the harmonic $\hat{\rho}(l_1, l_2, l_3)$ are given by

$$L_0 + l_1^* + l_2^* L_2 + l_3^* L_3$$

where

$$l_i^* = \begin{cases} l_i & \text{part even in } l_i \\ l_i + N/2 & \text{part odd in } l_i \end{cases}$$

and $0 \leqslant l_i \leqslant N/2$ for the even parts and $0 < l_i < N/2$ for the odd parts.

The eight parts of the potential harmonics (odd or even in each of its indices) are found by multiplying the corresponding elements of $\hat{\rho}$ by the influence function values $\hat{G}$ and are stored by overwriting elements of $\hat{\rho}$ in the workspace array CORE. It is at this stage that the mesh-calculated part of the potential energy [Eqs. (12-58) to (12-60)] is accumulated. Potential values are obtained by reversing the transformation and overwriting sequence used in obtaining charge density harmonics from charge density values.

A more detailed outline of the three stages of the potential solver algorithm stated at the start of this subsection is

1. Transform all N^2 lines of data parallel to x axis:

$$\rho(p_1, p_2, p_3) \supset \begin{cases} \tilde{\rho}_e(l_1, p_2, p_3) & 0 \leqslant l_1 \leqslant N/2 \\ \tilde{\rho}_o(l_1, p_2, p_3) & 0 < l_1 < N/2 \end{cases}$$

2. Transform all N^2 lines of data parallel to y axis:

$$\tilde{\rho}_e(l_1, p_2, p_3) \supset \begin{cases} \tilde{\rho}_{ee}(l_1, l_2, p_3) & 0 \leqslant l_2 \leqslant N/2 \\ \tilde{\rho}_{eo}(l_1, l_2, p_3) & 0 < l_2 < N/2 \end{cases}$$

$$\tilde{\rho}_o(l_1, p_2, p_3) \supset \begin{cases} \tilde{\rho}_{oe}(l_1, l_2, p_3) & 0 \leqslant l_2 \leqslant N/2 \\ \tilde{\rho}_{oo}(l_1, l_2, p_3) & 0 < l_2 < N/2 \end{cases}$$

3. Transform all N^2 lines of data parallel to z axis:

$$\tilde{\rho}_{ee}(l_1, l_2, p_3) \supset \begin{cases} \hat{\rho}_{eee}(l_1, l_1, l_3) & 0 \leqslant l_3 \leqslant N/2 \\ \hat{\rho}_{eeo}(l_1, l_2, l_3) & 0 < l_3 < N/2 \end{cases}$$

etcetera.

4. Compute potential harmonics:

$$\hat{\phi}_{eee}(l_1, l_2, l_3) = \hat{G}(l_1, l_2, l_3)\hat{\rho}_{eee}(l_1, l_2, l_3) \qquad 0 \leqslant (l_1, l_2, l_3) \leqslant N/2$$

$$\hat{\phi}_{eeo}(l_1, l_2, l_3) = \hat{G}(l_1, l_2, l_3)\hat{\rho}_{eeo}(l_1, l_2, l_3) \qquad 0 \leqslant l_1, l_2 \leqslant N/2, \quad 0 < l_3 < N/2$$

etcetera.

5. Transform all N^2 lines of data parallel to z axis:

$$\left.\begin{matrix} \hat{\phi}_{eee}(l_1, l_2, l_3) \\ \hat{\phi}_{eeo}(l_1, l_2, l_3) \end{matrix}\right\} \supset \tilde{\phi}_{ee}(l_1, l_2, p_3) \qquad 0 \leqslant p_3 \leqslant N-1$$

etcetera.

6. Transform all N^2 lines of data parallel to y axis:

$$\left.\begin{matrix} \tilde{\phi}_{ee}(l_1, l_2, p_3) \\ \tilde{\phi}_{eo}(l_1, l_2, p_3) \end{matrix}\right\} \supset \phi_e(l_1, p_2, p_3) \qquad 0 \leqslant p_2 \leqslant N-1$$

$$\left.\begin{matrix} \tilde{\phi}_{oe}(l_1, l_2, p_3) \\ \tilde{\phi}_{oo}(l_1, l_2, p_3) \end{matrix}\right\} \supset \tilde{\phi}_o(l_1, p_2, p_3) \qquad 0 \leqslant p_2 \leqslant N-1$$

7. Transform all N^2 lines of data parallel to x axis:

$$\left.\begin{matrix} \tilde{\phi}_e(l_1, p_2, p_3) \\ \hat{\phi}_o(l_1, p_2, p_3) \end{matrix}\right\} \supset \phi(p_1, p_2, p_3) \qquad 0 \leqslant p_1 \leqslant N\text{-}1$$

8-7-5 The Short-Range Force

The main practical problems associated with the short-range force calculation is the location of neighboring particles and ordering the operations to avoid the double computation of pair forces. Since these two topics have been treated in detail in Sec. 8-4, we shall not pursue them further here.

Values of short-range forces used in loading interparticle forces are obtained from Eq. (8-7). Note that only forces between an isolated pair of particles need be considered. Forces due to periodic images of the pair of particles are included in the PM part of the force calculation. Values of $f^{sr}(r)/r$ are stored in the tables (cf. Sec. 8-4-3) rather than values of $f^{sr}(r)$ to eliminate the divisions by r, and values are stored at uniform intervals of Δr^2 to avoid the square-root computation of $\sqrt{(r^2)}$ in the inner loop of the pairwise force sums. Tabular values $\mathscr{F}$ are given by

$$\mathscr{F}(r) = \frac{f^{sr}(r)}{r} = \frac{(f^{tot}_{(r)} - R(r))}{r} \tag{8-70}$$

where

$$R(r) = \mathbf{R}(\mathbf{x})\cdot\hat{\mathbf{r}} = \frac{\hat{\mathbf{r}}}{4\pi\epsilon_0}\cdot\int d\mathbf{x}'\int d\mathbf{x}''\, S(\mathbf{x}'')S(\mathbf{x}'-\mathbf{x}')\frac{(\mathbf{x}'-\mathbf{x}'')}{|\mathbf{x}'-\mathbf{x}''|^3} \tag{8-71}$$

For an $S2$-shaped particle, S is given by Eq. (8-3). Substituting Eq. (8-3) into Eq. (8-71) yields after some straightforward but lengthy manipulation

$$R(r) = \frac{1}{4\pi\epsilon_0}\times\begin{cases} \dfrac{1}{35a^2}(224\xi - 224\xi^3 + 70\xi^4 + 48\xi^5 - 21\xi^6) & 0 \leqslant \xi \leqslant 1 \\ \dfrac{1}{35a^2}\left(\dfrac{12}{\xi^2} - 224 + 896\xi - 840\xi^2 + 224\xi^3 + 70\xi^4 - 48\xi^5 + 7\xi^6\right) & 1 \leqslant \xi \leqslant 2 \\ \dfrac{1}{r^2} & \xi \geqslant 2 \end{cases} \tag{8-72}$$

where

$$\xi = \frac{2r}{a} \tag{8-73}$$

Tabular values of the short-range part of the pair-potential energy $\phi^{sr}(r)$ are similarly computed from the difference between the total pair-potential energy ϕ and the pair-potential energy ϕ^m arising from the mesh reference interparticle force $\mathbf{R}$ (cf. Sec. 12-3-4)

$$\phi^{sr}(r) = \phi(r) - \phi^m(r) \tag{8-74}$$

where, for an $S2$-shaped mesh reference particle,

$$\phi^m = \frac{1}{4\pi\epsilon_0}\times\begin{cases} \dfrac{1}{70a}[208 - 112\xi^2 + 56\xi^4 - 14\xi^5 - 8\xi^6 + 3\xi^7] & 0 \leqslant \xi \leqslant 1 \\ \dfrac{1}{70a}\left[\dfrac{12}{\xi} + 128 + 224\xi - 448\xi^2 + 280\xi^3 - 56\xi^4 - 14\xi^5 + 8\xi^6 - \xi^7\right] & 1 \leqslant \xi \leqslant 2 \\ \dfrac{1}{r} & 2 \leqslant \xi \end{cases} \tag{8-75}$$

To prevent instabilities in the time integration for ionic systems (arising when $\omega DT > 2$; cf. Chapter 4), an upper limit on the value of the interparticle force is set so that $\omega_{max} DT < 2$ (see also Secs. 12-2-2 and 12-3-2). If the timestep DT is chosen appropriately, then the probability of the separation of two ions being less than the radius b within which the force is set to its upper limit is negligible. A simple means of monitoring that interparticle separations less than b are not encountered is to set tabular values of ϕ^{sr} to some physically unattainable value. Then, if separations less than b do occur, they lead to "spikes" in the energy, indicating that startup transients are present or that a smaller timestep is needed.

8-7-6 Parameter Selection

The final step before physically useful results can be obtained from a simulation program is to specify the input data for the simulation. A poor selection of parameters can lead to an inefficient program or, worse still, physically meaningless results. In this section, we describe a procedure for parameter selection and evaluation that we have found useful in setting up P3M3DP for simulating ionic liquids.

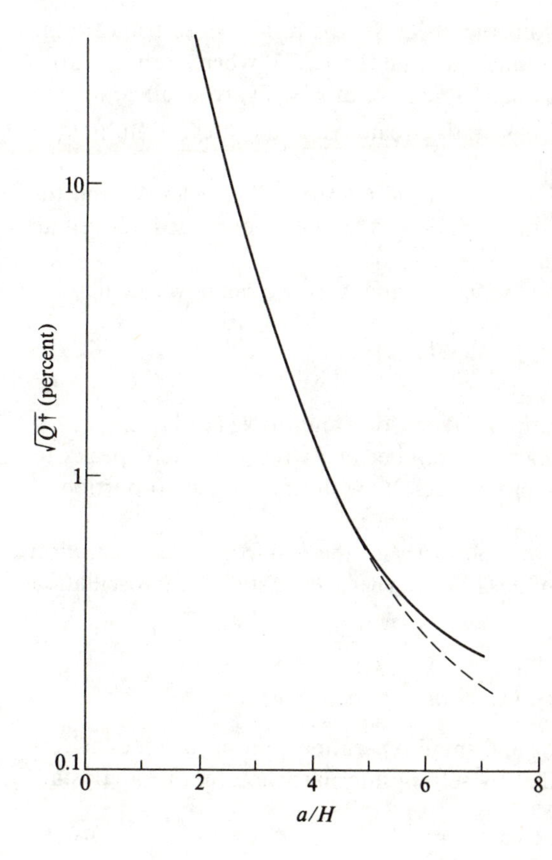

Figure 8-12 The quality measure, $\sqrt{Q^\dagger}$ as a function of the cutoff radius a. The dotted curve was obtained using a 16^3 mesh and the solid curve was obtained using a 32^3 mesh.

There are five main stages in defining the input for the program: specification of the physical system, the numerical scheme, initial conditions, output selection, control and diagnostics. The material under investigation is specified by the number of particle species and the attributes of those species; charge, mass, the form of the (short-range) core repulsion force, etc. The quantity of material is described by the number of ions of each species, and the volume of space it occupies is defined by the size of the computation box: Together, the number of ions and box volume prescribe the average density. The numerical scheme is specified by the mesh size N, the cutoff radius a, the effective cutoff radius, r_e, the timestep DT, and the force cutoff for the short-range force tables.

A suitable combination of physical and numerical parameters are obtained as follows:

1. Specify the material:
   —number of species
   —ion charge and mass
   —short-range force law: The range of the noncoulombic part leads to a minimum cutoff radius r_e^0.
2. Choose average density n_0.
3. Select $\sqrt{Q^\dagger}$ and thence (a/H) from Fig. 8-12. Set the ratio r_e/a to truncate the short-range force correction tables at radius (r_e/H) where table entries (expressed as percentages of the coulombic force at $r = H$) are small compared with $\sqrt{Q^\dagger}$, preferably where N is slightly greater than an integral multiple of (r_e/H). Typically, $0.7 \leqslant r_e/a \leqslant 1.0$.
4. Step 3 gave (r_e/H). Figure 8-13 gives the total number of particles N_p and the mesh size N. Some iteration on steps 3 and 4 may be needed to get an acceptable cost-quality compromise.
5. The preceding steps specify r_e^0, (r_e/H), n_0, N_p, and N, so we can now calculate

$$r_e = \left(\frac{r_e}{H}\right)\frac{1}{N}\left(\frac{N_p}{n_0}\right)^{1/3}$$

   If $r_e < r_e^0$, then we increase the ratio (r_e/a) and/or a and/or N_p until $r_e \geqslant r_e^0$.
6. Calculate $L = (N_p/n_0)^{1/3}$ and scale the number of particles of each species to obtain the chosen total number of particles N_p with the correct proportion of each species.
7. Choose DT to ensure stability and accuracy of the leapfrog time integration scheme. The stability criterion $\omega DT < 2$, where ω is the local oscillation frequency

$$\omega = \left|2\left(\frac{1}{m_1} + \frac{1}{m_2}\right)\frac{\partial F}{\partial r}\right|^{1/2}$$

   is most restrictive for lighter ions and small separations. Large ω values at very small ion separations are avoided by setting interparticle forces to a constant for small r (cf. Secs. 12-2-2 and 12-3-2).

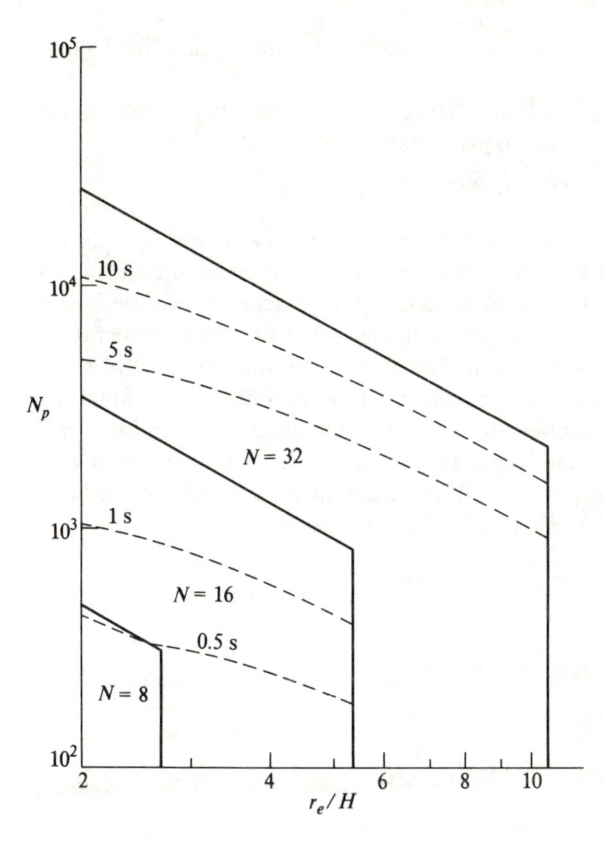

Figure 8-13 Contours of constant cycle time (broken curves) and regions of particle number/effective cutoff radius space where particular choices of grid size are most appropriate (solid lines). The vertical sections of the solid lines arise from the constraint $3(r_e/H) < N$. Cycle times are seconds as measured on an IBM 360/195 computer (shown on curves as 10 s, 5 s, etc.) and include the computation of pressure, temperature, energy, and radial distribution function.

The selection of initial conditions and output are specific to the particular experiment being performed. Examples of these are discussed further in later chapters. One problem with initialization of dense ionic systems not generally encountered in collisionless PM simulations is that of startup transients. The potential energy and kinetic energy in dense systems are comparable. Loading ion positions and momenta with specified first and second moments usually leads to a much hotter system than desired. This problem is overcome by rescaling momenta for the first one or two hundred timesteps to obtain the desired velocity moments (e.g., nondrifting maxwellians with temperature T).

Simple checks of the suitability of the combination of input parameters chosen by the above method are furnished by test runs performed with a system of just two ions—this provides results which can be checked analytically. Graphical output is invaluable as it allows the form of the force law, the force splitting, initial distributions, and orbits to be rapidly inspected for errors. A useful cross-check on the suitability of the effective cutoff radius is given by the energy and pressure truncation errors [see Sec. 12-3-4(4)]. Running checks as simulations proceed are furnished by the total linear momentum, angular momentum, and energy. These monitor both roundoff and truncation errors, and can also be used to monitor transients if the short-range energy correction tables are loaded as described above (Sec. 8-7-5).

The cycle time of the program P3M3DP on an IBM 360/195 has been found experimentally to be given (to within 10 percent) by

$$T = (50+2N_n)10^{-5}N_p+\beta(N) \text{ seconds} \tag{8-76}$$

where N_n is the number of particles in a sphere whose radius equals r_e, N_p is the total number of particles, and $\beta(N)$ is the time for the potential solution ($\beta = 0.3$, 2, and 12 seconds for $N = 16$, 32, and 64, respectively). T includes the time for the dynamical calculation and the time for the calculation of pressure, temperature, energies, and radial distribution function. For typical parameters used in simulating ionic liquids, a cycle time of $1\frac{1}{2}$ seconds per thousand particles is obtained. For example, cycle times of 6 seconds for 4,096 particles and 33 seconds for 21,952 particles have been measured (Hockney, 1980). These are to be compared to a cycle time of 1 second for 256 particles obtained for direct summation methods, which scales to approximately 2 hours for 21,952 particles!

CHAPTER

NINE

PLASMA SIMULATION

9-1 INTRODUCTION

A plasma is a hot, fully ionized gas which may be regarded as a collection of positive ions and negative electrons interacting through their mutual electric and magnetic fields (**E** and **B**). The fields are related to the charge and current density (ρ and **j**) by Maxwell's equations. In vacuum we have

$$\operatorname{div} \mathbf{B} = 0 \tag{9-1a}$$

$$\operatorname{curl} \mathbf{B} = \mu_0 \mathbf{j} + \frac{1}{c^2}\frac{\partial \mathbf{E}}{\partial t} \tag{9-1b}$$

$$\operatorname{div} \mathbf{E} = \rho/\epsilon_0 \tag{9-1c}$$

$$\operatorname{curl} \mathbf{E} = -\frac{\partial \mathbf{B}}{\partial t} \tag{9-1d}$$

where the velocity of light in free space $c = (\epsilon_0 \mu_0)^{-1/2}$.

The force on a charge q moving through these fields with a velocity **v** is

$$\mathbf{F} = q[\mathbf{E} + \mathbf{v} \times \mathbf{B}] \tag{9-2a}$$

and the motion of the charge is determined by Newton's law

$$m\frac{d\mathbf{v}}{dt} = \mathbf{F} \tag{9-2b}$$

If in a small volume of space V (for example a mesh cell) there are N charges with charge q_i and velocity $\mathbf{v}_i$ ($i = 1, \ldots, N$), the charge density and current density may be calculated by summation from

$$\rho = \sum_{i=1}^{N} q_i/V \tag{9-3a}$$

$$\mathbf{j} = \sum_{i=1}^{N} q_i \mathbf{v}_i/V \tag{9-3b}$$

A full three-dimensional simulation of the above system [Eqs. (9-1) to (9-3)] is just possible on computers such as the IBM 360/195 or CDC 7600 (Buneman, 1976; Langdon and Lasinski, 1976). However, with a timestep of several minutes the computer time required for a realistic computation is too long for most purposes. Parallel computers such as the CRAY-1 with speeds at least ten times that of an IBM 360/195 are necessary for such calculations (see Hockney, 1979; Hockney and Jesshope, 1981).

We are obliged therefore to consider simplifications to the above equations. A range of approximations is possible but here we consider the two extreme cases of magnetohydrodynamics (MHD) and the electrostatic approximation (ES). In both cases we consider timescales very much longer than the time for light to cross the system and the second term on the right-hand side of Eq. (9-1*b*), which is proportional to c^{-2}, is ignored. This eliminates electromagnetic waves and implies that the **E** and **B** fields adjust themselves instantaneously to the present distribution of charges and currents. We now consider the two approximations in more detail.

9-1-1 Magnetohydrodynamics

This is a low-density, low-frequency approximation in which we consider space scales much larger than the Debye length. The Debye length λ_D is the maximum distance over which charge separation can occur in a thermal plasma. Hence if we are only interested in averages over many λ_D, we can assume charge neutrality and ignore the right-hand side of Eq. (9-1*c*). Alternatively, we may regard all currents as flowing in closed circuits and therefore leading to no accumulation of charge. The MHD plasma may therefore be regarded as a neutral fluid through which currents flow. The approximation of charge neutrality eliminates the mechanism responsible for plasma oscillations and the high frequencies associated with them.

In the MHD plasma the magnetic fields are usually large and imposed (e.g., in a fusion machine) and the system is dominated by currents and the magnetic forces upon them. The currents themselves produce magnetic fields that are important, by Eq. (9-1*b*). The effect of the electric field is less important and arises by the induction effect of the time variation of **B**, by Eq. (9-1*d*).

With these approximations Maxwell's equations become

$$\operatorname{curl} \mathbf{B} = \mu_0 \mathbf{j} \tag{9-4a}$$

$$\operatorname{curl} \mathbf{E} = -\frac{\partial \mathbf{B}}{\partial t} \tag{9-4b}$$

$$\operatorname{div} \mathbf{B} = \operatorname{div} \mathbf{E} = 0 \tag{9-4c}$$

If σ is the conductivity the force equation becomes

$$\mathbf{j} = \sigma[\mathbf{E} + \mathbf{v} \times \mathbf{B}] \tag{9-5}$$

where $\mathbf{v}$ is the bulk velocity of the fluid.

If the conductivity is high and $\mathbf{j}$ finite then Eq. (9-5) implies

$$\mathbf{E} + \mathbf{v} \times \mathbf{B} = 0 \tag{9-6}$$

which on combining with Eq. (9-4*b*) leads to

$$\frac{\partial \mathbf{B}}{\partial t} = \operatorname{curl}(\mathbf{v} \times \mathbf{B}) \tag{9-7}$$

Equation (9-7) is the condition that the magnetic flux $\int \mathbf{B} \cdot \mathbf{ds}$ is constant through any surface moving with the velocity $\mathbf{v}$ of the fluid. Hence this equation can be interpreted as meaning that the magnetic lines of force are "frozen into", and thus constrained to move with, the fluid.

The MHD equations are completed by the equations of continuity and force balance for the fluid

$$\frac{\partial \rho_m}{\partial t} = -\operatorname{div}(\rho_m \mathbf{v}) \tag{9-8a}$$

$$\rho_m \frac{d\mathbf{v}}{dt} = -\operatorname{grad} p + \mathbf{j} \times \mathbf{B} \tag{9-8b}$$

where the substantive derivative, or rate of change as seen by a particle moving with the fluid, is

$$\frac{d}{dt} = \frac{\partial}{\partial t} + \mathbf{v} \cdot \operatorname{grad} \tag{9-8c}$$

and ρ_m is the mass density.

The pressure p is usually related to the density via the adiabatic law p/ρ_m^γ = constant, where γ is the ratio of specific heat at constant pressure to that at constant volume.

The MHD equations are usually solved by finite-difference methods which are familiar in fluid dynamics (e.g., the Lax–Wendroff method). These methods are well-described in "An Introduction to Computational Physics" (Potter, 1973) and in "Difference Methods for Initial-Value Problems" (Richtmyer and Morton, 1967). Such methods are outside the scope of this book and will not be described here.

Since the derivation of the MHD equations implies taking averages over many ion and electron orbits, this approximation might not seem suited to simulation by particle methods. However, in several special cases particle methods can be successfully applied. For example, Rathmann, Vomvoridis, and Denavit (1978), in the study of Langmuir and Whistler waves, divide the plasma into a high-density cold component which constitutes the main propagating medium, and a low-density energetic electron component which is responsible for resonant particle interactions. The cold component is then represented by a fluid model that calculates the time variation of the amplitudes and phases of a number of waves. The orbits of the high-energy electrons in these waves are then calculated analytically to first order in the electric field. In this way the timestep is related to the trapping frequency of the particle in the waves, and may represent an integration over many plasma periods. It is further assumed that the amplitude and phase of the waves varies slowly over a period of oscillation. This method of calculation is known as a "hybrid simulation" because it combines both a particle and fluid representation. It has the additional advantage of being economic in the use of particles because these are only used to represent the low-density component. The time integration is referred to as a "long-time scale" (LTS) simulation. We describe in Sec. 9-4 two other interesting examples of the use of particle methods in the MHD limit. They are applied to the study of the interaction of the solar wind with the earth's magnetic dipole, which leads to the formation of the magnetosphere and the geomagnetic tail.

9-1-2 Electrostatic Plasma

In contrast to the MHD plasma, the electrostatic (ES) plasma approximation deals with high frequencies (such as the plasma frequency) and small space scales of the order of the Debye length. In these circumstances charge separation can occur and we retain the right-hand side of Eq. (9-1*c*). However the currents are presumed small so that the self-magnetic field is small. This results in making zero the right-hand side of Eqs. (9-1*b*) and (9-1*d*). In this approximation Maxwell's equations reduce to

$$\operatorname{div} \mathbf{E} = \rho/\epsilon_0 \tag{9-9a}$$

$$\operatorname{div} \mathbf{B} = 0 \tag{9-9b}$$

$$\operatorname{curl} \mathbf{B} = \operatorname{curl} \mathbf{E} = 0 \tag{9-9c}$$

In these circumstances it is convenient to introduce the electrostatic potential ϕ such that

$$\mathbf{E} = -\operatorname{grad} \phi \tag{9-10a}$$

then

$$\operatorname{curl} \mathbf{E} = -\operatorname{curl} \operatorname{grad} \phi = 0 \tag{9-10b}$$

as required, and also

$$\operatorname{div} \mathbf{E} = -\operatorname{div} \operatorname{grad} \phi = -\nabla^2 \phi = \rho/\epsilon_0 \tag{9-10c}$$

Thus Maxwell's equations have been reduced to Eq. (9-10*c*), which is the familiar Poisson's equation.

Since the space scale of interest is now of the order of the Debye length and the orbits of individual ions or electrons are important, the electrostatic plasma is well suited to simulation by particle-mesh methods.

The complete set of equations for the electrostatic plasma is, for the fields,

$$\nabla^2 \phi = -\rho/\epsilon_0 \tag{9-11a}$$

$$\mathbf{E} = -\text{grad}\,\phi \tag{9-11b}$$

$$\rho = \sum_i q_i/V \tag{9-11c}$$

and for the motion of each particle i

$$m_i \frac{d\mathbf{v}_i}{dt} = q_i[\mathbf{E} + \mathbf{v}_i \times \mathbf{B}_0] \tag{9-11d}$$

$$\frac{d\mathbf{x}_i}{dt} = \mathbf{v}_i \tag{9-11e}$$

where $\mathbf{B}_0$ is retained in Eq. (9-11*d*) in order to include the effect of a fixed external magnetic field.

In Sec. 9-2 we discuss in detail the properties of the two-dimensional electrostatic particle-mesh model and in Sec. 9-3 we describe its application to the study of the diffusion of plasma across a strong magnetic field. Strong magnetic fields are used to contain hot-gas plasmas in devices designed to extract power from controlled thermonuclear reactions. The unexpectedly high (or anomalous) level of plasma diffusion across the containing fields has been one factor delaying the success of these devices and computer simulations have played an important role in extending our knowledge of the basic mechanisms causing the anomalous diffusion.

9-1-3 Historical Survey

By far the largest effort in computer simulation has been devoted to the study of hot-gas plasmas, spurred on by the investment in research aimed at the production of power from controlled thermonuclear fusion. Most of the work is the output of organized research groups in the United States [at Los Alamos Scientific Laboratory (LASL), the Magnetic Fusion Energy Computer Center (MFECC) at Livermore, Lawrence Livermore Laboratory (LLL), Naval Research Laboratory Washington (NRL), University of California, Berkeley (UCB) and Los Angeles (UCLA), and the Universities of Stanford, Princeton, and Columbia] and in Britain (at the UKAEA Culham Laboratory, Imperial College, and Reading University), although important contributions have been made elsewhere (e.g., the Max Planck Institute, Garching, West Germany).

The literature, starting about 1960, is now vast and we can only touch on the most significant contributions here. Our emphasis will be on papers concerned with the development of simulation techniques using particles. Early papers appeared principally in *Physics of Fluids* but now such papers appear in the *Journal of Computational Physics*—which specializes in computational techniques and gives details of models that are often only cursorily mentioned, if at all, in other journals. More comprehensive invited papers appear in a series of volumes "Methods in Computational Physics" (eds. B. Alder, S. Fernbach, and M. Rotenberg, Academic Press, New York and London), of which volumes 9 (1970) and 16 (1976) are devoted respectively to plasma physics and controlled fusion. These volumes, together with the proceedings of the specialist conferences on *Numerical Simulation of Plasma* (LASL 1968, NRL 1970, LLL 1973, and New York University 1975) form the main collected research material on plasma simulation methods. Specific reviews of the role of computer simulation in plasma physics have been given by Buneman and Dunn (1966) and by Birdsall and Dawson (1970).

Printed course notes have been available from UCB since 1975 under the title *Plasma Physics via Computer Simulation* (Birdsall and Langdon, fourth printing 1978). This deals in detail—including sample programs—with the techniques and theory of electrostatic plasma codes, with and without a fixed magnetic field, and has now been published commercially (Birdsall and Langdon, 1981). Tested and authenticated computer programs for some models have been published in *Computer Physics Communications* (North-Holland Publishing Co., Amsterdam). Programs published in this journal may be obtained at minimal cost from the CPC program library, Queen's University, Belfast.

The first particle models of an electrostatic plasma resulted from the pioneering work of Buneman (1959) at Stanford on the "dissipation of currents in ionized media" and of Birdsall and Bridges (1961) at Berkeley on "space-charge instabilities in electron diodes and plasma converters." These models were one-dimensional and did not use a mesh for the calculation of the fields. Similar models were developed by Dawson (1962) at Princeton and Eldridge and Feix (1962). The important step of introducing a mesh for the calculation of the fields, resulting in the first PM algorithms, was taken at Stanford independently by Burger (1965) in one dimension and Hockney (1966*b*) in two dimensions. The latter required the parallel development of rapid direct methods for the solution of Poisson's equation, that is to say the first RES algorithms (Hockney, 1965, and Chapter 6). The above PM algorithms used NGP charge assignment and field interpolation in order to economize on computer time. Over 3,000 particles could be simulated in one dimension and up to 2,000 in two dimensions on the IBM 7090 computer. Higher-order interpolation schemes (CIC) were first introduced by the Berkeley group (Birdsall and Fuss, 1969) in order to reduce the noise in simulations, and by Morse and Nielson (1969) using Harlow's (1964) PIC technique for fluid dynamics simulation. The PIC method applied to plasma physics is also discussed by Morse (1970). Work with one-dimensional electrostatic models has been reviewed by Dawson (1970), and methods for efficiently implementing codes have been considered by Boris and Roberts (1969), and Buneman (1976).

An alternative technique for calculating the fields— the multipole-expansion method—has been developed at Princeton and UCLA (Kruer, Dawson, and Rosen, 1973; Chen and Okuda, 1975). The method is based on the expansion of the charge distribution in each mesh cell in terms of the monopole, dipole, and quadrupole moments positioned at the center of each cell. These separate distributions are Fourier-analyzed and summed in order to obtain the Fourier transform of the monopole, dipole, and quadrupole strengths of the field. The Fourier synthesis of these gives the multipole strengths of the field at the mesh points and a Taylor expansion using these strengths gives the field at each particle position. Even if the multipole expansion is taken only to the dipole terms, substantially more computation and storage is required compared with the PM algorithms described in this book. This is particularly the case if the method is generalized to three dimensions, in which case there are four distributions to analyze, synthesize, and store (the monopole and three components of the dipole) compared with a single equivalent monopole distribution used in a PM algorithm. Nevertheless, successful one- and two-dimensional multipole codes are in use.

The theory of the PM algorithm was first developed in a coherent way by Birdsall, Langdon, and their coworkers at Berkeley in a series of important papers. The fact that a PM algorithm is a model of a plasma of finite-sized particles or clouds was recognized by Hockney (1966*b*) in describing the earliest PM model, and the theory of simulation using such finite-sized particles was developed by Langdon and Birdsall (1970). The effect on the dispersion relation of the model due to the finite spatial mesh (Langdon, 1970*a*,*b*) and finite timestep (Langdon, 1970*a*; Chen, Langdon, and Birdsall, 1974; Langdon, 1979*b*) was subsequently given. This work was extended to the energy-conserving algorithms of Lewis (1970*a*,*b*) by Langdon in 1973. The above papers form the basis of our Chapter 7. The kinetic theory of fluctuations and noise in such computer models has been given by Langdon (1979*a*).

The finite-size concept was used by Okuda and Birdsall (1970) to calculate the dramatic reduction of the collision rate resulting from the use of the PM method that inherently simulates particles comparable to a mesh cell in size. This effect was confirmed in an extensive empirical study of collision and heating times by Hockney (1971), which is presented here in Secs. 9-2-1 to 9-2-3, and by Okuda (1972*b*). The energy conservation and momentum conservation properties of a variety of PM codes have also been compared by Lewis, Sykes, and Wesson (1972), who showed that, because of the effect of a finite timestep, the so-called "energy-conserving" schemes (see Sec. 5-5) do not have noticeably better energy conservation properties than momentum-conserving codes.

Confirmation of Langdon's theories of mesh effects in PM algorithms has been found in the prediction and experimental verification of unphysical instabilities in a cold plasma due to motion through the computational mesh, and instabilities in a thermal plasma when $(H/\lambda_D) > \pi$ (Okuda, 1972*a*; Langdon, 1973; Chen, Langdon, and Birdsall, 1974; Abe, Miyamoto, and Itatani, 1975; Birdsall and Maron, 1980). Fortunately, neither of these instabilities is destructive. The cold-plasma instability saturates at a small amplitude, and the thermal-plasma instability ceases to grow when it has raised the temperature and thereby increased λ_D such that $(H/\lambda_D) < \pi$, thus stabilizing the plasma. Nevertheless, a good model should minimize these unwanted effects.

Efforts to reduce the effect of the spatial mesh have led both the Berkeley and Reading groups to the next-higher-order charge sharing and field interpolation scheme. This is the triangular-shaped density cloud (TSC, see Chapter 5) or quadratic-spline (QS) interpolation in the Berkeley notation. The Reading group has also shown that the Green's function used in the calculation of the potential may be shaped to compensate for the squareness of the mesh. The result is an algorithm in which the effect of the mesh has, for all practical purposes, been eliminated. This was first demonstrated in two dimensions by empirically shaping the Green's function (Hockney, Goel, and Eastwood, 1974; Eastwood and Hockney, 1974). The theory has been developed by Eastwood (1975; 1976*a*,*b*; and Chapters 7 and 8) who derives, by his *Q-minimization* procedure, expressions for the best choice of Green's function. These algorithms are at least one-hundred times less noisy than the original CIC method and are known as quiet particle mesh codes (QPM). The addition of direct particle-particle interactions with nearby particles (producing the P^3M code) enables any desired short-range interaction to be included, and subgrid resolution can be achieved. A three-dimensional computer code P3M3DP has been published incorporating these ideas and is available from the CPC program library (Eastwood, Hockney, and Lawrence, 1980).

The advent of lasers and the possibility of laser-induced fusion has led to the development of fully electromagnetic (in contrast to electrostatic) codes in which all Maxwell's equations [Eqs. (9-1)] for the electric and magnetic fields are solved self-consistently with the motion of the charged particles. Such models are reviewed by Langdon and Lasinski (1976) and, in the nonradiative limit, by Nielson and Lewis (1976). The advance from two-dimensional electrostatic codes to three-dimensional electromagnetic codes is discussed by Buneman (1976), who describes the Stanford electromagnetic code. This comprises a particle-moving phase using a relativistic extension of second-order time-centered differencing for the Lorentz acceleration (Sec. 4-7-1), and a field update phase for the $\mathbf{E}$ and $\mathbf{B}$ fields using a triple Fourier transform and solution in $\mathbf{k}$ space. There are ten variables per mesh point (ρ and three components of $\mathbf{E}$, $\mathbf{B}$, and $\mathbf{j}$), and six variables per particle (three components of $\mathbf{x}$ and $\mathbf{v}$). Because of their number, the particle and field values are stored on disk and are transferred to and from the fast store of the computer for updating. Considerable ingenuity is displayed in overlapping input/output operations between the two levels of storage with the arithmetic processing (Barnes, 1972). On the IBM 360/91 (between one-half and one-third the speed of the CDC 7600; see Hockney 1978*a*), the code uses a 64^3 mesh with quadratic-spline interpolation (i.e., 3D TSC). The particle moving phase takes 250 μs per particle and the field update 70 s. A timestep of $\sim$4 min is reported for 500,000 particles. A highly optimized CDC 7600 version of this code due to Nielsen (1978), Green, and Mirabella, called SPLASH, is available on the MFECC library at Livermore (Buneman et al., 1980), and has been used to study the dynamic evolution of a *Z*-pinch by Nielsen et al. (1978). The particle moving phase handles up to 300,000 particles at 130 μs per particle and the field updating on a 32^3 mesh takes 4 s. A vectorized version written by Buneman (1979) for the CRAY-1 used linear tetrahedral interpolation and a $(128 \times 128 \times 256)$ mesh (4 million mesh points). Tetrahedral interpolation (Buneman, 1980) is a form of 3D linear interpolation that requires data from only the four nearest mesh points, compared with the nearest eight points required by the trilinear interpolation of a three-dimensional CIC code. The six coordinates of each particle are packed into two CRAY-1 64-bit words in fixed point with a precision of 21 bits. The code is written in CAL assembler language and the particle-moving phase has been reduced to 7 μs per particle by the combined effects of the drop in the order of interpolation (reduction by factor $\frac{1}{4}$) and the increased speed of a carefully coded CRAY-1 (reduction by a factor $\frac{1}{5}$). Simulations with several million particles are feasible using this code with a timestep of about a minute.

The models described up to now, which calculate on the microscopic scale by following the orbits of individual plasma particles, are limited to the study of the fundamental properties of a plasma on time and space scales very much smaller than those that are important in actual fusion machines. In order to study the latter one must inevitably proceed to the macroscopic equations of magnetohydrodynamics [Eqs. (9-4) to (9-5)] or to a hybrid macroscopic/microscopic model. The role of the pure microscopic simulation is to provide understanding of the basic behavior of plasma and values of, for example, transport coefficients for use in the macroscopic model. The solution of the latter by conventional difference methods using both eulerian and lagrangian techniques is given in the reviews by Roberts and Potter (1970) and by Brackbill (1976).

Particle methods do find their role in macroscopic simulation of fluids, particularly where moving interfaces are involved. A derivative of the famous particle-in-cell (PIC) method, developed by Harlow (1964) for hydrodynamic calculations, has been successfully extended to MHD by Marder (1975) and used, for example, by Todd (1975) for the study of shock-heated plasmas. Another approach is the "waterbag" model, in which the system is defined by storing and moving the coordinates of particles defining the contours of some quantity. This was first used in one space dimension by Berk and Roberts (1970) to follow density contours in (x, v) phase space but has since been extended by Potter (1976) to the study in two space dimensions of magnetic field contours in realistic toroidal fusion devices.

Very often the parameters of a plasma lie between those that can be modeled satisfactorily by a microscopic model and those that can be modeled satisfactorily by a macroscopic model. Such is the case in a strong magnetic field when the electrons' larmor radius is small and can be ignored, but the effect of the larmor radius of the ions must be included. In this case it is appropriate to treat the electrons as a fluid and to calculate the orbits of the ions by a collisionless particle technique. Such hybrid methods are discussed by Hewett and Nielson (1978) and by Byers, Cohen, Condit, and Hanson (1978).

9-2 TWO-DIMENSIONAL ELECTROSTATIC MODEL

We have already used a one-dimensional model of an electrostatic plasma as an illustrative example of the particle-mesh method in Chapters 2 and 3. In higher dimensions the basic cycle remains the same but different methods must be used for the solution of Poisson's equation. The choice of suitable methods has already been fully discussed in Chapter 6. Computer-time constraints usually encourage the use of rapid transform methods (RES algorithms, Secs. 6-5-1 to 6-5-7) if the

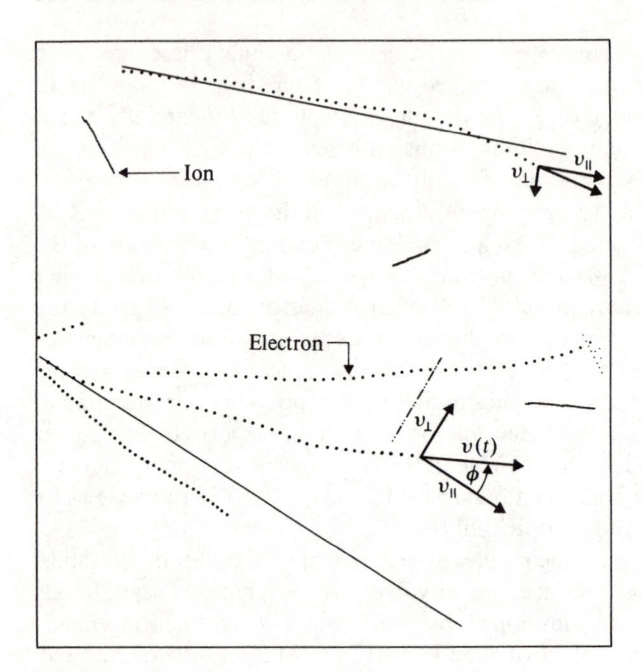

Figure 9-1 Typical orbits in a two-dimensional particle-mesh model of an electrostatic plasma. (*After Hockney, 1971, courtesy of Journal of Computational Physics, © Academic Press, Inc.*)

geometry of the problem can be simplified sufficiently. Two properties that characterize the behavior of such models are the collision time and the heating time. These have been studied empirically by Hockney (1971) and Hockney, Goel, and Eastwood (1974). The results are presented below.

9-2-1 Collision Time

In order to study the fundamental properties of an electrostatic plasma we consider a square region with doubly periodic boundary conditions. Figure 9-1 shows some typical particle orbits in such a model. In an ideal, uniform, collisionless plasma the orbits of all ions and electrons are straight lines. The fact that the observed orbits in the model are not straight can be attributed to the presence of collisional effects in the model. This can be expressed quantitatively by defining a collision time as follows.

Suppose the orbit of the ith particle has been deflected from its initial direction by an angle $\phi_i(t)$ after a time t; then one can define the root-mean-square (RMS) average deflection as

$$\langle \phi^2(t) \rangle^{1/2} = \left(\frac{1}{N} \sum_{i=1}^{N} \phi_i^2(t) \right)^{1/2} \tag{9-12a}$$

The collision time τ_c is the time for this average deflection to reach 90°. That is to say

$$\langle \phi^2(\tau_c) \rangle^{1/2} = \pi/2 \tag{9-12b}$$

Throughout this section we use the angled brackets to denote the value of a quantity averaged over the ensemble of particles under consideration. For example, if a_i is the value of a quantity for the ith particle then

$$\langle a \rangle = \frac{1}{N} \sum_{i=1}^{N} a_i \tag{9-12c}$$

We shall also need some results for a random walk of n steps with vector step length $\delta \mathbf{a}$. We assume the step length to have constant magnitude $|\delta \mathbf{a}|$ but random direction. After n steps, the ith particle has moved a distance $\Delta \mathbf{a}_i$ and the following averages can be calculated

$$\langle \Delta \mathbf{a} \rangle = 0 \tag{9-13a}$$

and

$$\langle |\Delta \mathbf{a}|^2 \rangle = n |\delta \mathbf{a}|^2 \tag{9-13b}$$

The first result follows from symmetry. The second is more subtle and states that the squared total change can be obtained by adding up the squares of the individual steps. Thus the mean-squared value of the quantity grows linearly with the number of steps. If the quantity $\delta \mathbf{a}$ is a displacement then Eq. (9-13b) is the equation for the diffusion of a cloud of particles and $\langle |\Delta \mathbf{a}|^2 \rangle^{1/2}$ is a measure of the width of the cloud.

If the total deflection ϕ is the result of a large number of small deflections $\delta \phi$

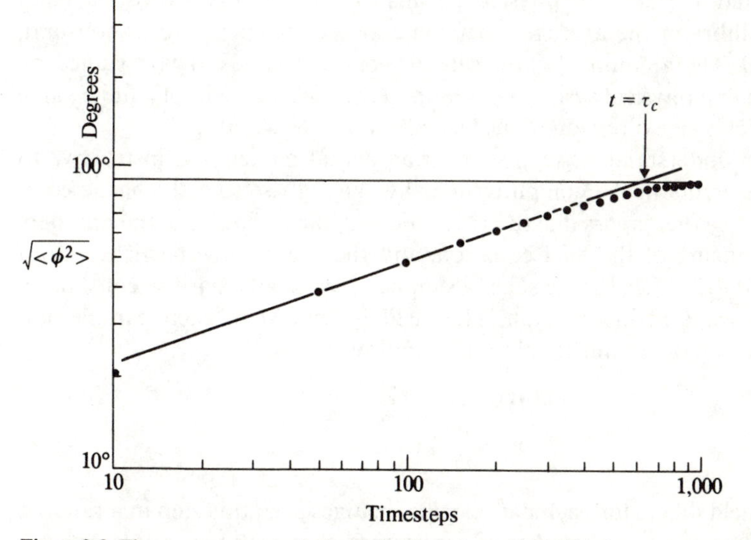

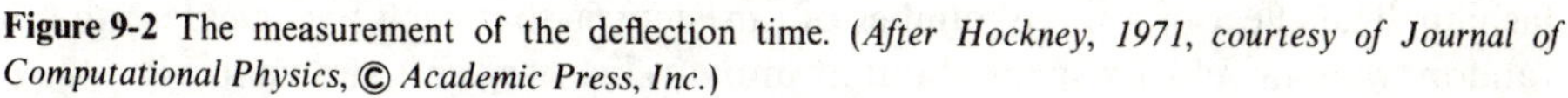

Figure 9-2 The measurement of the deflection time. (*After Hockney, 1971, courtesy of Journal of Computational Physics, © Academic Press, Inc.*)

that are stochastic in origin, we can consider the scattering process as a random walk in angle and apply Eq. (9-13*b*) to obtain the squared total deflection after n steps

$$\langle \phi^2 \rangle = n\,(\delta\phi)^2 \tag{9-14a}$$

or

$$\langle \phi^2 \rangle^{1/2} = \delta\phi\, n^{1/2} \tag{9-14b}$$

Thus the root-mean-squared deflection is proportional to the square root of the number of steps. Figure 9-2 shows the measurement of $\langle \phi^2 \rangle^{1/2}$ in an electrostatic plasma model. The root-mean-squared deflection is seen to be proportional to $n^{1/2}$ in agreement with the stochastic theory given above. The collision time is determined quite precisely.

9-2-2 Heating Time

Another quantity of importance in judging the quality of a plasma model is the heating time. If $h_i(t)$ is the deviation of the kinetic energy of the ith particle from its initial value, then we can define the average change in kinetic energy per particle over the ensemble

$$\langle h(t) \rangle = \frac{1}{N} \sum_{i=1}^{N} h_i(t) \tag{9-15a}$$

The heating time τ_H is defined as the time for the average kinetic energy per particle to increase by $\frac{1}{2}k_B T$. That is to say

$$\langle h(\tau_H) \rangle = \tfrac{1}{2}k_B T \tag{9-15b}$$

where k_B is Boltzmann's constant and T the absolute temperature.

In an isolated nonradiating physical plasma the total energy is conserved and in thermal equilibrium the average kinetic energy (i.e., temperature) is constant, thus $\langle h(t) \rangle \equiv 0$. The amount of energy lost by one particle is always gained by another. Hence in a physical system $\tau_H = \infty$ and the measurement of a finite value for τ_H in a model is an expression of the lack of energy conservation.

In order to understand why $\tau_H < \infty$ in an actual model, it is instructive to consider all the errors in the computer model as giving rise to a stochastic error field. We lump together errors due to arithmetic rounding, the size of the timestep, the finite-differencing of the field equation, and the use of superparticles into a single error field $\delta\mathbf{E}$ which, for the sake of simplicity, we shall regard as constant in magnitude but random in direction. This field is applied to each particle and causes a change of momentum equal to the impulse. Hence

$$m\,\delta\mathbf{v} = q\,\delta\mathbf{E}\,DT \tag{9-16a}$$

and

$$\delta\mathbf{v} = \frac{q}{m}\,DT\,\delta\mathbf{E} \tag{9-16b}$$

The error field differs for each particle and changes each timestep in a random fashion. The effect of a large number of timesteps is that each particle describes a random walk in velocity space about its initial velocity $\mathbf{v}_0$ as is shown in Fig. 9-3.

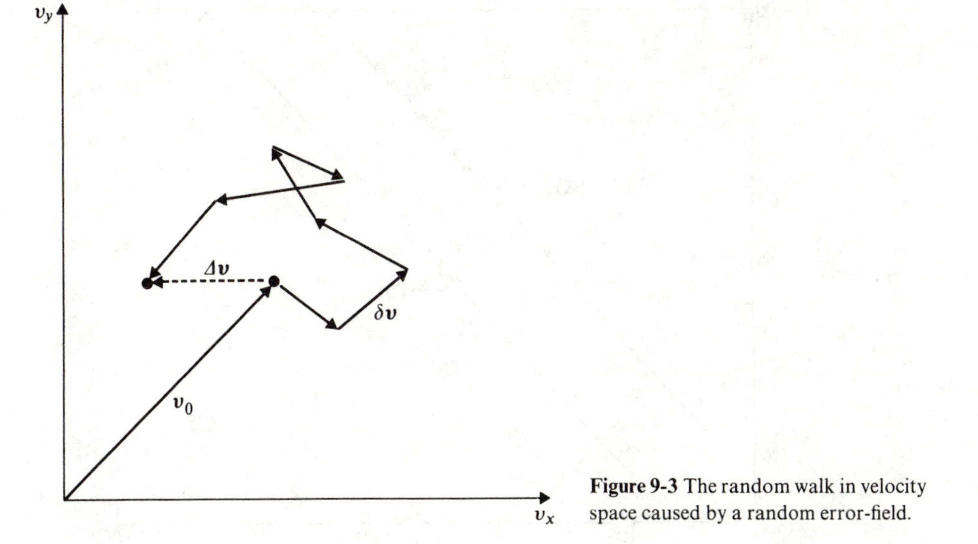

Figure 9-3 The random walk in velocity space caused by a random error-field.

If we have an ensemble of particles initially all with velocity $\mathbf{v}_0$, then after n steps we have, using Eq. (9-13),

$$\langle \Delta \mathbf{v} \rangle = 0 \tag{9-17a}$$

$$\langle |\Delta \mathbf{v}|^2 \rangle = n \frac{q^2}{m^2} DT^2 |\delta \mathbf{E}|^2 \tag{9-17b}$$

and the average change in kinetic energy per particle is

$$\begin{aligned} \langle h \rangle &= \tfrac{1}{2} m \langle |\mathbf{v}_0 + \Delta \mathbf{v}|^2 \rangle - \tfrac{1}{2} m \langle |\mathbf{v}_0|^2 \rangle \\ &= \tfrac{1}{2} m (2\mathbf{v}_0 \cdot \langle \Delta \mathbf{v} \rangle) + \tfrac{1}{2} m \langle |\Delta \mathbf{v}|^2 \rangle \end{aligned}$$

which, using Eq. (9-17a), becomes

$$\langle h \rangle = \tfrac{1}{2} \frac{q^2}{m} DT^2 |\delta \mathbf{E}|^2 n \tag{9-18}$$

Equation (9-18) applies to any group of particles independent of their initial velocity $\mathbf{v}_0$, and hence applies to a system with a distribution of velocities, such as a maxwellian plasma. The important result is that the average kinetic energy increases linearly with the number of timesteps, provided the error fields are random. This effect is called "stochastic heating" and is present to some degree in all computer models. Since this nonphysical effect cannot be eliminated, it is important to know its magnitude and how to control it.

Figure 9-4 shows the measurement of $\langle h \rangle$ as a function of time on a log–log plot. The graph clearly shows the linear increase of kinetic energy with time, in agreement with the simple stochastic theory given above. The measurements for the NGP and CIC models also show that stochastic heating is significantly less in the CIC model than it is in the NGP model.

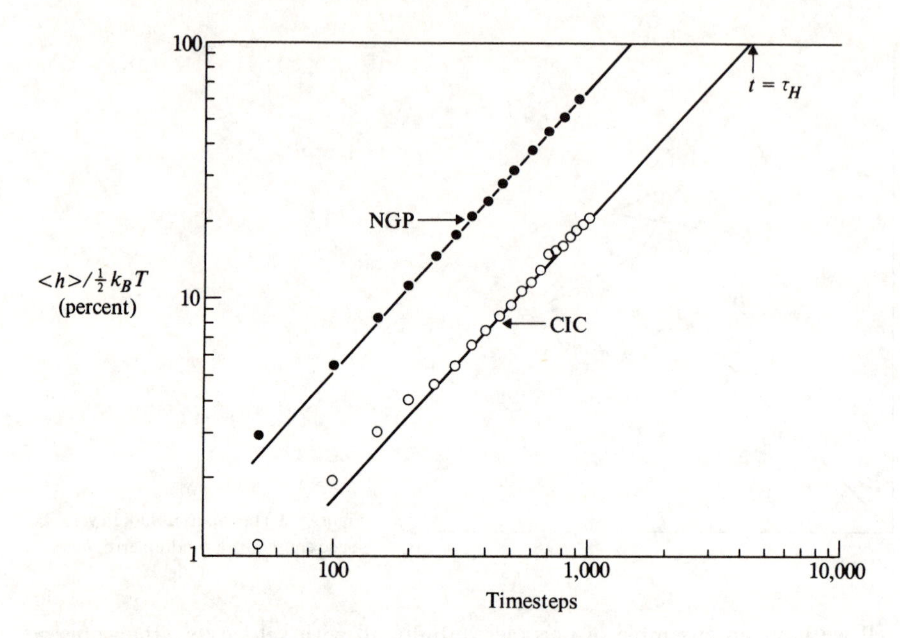

Figure 9-4 The measurement of the heating time τ_H for the NGP and CIC models. (*After Hockney, 1971, courtesy of Journal of Computational Physics, © Academic Press, Inc.*)

In a plasma the potential energy is small and may be neglected compared with the kinetic energy. Further, Eq. (9-18) shows that the rate of stochastic heating is inversely proportional to the mass of the particles being considered. Stochastic heating therefore shows up primarily as an increase in temperature of the lighter electrons. The rate of increase of the ion temperature is much slower and can usually be neglected. For this reason the quantity $\langle h \rangle$ is measured for the electrons only. Since, in two dimensions, the average energy of an electron is k_BT ($\frac{1}{2}k_BT$ for each degree of freedom) and initially the kinetic energy of the ions equals the kinetic energy of the electrons, the heating time is the time for the total energy of the system to increase by 25 percent.

9-2-3 Empirical Correlations

A computer plasma differs from a real plasma in the density n of particles, and in the finite discretization of space and time (with space-mesh H and timestep DT). The following dimensionless forms of these variables are used to correlate the experimental measurements of the collision and heating times. These are:

1. $N_C = N_D\{1+(W/\lambda_D)^2\}$ where W is the width of the particle, a number that characterizes the collision time. It varies from the number of particles per Debye square N_D $(=n\lambda_D^2)$ for small particles with $W \ll \lambda_D$, to the number of particles within a particle for large particles with $W \gg \lambda_D$.

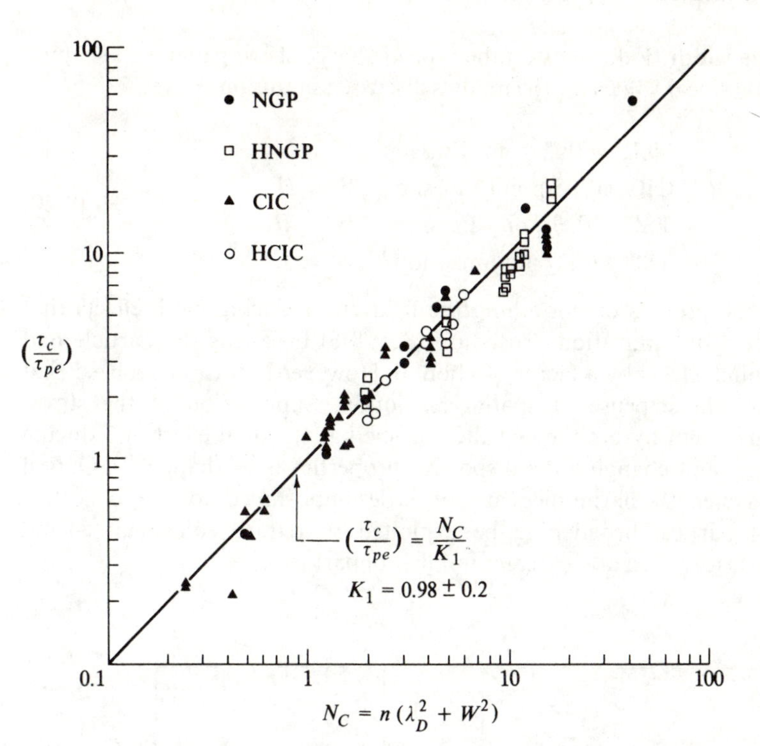

Figure 9-5 Correlation of the collision time with particle number. (*After Hockney, 1971, courtesy of Journal of Computational Physics, © Academic Press, Inc.*)

2. H/λ_D, the number of Debye lengths per cell in the spatial grid.
3. $\omega_{pe}DT$, the electron plasma frequency multiplied by the timestep of the fraction of a Debye length traveled by a thermal electron in a timestep.

Figure 9-5 shows the correlation of the collision time for a wide range of computer models with the single variable N_C. To within 20 percent, all the measurements can be fitted by the relation

$$\left(\frac{\tau_c}{\tau_{pe}}\right) = N_C = N_D\left\{1 + \left(\frac{W}{\lambda_D}\right)^2\right\} = n(\lambda_D^2 + W^2) \tag{9-19a}$$

or alternatively the collision frequency ν_c $(=\tau_c^{-1})$ is given by

$$\left(\frac{\nu_c}{\omega_{pe}}\right) = \frac{1}{2\pi N_C} \tag{9-19b}$$

Equation (9-19*a*) shows that the collision time is primarily determined by the density of particles through the variable *n*, and only slightly by the type of model used (e.g., NGP or CIC) through the particle width *W*. For example, if the model contains only one-millionth of the particles in the real system, then the collisional effects will be enhanced and the collision time will be shortened by the same factor.

The particle width W depends on the type of charge sharing and field shaping that is used and typical values for the models discussed in this book are:

$$\begin{array}{ll} \text{NGP with 5-point Poisson} & W = H \\ \text{CIC with 5-point Poisson} & W = H \\ \text{TSC with 9-point Poisson} & W = 2H \\ \text{TSC with } Q \text{ minimization} & W = 3H \end{array} \qquad (9\text{-}20)$$

We note how the process of smoothing the field and reducing mesh effects that occurs with the Q-minimization procedure (Sec. 8-6) broadens the particle and reduces collisional effects by a factor of about 3. However it must be realized that this is done at the expense of spatial resolution. Suppression of the short-wavelength harmonics by broadening the particle has the dual effect of reducing the collision rate and changing the dispersive properties (see Chapter 7). Only if these short-wavelength harmonics are of little importance to the collective properties can particle broadening be exploited to reduce collisional effects. Otherwise, one is forced to use a greater number of particles.

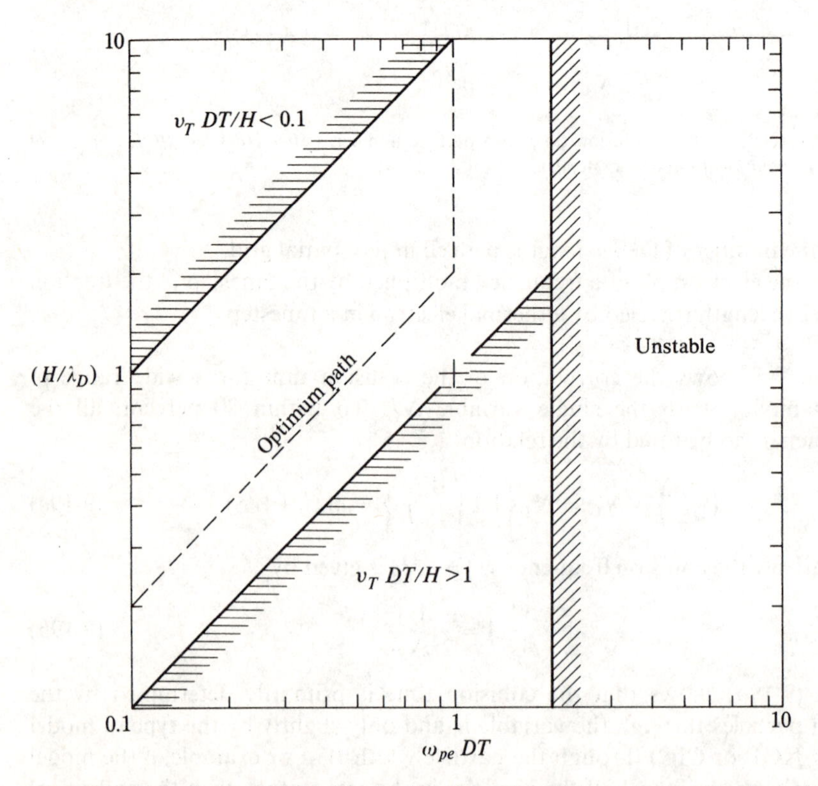

Figure 9-6 The optimum path for the selection of the timestep in an electrostatic plasma model. (*After Hockney, 1971, courtesy of Journal of Computational Physics, © Academic Press, Inc.*)

Unlike the collision time, the heating time is strongly dependent on H and DT. The dependence on N_C is the same as that found above for the collision time and we eliminate this variation by concentrating attention on the ratio of heating time to collision time (τ_H/τ_c). It is in any case the ratio of these quantities that is important, since this determines the number of collision times that can be computed before the violation of energy conservation becomes serious. For example, if we select model parameters for which $(\tau_H/\tau_c) = 10$, then we may compute for a collision time with 2.5 percent conservation of energy.

The heating–collision time ratio is a complicated function of (H/λ_D) and $(\omega_{pe}DT)$ and its variation has been measured in the parameter plane of these two variables. The region of interest is shown in Fig. 9-6. For $\omega_{pe}DT > 2$ the explicit finite-difference approximation to Newton's laws of motion is unstable (Sec. 4-4-1) and this region of the plane cannot be used. Finite-difference truncation errors will become serious if a thermal electron moves more than about one cell per timestep, i.e., if $v_T DT/H > 1$, since in this case many particles are jumping over field variations. Since the resolution of the field description is H, there is nothing to be gained by taking a timestep shorter than, say, one-tenth of the time for a thermal particle to cross a cell. Hence one may disregard the upper left of the plane for which $v_T DT/H < 0.1$.

Within the region of interest we define an optimum path, shown by the dotted line in Fig. 9-6 and given by

$$(\omega_{pe}DT)_{\text{opt}} = \min\left[\tfrac{1}{2}H/\lambda_D, 1\right] \tag{9-21}$$

This line represents the most sensible choice of timestep for a given (H/λ_D) in the sense that a reduction in DT causes relatively little increase in the heating time whereas an increase causes a rapid degradation of the model (i.e., decrease in τ_H).

If we assume that the timestep is selected along the optimum path then we can correlate the heating-time–collision-time ratio with the single variable (H/λ_D) and (see Fig. 9-7) obtain

$$\frac{\tau_H}{\tau_c} = \frac{K_H}{(H/\lambda_D)^2} \tag{9-22}$$

where the heating constant K_H depends strongly on the type of model used:

$$\begin{aligned} K_H &= 2 \text{ for NGP with 5-point Poisson} \\ &= 40 \text{ for CIC with 5-point Poisson} \\ &= 150 \text{ for TSC with 9-point Poisson} \\ &= 600\dagger \text{ for CIC with } Q\text{-minimization} \\ &= 3{,}000\dagger \text{ for TSC with } Q\text{-minimization} \end{aligned} \tag{9-23}$$

These results show the striking improvement in the quality of the model achieved by the higher-order charge sharing and the Q-minimization force-shaping

† Numbers taken from Hockney, Goel, and Eastwood (1974) using an early empirical method of minimization. The analytic methods of Chapter 8 are expected to give higher values of K_H.

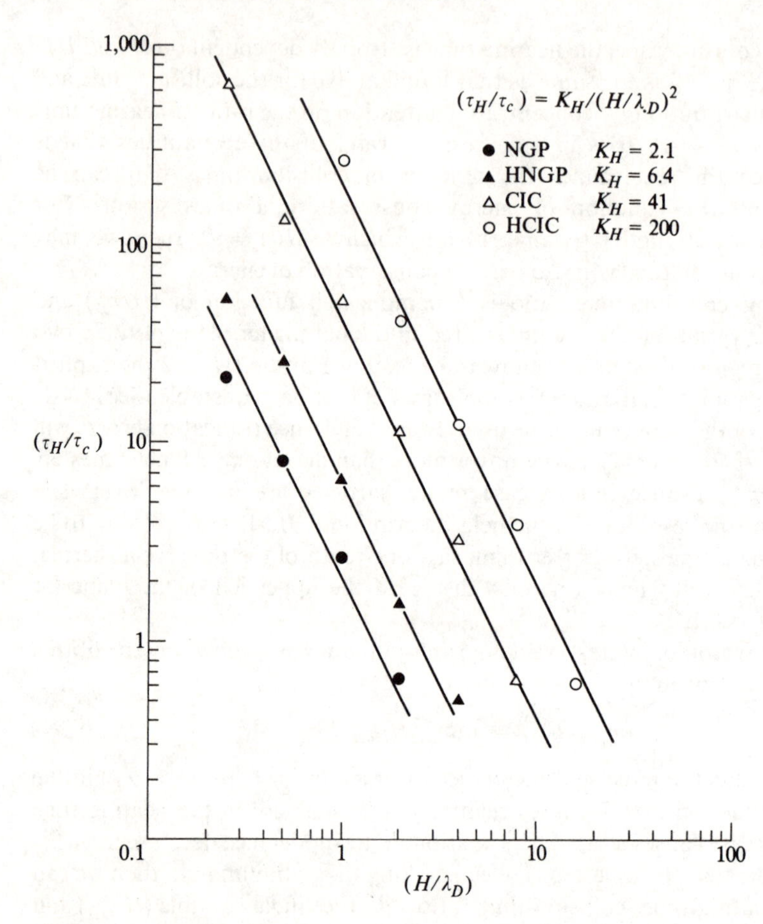

Figure 9-7 Correlation of the heating time along the optimum path. (*After Hockney, 1971, courtesy of Journal of Computational Physics, © Academic Press, Inc.*)

procedures. They also show that substantial improvements can be obtained by adding Q-minimization to a CIC model. However, it must be remembered that the use of higher-order schemes introduces a cost in computer time. Some typical CPU times on the IBM 360/195 for the particle-pushing phase of a two-dimensional electrostatic plasma model are:

NGP	1-point scheme	23 μs/particle	
CIC	4-point scheme	44 μs/particle	(9-24)
TSC	9-point scheme	104 μs/particle	

The use or not of Q-minimization does not affect these times since it only changes the values of the Green's function used in the solution of Poisson's equation. The figures are for unoptimized FORTRAN code and substantial improvements (perhaps by a factor of two) are likely to be obtained by careful hand coding in assembler language.

9-3 ANOMALOUS DIFFUSION

The two-dimensional electrostatic particle-mesh model was first applied in the study of the anomalously high diffusion of plasma across a magnetic field (Hockney, 1966*a*,*b*). Most devices planned for the production of power from the deuterium in sea water will operate by confining a very hot ($\sim 10^8$ K) deuterium-tritium plasma in the interior of a toroidal vacuum vessel with a strong magnetic field. In order to avoid radioactivity, experimental laboratory devices for the study of magnetic confinement usually use pure deuterium or hydrogen plasmas. In order to simulate any of these devices one must model the motion of charged particles in a strong magnetic field. The orbit of such a particle is a helix with its axis parallel to the magnetic field lines. The motion along the field is unaffected by the magnetic field, but the motion projected in the plane perpendicular to the field is a circle with radius inversely proportional to the strength of the magnetic field. This gyration of the plasma particles about the field lines is additionally subject to drifts arising from gradients and curvatures of the magnetic field, and from the presence of electric fields (see, e.g., Longmire, 1963, Chapter VI).

The basic problem in magnetic confinement is to design a field configuration that will confine the plasma long enough for fusion to take place. In addition to the temperature requirement of $\sim 10^8$ K, the condition for a viable fusion reactor is expressed by the Lawson criterion which states that the product of plasma density and confinement time should exceed $\sim 10^{20}\ \mathrm{m}^{-3}$ s. If the magnetic field lines are bent to follow the interior of a toroidal device, the motion of plasma along the field lines does not lead to the loss of plasma from the device. Loss occurs and the confinement time is reduced, owing to motion of plasma particles perpendicular to the field lines, and we therefore now confine our attention to this two-dimensional motion in the plane perpendicular to the magnetic field lines. In addition, we will assume a plasma comprising electrons with charge $-e$ and mass m_e, and a single species of positive ion with charge $+e$ and mass m_i. The modulus of the electronic charge e is equal to 1.60210×10^{-19} C.

The orbit of a particle with charge $\pm e$, mass m, and speed v, perpendicular to a magnetic field of strength B, is a stationary gyration in a circle of gyro (or Larmor) radius

$$a = \frac{mv}{eB} \tag{9-25a}$$

with a gyrofrequency

$$\Omega = \frac{eB}{m} \tag{9-25b}$$

Hence, if there were no interactions with other particles of the plasma (i.e., no collisions and no collective instabilities) there would be perfect containment of the plasma. The effect of a collision is to move the center of gyration in a random direction through a vector distance $\delta\mathbf{a}$, such that $0 \leqslant |\delta\mathbf{a}| \leqslant 2a$. We therefore take on average $|\delta\mathbf{a}| = a$. There will be a diffusion of plasma across the magnetic field given by the random-walk equation (9-13). The mean-squared distance diffused after n steps will be

$$\langle|\Delta\mathbf{a}|^2\rangle = n|\delta\mathbf{a}|^2 = n\left(\frac{m^2v^2}{e^2B^2}\right) \tag{9-26a}$$

$$= \frac{\nu_c m^2 v^2}{e^2B^2}t \tag{9-26b}$$

where vertical bars denote the magnitude of a vector, and $\nu_c = n/t$ is the physical collision frequency. The diffusion coefficient D_c is, in general, related to the mean-squared displacement by the equation [see Sec. 12-3-4(6), Eq. (12-82)].

$$\langle|\Delta\mathbf{a}|^2\rangle = 2dD_c t \tag{9-27a}$$

where d is the dimensionality. In our case we are considering two-dimensional diffusion in the plane perpendicular to the magnetic field, therefore $d = 2$. Hence

$$D_c = \frac{\nu_c}{2d}\frac{m^2v^2}{e^2B^2} = \frac{\nu_c}{4}\frac{v^2}{\Omega^2} \tag{9-27b}$$

If realistic values are inserted for the physical collision rate, one predicts very slow diffusion and excellent containment of the plasma. The first fusion machines were built on the promise of such estimates and it was a major setback to the development of fusion as a source of controlled power when the diffusion was observed to be many orders of magnitude larger than the estimated collisional diffusion. Consequently, this phenomenon became known as "anomalous diffusion." As a result, the machines failed to contain the plasma long enough for fusion to take place.

A useful scale to use in the discussion of anomalous diffusion is given by the Bohm diffusion coefficient

$$D_B = \frac{1}{16}\frac{k_BT}{eB} = \frac{1}{16}\frac{v_T^2}{\Omega} \tag{9-28}$$

which corresponds to the observed anomalous diffusion in some parameter ranges (see Sec. 9-3-5). We note that Bohm diffusion scales like B^{-1}, whereas collisional diffusion [Eq. (9-27b)] scales like B^{-2}. Since anomalous diffusion is observed when all MHD instabilities have been suppressed it is natural to look for its explanation in effects, such as charge separation, that are neglected in the MHD approximation. A two-dimensional electrostatic model was therefore set up to study the problem (Hockney, 1966a,b). An NGP model was used with Newton's equations integrated by the methods of Sec. 4-7-1. Poisson's equation was solved on a (48×48) mesh using the FACR(1) algorithm (see Sec. 6-5-3 and Hockney, 1965).

9-3-1 Diffusion Experiment

In order to eliminate end effects, most fusion devices are toroidal and in Fig. 9-8 (left) we show a cross section through the minor radius of such a device. The plasma is generated near the center and confined by a strong magnetic field

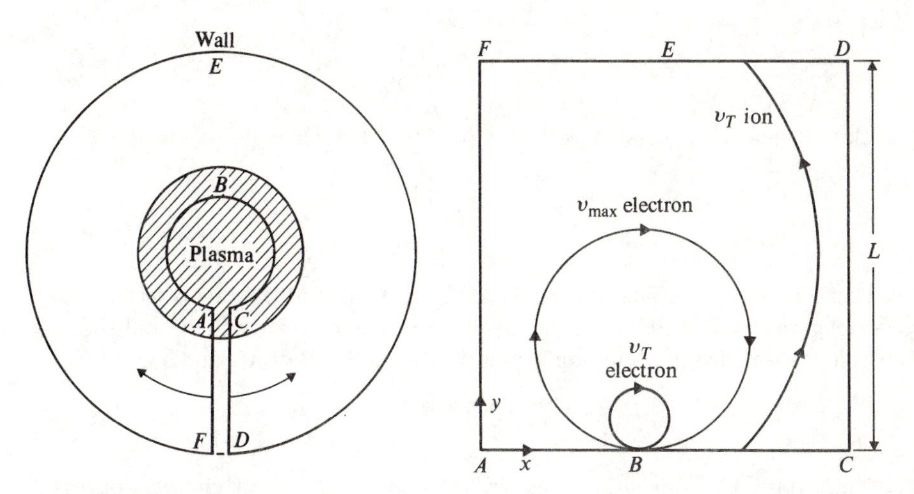

Figure 9-8 *Left*, the physical geometry of a typical fusion device. *Right*, the idealization used for computational purposes. (*After Hockney, 1966a.*)

directed into the paper. We wish to model the diffusion of plasma across this magnetic field from the center to the wall. The experiment we are about to describe was performed about 1964–5 on the IBM 7090 computer and, in order to permit a simulation at all, considerable simplifications had to be made to the geometry.

The geometry actually solved is shown in Fig. 9-8 (right). This is a square region with plasma particles injected from the bottom surface ABC towards the wall, represented by the top surface FED. The model is assumed to be periodic in the x direction so that the surfaces AF and CD act as though they are adjacent. The computed geometry (right) can be obtained from the actual geometry (left) by making a cut along the contour $FABCD$ and opening out the cylinder. The periodicity in azimuth in the real problem then becomes the periodicity in x in the model.

The scaling of the problem is also shown in Fig. 9-8 (right). An electron with a typical thermal velocity $v_T = (k_B T/m)^{1/2}$ has a gyroradius less than the distance L between the plasma edge and the wall. Even the maximum velocity assigned initially to the electrons, $v_{\max} \cong 3v_T$, leads to a gyroradius less than L. Hence in the absence of other effects no electrons can reach the wall. The ions, on the other hand, are heavier than the electrons and by Eq. (9-25a) have much larger gyroradii. Thus in the presence of only the external B field the ions have no difficulty in reaching the wall. The condition at the wall is that any charges arriving stick at their point of impact. This wall charge disappears only if it is discharged by the arrival of oppositely charged electrons.

If the region is initially empty and plasma injection starts from the bottom surface, ions travel to the wall until the potential on the wall ϕ_W is sufficiently positive to repel further ions. This will occur when the potential energy barrier to be climbed by a thermal ion is equal to its available energy, i.e.,

$$e\phi_W \cong k_B T \tag{9-29a}$$

$$\phi_W \cong \left(\frac{k_B T}{e}\right) \tag{9-29b}$$

The electric field is associated with this potential barrier is perpendicular to the wall and of value

$$E_\perp \cong \frac{k_B T}{eL} \tag{9-30}$$

The effect of an electric field on the gyration of a charged particle is to cause the center of gyration to drift with a velocity perpendicular to both the **E** and **B** fields. This velocity is called the "guiding-center drift" (or "**E** × **B** drift") and has the value

$$\mathbf{v}_g = \frac{\mathbf{E} \times \mathbf{B}}{B^2} \tag{9-31}$$

It is important to notice that this drift is the same for all charged particles regardless of their sign, charge, or mass. The latter three quantities effect the direction of gyration, the gyroperiod, and the gyroradius but *not* the drift velocity of the center of gyration.

Figure 9-9 shows the orbits of typical electrons and ions in the simulation after

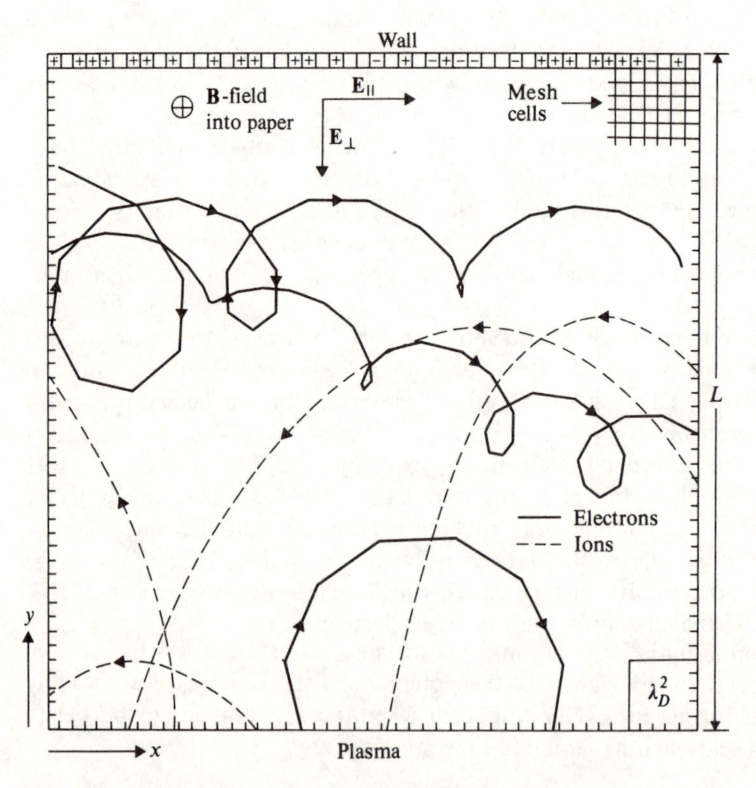

Figure 9-9 Typical particle orbits after the wall has charged positive. Note the **E** × **B** drift of the electrons. (*After Hockney, 1966b, courtesy of Physics of Fluids, © American Institute of Physics.*)

the wall charge has been established. The heavy ions describe roughly parabolic orbits as they climb and fall on the potential barrier. The electrons gyrate and drift from left to right according to Eqs. (9-25) and (9-31). In this situation we still have containment of the plasma because, although the particles drift, the direction of this drift is always parallel to the wall and never towards it.

Diffusion to the wall will only occur if there is a component of electric field parallel to the wall, $E_{\parallel}$. Such a component arises due to the growth of instabilities in the plasma. Figure 9-10 shows the situation in the simulation after the instability has grown to saturation amplitude. The perturbation of the potential due to the instability is of the form

$$\phi' = A\frac{k_B T}{e}\cos\left(\frac{2\pi mx}{L} + \delta\right) \tag{9-32}$$

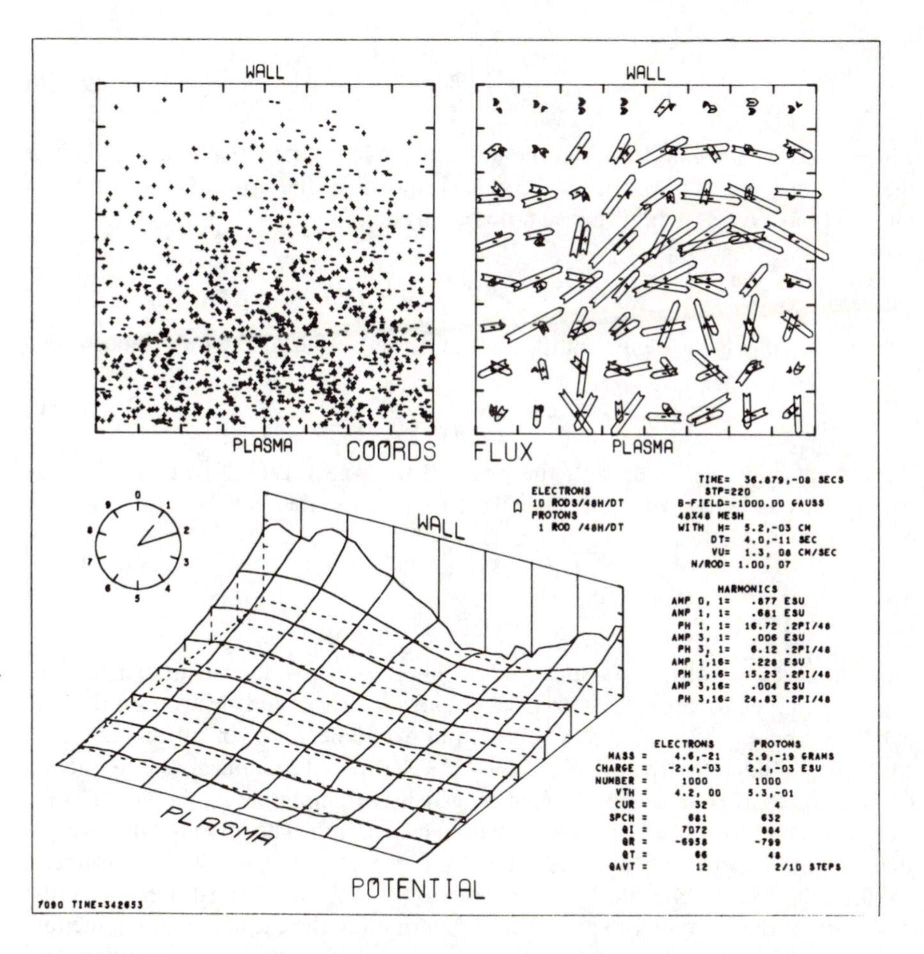

Figure 9-10 Situation after the growth of a strong $m = 1$ instability in the plasma. The flux arrows (*top right*) show the flow of plasma to the wall due to the **E** field of the wave. (*From Hockney, 1966b, courtesy of Physics of Fluids, © American Institute of Physics.*)

where A is the amplitude (in units of $k_B T/e$), m the wavenumber, and δ the phase. It is clear from Fig. 9-10 that the dominant instability is for $m = 1$. The amplitude and phase of the $m = 1$ wave, on the wall and at $y = \frac{2}{3}L$, are a byproduct of the solution of the potential by FFT techniques. They are plotted in Fig. 9-11 together with the arrival of electrons and ions at the wall.

The amount of diffusion to the wall caused by the wave can be calculated as follows. The drift velocity of particles to the wall is (taking $\delta = 0$ for convenience)

$$v_y = \frac{E_\parallel}{B} = -\frac{1}{B}\frac{\partial \phi'}{\partial x} = \frac{A}{B}\frac{k_B T}{e}\frac{2\pi}{L}\sin \cdot \frac{2\pi x}{L} \tag{9-33}$$

and the associated average flux at the wall is

$$\Gamma = \frac{n_W}{L}\int_{x=0}^{L/2} v_y \, dx \tag{9-34a}$$

$$= 2\frac{A}{B}\frac{k_B T}{e}\frac{n_W}{L} \tag{9-34b}$$

where n_W is the plasma density at the wall and we have integrated over that half of the wave for which the drift velocity v_y is positive and towards the wall. To interpret this flow as a diffusion we note that by definition

$$\Gamma = -D \operatorname{grad} n \simeq D\frac{n_p}{L} \tag{9-35}$$

where n_p ($\gg n_W$) is the density at the plasma (lower surface of region). Hence

$$D = 2A\frac{n_W}{n_p}\frac{k_B T}{eB} \tag{9-36}$$

Comparing Eq. (9-36) with the physical formula Eq. (9-28) we see that the dependence on temperature and B field is the same and therefore

$$D = 32A\frac{n_W}{n_p}D_B \tag{9-37a}$$

$$= 1.2D_B \tag{9-37b}$$

after the insertion of the measured values of A, n_W, n_p from the simulation. The measured diffusion rate calculated from the arrival of ions and electrons at the wall is $0.9D_B$, and Eq. (9-37*b*) shows that within experimental error all this diffusion can be attributed to the effect of the $m = 1$ wave. To confirm this a second computer experiment was performed in which the amplitude of the $m = 1$ wave was artificially set to zero after the potential calculation. The observed diffusion in this second experiment was less than 4 percent of the original experiment, confirming directly the above conclusion. This procedure illustrates the great flexibility of the computer experiment in determining the cause of a phenomenon. In contrast there would be great difficulty in suppressing this wave in a laboratory experiment.

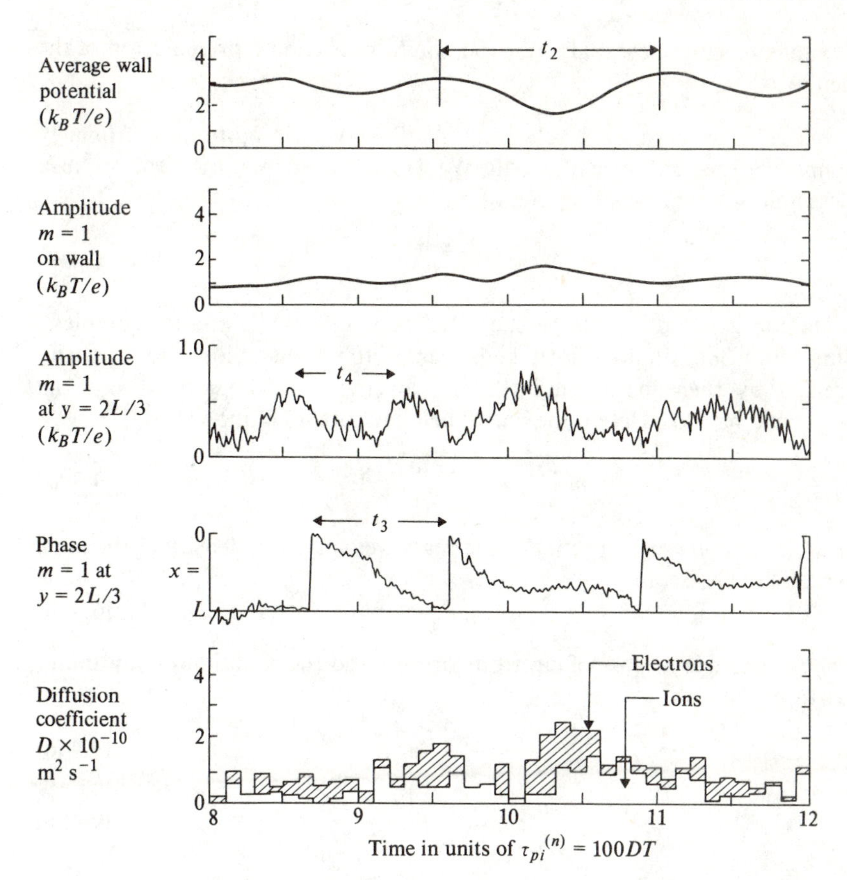

Figure 9-11 The amplitude and phase of the $m = 1$ instability and the arrival of plasma at the wall, during fairly steady diffusion. Time unit $\tau_{pi}^{(n)} = 1.4\tau_{pi}^{(0)} = 100DT$. (*After Hockney, 1966b, courtesy of Physics of Fluids, © American Institute of Physics.*)

The two-dimensional model described above neglects the effect of shear in the magnetic field and will give a diffusion larger than that in a good physical containment experiment. A computer code including this effect is described in Sec. 9-3-5.

9-3-2 Supporting Theory

One role of the computer experiment is to suggest the manner in which simpler analytic theories might be constructed. The experiment itself has demonstrated that instabilities do grow in a confined electrostatic plasma and that they give rise to diffusion across the field similar in magnitude to the anomalously high diffusion observed in the laboratory. The purpose of the supporting theory is to try to understand the origin of the instability from simpler considerations. The observation of the behavior of ions and electrons in the computer model indicates that the following simplified analytic model might be a good representation of the phenomenon:

1. *Electrons.* The gyroradii are relatively small and their motion is primarily determined by the guiding-center drift. We will therefore treat the electrons as a massless fluid with the equation of motion

$$\mathbf{v} = \frac{\mathbf{E} \times \mathbf{B}}{B^2} \tag{9-38}$$

2. *Ions.* The ion gyroradii are large such that their gyroorbits are not complete. Guiding-center motion does not take place and the orbits of ions are primarily determined by their inertia and the electrostatic field. That is to say, the magnetic field has no effect on the ions. Their equation of motion is

$$m_i \frac{d\mathbf{v}_i}{dt} = -\frac{1}{n_i} \frac{\partial p}{\partial \mathbf{x}} + e\mathbf{E} \tag{9-39a}$$

where m_i, $\mathbf{v}_i$, and n_i are, respectively, the mass, velocity, and density of the ions, and we take the pressure

$$p = n_i k_B T \tag{9-39b}$$

3. *Continuity.* There is no loss of electrons or ions and the equation of continuity applies to both fluids

$$\frac{\partial n_i}{\partial t} = -\frac{\partial}{\partial \mathbf{x}}(n_i \mathbf{v}_i) \tag{9-40a}$$

$$\frac{\partial n_e}{\partial t} = -\frac{\partial}{\partial \mathbf{x}}(n_e \mathbf{v}_e) \tag{9-40b}$$

where the subscript e refers to electron values.

4. *Fields.* The magnetic field is constant in the z direction and the model neglects the self-magnetic field of the moving charges. Hence

$$\mathbf{B} = (0, 0, B_z) \tag{9-41}$$

Charge separation does take place and the electric field is determined by Poisson's equation

$$\nabla^2 \phi = -\rho/\epsilon_0 \tag{9-42a}$$

and

$$\mathbf{E} = -\text{grad}\, \phi \tag{9-42b}$$

The above description specifies a two-fluid model and it only remains to define an initial equilibrium (indicated by starred quantities) and perform a stability analysis about this. Inspection of the simulation results again provides the clue. Early in the experiment after the wall has charged and before the $m = 1$ instability has grown the potential distribution is as shown in Fig. 9-12. This state is characterized by

1. A constant field E_y^* perpendicular to the wall

$$\phi^* = -E_y^* y \tag{9-43a}$$

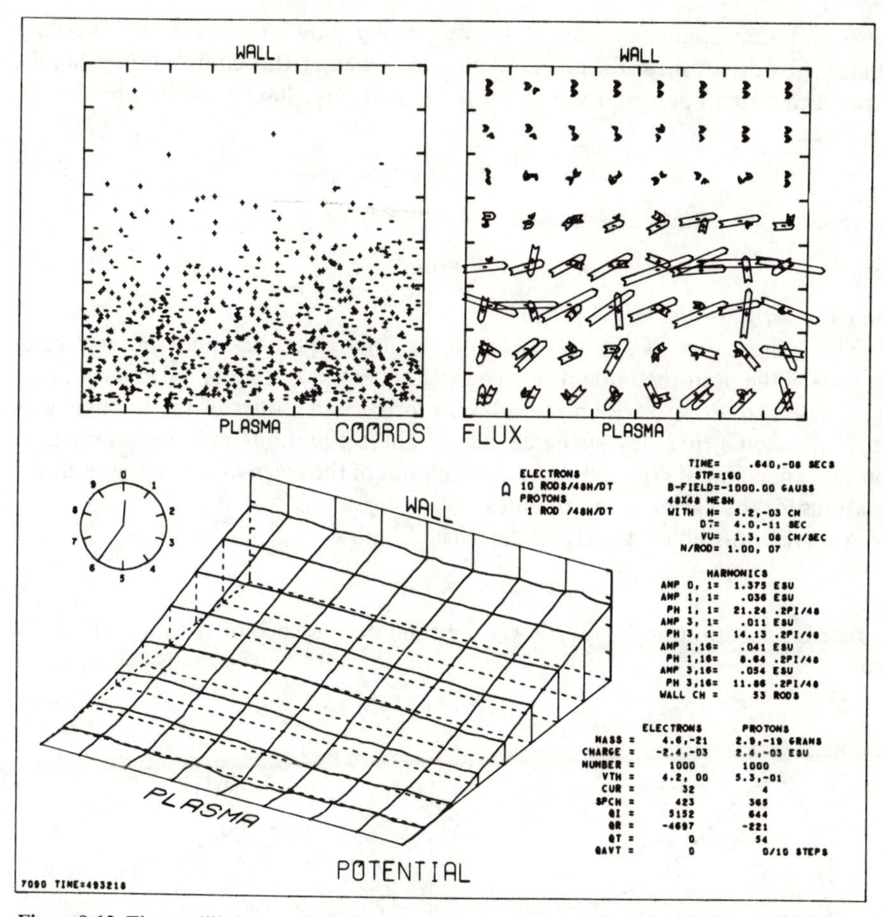

Figure 9-12 The equilibrium state, before the growth of the $m = 1$ instability. The potential surface shows a constant **E** field perpendicular to the wall, and the arrows the resulting **E** × **B** drift of the electrons parallel to the wall. This is the initial state for the instability analysis. (*From Hockney, 1966b, courtesy of Physics of Fluids,* © *American Institute of Physics.*)

2. A guiding-center drift of the electron fluid given by

$$v_x^* = \frac{E_y^*}{B_z} \tag{9-43b}$$

3. An ion density gradient determined by the Boltzmann relation

$$n_i^*(y) \propto \exp\left(-e\phi^*/k_B T\right) \tag{9-44a}$$

$$\propto \exp\left(eE_y^* y/k_B T\right) \tag{9-44b}$$

4. Charge neutrality in the interior

$$n^*(y) = n_e^*(y) = n_i^*(y) \tag{9-45}$$

and we henceforth drop the subscripts on the density variable.

The equilibrium quantities defined above may be shown to satisfy the two-fluid equations by direct substitution. To determine whether this equilibrium is stable, we perform a small perturbation about the equilibrium state, indicated by dashed quantities

$$n(x, y, t) = n^*(y) + n'(x, y, t) \tag{9-46}$$

and expand the perturbed quantities

$$n'(x, y, t) = \hat{n}(y)e^{i(kx-\omega t)} \tag{9-47}$$

where $k = 2\pi/L$ for the $m = 1$ mode.

The substitution of Eq. (9-47) in the original equations and the neglect of products of the perturbed quantities lead to a set of linear differential equations for $\hat{n}(y)$. These equations have only a solution for certain values of the frequency ω, which are the eigenvalues of the equation. These equations may be solved by a Fourier expansion of $\hat{n}(y)$ and the determination of the eigenvalues of the resulting equations for the harmonic amplitudes.

A wave is unstable if the eigenvalue ω is complex

$$\omega = \omega_r + i\gamma \tag{9-48}$$

because on substitution in Eq. (9-47) the time dependence of the mode is of the form

$$e^{-i\omega t} = e^{\gamma t}e^{-i\omega_r t} \tag{9-49}$$

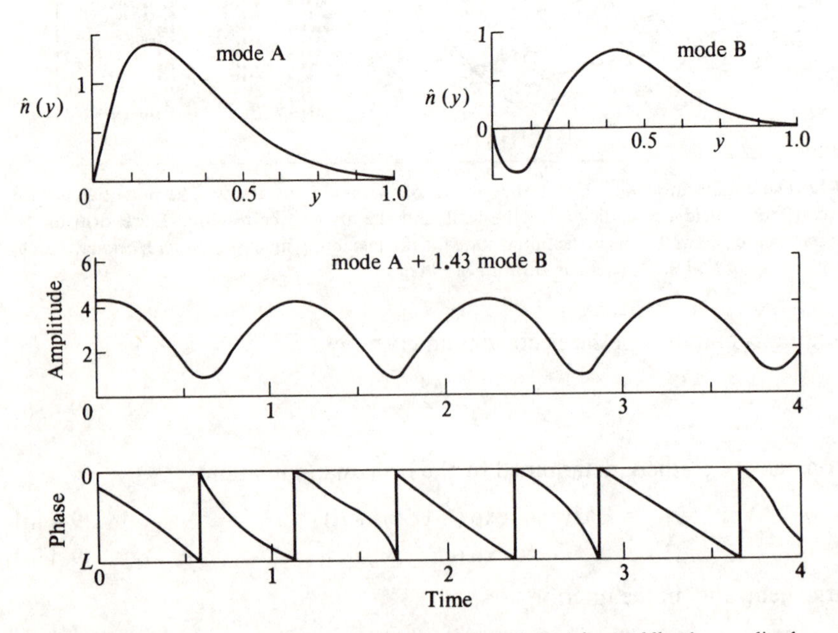

Figure 9-13 *Top*, the *y*-dependence of modes *A*, *left*, and *B*, *right*. *Middle*, the amplitude and *bottom*, phase variation resulting from the interference of modes *A* and *B* in the ratio 1 to 1.43. Time unit $\tau_{pi}^{(n)} = 1.4\tau_{pi}^{(0)} = 100DT$. (*After Hockney, 1966a.*)

and the amplitude of the disturbance grows exponentially in time with the growth rate γ.

The result of this analysis shows that the two most unstable $m = 1$ modes have a y variation as shown in Fig. 9-13 (*top*). After allowing for the stabilizing effect of Landau damping, the frequency ω, transit time t_3, and growth rate γ of these waves are:

$$\text{mode } A\text{:} \quad \omega = 0.52\omega_{pi}^{(0)}, \quad t_3 = 1.9\tau_{pi}^{(0)}, \quad \gamma = 0.34\omega_{pi}^{(0)} \tag{9-50a}$$

$$\text{mode } B\text{:} \quad \omega = 1.15\omega_{pi}^{(0)}, \quad t_3 = 0.87\tau_{pi}^{(0)}, \quad \gamma = 0.14\omega_{pi}^{(0)} \tag{9-50b}$$

where $\omega_{pi}^{(0)}$ and $\tau_{pi}^{(0)}$ are the ion plasma frequency and period at the injection plane.

In the computer experiment (see Fig. 9-11) we observe that the phase of the $m = 1$ unstable wave moves across the region with a transit time varying between $0.7\tau_{pi}^{(0)}$ and $\tau_{pi}^{(0)}$. At the same time the amplitude of the wave oscillates with period $t_4 \approx 0.7\tau_{pi}^{(0)}$. A single unstable mode would move across the system with constant amplitude and transit time. The variation in amplitude seen in the experiment and the variation in the transit time can be explained only by the interference of two waves, traveling with slightly different velocities. Figure 9-13 (*middle, bottom*) shows the total amplitude and phase variation obtained from the interference of modes A and B in the ratio 1:1.43. This is in excellent agreement with the computer experiment of Fig. 9-11.

As a further confirmation of the supporting theory the initial growth of the instability shows evidence of two growth rates that can be identified as belonging to modes A and B. This is shown in Fig. 9-14.

9-3-3 Choice of Timestep and Mesh Size

The choice of timestep and mesh size must be related to the characteristic physical frequencies and distances of the problem. These are:

Electron plasma frequency

$$\omega_{pe} = \left(\frac{n_e e^2}{\epsilon_0 m_e}\right)^{1/2} \tag{9-51a}$$

Electron gyrofrequency

$$\Omega_e = \frac{eB}{m_e} \tag{9-51b}$$

Ion plasma frequency

$$\omega_{pi} = \left(\frac{n_i e^2}{\epsilon_0 m_i}\right)^{1/2} \tag{9-51c}$$

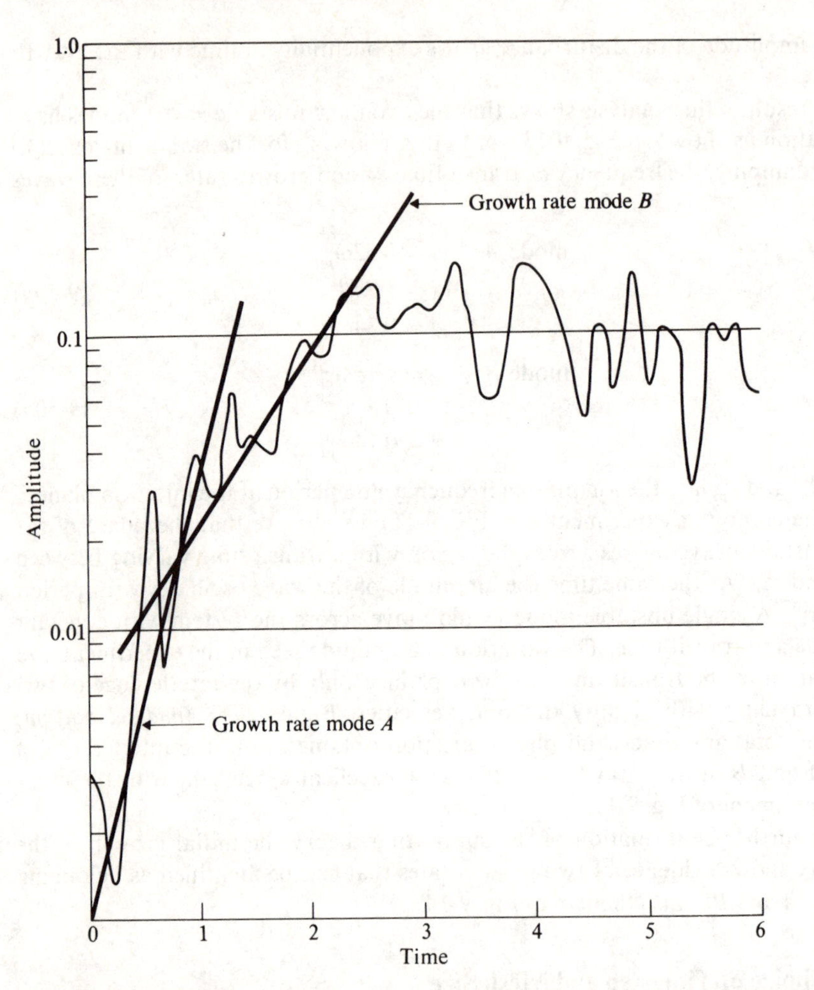

Figure 9-14 The initial growth of the instability analyzed in terms of modes A and B. The time unit $\tau_{pi}^{(n)} = 100DT$. (*From Hockney, 1966a.*)

Ion gyrofrequency

$$\Omega_i = \frac{eB}{m_i} \tag{9-51d}$$

Debye length

$$\lambda_D = \left(\frac{k_B T \epsilon_0}{ne^2}\right)^{1/2} \tag{9-51e}$$

Electron gyroradius

$$a_e = \frac{mv_{Te}}{eB} = \frac{v_{Te}}{\Omega_e} \tag{9-51f}$$

where the characteristic thermal velocity is defined as

$$v_{Te} = (k_B T/m_e)^{1/2} \tag{9-51g}$$

Ion gyroradius

$$a_i = \frac{mv_{Ti}}{eB} = \frac{v_{Ti}}{\Omega_i} \tag{9-51h}$$

$$v_{Ti} = (k_B T/m_i)^{1/2} \tag{9-51i}$$

The above formulae show that the ratio of electron to ion plasma frequency is in the inverse ratio of the square root of the electron to ion mass, and that the ratio of electron to ion gyrofrequency is in the inverse ratio of the electron to ion mass. Since the physical ion to electron mass for a deuterium plasma is about 4,000, the ratio of frequencies in the physical problem is very large. The timestep must be chosen such that $\omega DT < 2$ for the highest frequency in the problem. This means that if we used the physical mass ratio it would require several thousand timesteps to describe just one oscillation of the lowest frequency. This is the basic dilemma of the microscopic particle-mesh simulation and prevents the use of the physical mass ratio in practical calculations. However, it is believed that if a sufficiently large mass ratio is used to separate the frequencies, and the results are interpreted as dimensionless ratios, then the results can be validly interpreted in terms of the physical mass ratio. A typical choice used in this experiment is

$$\left(\frac{m_i}{m_e}\right) = 64 \tag{9-52a}$$

hence

$$\left(\frac{\omega_{pe}}{\omega_{pi}}\right) = 8 \tag{9-52b}$$

and

$$\left(\frac{\Omega_e}{\Omega_i}\right) = 64 \tag{9-52c}$$

The highest frequencies in the problem are the electron plasma and gyrofrequencies. The timestep must be chosen to satisfy the leapfrog stability criteria of $\omega DT < 2$. For reasonably accurate integration of the electron orbits a timestep significantly less than this is required. The timestep chosen in the simulation satisfies:

$$\omega_{pe} DT = 0.72 \tag{9-53a}$$

$$\Omega_e DT = 0.7 \tag{9-53b}$$

$$\omega_{pi} DT = 0.09 \tag{9-53c}$$

$$\Omega_i DT = 0.01 \tag{9-53d}$$

With this choice of timestep the ion orbits are being accurately calculated. On the other hand the electron orbits have only ~ 9 steps per gyroperiod. Although this is stable it is not an accurate integration of the gyroorbit. The finite-difference scheme used is that given in Chapter 4 [Eqs. (4-96) and (4-97)] and it is shown there that the guiding center drift is determined accurately even for large timesteps [Eq. (4-98)]. Our theory of the instability has shown that the drift velocity and not the details of the gyroorbit causes the instability. Hence the inaccuracy of the electron gyroorbit is not a serious error.

The space mesh distance H must be chosen smaller than any physical distances it is required to resolve. The value chosen satisfies:

$$(H/\lambda_D) = 0.23 \tag{9-54a}$$

$$(H/a_e) = 0.12 \tag{9-54b}$$

$$(H/a_i) = 0.002 \tag{9-54c}$$

$$(H/L) = 1/48 = 0.02 \tag{9-54d}$$

$$(H/\lambda_1) = 0.02 \tag{9-54e}$$

where λ_1 is the wavelength of the unstable $m = 1$ wave.

We can examine the above choice of timestep and space mesh in the light of the collision and noise measurements given in Secs. 9-2-1 to 9-2-3. We know that collisional effects cause diffusion across a magnetic field and that our computational models greatly enhance collisional effects compared with physical plasmas. It is therefore important to estimate, reasonably accurately, the collisional effects in the model and to show that these are not the cause of any observed diffusion. If they were, then we would have set up an inaccurate model and measured a diffusion in it due to modeling errors that had nothing whatsoever to do with processes in a physical plasma. In our model it is the transport of electrons across the magnetic field that controls the total plasma flow to the wall. The ions are almost unaffected by the magnetic field, and an equal number will always reach the wall to neutralize the charge of any electrons that arrive. Consequently in the discussion that follows we need consider only the electron diffusion.

Taking the ratio of the diffusion due to collisional effects D_c [Eq. (9-27b)] and the observed anomalous diffusion D_B [Eq. (9-28)] we obtain

$$\frac{D_c}{D_B} = 16\frac{\Omega_e}{v_{Te}^2}\frac{\nu_c}{4}\frac{v_{Te}^2}{\Omega_e^2} = 4\left(\frac{\nu_c}{\Omega_e}\right) \tag{9-55}$$

The collision frequency ν_c in the model can be obtained from Eq. (9-19b) and finally

$$\frac{D_c}{D_B} = \frac{4}{2\pi N_C}\left(\frac{\omega_{pe}}{\Omega_e}\right) \tag{9-56a}$$

At the plasma injection plane we have

$$N_D = n\lambda_D^2 \simeq 10 \tag{9-56b}$$

$$\left(\frac{W}{\lambda_D}\right) = 0.2 \tag{9-56c}$$

$$\left(\frac{\omega_{pe}}{\Omega_e}\right) \simeq 1 \tag{9-56d}$$

therefore $$N_C \simeq N_D(1+(W/\lambda_D)^2) \simeq 10 \tag{9-56e}$$

and $$\left.\frac{D_c}{D_B}\right|_{\text{plasma}} = 0.06 \tag{9-56f}$$

We conclude therefore that, at the injection plane, about 6 percent of the observed diffusion can be attributed to unrealistic collisional effects of the model. In order to estimate the collisional diffusion elsewhere in the plasma, it is necessary to consider its variation with density.

In a constant-temperature, two-dimensional plasma the number of particles per Debye square is independent of density because

$$N_D = n\lambda_D^2 = \frac{k_B T \epsilon_0}{q^2} \tag{9-57}$$

where q is the charge per unit length of the superparticles of the model. Furthermore, since Ω_e is independent of density [see Eq. (9-51b)] and ω_{pe} is proportional to $n^{1/2}$ [see Eq. (9-51a)], we conclude

$$\frac{D_c}{D_B} \propto n^{1/2} \tag{9-58}$$

The density of the wall is observed to be $\frac{1}{100}$ of the plasma density, hence

$$\left.\frac{D_c}{D_B}\right|_{\text{wall}} = 0.6 \text{ percent} \tag{9-59}$$

It is evident, therefore, that near the wall very little of the observed diffusion can be attributed to modeling errors. The above estimates are nicely confirmed by the fact that the measured diffusion when the $m = 1$ wave is artificially suppressed (4 percent D_B) lies between the two calculated values.

9-3-4 Criticism of the Experiment

The computer experiment we have described was the earliest simulation of anomalous diffusion and can be criticized on a number of counts:

1. *Geometry*. The idealized cartesian geometry ignores some important aspects of actual devices, for example the curvature of the field lines and the variation of field strength with radius. Furthermore, the variation in direction of the field lines with radius (or shear), an important stabilizing mechanism in fusion machines, is absent.
2. *Size*. The linear size of the computed system ($\sim 10\lambda_D \approx 0.2 \times 10^{-2}$ m) is at least one hundred times smaller than a physical device. Furthermore, the ion gyroradius is greater than the dimensions of the system, whereas in most real devices the reverse is the case.
3. *Boundary conditions*. The condition at the wall was chosen to represent an insulating container such as glass. The experiment shows that the potential distribution on the wall has an important, even dominating, influence on the

behavior. Since most fusion devices have a metal wall, a more realistic condition might have been a conducting wall at constant potential. The plasma density drops off rapidly as one travels away from the bottom surface, showing that the density profile is determined by the position of the injecting surface, rather than by the physical condition of the plasma. It could be that the observed results are strongly influenced by these rather artificial boundary conditions.

4. *Dynamics.* In a physical plasma, motion in the z direction, parallel to the lines of B field, is an important mechanism for neutralizing charge fluctuations. Since the diffusion is a result of such charge fluctuations, the results could be distorted by the neglect of z motion in the model.
5. *Timestep.* The investigations of the two-dimensional electrostatic plasma in Sec. 9-2-3 showed that the optimal choice of timestep is given by

$$\omega_{pe}DT|_{\text{opt}} = \min\left[\tfrac{1}{2}(H/\lambda_D), 1\right] = 0.12 \tag{9-60}$$

in our scaling [(Eq. (9-54a)]. The actual value of timestep used is six times larger, namely $\omega_{pe}DT = 0.72$ [see Eq. (9-53a)]. In this case the average number of mesh cells traveled by a thermal electron in a timestep is

$$\frac{v_{Te}DT}{H} = \frac{(\omega_{pe}DT)}{(H/\lambda_D)} = 3 \tag{9-61}$$

The use of such a large value of $v_{Te}DT/H$, so far from the optimal value of $\frac{1}{2}$, is clearly unsatisfactory and will lead to significant nonphysical heating effects.

9-3-5 Two-and-a-Half- and Three-Dimensional Models

In order to overcome some of these objections Dawson, Okuda, and Rosen (1976) have generalized the electrostatic model as follows and conducted an extensive series of experiments. In their, so called, $2\frac{1}{2}$-dimensional model, motion in the z direction is allowed and five coordinates are carried for each particle, namely (x, y, v_x, v_y, v_z). The inclusion of the velocity component v_z allows a crude representation of magnetic mirroring in an actual device. The upper boundary of the square calculation represents a magnetic mirror. As particles approach this boundary the ratio of the velocity parallel and perpendicular to **B** is computed. If this ratio is less than $[B_{\max}/B_{\min} - 1]^{1/2}$, where $B_{\max}/B_{\min}$ is the mirror ratio, then reflection takes place and the component of velocity parallel to **B** is reversed. Otherwise the particle proceeds through the boundary and reenters the system on the other side.

The model also allows the **B** field to lie at an angle to the z axis, hence all components of $\mathbf{B} = (B_x, B_y, B_z)$ exist. Gyration about such a field line involves all components of velocity $\mathbf{v} = (v_x, v_y, v_z)$ and the inclusion of v_z is necessary to represent this more general gyration. Since this model goes partway (it does not contain z) to a three-dimensional model, it is called a "two-and-a-half-dimensional" model. Magnetic shear can be represented in this model by allowing the direction of **B** to vary linearly with either x or y.

In order to avoid artificial effects due to unrealistic boundary conditions, the diffusion is measured in a uniform plasma with no net drifts, by observing the motion of test particles from their initial positions. If $\Delta\mathbf{r}$ is the displacement after a time t, then since the diffusion is two-dimensional ($d = 2$), Eq. (9-27*a*) shows that

$$D = \tfrac{1}{4} \lim_{t \to \infty} \langle |\Delta\mathbf{r}|^2 \rangle / t \tag{9-62}$$

where the angle brackets denote an average over the particles of the system. Using this method of measurement the authors find anomalously high diffusion in a uniform, stable thermal plasma due to $\mathbf{E} \times \mathbf{B}$ convective effects arising from thermal fluctuations in the $\mathbf{E}$ field. We contrast this with the experiments described earlier in this chapter in which the anomalous diffusion arose from an electrostatic plasma instability. It is clear that anomalous diffusion can arise from a variety of causes.

The computer experiments of Dawson, Okuda, and Rosen (1976) were conducted for a range of $\mathbf{B}$ field values and three regions were identified. For weak magnetic fields, $(\omega_{pe}/\Omega_e) > 3$, the diffusion varied as B^{-2} and was in agreement with classical collisional theory [Eq. (9-27*b*)]. For an intermediate range, $0.3 < (\omega_{pe}/\Omega_e) < 3$, the diffusion is approximately independent of B, and for strong fields, $(\omega_{pe}/\Omega_e) < 0.3$, the diffusion is anomalous and agrees with Bohm diffusion [αB^{-1}, Eq. (9-28)].

Further experiments were performed in an inhomogenous plasma with a density gradient. In this case the plasma is unstable to a flutelike drift wave and the diffusion is further enhanced ($\times 3$) by the $\mathbf{E} \times \mathbf{B}$ drifts associated with the $\mathbf{E}$ field of this wave. As in the experiments of this chapter, this example of anomalous diffusion is due to a plasma instability. A limited number of three-dimensional experiments are reported in which the six coordinates (x, y, z, v_x, v_y, v_z) are recorded for each particle. These show the same three regions of diffusion but the transition to collisional diffusion occurs for higher fields at $(\omega_{pe}/\Omega_e) \simeq 1$.

The generalization that $\mathbf{B}$ may be inclined to the z axis allows for three different cases that have analogies in a toroidal fusion device. If $\mathbf{B}$ is parallel to z the field lines close on themselves after one transit of the system. For other inclinations the $\mathbf{B}$ field may either close after a finite number of passes or continue to fill the system, ergodically, without closure.

It should be clear now that the relatively trivial inclusion of extra coordinate v_z (and its accompanying equation of motion) has greatly extended the application of the model. However, it must be pointed out that the $2\tfrac{1}{2}$D model does not overcome the objection to the inadequacy of representing a toroidal device in cartesian geometry or the objection to the small size of the simulated system.

9-3-6 Diagnostics and Display

It is a common feature of computer experiments that the time spent in programming diagnostic and display routines often exceeds that spent on the basic timestep cycle. This is even the case if the program is written under the OLYMPUS system (see Chapter 3), that provides some standard input and output packages. On the other hand, the usefulness of a code depends primarily on the way the input has to be given and on the way the results are analyzed and presented. These aspects of the programming differ radically between different applications and very little standardization has been achieved. There are few utility programs available, except perhaps for the printing of tables of numbers.

The interpretation of results is greatly facilitated by their being presented in graphical form and several examples are shown in the figures in this chapter. Unfortunately there is, as yet, no standard language for describing graphical output, although there are several candidates (e.g., the Culham GHOST system, GINO, and some commercial products). The consequence is that graphical routines must often be written at a very low level and are usually not transferred readily from one computer installation to another. We shall now examine some of the methods of diagnosis and display that are suitable for particle-mesh simulations.

It is never wise to compute blindfolded, without being able to see what is happening in detail in the simulation. At the lowest level it is essential to be able to examine the orbits of a few (say 10) of the particles in the simulation. For this purpose the positions of the selected particles may be stored for, say, 100 steps and then plotted out on a Calcomp pen recorder. Such a plot is shown in Fig. 9-9. Gross errors in the choice of timestep and other misconceptions in scaling show up immediately in such a display. If a storage tube display (e.g., Tektronix 4010) or microfilm recorder (e.g., III FR80) is used, one need only plot the positions of the selected particles at every timestep and the orbit will be gradually built up as a series of disconnected dots. Such a plot is shown in Fig. 9-1. If the orbit is difficult to discern the timestep is probably too long.

Figure 9-10 shows an extreme case in which all the qualitative features of an experiment together with the key input data are plotted. Such a display contains

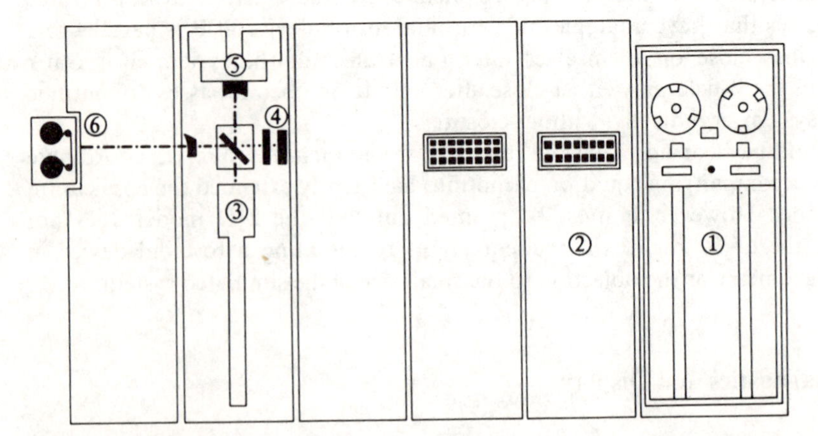

Figure 9-15 The Stromberg-Carlson 4020 microfilm recorder: magnetic tape drive (1), electronics (2), high-precision cathode ray tube (3), forms slide (4), movie camera (5), and hardcopy paper camera (6). (*Courtesy of Stromberg-Carlson, a division of General Dynamics.*)

about 10,000 dots and lines and can only be plotted on a high-speed microfilm recorder. On such a device one picture can be plotted in a few seconds. The device used for this figure was the Stromberg-Carlson 4020 microfilm recorder (Fig. 9-15). This device will plot characters at any of a (1024×1024) raster of addressable points. Lines may also be drawn between pairs of such points. The plotting instructions generated during the execution of the simulation program are generally stored on magnetic tape, which is subsequently plotted off line on the SC4020. In some systems, however, the recorder may be more closely linked to the main computer system and plotting may take place from instructions stored in temporary disk files.

The plotting instructions are read from tape by the tape drive (1) and they control, with the electronics (2), the electron beam of a high-precision cathode ray tube (3). The image on the screen is divided by a half-silvered mirror and is directed both to a movie camera (5) and a hardcopy camera (6). The former produces an image on 16 or 32 mm movie film and the latter on 10-inch photosensitized paper. If a fixed background image is desired, a forms slide (4) can be inserted containing this image. The image on either of the cameras is accumulated (or painted) on the photosensitive medium until an instruction for the advance of film is received. A mask is built into the CRT to shape the electron beam into one of 64 possible hardware character shapes. Alternative character sets and fonts can be composed out of straight-line segments and are available as "software" characters.

At the top left of Fig. 9-10 every particle in the plasma simulation is recorded using the hardware plus and minus characters for the ions and electrons, respectively. The lettering WALL and PLASMA on the outside of the box are examples of software lettering from straight-line segments. In the top right the average velocity of each type of particle has been computed in an (8×8) array of squares. An arrow is then drawn whose length and direction are proportional to the mean velocity vector. This display shows immediately the main characteristics of the flow, which is a strong flow of electrons to the wall near the center of the box. Below we have a clock face indicating the flow of time and an isometric projection of the electrostatic potential surface. The display is completed with a record in small hardware characters of the principal numerical constants of the simulation (timestep, space mesh, **B** field, charge, mass of particles, and amplitudes of waves).

If the record such as Fig. 9-10 is plotted on 16 mm movie film with one frame corresponding to each timestep, and projected at 16 frames per second, one usually obtains a satisfactory dynamic display of the time evolution of the experiment. Such a display is invaluable in identifying the qualitative features of any instabilities that may not be evident from the examination of static displays taken, for example, every 100 timesteps. Movie making using general-purpose graphics software is likely to be costly both in the use of the main computer and in the use of the microfilm recorder. For this reason it is seldom possible to make films of all the cases that are computed and standard output may have to be limited to static displays. However, it is wise to make a movie whenever a major change is made in scaling or geometry. Failure to do so may mean that important physical phenomena or programming errors go undetected. If, on the other hand, a simple optimized assembler code subroutine is written (e.g., for plotting dots only) several authors have found that movie making can be made a small fraction of the cycle time and become a routine form of output (Levy and Hockney, 1968; Boris and Roberts, 1969).

The purpose of the static and movie displays is to highlight the qualitative phenomenon occuring in the model, such as the $m = 1$ wave in Fig. 9-10. We also learn whether it is a standing or traveling wave. Quantitative measurements can rarely be made from such displays, and once the nature of the phenomenon is recognized it is essential to write special purpose diagnostic programs to compute numerically the key features of the phenomenon. In our case it is clear that there is a nearly sinusoidal $m = 1$ wave on the potential surface. The wave is quantified by calculating its amplitude and phase and these are recorded in a display such as Fig. 9-11, which is produced on a Calcomp pen recorder. Data on the wave are recorded at the wall as well as in the interior of the plasma. The arrival of electrons and ions at the wall are also recorded. Two-dimensional graphs on paper from which measurements can be made are most convenient for quantitative work. In this example the transit time t_3 can be measured from the phase graph and the period t_4 from the amplitude graph.

The Stromberg-Carlson 4020 was probably the most widely used and successful microfilm recorder and was available to computational scientists between the early 1960s and the mid 1970s. Such a device is obviously an essential part of any computer installation engaged in large-scale computational science. More modern devices (circa 1975), such as the III FR80, work on the same principles but provide greater precision [(16,000 × 16,000) addressable raster] and other facilities (e.g., the generation of color movies). The introduction of color has obvious real applications to the recording of the results from simulations. These include distinguishing particle types and contour levels by color.

It is sometimes possible to use a microfilm recorder to draw an image directly onto the sound track of a 16- or 35-mm movie film, and thereby generate different pitches of sound. This could find use in recording the arrival of simulated particles at a wall, a different pitch of sound being used for the different types of particle. However, microfilm recorders were not designed for this type of use and the potential of artificially generated sound has not been exploited to any extent.

9-4 THE MAGNETOSPHERE

The visible sun is surrounded by a proton plasma, called the corona, at a temperature of about 10^6 K. A stream of neutral plasma, called the "solar wind", is ejected from the corona and continually bombards the earth and the other planets. The velocity of the plasma at the earth is typically 7.5×10^5 m s^{-1}. The frozen-in approximation is valid and the solar wind carries with it a magnetic field from the sun. This interplanetary magnetic field is observed to be divided into sectors, two or four in number, in which the frozen-in magnetic field is alternately in the northward and southward directions. The problem of the magnetosphere refers to

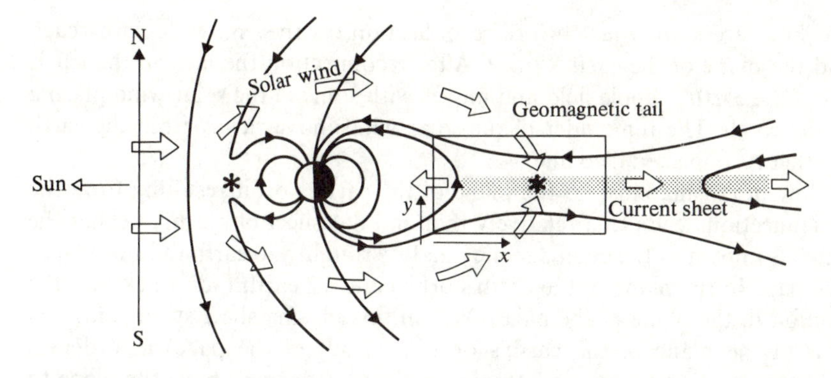

Figure 9-16 The Dungey model of the magnetosphere with a southward-directed magnetic field in the solar wind. Plasma from the solar wind (large arrows) flows over the poles to the neutral point on the downwind side of the earth whence it either returns to earth at lower latitudes or passes to infinity. The plasma behind the earth forms into a relatively thin current sheet. The rectangle shows the region studied in Fig. 9-19. (*After Dungey, 1961, courtesy of Physical Review Letters, © American Institute of Physics.*)

the whole calculation in three dimensions of the interaction of the solar wind with the magnetic field of the earth. It is important also to have a time-dependent simulation because the auroral lights and the associated magnetic storms observed on the earth are all thought to be related to changes in the plasma and magnetic field patterns in the magnetosphere, triggered by changes in the solar wind and its magnetic field.

Figure 9-16 is a sketch of the magnetic field and plasma flow patterns as envisaged by Dungey (1961) for the case that the solar wind carries a southward magnetic field. The earth-sun line lies in the plane of the page with the sun to the left. The drawing is in a plane perpendicular to the orbit of the earth and the geomagnetic dipole is drawn for the case that it is perpendicular to the earth-sun line, although in general the dipole axis may deviate by up to about 45° from this direction, further complicating the problem by removing all symmetries.

For the case of a southward (i.e., southward-directed) solar wind field there are two neutral points at which the magnetic field of the sun and the earth cancel. These are shown as two stars on the upwind and downwind side of the earth. At these points the magnetic field lines carried by the moving plasma move together and touch in the shape of an "X," then break and reconnect in the opposite sense. This important phenomenon of reconnection is demonstrated by simulation in Sec. 9-4-4. The physics of the problem is reviewed by Sagdeev (1979).

Figure 9-16 should therefore be viewed dynamically with the solar wind plasma (large arrows) and its frozen-in magnetic field lines continually hitting the earth's dipole field. The incident solar-wind magnetic field lines meet and reconnect at the neutral point on the sun side of the earth with field lines moving out from the earth. One pair of lines is shown just after reconnection, slightly to the right of the neutral point. The reconnected lines with the solar-wind plasma then flow over the poles to the neutral point on the downwind side of the earth.

Here two lines are shown just before reconnection, as they move towards each other and the plane of the earth's orbit. After reconnection the line on the left is now part of the earth's dipole field and moves with some of the solar wind plasma towards the earth. The remainder of the solar wind moves away from the earth with the other reconnected field line.

The flow of plasma close to the plane of the earth's orbit resulting from the above reconnection, leads to a relatively thin and flat sheet of plasma behind the earth which is known as the geomagnetic tail. It is roughly 40 earth radii wide (i.e., out of the page in the plane of the earth's orbit) and 12 earth radii thick (i.e., the N-S direction in the plane of the page). Within this plasma sheet strong currents must flow in the plane of the earth's orbit (i.e., out of the page) in order to produce the magnetic field reversal from a sunward direction above the plane to the antisunward direction below the plane. Owing to the fact that like current elements attract each other, these currents attract and are confined to a very thin current sheet about 0.1 earth radii thick. The formation of this current sheet in the region of the rectangle drawn in Fig. 9-16 is the subject of a simulation by particles described in Secs. 9-4-3 and 9-4-4. The formation of the overall plasma flow and field pattern including the simulation of a magnetic substorm is the subject of the next two sections (Secs. 9-4-1 and 9-4-2).

9-4-1 Magnetohydrodynamic Particle Model

The ideal magnetohydrodynamic equations for infinite conductivity were derived in Sec. 9-1, Eqs. (9-4) to (9-8). These may be written

$$\frac{\partial \rho_m}{\partial t} = -\operatorname{div}(\rho_m \mathbf{v}) \tag{9-63a}$$

$$\rho_m \frac{d\mathbf{v}}{dt} = -\nabla p - \frac{1}{2\mu_0}\nabla B^2 + \frac{1}{\mu_0}\nabla \cdot \mathbf{BB} \tag{9-63b}$$

$$\frac{\partial \mathbf{B}}{\partial t} = \operatorname{curl}(\mathbf{v} \times \mathbf{B}) \tag{9-63c}$$

$$p \propto \rho_m^\gamma \tag{9-63d}$$

where the $\mathbf{j} \times \mathbf{B}$ term on the right-hand side of Eq. (9-8*b*) has been expressed equivalently in terms of the magnetic pressure tensor.

In traditional methods these equations are all differenced on a fixed eulerian mesh and solved in the order shown for ρ_m, $\mathbf{v}$, and $\mathbf{B}$ at the next timelevel. These methods encounter difficulties when applied to the magnetosphere because the low densities encountered lead to instabilities, and the high fluid velocities can produce unphysical negative densities during the solution of the continuity equation (9-63*a*). One way of overcoming these difficulties has been given by Leboeuf, Tajima, and Dawson (LTD) (1979), who represent the MHD plasma as a collection of particles. Particles cannot be lost in such a representation, so that the continuity equation is automatically obeyed and negative densities cannot arise.

Equation (9-63*b*) becomes the equation of motion for the particle of plasma. Hence for the *i*th particle we have

$$\frac{d\mathbf{v}_i}{dt} = m^{-1}\left(-\nabla p - \frac{1}{2\mu_0}\nabla B^2 + \frac{1}{\mu_0}\nabla\cdot\mathbf{BB}\right) \tag{9-64a}$$

$$\frac{d\mathbf{r}_i}{dt} = \mathbf{v}_i \tag{9-64b}$$

where m is the mass of the particle. The derivation of field quantities, such as the number density n, the mass density ρ_m, and plasma velocity $\mathbf{v}$, from the particle values is performed by assignment to a fixed eulerian mesh using a weighting function w in the same way as charge is assigned in an electrostatic code, thus

$$n(\mathbf{r}) = \sum_i w(\mathbf{r}-\mathbf{r}_i)$$

$$\rho_m(\mathbf{r}) = mn(\mathbf{r}) \tag{9-65a}$$

and

$$\mathbf{v}(\mathbf{r}) = \sum_i \mathbf{v}_i w(\mathbf{r}-\mathbf{r}_i)/n(\mathbf{r}) \tag{9-65b}$$

The weighting function used by LTD corresponds to treating each particle as a cloud of mass with a gaussian-shaped density distributed symmetrically about the coordinate of the particle. The pressure at each mesh point is obtained from ρ_m using Eq. (9-63*d*) and the pressure force $-\nabla p$ obtained by local differencing or by Fourier transform techniques (Kruer et al., 1973). The magnetic pressure-tensor terms on the right-hand side of Eq. (9-64*a*) are evaluated by conventional finite-difference methods.

The magnetic field is advanced in time by Eq. (9-63*c*) using the conservative Lax method of differencing (Potter, 1973). The use of this method introduces a numerical diffusion because one is effectively solving the equation

$$\frac{\partial\mathbf{B}}{\partial t} - \text{curl}\,(\mathbf{v}\times\mathbf{B}) = (H^2/4DT - DTc^2/2)\nabla^2\mathbf{B} \tag{9-66}$$

where c is the fastest propagation velocity on the mesh. The terms on the right-hand side arise from the finite size of the spatial mesh interval H and the timestep DT. These error terms cannot be made to cancel everywhere because the propagation velocity varies with position. It is necessary, however, that their sum be positive in order to avoid the introduction of a negative diffusion that causes numerical instability. This leads to a Courant–Friedrichs–Lewy condition for the maximum size of the timestep.

$$DT < 2^{-1/2}H/c \tag{9-67}$$

The presence of a positive numerical diffusion means that, although we started by approximating the equations for infinite conductivity, Eq. (9-63*c*), we have produced a simulation with finite conductivity. The result is that magnetic field lines do diffuse across the plasma and the frozen-in property is only approximately obeyed.

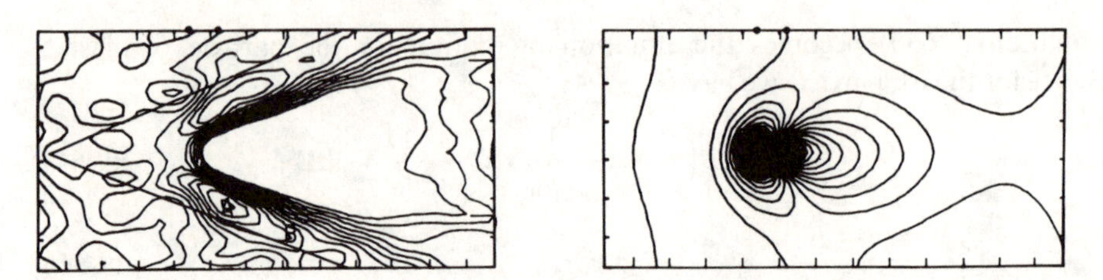

Figure 9-17 Simulation of the solar wind carrying no magnetic field with a magnetic dipole representing the earth. *Left*, average-density contours showing the mach cone and magnetopause. *Right*, magnetic field lines showing the magnetosphere extended in the direction of the solar wind. (*From Leboeuf, Tajima, Kennel, and Dawson, 1979, courtesy of Geophysical Monograph Series,* © *American Geophysical Union.*)

9-4-2 Overall Magnetosphere

In the study of the magnetosphere a two-and-a-half-dimensional version of the above code was used with two spatial dimensions and three velocity and field dimensions (Leboeuf, Tajima, Kennel, and Dawson, 1978, 1979). In the experiment shown in Fig. 9-17 the solar wind enters from the left carrying no magnetic field and impinges on a dipole magnetic field that represents the earth. The wind is supersonic and the Mach cone and density contours are shown on the left. Contours obtained for densities less than the upstream solar wind are suppressed for visual clarification. On the right the magnetic field lines are shown forming an essentially closed "teardrop" magnetosphere similar to that predicted by Johnson (1960). The shape of the outermost field lines is influenced by the doubly periodic boundary conditions imposed on the rectangle.

In order to simulate the conditions envisaged by Dungey (1961) and sketched in Fig. 9-16, a simulation was performed in which the solar wind magnetic field was switched from an E-W orientation to a southerly direction, as would occur during a magnetic storm caused by the earth entering a segment of the solar wind with a southward field. On the left of Fig. 9-18 the southward field has not quite reached the earth and a neutral point has formed on the sun side of the earth with

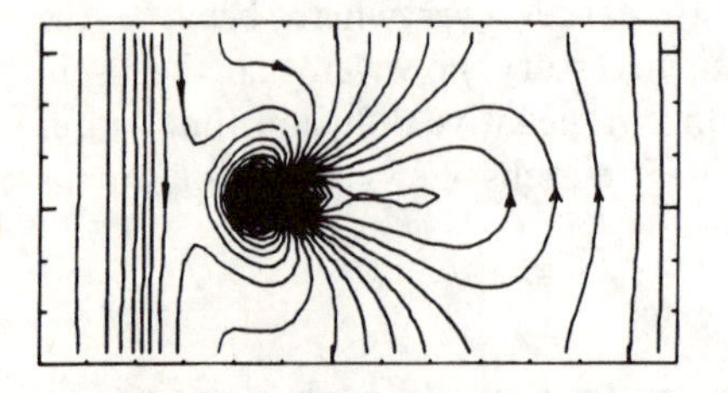

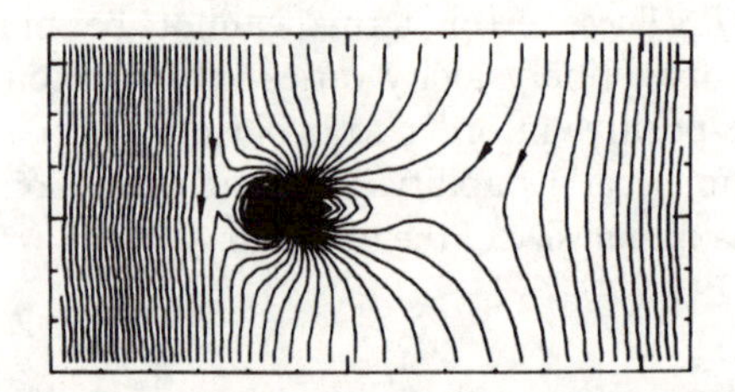

Figure 9-18 A simulated magnetic substorm. The magnetic field carried by the solar wind is switched from E-W to southward, and passes over the earth's dipole. *Left*, the southward field change has not quite reached the earth and, *right*, it has passed completely across the system leading to a field distribution closely similar to the Dungey (1961) model of Fig. 9-16. (*From Leboeuf, Tajima, Kennel, and Dawson, 1979, courtesy of Geophysical Monograph Series,* © *American Geophysical Union.*)

field lines breaking and reconnecting at this point. On the right the distribution is shown when the field switch has passed completely across the system and a field profile similar to the Dungey model is obtained including a geomagnetic tail, although this is much shorter than it should be owing to the finite conductivity of the model. Nevertheless, the simulation has shown all the qualitative features of the Dungey model. A fully three-dimensional model and a significant reduction in the numerical diffusion are required before the simulation can be regarded as satisfactory. Such simulations should be possible on computers like the CRAY-1 with speeds some ten times the IBM 360/195.

The above techniques for the modeling of the MHD fluid can clearly be applied to the modeling of other fluids by particle methods. In hydrodynamics the magnetic pressure-tensor terms in Eq. (9-63*b*) are replaced by viscous terms related to derivatives of the velocity field, and the magnetic field equation (9-63*c*) is absent. However, the principle of moving the fluid as particles in order to ensure positive density and mass conservation, and calculating the force on the fluid particles from field quantities derived on a fixed mesh, is sound and is likely to have wide application in fluid dynamics. Such particle methods have not yet received the attention they deserve.

9-4-3 Ampere Particle Model

A different type of particle model has been developed by Eastwood and Hamilton for the detailed study of reconnection and the formation of the current sheet. This model (Hamilton, 1976) is entirely two-dimensional and is applicable to the study of the geomagnetic tail in the region indicated by the rectangle in Fig. 9-16, where we have already noted that the depth of the tail out of the paper (z direction) is many times its thickness in the plane of the paper (x, y plane).

In this region the magnetic field may be assumed to be confined to the (x, y) plane, hence

$$\mathbf{B} = (B_x, B_y, 0) \tag{9-68}$$

and currents, electric field, and vector potential are confined to the z direction:

$$\mathbf{j} = (0, 0, j_z) \tag{9-69a}$$

$$\mathbf{E} = (0, 0, E_z) \tag{9-69b}$$

$$\mathbf{A} = (0, 0, A_z) \tag{9-69c}$$

The magnetic field is related to the vector potential $\mathbf{A}$ by

$$\mathbf{B} = \operatorname{curl} \mathbf{A} \tag{9-70a}$$

whence

$$B_x = \frac{\partial}{\partial y} A_z \qquad \text{and} \qquad B_y = -\frac{\partial}{\partial x} A_z \tag{9-70b}$$

The Ampere equation (9-4*a*)

$$\operatorname{curl} \mathbf{B} = \mu_0 \mathbf{j} \tag{9-71a}$$

leads to

$$j_x = j_y = 0$$

and

$$\frac{\partial}{\partial x} B_y - \frac{\partial}{\partial y} B_x = \mu_0 j_z \tag{9-71b}$$

or on substituting Eq. (9-70*b*)

$$\nabla^2 A_z = -\mu_0 j_z \tag{9-71c}$$

The electric field is obtained by induction from Eq. (9-4*b*)

$$\text{curl}\,\mathbf{E} = -\frac{\partial \mathbf{B}}{\partial t} \tag{9-72a}$$

which on substituting Eqs. (9-70*a*) and (9-69*b*,*c*) can be expressed in terms of the vector potential

$$E_z = -\frac{\partial A_z}{\partial t} \tag{9-72b}$$

In order to obtain an expression for the current j_z one must consider the orbits of the ions which, because of their large Larmor radius, may be shown to carry most of the current (Eastwood, 1972, 1974). The equation of motion for an ion of mass m and charge q in electric and magnetic fields is

$$\frac{d\mathbf{v}}{dt} = \frac{q}{m}(\mathbf{E} + \mathbf{v} \times \mathbf{B}) \tag{9-73}$$

which in coordinate form, using the simplifying assumptions of Eqs. (9-68) and (9-69), becomes

$$m\frac{dv_x}{dt} = -qB_y v_z \tag{9-74a}$$

$$m\frac{dv_y}{dt} = qB_x v_z \tag{9-74b}$$

and

$$m\frac{dv_z}{dt} = q(E_z + v_x B_y - v_y B_x) \tag{9-74c}$$

The last equation (9-74*c*) can be integrated by substituting Eqs. (9-70*b*) and (9-72*b*), when one obtains

$$m\frac{dv_z}{dt} = -q\left(\frac{\partial A_z}{\partial t} + v_x\frac{\partial A_z}{\partial x} + v_y\frac{\partial A_z}{\partial y}\right) \tag{9-75a}$$

The expression in parenthesis on the right-hand side is dA_z/dt, the time derivative as seen by the moving particle, hence

$$\frac{d}{dt}(mv_z + qA_z) = 0 \tag{9-75b}$$

or on integrating

$$p_z = mv_z + qA_z = \text{constant} \tag{9-75c}$$

The quantity p_z, called the canonical momentum, is therefore a constant of the motion of an ion. The value of p_z is fixed by the initial conditions and is generally different for each ion. The velocity of an ion is then known from the value of the vector potential

$$v_z = m^{-1}(p_z - qA_z) \tag{9-76a}$$

and the current is given by

$$j_z = nq\bar{v}_z = nqm^{-1}(\bar{p}_z - qA_z) \tag{9-76b}$$

where n is the density of particles, and $\bar{v}_z$ and $\bar{p}_z$ the local average particle velocity and momentum.

Substituting Eq. (9-76*b*) into the Ampere equation (9-71*c*), one obtains the field equation for the problem

$$\left(-\nabla^2 + \frac{\mu_0 q^2 n}{m}\right)A_z = \frac{\mu_0 q}{m}(n\bar{p}_z) \tag{9-77}$$

The above equation is discretized and values of A_z calculated on a regular set of mesh points s, with coordinates $\mathbf{r}_s$. The values of density and momentum at these mesh points are then obtained by assigning particle values to the mesh points using a weighting function w, in a way analogous to the assignment of charge in the electrostatic plasma. Thus

$$n_s = n_z H^{-2} \sum_i w(\mathbf{r}_s - \mathbf{r}_i) \tag{9-78a}$$

and

$$(n\bar{p}_z)_s = n_z H^{-2} \sum_i w(\mathbf{r}_s - \mathbf{r}_i) p_{zi} \tag{9-78b}$$

where the summations are over all particles i at position $\mathbf{r}_i$. The mesh is assumed to be square with spacing H and n_z is the number of protons per metre in the z direction that is represented by one particle in the simulation. If the region simulated is $L_x \times L_y$ and there are N particles representing an average three-dimensional proton density n_0 then

$$n_z = L_x L_y n_0 / N \tag{9-78c}$$

Because the density n is an arbitrary function of position and of the distribution of particles through Eq. (9-78*a*), the field equation (9-77) is a nonseparable Poisson type problem. Since we are satisfied with solution within a rectangle and require a large number of mesh points, solution by Concus and Golub iteration is recommended (see Chapter 6, Fig. 6-1).

In the Concus and Golub method we introduce a shift constant k to both sides of Eq. (9-77) and define the scheme

$$(-\nabla^2 + k)A_z^{(i+1)} = (k - \mu_0 q^2 m^{-1} n)A_z^{(i)} + \mu_0 q m^{-1}(n\bar{p}_z) \tag{9-79}$$

where the superscript i denotes the iteration number. The iteration obviously converges in one step if the density is constant and $k = \mu_0 q^2 m^{-1} n$. If the density varies, as is normally the case, then the best choice is

$$k = \tfrac{1}{2}\mu_0 q^2 m^{-1}(n_{\max} - n_{\min}) \tag{9-80}$$

where n_{max} and n_{min} are the maximum and minimum values of the density n assigned to the mesh points by Eq. (9-78*a*). Convergence is most rapid for near-uniform plasmas with a low level of density fluctuations on the mesh. For a given physical variation the fluctuations are less and the convergence is better the larger the number of particles, the coarser the mesh and the higher the order of the assignment scheme (i.e., CIC better than NGP).

Equation (9-79) is separable and $A_z^{(i+1)}$ may be found using any rapid elliptic solver (see Secs. 6-5-1 to 6-5-4). In the program COIRCE (Hamilton, 1977), convolution methods are used (Sec. 6-5-4) based on a double Fast Fourier Transform. Using CIC assignment on a (32×32) mesh, one iteration of Eq. (9-79) takes 0.046 s on the IBM 360/195. With 8192 particles representing the plasma, the solution to the field equation (9-77) is found to an accuracy of one part in 10^3 in 10 iterations or 0.5 s.

The equations of motion, Eqs. (9-74), are integrated using the leapfrog differencing scheme (Sec. 2-2-1).

$$v_{xi}^{(t+1/2)} = v_{xi}^{(t-1/2)} - B_{yi}^{(t)} v_{zi}^{(t)} \frac{q}{m} DT \tag{9-81a}$$

$$x_i^{(t+1)} = x_i^{(t)} + v_{xi}^{(t+1/2)} DT \tag{9-81b}$$

and similarly for v_y and y from Eq. (9-74*b*) also

$$p_{zi}^{(t+1)} = p_{zi}^{(t)} \tag{9-81c}$$

by conservation of canonical momentum. The values of B_y and v_z on a particle are obtained by interpolation from mesh values with the same weighting function as used in assignment. Thus the field B_{yi} on the ith particle at $\mathbf{r}_i$ is

$$B_{yi}^{(t)} = \sum_s w(\mathbf{r}_i - \mathbf{r}_s) B_y^{(t)}(\mathbf{r}_s) \tag{9-82a}$$

where the summation is over mesh cells s and

$$\begin{aligned} B_y^{(t)}(\mathbf{r}_s) &= -\frac{\partial}{\partial x} A_z \\ &\simeq -[A_z^{(t)}(\mathbf{r}_s + H) - A_z^{(t)}(r_s - H)]/2H \end{aligned} \tag{9-82b}$$

and similarly for B_{xi}. The z velocity in Eq. (9-81*a*) is obtained from Eq. (9-76*a*)

$$v_{zi}^{(t)} = m^{-1}(p_{zi}^{(t)} - qA_{zi}^{(t)}) \tag{9-83a}$$

where

$$A_{zi}^{(t)} = \sum_s w(\mathbf{r}_i - \mathbf{r}_s) A_z^{(t)}(\mathbf{r}_s) \tag{9-83b}$$

In the program COIRCE, CIC weighting is used for interpolation between mesh values in Eqs. (9-82*a*) and (9-83*b*).

The timestep loop of the Ampere model may be summarized as follows.

Store for each particle i the coordinates

$$(x, y, p_z)_i^{(t)} (v_x, v_y)_i^{(t-1/2)} \tag{9-84}$$

1. Assign density and z momentum to mesh by Eqs. (9-78) using $(x, y, p_z)^{(t)}$.
2. Solve field equation Eq. (9-77) for $A_z^{(t)}$.
3. Integrate equations of motion to advance to $(v_x, v_y)^{t+1/2}$ and $(x, y, p_z)^{(t+1)}$ using Eqs. (9-81).

The model described above calculates explicitly the three-dimensional motion of the heavy protons, taking into account their inertia. The effect of finite gyroradius and gyroperiod, which are absent from the MHD particle model of Sec. 9-4-1, are correctly taken into account. This enables the model to be used near the field reversal in the geomagnetic tail. Sound and Alfven waves are also represented. The role of the electrons, which are not explicitly represented in the model, is to maintain charge neutrality and to cancel any proton currents in the (x, y) plane. This explains why we have set $j_x = j_y = 0$, although we have charged particle velocities v_x and $v_y \neq 0$. The small gyroradius of the electrons, however, prevents their canceling the proton currents in the z direction. The motion of the particles is therefore the projection of the proton orbits onto the (x, y) plane. They may also be regarded as rods of neutral plasma moving across the (B_x, B_y) field and carrying induced currents along their length in the z direction.

9-4-4 Geomagnetic Tail

The Ampere model has been used to study the reconnection of magnetic field lines at the neutral point on the downwind side of the earth and the stability of the current sheet in the geomagnetic tail.

Figure 9-19 shows a model of reconnection. A neutral point at the center is produced by two stationary rods carrying current in the same direction on the left

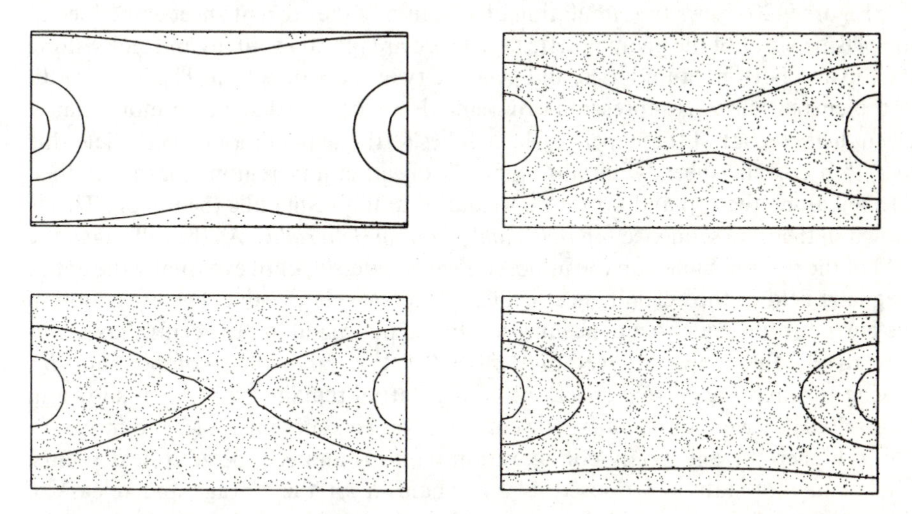

Figure 9-19 Simulation of the breaking and reconnection of magnetic field lines at a neutral point. (*Courtesy of J. E. M. Hamilton and J. W. Eastwood, unpublished.*)

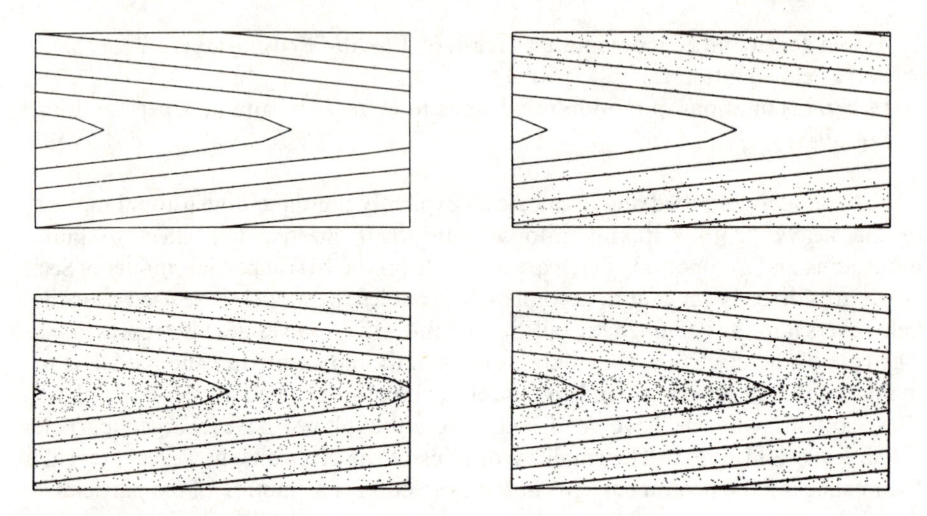

Figure 9-20 Formation of a current sheet on the left side of a neutral point. (*Courtesy of J. E. M. Hamilton and J. W. Eastwood, unpublished.*)

and right of the computational region. Plasma is injected from the top and bottom into an initially empty system. Field lines are carried into the system with the plasma as can be seen in time sequence between the top left and right frames. A little later in the bottom left frame the magnetic field lines can be seen to have broken and reconnected in the opposite sense. At the bottom right the reconnected lines are moving into the two fixed-current rods and a new pair of lines are entering from the top and bottom surfaces. During this process the plasma flows in from the top and bottom and out at the left and right sides.

Figure 9-20 shows the simulation of a section of the current sheet embedded in the plasma sheet of the geomagnetic tail. There are plasma sources and sinks along the top and bottom boundaries, and periodicity in the x-direction. The finite width of the current sheet is represented by removing particles that move more than a given distance out of the paper. The earth lies to the left, and a magnetic field due to a thin current sheet is initially imposed on an empty region. Plasma is then injected along the top and bottom boundaries until plasma fills the region. This is shown in the time sequence *top* (*left*, *right*), *bottom* (*left*, *right*). As the self-magnetic field of the plasma builds up the imposed field is reduced, until eventually the entire magnetic field of the current sheet is due to the moving plasma. The attraction of the current rods leads to a concentration of the plasma along the central axis into a relatively thin current sheet (*bottom*, *right*). For large field components, B_y, linking through the current sheet, the configuration remains stable as is shown in Fig. 9-20. However, for small B_y, the current layer becomes unstable, and moves to a new higher-current steady configuration with O- and X-type zero field points. This change of state is proposed as a mechanism for the storage and release of magnetic energy in geometric substorms (Hamilton, 1980).

CHAPTER
TEN
SEMICONDUCTOR DEVICE SIMULATION

10-1 INTRODUCTION

10-1-1 Purpose of Simulation

One of the most promising applications of computer simulation is to the design of semiconductor devices. If a reliable computer model of a semiconductor device can be made it can be used as a design tool. In certain circumstances the characteristics of a new device can be found more economically by computer simulation than by traditional laboratory techniques. In any case, a comprehensive program of computer simulation is a sensible accompaniment to any program of laboratory experiments. In this chapter we discuss the application of computer simulation to the design of one class of semiconductor devices, namely gallium arsenide (GaAs) field effect transistors (FETs). The procedures, however, apply more generally to the design of other devices.

The advantages of computer simulation may be itemized as follows:

1. *Parameters.* The parameters used in any computer simulation are precisely known because they are determined by the input data used. In contrast the parameters, such as doping level and profile, which exist in any laboratory device are only known approximately because of the difficulty of measurement.
2. *Parameter range.* The range of parameters (e.g., electrode spacing) accessible to laboratory measurement is limited by the existing manufacturing technology. In the computer simulation, however, a much wider range of parameters may be used. The investigation of parameter ranges on the computer that cannot presently be reached in the laboratory may indicate the areas in which it is worth investing in the development of better laboratory techniques. On the other hand, computer experiments may show the parameter areas for which it is

not worth developing new laboratory techniques. In either case, the large capital sums involved show that the manufacturer of electronic devices ignores computer simulation at his peril.

3. *Speed.* If the laboratory measurement on a device requires any manufacturing process, such as mask making, then it is almost certain that the characteristics of the device can be obtained more rapidly by computer simulation than they can in the laboratory.

Having enumerated the advantages of computer simulation, it is prudent to remember that the simulation is only as good as the computer model used. The computer simulation is, after all, only an approximate model of the real world and in the final analysis its results must yield to those obtained in the laboratory.

10-1-2 Defining the Problem

The first stage in simulation is to produce a simplified model of the actual device. In Fig. 10-1 we show two GaAs FETs manufactured by Plessey Ltd. They are used as high-frequency amplifiers in aerospace applications and can be used in the 200 MHz to 12 GHz frequency range. They have the advantage of low noise, high gain, and large dynamic range. The devices themselves are about 0.3 mm by 0.3 mm and are mounted on holders. On the left is a leadless inverted device 2 mm by 1 mm, used up to 8 GHz. On the right is a microstrip package 7 mm long and

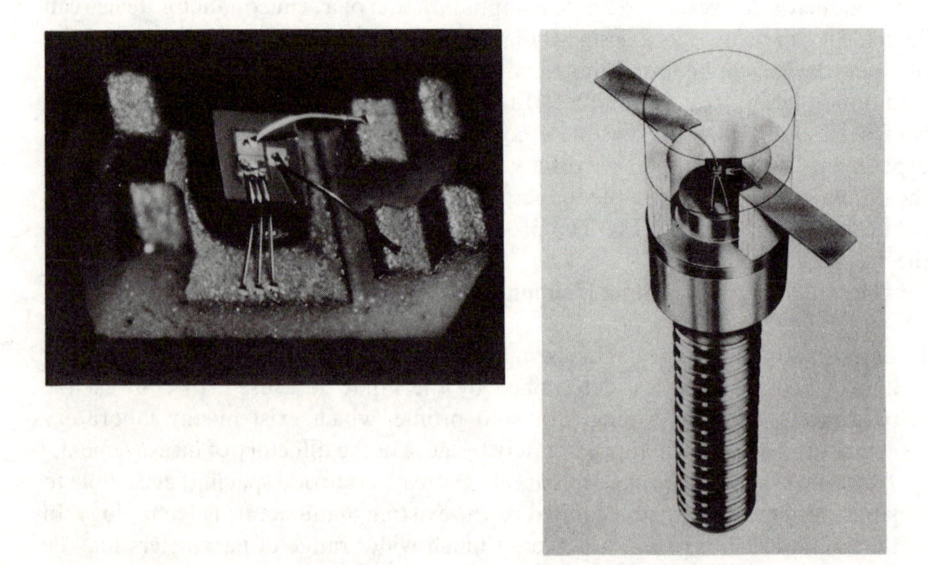

Figure 10-1 Two gallium arsenide field effect transistors made by Plessey Ltd. These combine the properties of low noise, high gain, and large dynamic range. They are particularly suitable as high-frequency amplifiers in the frequency range 200 MHz to 12 GHz. (*Photographs courtesy of R. Butlin and Plessey Research Ltd, Allen Clark Research Centre, Caswell.*)

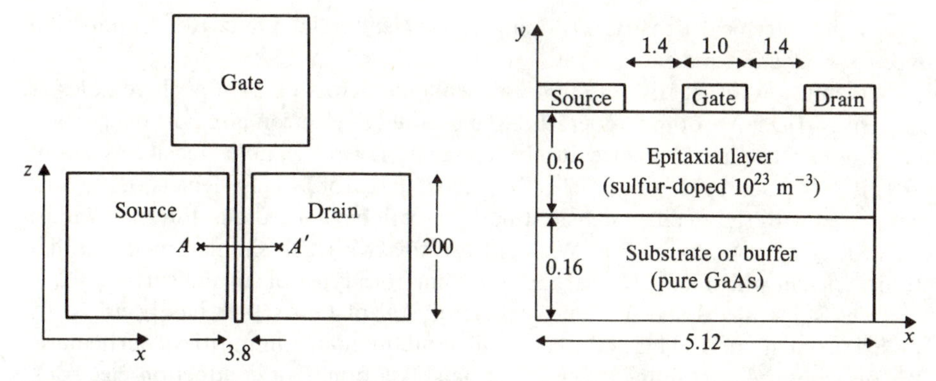

Figure 10-2 *Left*, plan view of device in Fig. 10-1 and, *right*, idealized section in depth below A–A' showing the active region simulated on the computer. Dimensions in μm, not to scale.

2 mm maximum diameter. This package is normally enclosed, but the top has been removed to reveal the transistor chip. There are three electrodes, the source S, the gate G, and the drain D. Current flows through the device from the source to the drain under the control of the voltage on the gate. For those familiar with vacuum electronics, the device acts like a triode valve, the source, drain, and gate corresponding respectively to the cathode, anode, and grid.

A plan view of the FET is shown in Fig. 10-2 (*left*). Three large pads of approximately $(0.2 \times 0.2)\,\text{mm}^2$ form the electrode connections. The active region of the device lies in the small channel ($\sim 4\,\mu\text{m}$) between the source and drain pads. The gate itself is a long narrow strip 1 μm wide lying in the center of the source–drain gap. The geometry is such that, except within a few micrometers of the ends (only 1 percent of the gap), there will be no variation of quantities in the z direction which is along the length of the gap. We will therefore assume in our model that we need only consider variations across the gap (the x direction) and upwards towards the surface (the y direction). It is also found that the majority of the current leaves the source and enters the drain within about 1 μm of the source–drain gap. Consequently, although the device occupies an area of about $(400 \times 400)\,\mu\text{m}^2$, only the small region represented by the section A–A' plays an active part in determining the characteristics of the device.

Figure 10-2 (*right*) shows the section below A–A' that forms the region to be modeled on the computer. Field effect transistors are made by growing a thin epitaxial layer ($\sim 0.16\,\mu\text{m}$) of doped GaAs onto a wafer of crystalline insulating GaAs ($\sim 10^5\,\Omega\,\text{m}$) called the substrate. The epitaxial layer is grown in the laboratory under carefully controlled conditions from vapor containing free arsenic and a suitable gallium compound with an addition of sulfur. The sulfur atoms, after releasing one of their outer electrons as a free conduction electron, are isoelectronic with arsenic and may replace arsenic in the GaAs lattice. The result is that fixed positive sulfur ions are rigidly bound as part of the crystal lattice and if they are fully ionized an equal number of free electrons are released as conduction electrons. As these free carriers are negatively charged, the layer is said to be doped as *n* type. A typical density of doping is 10^{23} sulfur ions and 10^{23} conduction electrons per cubic meter.

In addition to negative electrons all semiconductors contain positive holes as current carriers. At room temperature the product of electron and hole density is a constant ($\approx 10^{26}\,\mathrm{m}^{-6}$), hence in the epitaxial layer the hole density is about $10^{3}\,\mathrm{m}^{-3}$ (Kittel, 1971, page 369; Sze, 1969, page 58). This density is so low compared with the electron density that holes will be ignored and the FET will be treated as a unipolar device. We contrast this with the simulation of bipolar transistors, in which it is necessary to represent both types of current carrier.

The substrate is sliced from a single crystal of GaAs that has been slowly pulled from a melt. This substrate will contain many impurities, particularly silicon from the containing vessels that acts as a donor of conduction electrons. Chromium, which captures electrons, is therefore added to reduce the conduction electrons to a low level and to produce thereby a semi-insulating material. The chromium is said to compensate the original impurities and the net result is a material with a small net level of doping charge. In this work we will assume that compensation is exact and treat the semi-insulating substrate as pure, undoped GaAs.

A typical substrate wafer is 250 μm thick; however, it is only the substrate immediately adjacent to the epitaxial layer that affects the conduction between source and drain. Consequently it is usual to model only a substrate of thickness approximately equal to the epitaxial thickness, as is shown in Fig. 10-2 (*right*). In some devices an undoped layer of pure GaAs is grown on the semi-insulating substrate before the doped layer is grown. This buffer layer acts as a barrier against the diffusion of impurities from the substrate to the active doped layer during manufacture. In this case, what we have described as the substrate in the computer model is actually a model of the buffer layer and our treatment of it as pure GaAs is particularly apt. However, in practice, quite a lot of chromium may diffuse into the buffer layer.

The source and drain electrodes are made of a gold-indium-germanium alloy

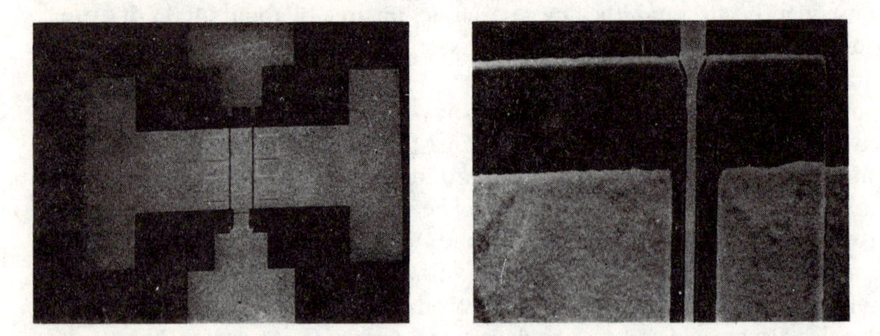

Figure 10-3 *Left*, scanning electron microscope photograph ($\sim \times 120$) of a pair of FETs with a common drain and overall size about 1 mm. *Right*, scanning electron microscope photograph ($\sim \times 4500$) of the ~ 3 micrometer source–drain region simulated on the computer. (*Photographs Crown Copyright, 1980.*)

which is sintered on to the active layer to produce low-resistance ohmic contacts. The voltage applied to these electrodes is therefore also the voltage applied to the active GaAs layer. The gate electrode is a layer of aluminum which forms a good Schottky barrier diode to the GaAs. This means that the voltage in the GaAs adjacent to the gate is 0.8 V less than the voltage applied to the gate electrode.

The geometry of a typical FET can be seen in photographs of actual devices under magnification. Figure 10-3 (*left*) is a scanning electron microscope photograph ($\sim \times 120$) of a pair of FETs with a common drain and interconnected gates, corresponding to the drawing in Fig. 10-2 (*left*), and its image mirrored about the center of the drain. The two source–drain gaps appear as two black vertical lines and at this magnification the gates cannot be seen. Increasing the magnification to $\sim \times 4{,}500$ at the right-hand gap reveals the gate (Fig. 10-3, *right*). Except for the fact that the source and drain are interchanged, this corresponds exactly to the diagram in Fig. 10-2 (*left*). The gate shows as a white vertical line, approximately 1 μm wide, running approximately down the center of the black ~ 3 μm gap between the rectangular source and drain pads.

Having defined the geometrical region of importance, we must now define the conditions to be applied on the boundary of the region. The electrons flow from the source to the drain under the influence of their own space charge and the external voltages applied at the electrodes. As regards the electrostatic potential, it is clear that the potential must take on given values on the boundary adjacent to the three electrodes. Elsewhere on the boundary, a zero value of the electric field normal to the boundary is prescribed. That is to say, the normal gradient of the potential is zero. Such conditions round the surface of a rectangle are ideally suited to solution by rapid elliptic solvers (RES) and computer programs such as POT4 (see Sec. 6-5-6) have been written for just such conditions. The justification of zero-field conditions on the left, right, and bottom boundaries is the natural tendency of free charge to neutralize any field variation within a Debye length (see Sec. 9-3-3) of any imposed change (e.g., the edge of an electrode). On the top surface the zero-field condition is a good approximation because of the large change in dielectric constant from the epitaxial layer ($\epsilon_e \sim 12$) to the surrounding air ($\epsilon_a \sim 1$). At such a discontinuity the normal value of the displacement $\mathbf{D} = \epsilon\mathbf{E}$ is continuous. Hence

$$\epsilon_e \mathbf{E}_e = \epsilon_a \mathbf{E}_a \tag{10-1a}$$

$$\mathbf{E}_e = \frac{\epsilon_a}{\epsilon_e} \mathbf{E}_a \approx 0 \tag{10-1b}$$

where the subscripts "e" and "a" refer respectively to the epitaxial and air regions. Since $\epsilon_e \gg \epsilon_a$ and $\mathbf{E}_a$ is a finite value, the electric field $\mathbf{E}_e$ just in the interior of the epitaxial layer must be small. A more sophisticated calculation would in fact include a calculation of the electric field in the air region. However, it is clear that this is a refinement which need not be considered in the first instance.

We now consider the conditions to be applied to the motion of the electrons at the boundaries. The requirement that there be no current flowing through the boundaries except at the electrodes is best included by reflecting any electron that hits the boundary. That is to say, the velocity of the electron normal to the boundary is reversed. At the source and drain electrodes we represent an ohmic contact by absorbing all electrons which hit the electrodes and by injecting, at each timestep, sufficient electrons to maintain a density of $10^{23}\,\mathrm{m}^{-3}$ in the row of potential mesh cells adjacent to the electrodes. The injection takes place with a half-maxwellian velocity distribution in the same way as for the plasma physics example (see Sec. 9-3-1). At the gate electrode we represent a Schottky barrier contact by applying a potential equal to the applied potential minus the barrier height of 0.8 V. There is absorption of particles but no injection. These latter conditions on the gate are in practice unimportant because under normal bias conditions very few electrons will reach the gate.

10-1-3 Types of Model

The remainder of this chapter is devoted to the use of particle techniques in the simulation of field effect transistors. To set this in context it is appropriate first to review the different types of model which have been used in semiconductor simulation. An extensive literature now exists and 486 references up to 1974 are given in a bibliography by Agajanian (1975). The reader is also referred to the book on semiconductor device modeling by Hamilton, Lindholm, and Marshak (1971).

The simplest model represents the device by an equivalent lumped-parameter circuit such as that described in Sec. 10-4-4. The properties of the device are then derived by standard circuit analysis methods (Faulkner, 1966). Drangeid and Sommerhalder (1970) have studied the dynamic performance of FETs by this method. Such calculations take little computer time and extensive parameter surveys are possible; however, this is achieved by ignoring the distributed nature of the parameters (e.g., resistance and capacitance) and the complex geometry of the device. The method therefore has only limited validity but, provided these limits are well known and understood by comparison with more sophisticated models, it is attractive because of the minimal use of computer resources. It is also popular because it expresses the behavior of the device in terms of circuit elements whose properties are familiar to electronic circuit engineers.

The next most complex model expresses the current flow by the fluid equations:

1. Poisson equation for the electric field **E**

$$\nabla^2\phi = \frac{e}{\epsilon\epsilon_0}(n-N) \qquad \mathbf{E} = -\nabla\phi \tag{10-2a}$$

2. The flux equation

$$\mathbf{\Gamma} = n\mathbf{v} - D\nabla n \qquad \mathbf{v} = -\mu\mathbf{E} \tag{10-2b}$$

3. The continuity equation

$$\frac{\partial n}{\partial t} = -\nabla\cdot\mathbf{\Gamma} \tag{10-2c}$$

In the above equations the current is carried by charges, $-e$,† with number density n and average velocity $\mathbf{v}$. The flux of carriers is $\boldsymbol{\Gamma}$ and the number density of doping charges $+e$ is N. We write the equations for a single type of carrier (electrons) as is appropriate in the FET. If more than one type of carrier is important Eqs. (10-2*b*) and (10-2*c*) are repeated for each type and extra terms are added to the right-hand side of Poisson's equation. The physics of the material enters the above equations only through the mobility μ, diffusivity D, and dielectric constant ϵ.

Reiser (1971, 1972, 1973) discusses the two-dimensional finite-difference approximations to the Eqs. (10-2) and the timestep instabilities that can arise. Applications to the study of FETs are also given by Reiser (1970, 1972) and Yamaguchi, Asai, and Kodera (1976). The procedure starts from a knowledge of the density distribution at a given time, computes the electrostatic potential and field from Eq. (10-2*a*), and then advances the density to the next time level using Eqs. (10-2*b*) and (10-2*c*). The timestep cycle then repeats. The Poisson equation is linear in ϕ and may be solved in most cases by rapid elliptic solvers (see Secs. 6-5-1 to 6-5-7), although many authors have used iterative schemes (see Secs. 6-3-1 to 6-3-6), finite-element methods (Barnes and Lomax, 1974; Barnes, Lomax and Haddad, 1976; and Sec. 6-4-3) or strongly-implicit procedures (Jesshope, 1975; and Sec. 6-4-5). Implicit time-differencing of the continuity equation is found necessary in order to avoid an excessively short timestep. The generalized Crank–Nicolson (1947) scheme with equal weights given to the new and old time levels is usually used.

The solution of the two-dimensional steady-state equations [Eq. (10-2) with $\partial/\partial t = 0$] has been obtained by Kennedy and O'Brien (1970) using the Picard iteration. The finite-difference approximations to Eqs. (10-2) were solved using mesh relaxation and an analysis was made of the operation of junction FETs in silicon. A similar model was used by Himsworth (1972) in the study of GaAs junction FETs. Silicon, GaAs, and InP metal semiconductor FETs have been simulated by Wada and Frey (1979) using the rapid elliptic solver POT4 (see Sec. 6-5-6) for the solution of Poisson's equation. The continuity equation is solved by under-relaxation using the method of Scharfetter and Gummel (1969) to prevent numerical instabilities. Metal oxide silicon FETs (MOST) have been simulated by Loeb, Andrew, and Love (1968) and bipolar transistors, for example, by Kilpatrick and Ryan (1971) using finite-difference approximations and Newton iteration, and by Slotboom (1973).

The density of electrons in a semiconductor device varies by many orders of magnitude because in thermodynamic equilibrium the density is proportional to $\exp(e\phi/k_B T)$, and small changes in potential ϕ lead to large changes of density. Slotboom (1969) suggested that the density variable be replaced by

$$\Phi = n \exp(-e\phi/k_B T) \tag{10-3}$$

Because of the introduction of the exponential factor into the definition, the new

† $e = +1.60210 \times 10^{-19}$ C is the absolute value of the electronic charge.

variable does not change by large amounts. Consequently, the finite-difference errors are reduced. In terms of the new Slotboom variable the Eqs. (10-2) become

$$\nabla^2\phi = \frac{e}{\varepsilon\varepsilon_0}[\Phi \exp(e\phi/k_B T) - N] \tag{10-4a}$$

and

$$\frac{\partial}{\partial t}[\Phi \exp(e\phi/k_B T)] = -\nabla\cdot[D\exp(e\phi/k_B T)\nabla\Phi] \tag{10-4b}$$

where we have used the Einstein relation $\mu = eD/k_B T$ to relate the mobility and diffusivity. The procedure for solution is to solve Eq. (10-4*a*) for the potential ϕ given the value of Φ and then compute Φ at the new time level from Eq. (10-4*b*) given the value of ϕ. We note that in the new variables the Poisson equation (10-4*a*) is nonlinear and can only be solved by iteration. The solution is usually obtained by Newton's method (see Sec. 6-2-1). The Eq. (10-4*b*) is as before a linear equation in Φ with variable coefficients and may be solved by mesh-relaxation or matrix methods (see Secs. 6-3-1 to 6-4-6). The cost therefore of introducing a better-behaved variable is the increased effort required for the solution of the Poisson equation.

In the steady-flow condition $\partial/\partial t = 0$ and Eq. (10-4*b*) becomes

$$\nabla\cdot[D\exp(e\phi/k_B T)\nabla\Phi] = 0 \tag{10-5}$$

which may be regarded as a variable coefficient Laplace equation for the Slotboom variable Φ and solved by mesh relaxation, matrix methods, or by the Concus and Golub method (see Sec. 6-5-7). The complete solution of the steady-flow problem requires the simultaneous solution of Eqs. (10-4*a*) and (10-5). Many authors use an algorithm described by Gummel (1964) for one dimension. The method is based on Newton iteration and has been reformulated and improved by Andrew (1972) for two dimensions, using a variation on the Slotboom variables.

In the above discussion we have assumed that the temperature of the device remains constant. In the simulation of thermal runaway in power devices it is also important to include the effects of heat generation within the device and the variation of the mobility and diffusivity with temperature. This requires the solution of the heat conduction equation

$$\nabla\cdot(K\nabla T) = \rho c\frac{\partial T}{\partial t} + H \tag{10-6}$$

for the temperature T. In this equation K is the thermal conductivity, ρ the mass density, c the heat capacity, and H the heat source. The steady-flow solution of Eq. (10-6) in conjunction with the Poisson and continuity equations (10-2) or (10-4) has been extensively studied by Gaur and Navon (1976). The carrier densities are used as the variables and the nonlinear equations are solved by Newton iteration. Stone's method (see Sec. 6-4-5) is used for the solution of the variable-coefficient Poisson problem. The time-dependent solution of these equations has been found in one dimension by Chryssafis and Love (1979) and in two dimensions by Alwin et al. (1977) using the following cycle:

1. Solve for potential from Eq. (10-2*a*).
2. Calculate **E**, μ, D as functions of field and temperature.
3. Solve continuity Eq. (10-2*c*) for density.
4. Calculate heat generation and solve Eq. (10-6) for the new temperature.

We note that in the above cycle the solution of a variable-coefficient, Poisson-like equation is required in three places, during (1), (3), and (4).

The models described above assume that the velocity of the charge carriers adjusts instantaneously to the electric field via the mobility relation $v = \mu E$. No account is taken of the time required to accelerate electrons to their terminal velocity or of the time delays associated with the scattering of electrons to higher energy states. A complete description of this effect requires a Monte-Carlo particle model, the subject of the rest of this chapter. Such time delays are not important in single-energy-band materials like silicon provided the period of oscillation is much longer than the scattering time, as is usually the case. For other materials like gallium arsenide the electrons may exist in two energy bands (see Sec. 10-2-2) and it is necessary to take into account the time delays associated with scattering between these bands.

Brewitt-Taylor (1979) and Robson of Sheffield University and Shur (1976) have shown how these effects can be partially accounted for by introducing first-order differential equations for the average drift velocity **v** and random energy ξ (i.e., the temperature) of the electrons

$$\frac{d\mathbf{v}}{dt} = -\frac{e\mathbf{E}}{m^*} - \frac{\mathbf{v}}{\tau_m} \tag{10-7a}$$

$$\frac{d\xi}{dt} = -e\mathbf{v}\cdot\mathbf{E} - \frac{\xi - \xi_L}{\tau_e} \tag{10-7b}$$

where τ_m and τ_e are, respectively, the time constants for momentum and energy relaxation, and ξ_L is the random thermal energy of the lattice (i.e., $\frac{3}{2}k_B T$). The effective mass m^* must take into account the proportion of the electrons in each energy band

$$m^* = m_1(1-p_2)+m_2 p_2 \tag{10-8}$$

where m_1 and m_2 are the masses of electrons in Band I and Band II, respectively, and p_2 is the proportion of electrons in Band II.

The time constants are found by considering the solution of Eqs. (10-7) in the steady state ($d/dt = 0$), whence

$$\tau_m = \frac{m^*}{e}\frac{v}{E} = \frac{m^*}{e}\mu \tag{10-9a†}$$

$$\tau_e = -\frac{\xi - \xi_L}{e\mathbf{v}\cdot\mathbf{E}} = \frac{\xi - \xi_L}{e\mu E^2} \tag{10-9b†}$$

† Vector quantities (e.g., **E**) are set in bold type and their magnitude (e.g., E) in italic.

where $\mu = v/E$ is the mobility. The quantities $\mathbf{v}$, p_2, and m^* are found from Monte-Carlo simulations (see Fawcett, Boardman, and Swain, 1970; Hill, Robson, Majerfeld, and Fawcett, 1977) as a function of electron random energy ξ, and the time constants are evaluated from Eqs. (10-9) also as a function of ξ. It is found that the above time constants when used in Eqs. (10-7) give remarkably good agreement with transient Monte-Carlo simulations of velocity overshoot in a uniform electric field. One has thus successfully predicted time-dependent behavior using parameters obtained from steady-state data. This leads to confidence in the assumptions of this approach. In more complex situations of actual device geometries, the agreement is less satisfactory.

The advantage of the modified fluid model is that it requires less computer resources than a full Monte-Carlo particle model. However, one cannot expect to represent the full complexities of electron scattering with two time constants and such a model must always be validated against the later type of model, the historical development of which we now describe.

Semiconductor simulation using particles was made possible by techniques developed originally for vacuum tubes and plasmas, in which Monte-Carlo scattering with a background medium is absent. These techniques were pioneered during World War II by Professor D. R. Hartree when he headed the Manchester University Group for the Committee on Valve Development (CVD) of the British Admiralty. Around 1943 Professor Hartree and Phyllis Nicolson (1941–1944) computed by hand on a desk calculator the orbits of about 30 interacting electrons in a magnetron using a one-dimensional approximation for the electric field (see Buneman and Dunn, 1966, p. 36). Their model included the effect of space charge and the electric and magnetic fields, but because of its restriction to one-dimension was not able to represent fully the two-dimensional charge flow in a magnetron. Later Hartree (1950) made a study of electron flow in a one-dimensional diode in a similar way, and observed the initial transient that occurs as an empty diode fills with electrons. At about this time the first digital computers became available, and perhaps the earliest simulation on such computers was the study of noise in a high-frequency diode by Tien and Moshman (1956) of Bell Laboratories. They used a Univac I to simulate a one-dimensional diode with about 360 electron sheets and a physical timestep of two picoseconds. About 3,000 timesteps were taken, each of which required between 25 and 40 seconds of computer time. The total computing task must rank as one of the most impressive achievements of early device modeling, requiring as it did some 25 hours of reliable computing on a first generation vacuum-tube computer with a mercury delay-line store of only about 1,000 numbers.

Hartree's early work was adapted for automatic computation on the EDSAC2 at Cambridge University, and extended to much longer times, by his student Lomax (1960) who observed quite unexpected oscillations in the space charge about the expected static solution to the equations. These plasma oscillations were also observed about the same time at the University of California, Berkeley, by Birdsall and Bridges (1961, 1966) with a similar model. Another early one-dimensional model was developed at Stanford University by Buneman (1959), who himself had been a student of Hartree and had worked in the Manchester CVD group. This model with 512 sheets was programmed by D. Thoe of Lockheed on a Univac 1103AF, and used to demonstrate the unexpectedly rapid dissipation of currents when entering ionized media. Later Burger, working as a student of Buneman, refined the one-dimensional model by using integer arithmetic and a mesh for the calculation of the fields. This model permitted the simulation of 10,000 electron sheets on an IBM 7090, and was used to study the operation of cesium thermionic converters and plasma diodes (Burger 1964, 1965; Burger, Dunn and Halstead, 1966).

The extension of the simulations to two-dimensions was made possible about 1964 by a new Poisson-solver developed by Hockney (1965, the FACR algorithm of Chapter 6), also working as a student of Buneman but on plasma simulation. This allowed the simulation of about 2,000 rods of charge and was successfully used to demonstrate anomalous diffusion in plasmas (Hockney, 1966*b* and Chapter 9); the full two-dimensional charge flow in magnetrons (Yu, Kooyers and Buneman, 1965)—thus completing, some twenty years later, Hartree's work in the war; and the neutralization of ion-beams intended for space-ship propulsion (Wadhwa, Buneman and Brauch, 1965).

Scattering with a background medium was first introduced into the above collisionless models by Burger (1967) in order to include electron-neutral collisions in his simulation of a one-dimensional plasma diode. Random numbers were used to select the mean-free-path for hard sphere elastic collisions. The same general technique was used by Lebwohl and Price (1971) to represent electron flow in a gallium arsenide semiconductor and the formation of Gunn domains. Two-dimensional models with scattering were introduced by Hockney, Warriner and Reiser (1974), and further developed by Warriner (1976, 1977*a*,*b*,*c*). They are the subject of the rest of this chapter.

10-2 ELECTRON FLOW IN SEMICONDUCTORS

10-2-1 Equations of Motion

Before one can consider how to model the flow of electrons in a semiconductor, it is necessary to review the basic physical laws. The motion of electrons in a semiconductor is described in terms of the band structure (Fig. 10-4) which is the relationship $\mathscr{E}(\mathbf{k})$ between the energy $\mathscr{E}$ of the electron and its wavenumber $\mathbf{k}$ (sometimes called its $\mathbf{k}$ vector). The band structure is found from quantum-mechanical calculations which need not concern us here (see for example Kittel, 1971, Chapters 9 and 10). From the point of view of device modeling we can take the band structure as given, and need only know how to use it to determine the motion of our simulated electrons.

An electron with wavenumber $\mathbf{k}$ has momentum $\mathbf{p}$ and energy $\mathscr{E}$ given by

$$\mathbf{p} = \hbar\mathbf{k} \qquad \mathscr{E} = \mathscr{E}(\mathbf{k}) \tag{10-10a}$$

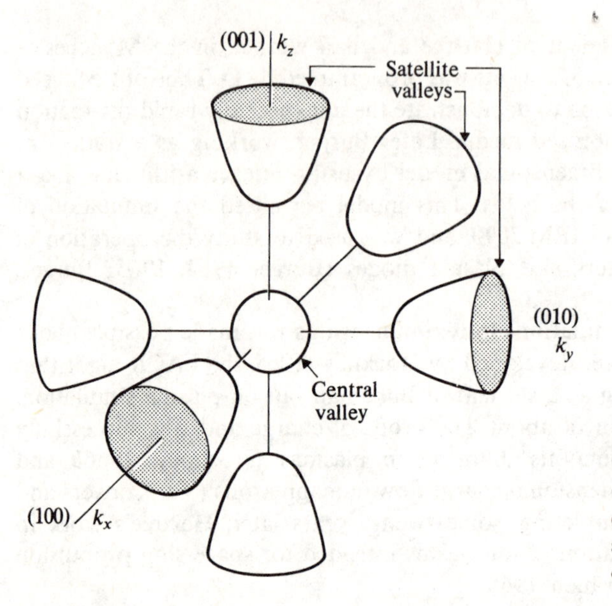

Figure 10-4 The three-dimensional band structure of gallium arsenide. One constant-energy surface in the central valley and each of the six satellite valleys is shown. (*Courtesy of R. A. Warriner, 1976.*)

where $\hbar$ is Planck's constant divided by 2π. The electron moves with the group velocity of a wave packet

$$\mathbf{v} = \frac{d\mathbf{x}}{dt} = \frac{1}{\hbar}\frac{d\mathscr{E}}{d\mathbf{k}} = \frac{1}{\hbar}\left(\frac{\partial\mathscr{E}}{\partial k_x}, \frac{\partial\mathscr{E}}{\partial k_y}, \frac{\partial\mathscr{E}}{\partial k_z}\right) \tag{10-10b}$$

The rate of change of momentum is always equal to the applied force **F**, hence

$$\frac{d\mathbf{k}}{dt} = \frac{1}{\hbar}\mathbf{F} \tag{10-10c}$$

The above equations, (10-10*b*) and (10-10*c*), determine completely the motion of the electron, provided the band structure $\mathscr{E}(\mathbf{k})$ is known. We store for each electron the coordinates (x, y, k_x, k_y, k_z) and advance the **k** vector in time by Eq. (10-10*c*) and the positions by Eq. (10-10*b*). We note also that the velocity is in the direction of the gradient of the energy surface in **k** space. Hence the velocity is only in the same direction as the **k** vector if the constant energy surfaces are spherical about the origin of the **k** vector.

In order to relate the motion of an electron to that of a classical particle and to the familiar notion of mass, the Eq. (10-10*c*) may equivalently be expressed in terms of the acceleration (see Blakemore, 1969, page 208)

$$\frac{dv_i}{dt} = F_j/m_{ij}^* \tag{10-11a}$$

$$\frac{dx_i}{dt} = v_i \tag{10-11b}$$

where $m_{ij}^* = \hbar^2 \left(\frac{\partial^2 \mathscr{E}}{\partial k_i \partial k_j} \right)^{-1}$ (10-11c)

and i and j stand for the coordinate direction x, y, z. We store for each particle the usual coordinates (x, y, v_x, v_y, v_z) and advance the velocities by Eq. (10-11*a*) and the positions by Eq. (10-11*b*).

By analogy with Newton's law of motion for a classical particle with mass m, that is to say

$$\frac{d\mathbf{v}}{dt} = \frac{\mathbf{F}}{m} \qquad \frac{d\mathbf{x}}{dt} = \mathbf{v} \tag{10-12}$$

we see that m_{ij}^*, which is inversely proportional to the curvature of the energy surface, plays the role of a mass and is called the "effective mass" of the electron. In general it depends on the energy of the electron and differs significantly from the rest mass m_e of a stationary electron in free space. Depending on the curvature $\partial^2 \mathscr{E} / \partial k_i \partial k_j$ of the energy band, the mass may even be negative. The effective mass has nine components (i.e., it is a tensor)

$$m^* = \begin{bmatrix} m_{xx}^* & m_{xy}^* & m_{xz}^* \\ m_{yx}^* & m_{yy}^* & m_{yz}^* \\ m_{zx}^* & m_{zy}^* & m_{zz}^* \end{bmatrix} \tag{10-13}$$

and Eq. (10-11*a*) shows that if all components of m^* are nonzero, a force in one coordinate direction produces accelerations in all three coordinate directions. In other words, the acceleration is not in the same direction as the force causing it.

The above formulation in terms of the effective mass is as general as Eqs. (10-10*b*) and (10-10*c*), and shows that the motion of an electron in a semiconductor can be regarded as that of a classical particle with an energy-dependent mass of tensor form. The strange behavior of the effective mass tensor m_{ij}^* simply represents the quantum-mechanical effects of the semiconducting medium on the motion of the electron. This formulation is particularly useful in providing a qualitative feel for the motion of the electron through the introduction of the concept of effective mass. However, the **k** vector is not computed in Eq. (10-11) and the difficulty of expressing the effective mass in terms of the available velocity coordinate makes the formulation unsuitable for calculational purposes, except in the simple case when the effective mass is independent of energy. This occurs in socalled parabolic energy bands in which the energy is proportional to the square of the **k** vector.

Fortunately, at room temperature most electrons will lie near the bottom (i.e., within 0.025 eV) of minima in the band structure, of which there may be several. In this case the band structure near each minimum may be approximately represented as parabolic. If, in addition, we assume that the band is spherically symmetric about the minimum, we have, measuring **k** from the minimum

$$\mathscr{E}(\mathbf{k}) = ck^2 = c(k_x^2 + k_y^2 + k_z^2) \tag{10-14a}$$

where c is a constant, and hence

$$\mathbf{v} = \frac{2c}{\hbar} \mathbf{k} = \frac{2c}{\hbar} (k_x, k_y, k_z) = \frac{\hbar}{m^*} \mathbf{k} \tag{10-14b}$$

$$m_{ij}^* = \frac{\hbar^2}{2c}\,\delta_{ij} = m^* \tag{10-14c}$$

Thus we find that the mass is a scalar quantity, m^*, independent of both direction and energy. Furthermore the off-diagonal elements of m_{ij}^* are zero, which means that the acceleration is in the direction of the force. The motion is therefore identical with that of a classical particle of mass m^* obeying Newton's laws of motion

$$\frac{d\mathbf{x}}{dt} = \mathbf{v} \tag{10-15a}$$

$$\frac{d\mathbf{v}}{dt} = \frac{\mathbf{F}}{m^*} \tag{10-15b}$$

where $m^* = \hbar^2/2c$ and the energy is seen to be the normal kinetic energy

$$\mathscr{E}(\mathbf{k}) = \frac{\hbar^2}{2m^*}\,k^2 = \tfrac{1}{2}m^* v^2 \tag{10-15c}$$

If the spherical parabolic approximation is satisfactory, and it will certainly be as a first approximation, the acceleration and move subroutines written for a plasma simulation (see Chapter 9) can be used for the free-flight parts of a semiconductor code, provided the correct effective mass is used. The effective mass formulation may also be used if the energy surfaces are ellipsoids and vary parabolically along each axis. In this case the mass is still independent of energy but is no longer a scalar quantity. There are different effective masses along the three axes of the ellipsoids and the tensor form of the acceleration, Eq. (10-11a), must be used.

Away from the minimum of a band, the effects of nonparabolicity become important, and the next approximation is usually to represent the band structure by

$$\frac{\hbar^2 k^2}{2m^*} = \mathscr{E}(1 + \alpha\mathscr{E}) \tag{10-16a}$$

where

$$\alpha = \frac{1}{E_g}\left(1 + \frac{m^*}{m_e}\right)^2 \tag{10-16b}$$

and E_g is the energy gap between the conduction and valence bands. The relation between velocity and wavenumber is obtained from Eq. (10-10b)

$$\mathbf{v} = \frac{\hbar\mathbf{k}}{m^*(1 + 2\alpha\mathscr{E})} \tag{10-16c}$$

In low electric fields, when the energy of the electrons $\mathscr{E}$ is small, the effect of nonparabolicity is small. However in the high electric fields that occur between the gate and drain of a FET, electron energies up to about $0.24E_g$ are attained and the nonparabolicity term $2\alpha\mathscr{E}$ in the denominator of Eq. (10-16c) is approximately 0.5. In these circumstances it is obviously necessary to use Eq. (10-16c) in place of Eq. (10-14b). Since the effective mass is now dependent on energy, the formulation of the equations of motion in terms of the $\mathbf{k}$ vector must be used, Eqs. (10-10b,c).

10-2-2 Band Structure of Gallium Arsenide

The three-dimensional band structure of gallium arsenide is shown in Fig. 10-4. Surfaces of constant energy $\mathscr{E}(\mathbf{k})$ are sketched in three-dimensional **k** space. At normal temperatures and low electric fields all the conduction electrons reside in the central (000) valley near its minimum at $\mathbf{k} = 0$. The constant-energy surfaces are spheres centered on the origin, and one such sphere is shown. Electrons in the central valley act like classical particles of effective mass $m_1^* = 0.067 m_e$, although at high energies allowance should be made for the nonparabolicity of the band by using Eq. (10-16*c*) with $\alpha = 0.576\,\text{eV}^{-1}$. Electrons in this band are therefore very light and mobile. If the electric field is high enough to give electrons an energy $\sim 0.36\,\text{eV}$, there is a strong probability that they will scatter into one of the six

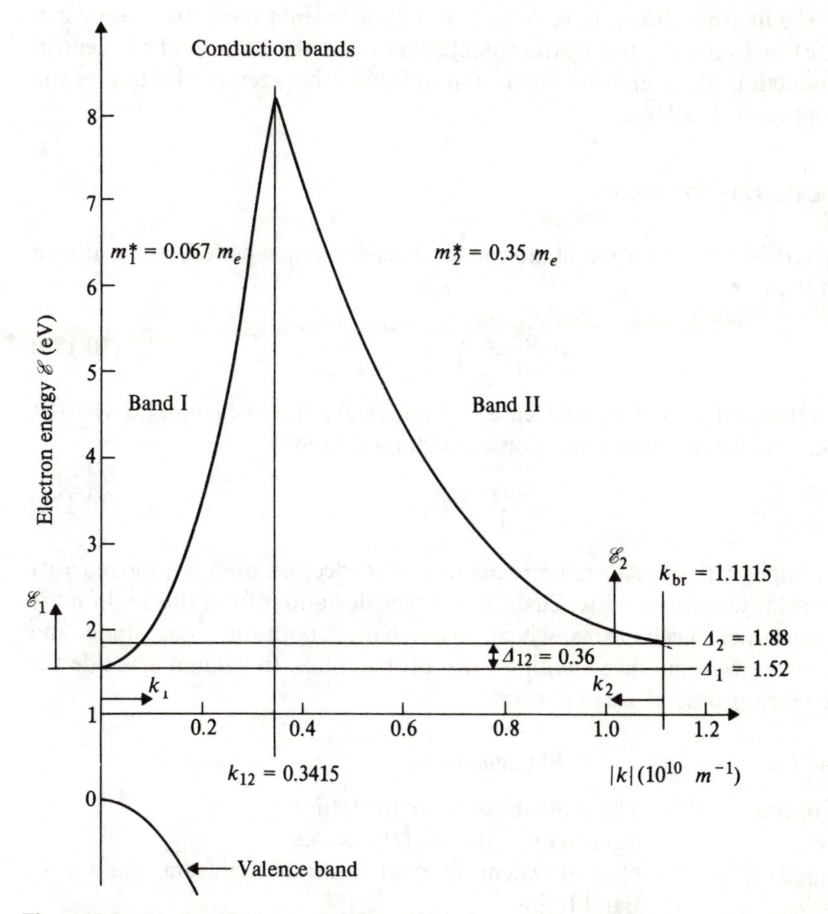

Figure 10-5 The simplified one-dimensional band structure of gallium arsenide, as used in the computer model. The six satellite valleys of Fig. 10-4 are combined into a single composite Band II. Energies are measured from the top of the valence band in electronvolts ($1\,\text{eV} = 1.6 \times 10^{-9}\,\text{J}$). (*Courtesy of R. A. Warriner, 1976.*)

satellite (or upper) valleys that lie in the (100), (010) and (001) directions in the reciprocal lattice (see Kittel, 1971, Chapter 2). Although the satellite valleys are ellipsoidal and have tensor effective masses, they will be treated as spherical bands with a scalar effective mass $m_2^* = 0.35m_e$. In this state the electron is five times more massive than it is in the central valley. Hence electrons in the satellite valleys are sluggish and have a low mobility.

In the computer model we do not attempt to simulate the full complexity of the three-dimensional band structure, although in principle this could be done. The computational cost of such a procedure would probably not be justified. Instead we use a simplified one-dimensional band structure comprising two bands. This is shown in Fig. 10-5. Band I represents the central valley, while the six satellite valleys are combined into a composite Band II. This simplified band structure is easier to compute and contains the main physical features of the material. The figure is drawn to scale and shows the valence band, the energy gap $E_g = 1.52\,\text{eV}$ between the top of the valence band and the bottom of the central valley conduction band, and the separation of 0.36 eV between the bottom of the satellite and central valleys.

10-2-3 Scattering Processes

From Eq. (10-10*c*) the equation of motion for an electron in a perfect lattice at zero degrees Kelvin is

$$\frac{d\mathbf{k}}{dt} = \frac{-e}{\hbar}\mathbf{E} \qquad (10\text{-}17a)$$

where **E** is the electric field. For the case of a spherical parabolic band, Eq. (10-15), this reduces to Newton's law of motion in its familiar form

$$m^* \frac{d\mathbf{v}}{dt} = -e\mathbf{E} \qquad (10\text{-}17b)$$

In an imperfect lattice at finite temperature an electron does not move freely through the lattice according to Eqs. (10-17). The deviations from this motion are due to scattering mechanisms which may change both the magnitude and direction of the wavenumber **k** with known probabilities. In gallium arsenide the following seven processes are important:

Band I	*Physical origin*
1. Polar optical	Thermal vibrations of the lattice
2. Acoustic	Thermal vibrations of the lattice
3. Interband	Nonequivalent intervalley scattering from Band I to Band II
Band II	
4. Polar optical	Thermal vibrations of the lattice
5. Acoustic	Thermal vibrations of the lattice

6. Interband	Nonequivalent intervalley scattering from Band II to Band I
7. Intraband	Equivalent intervalley scattering between the six satellite valleys of Band II

Each scattering process is described by a rule for computing a new wavenumber $\mathbf{k}'$ from the old wavenumber $\mathbf{k}$, and a formula for the scattering rate of the process. Let

$\lambda_i(\mathscr{E})$ = scattering rate of the ith process
$p_i(\theta)d\theta$ = probability of the angle between $\mathbf{k}$ and $\mathbf{k}'$ being between θ and $\theta + d\theta$.

The new energy after scattering will be

$$\mathscr{E}' = \mathscr{E} \pm \hbar\omega_i \tag{10-18}$$

where ω_i is the phonon frequency. The positive sign is used if the lattice gives energy to the electron (i.e., a phonon is absorbed) and the negative sign is used if the electron gives energy to the lattice (i.e., a phonon is emitted). In this way the process is considered as a "collision" between a "particle" of lattice vibration, called a "phonon," and the electron. With this information it is possible to calculate the new wavenumber as follows:

1. Calculate the old energy from the old wavenumber

$$\mathscr{E} = \frac{\hbar^2 k^2}{2m^*} \tag{10-19}$$

2. Calculate the scattering rates and select scattering process (see Sec. 10-3-2).

$$\lambda_i(\mathscr{E}) \tag{10-20}$$

3. Calculate new energy for selected process

$$\mathscr{E}' = \mathscr{E} \pm \hbar\omega_i \tag{10-21}$$

4. Calculate modulus of new wavenumber using Eq. (10-16a)

$$k' = [2m^*\mathscr{E}'(1+\alpha\mathscr{E}')]^{1/2}/\hbar \tag{10-22}$$

5. Select direction of new wavenumber by evaluating $p_i(\theta)$.

It would be inappropriate to give the formulae for λ_i and p_i for all the scattering processes. The formulae used in the computer simulation being described here were taken from Fawcett, Boardman, and Swain (1970).

Figure 10-6 shows the magnitude of the different scattering rates and the total scattering rate for electrons in Band I and Band II. In Band I acoustic scattering is unimportant and, at low energies, the predominant process is polar optical scattering. Above a threshold around 0.36 eV, or $k = 0.08 \times 10^{10}\,\mathrm{m}^{-1}$, intervalley scattering of electrons to Band II rapidly dominates the scattering. For electrons in Band II, there is always a high probability of scattering back into Band I. Acoustic

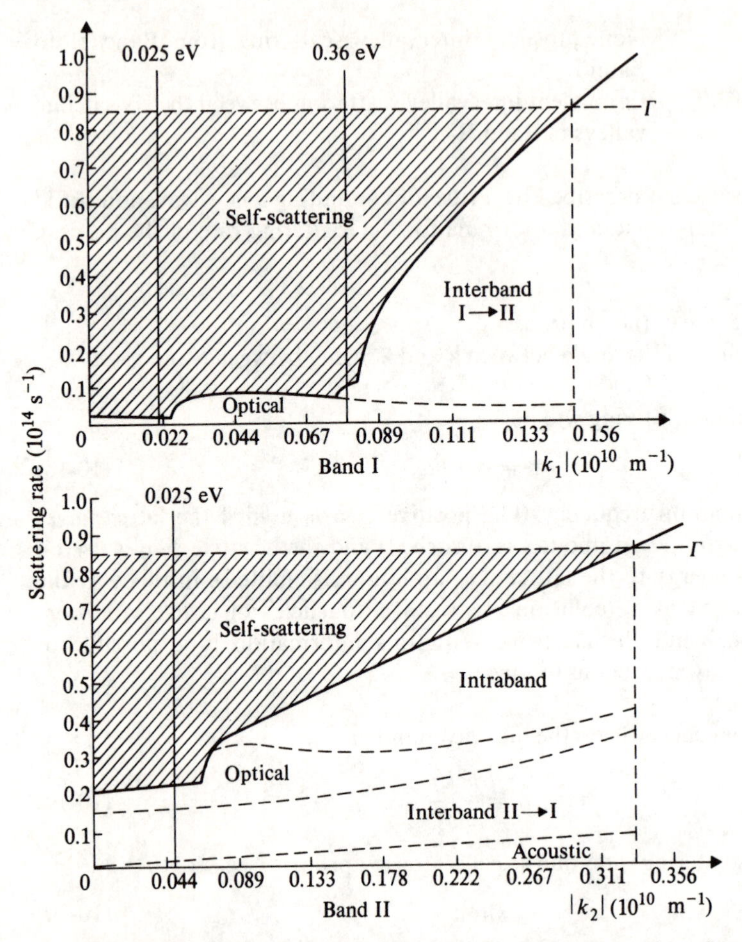

Figure 10-6 The total scattering rate (solid line) for the electrons in Band I and Band II as a function of wavenumber. Acoustic scattering is negligible in Band I. Wavenumbers k_1 and k_2 are measured from the center of each band, as in Fig. 10-5. The energy of an electron at room temperature (0.025 eV) and at the energy difference between bands (0.36 eV) is shown. (*Courtesy of R. A. Warriner, 1976.*)

scattering must be included throughout the energy range, because although its probability is relatively small, a scatter may produce a large change in the **k** vector. Above about $k \simeq 0.05 \times 10^{10}\,\text{m}^{-1}$ polar scattering becomes important. At higher k values, intraband scattering between the six constituent valleys of Band II begins to dominate the scattering.

10-2-4 Mobility

Equation (10-17a) shows that, in the presence of a constant electric field and in the absence of any scattering, the **k** vector (and the velocity) of an electron would increase linearly with time. The effect of scattering is to limit the drift velocity that is acquired, on average, by an electron to

$$\bar{\mathbf{v}} = -\mu(E)\mathbf{E} \tag{10-23a}$$

where μ is defined as the mobility. The mobility is related to the collision rate by

$$\mu = \frac{1}{2}\frac{e}{m^*}\tau = \frac{1}{2}\frac{e}{m^*\sigma} \tag{10-23b}$$

where τ is the mean scattering time and σ is the total scattering rate. The above formula is only approximate, but serves to show the influence of the scattering rate and effective mass on mobility. As expected, the greater the mass and scattering rate the lower the mobility. The formula is derived by assuming that the velocity of an electron is, on average, reduced to zero at every collision The mean velocity acquired is therefore half the acceleration in the electric field multiplied by the time between scatterings (or the reciprocal of the scattering rate).

Figure 10-7 shows the relation between the steady-state mean drift velocity of a cloud of electrons and the applied electric field. For low applied fields, less than a threshold value of about 4×10^5 V m^{-1}, the drift velocity is proportional to the field and GaAs is acting as a simple resistance with a roughly constant mobility of about 0.6 m^2 V^{-1} s^{-1}. Actual GaAs samples doped at 10^{23} m^{-3} have low-field mobilities in the range 0.4 to 0.5 m^2 V^{-1} s^{-1} and threshold fields of 3 to 4×10^5 V m^{-1}. This discrepancy is accounted for by the omission of ionized-impurity scattering from the computer model presently being described. Physically, this arises from the scattering of the conduction electrons by the fields of the fixed ionized sulfur atoms in the lattice. At doping densities above 10^{21} m^{-3} this becomes an important mechanism. Ionized-impurity scattering has been

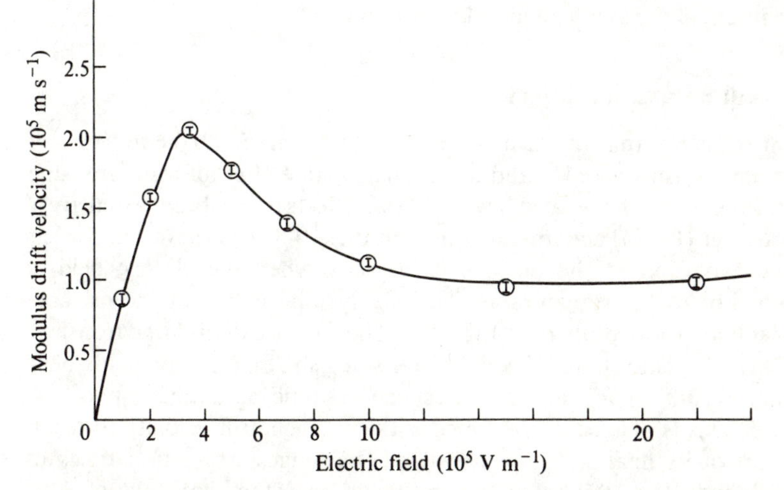

Figure 10-7 The steady-state mean drift velocity versus applied electric field. Solid line Fawcett et al. (1970), circles Warriner. (*From Warriner, 1977a, courtesy of Solid-State and Electron Devices,* © *Institution of Electrical Engineers.*)

included in subsequent work and the resulting mobilities and resistances are then in good agreement with experiment (see Table 10-2, page 400).

For applied fields above $4 \times 10^5\,\mathrm{V\,m^{-1}}$ an increasing number of electrons acquire an energy of $\sim 0.36\,\mathrm{eV}$, the energy gap between the satellite and central valleys. Figure 10-6 shows that, at these energies, the probability of an electron scattering to Band II becomes high and as it does so its effective mass increases fivefold. At the same time, most of its kinetic energy is converted to the potential energy of Band II and the velocity of the electron decreases to a small value. The overall result is that the drift velocity drops with increasing field. In this region between 4×10^5 and $12 \times 10^5\,\mathrm{V\,m^{-1}}$, although the total mobility $\mu = \bar{v}/E$ is still positive, the differential mobility $\mu_D = d\bar{v}/dE$ is negative. This means that the material acts as a negative resistance for any perturbations of the steady-flow condition. A decrease in electric field causes an increase, rather than a decrease, in velocity and current. Such a condition is clearly unstable and gives rise to the formation of charge accumulations and depletions.

It is possible to exploit this unusual property in certain electronic devices. For example, Gunn (1963) found that a simple GaAs diode will repeatedly generate dipole regions of charge, called Gunn domains, which travel from cathode to anode. These devices thus convert a simple dc bias into an ac microwave current, the frequency of which depends on the transit time of a Gunn domain. The properties of such domains have been studied by Warriner (1977*b*) using the computer model described in this chapter. An excellent review of the Gunn effect in gallium arsenide is given by Bott and Fawcett (1968).

For fields above about $20 \times 10^5\,\mathrm{V\,m^{-1}}$ the majority of electrons have been scattered into Band II and the material acts as a normal positive resistance. Because of the fivefold increase of mass and approximately fourfold higher scattering rates in Band II, Eq. (10-23*b*) shows that the Band II mobility is only about one-twentieth of the low-field mobility in Band I.

10-2-5 Transient Relaxation Effects

It is important to realize that the drift-velocity curve described above in terms of the scattering mechanisms is only valid in the steady state after all transients due to finite scattering times have died away. These effects have been extensively studied by Warriner (1977*a*) and are illustrated in Fig. 10-8. The curves on the left show the time evolution of the mean drift velocity when the electric field is increased from zero to the stated value. The steady-state values at the extreme right of the graph are those plotted on Fig. 10-7. The curves exhibit large transient velocities for high fields around $10^6\,\mathrm{V\,m^{-1}}$ and sluggish response for fields near threshold. Figure 10-8 (*right*) shows the relaxation time as a function of the increase in field. This is defined as the time for the average drift velocity to reach within 5 percent of its final steady-state value. The longest relaxation times of about 5 picoseconds ($5 \times 10^{-12}\,\mathrm{s}$) occur for fields near the threshold of 3 to $4 \times 10^5\,\mathrm{V\,m^{-1}}$. The time delays occur because of the finite acceleration of the electrons and because it takes several scattering times before steady-flow conditions are established.

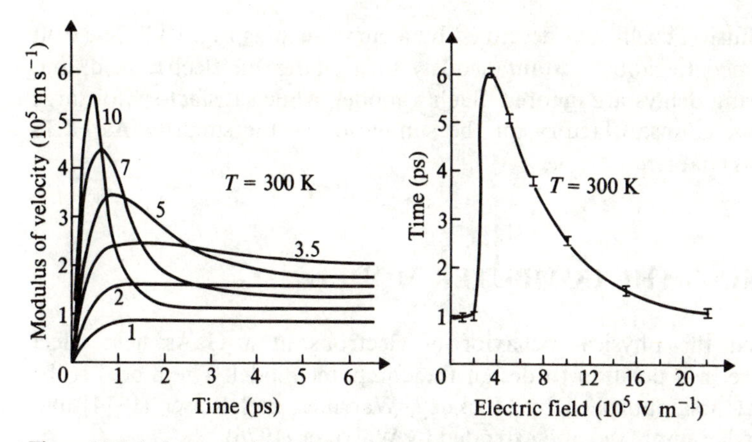

Figure 10-8 *Left*, mean drift velocity of electrons versus time in response to a step change in electric field. The initial field is zero and the curves are labeled with the field change in units of 10^5 V m^{-1}. *Right*, time to reach within 5 percent of final steady-state value, as a function of the field change. (*From Warriner, 1977a, courtesy of Solid-State and Electron Devices, © Institution of Electrical Engineers.*)

The existence of transient effects has several consequences. The most important of these concerns the dimensions and operating frequency of the device. A typical GaAs FET (circa 1978) operates with about 3 volts across a gate to drain gap of one micrometer. This corresponds to an average electric field of 30×10^5 V m^{-1}. At these fields an average velocity of 10^6 m s^{-1} is maintained for about one picosecond before the steady state is established. During this time a typical electron will have moved a distance of about one micrometer. That is to say, the distance required to establish the steady-state flow condition is equal to the gate-drain gap. Consequently, there is no part of this region where the steady-state $\bar{v}$ vs. E curve of Fig. 10-7 can be used. In order to represent correctly the flow of electrons across the gate-drain gap it is essential to take into account the scattering delays correctly. The maximum operating frequency can be estimated by allowing 5 ps for every quarter period of the oscillation. At this frequency the transients have decayed before the next change in direction of the field. This estimate gives 20 ps for the period of oscillation or a frequency of 50 GHz. As operating frequencies of up to 30 GHz are quoted for GaAs FETs, scattering delays must be included in the simulation of such devices. It is clear from the above considerations that the inclusion of scattering delays is important even for one-micrometer devices. With improvements in technology smaller and higher-frequency devices will be made for which the above statements will be even more valid.

We are describing in this chapter a simulation method which takes into account the individual scatters and associated delays of the electrons with the lattice. Such a microscope scattering model can correctly describe the motion of electrons in GaAs. We contrast this model with the traditional diffusion model for electron transport in transistors which has been used by many authors including Reiser (1972). In the latter the electrons drift and diffuse through the lattice with a mobility and diffusion coefficient described by a curve such as Fig. 10-7. The drift velocity is assumed to adjust instantaneously to a change in electric field, and hence all scattering delays are ignored. Such a model, while satisfactory for large and slow devices, is unsatisfactory for the simulation of the small GaAs FETs considered in this chapter.

10-3 DESIGNING THE COMPUTER MODEL

Having reviewed the physical behavior of electrons in a GaAs field effect transistor, we are in a position to design the computer model. The model to be described is that first published by Hockney, Warriner, and Reiser (1974) and subsequently further improved and extended by Warriner (1976).

If it were not for the scattering of electrons by the lattice, the electrons would behave like a collisionless electron plasma, for which the particle-mesh simulation techniques of Chapter 9 are appropriate. Apart from the boundary conditions already discussed (Sec. 10-1-2), all that need be considered in extending the plasma model to the FET problem is the inclusion of lattice scattering. The scattering processes will be chosen on a statistical basis using random numbers and for this reason the procedure is often referred to as a Monte-Carlo selection. We will now consider separately the particle-mesh and Monte-Carlo parts of the simulation and finally how they are joined together in a modified timestep cycle.

10-3-1 Particle-Mesh Calculation

In Sec. 9-2-3 we examined the influence of the timestep and spatial mesh on the properties of a two-dimensional particle-mesh model of a plasma. In particular, we noted that the timestep and mesh spacing must be related to the physical plasma frequency and Debye length of the plasma. Failure to so relate them could result in an unstable or highly collisional plasma. In the FET simulation we wish to superimpose physical collisional effects on a particle-mesh model that is, for practical purposes, collisionless. It is important therefore to choose parameters that ensure that the collision rate inherent in the particle-mesh model is less than the physical mechanisms we wish to superimpose.

We consider first the question of stability and the requirement that $\omega_p DT < 2$, where ω_p is the plasma frequency. For the worst case of the light Band I electrons we have

$$\omega_{p1} = \left(\frac{ne^2}{\epsilon\epsilon_0 m}\right)^{1/2} = 2.12 \times 10^{13}\,\mathrm{s}^{-1} \tag{10-24a}$$

where

$$\begin{aligned} n &= 10^{23}\,\mathrm{m}^{-3} \\ \epsilon &= 10, \text{ dielectric constant} \\ m &= 0.067 m_e \end{aligned} \tag{10-24b}$$

The timestep chosen for the particle-mesh calculation will be called the "field-adjusting" timestep in order to distinguish it from the free-flight times used later in the selection of scattering processes. The value chosen in the simulation was

$$DT = 5 \times 10^{-14}\,\mathrm{s} \tag{10-25a}$$

leading

$$\omega_{p1} DT = 1.06 \tag{10-25b}$$

and hence the plasma period

$$\tau_{p1} = \frac{2\pi}{\omega_{p1}} = 6DT = 30 \times 10^{-14}\,\mathrm{s} \tag{10-25c}$$

Similarly, for the heavier electrons ($m = 0.35m_e$) in Band II we have

$$\omega_{p2} DT = 0.464 \tag{10-26a}$$

$$\tau_{p2} = 13.5DT \tag{10-26b}$$

The choice of DT therefore satisfies the stability criterion for both species of electron.

In order to calculate the collision rate in the model, we must calculate the Debye length λ_D and the number of model particles per Debye square N_{DM}. The Debye length is independent of mass and is therefore the same for both species of electron

$$\lambda_D = \left(\frac{\epsilon\epsilon_0 k_B T}{ne^2}\right)^{1/2} = \frac{v_T}{\omega_p} = 1.2 \times 10^{-8}\,\mathrm{m} \tag{10-27a}$$

where the thermal velocity is defined as

$$\begin{aligned} v_T = \left(\frac{k_B T}{m}\right)^{1/2} &= 2.6 \times 10^5\,\mathrm{m\,s^{-1}} \quad \text{(Band I)} \\ &= 1.1 \times 10^5\,\mathrm{m\,s^{-1}} \quad \text{(Band II)} \end{aligned} \tag{10-27b}$$

The epitaxial layer of the model (see Fig. 10-2) is a region $5.12 \times 0.16\,\mu\mathrm{m}^2 = 430 \times 13.3\lambda_D^2$. This is occupied initially by 8,000 simulated electrons and hence

$$N_{\mathrm{DM}} = \frac{8{,}000}{430 \times 13.3} = 1.4 \tag{10-27c}$$

The collision time of the model τ_{CM} is given by Eq. (9-19a)

$$\frac{\tau_{\mathrm{CM}}}{\tau_{p1}} = N_{\mathrm{DM}}\left\{1 + \left(\frac{W}{\lambda_D}\right)^2\right\} \tag{10-28}$$

where W is the width of the particle. The FET model uses hollow-particle nearest-grid-point assignment for which $W = 2H = 4 \times 10^{-8}\,\mathrm{m} = 0.04\,\mu\mathrm{m}$, twice the spatial-mesh separation. Inserting these values, we obtain

$$\frac{W}{\lambda_D} = 3.34 \tag{10-29a}$$

$$\tau_{\mathrm{CM}} = 17.0\tau_{p1} = 510 \times 10^{-14}\,\mathrm{s} \tag{10-29b}$$

or, expressed as a collision rate,

$$\nu_{CM} = \tau_{CM}^{-1} = 0.0020 \times 10^{14}\,\text{s}^{-1} \tag{10-29c}$$

The weakest total physical scattering rate is that for low-energy Band I electrons and has a rate of $0.02 \times 10^{14}\,\text{s}^{-1}$ (see Fig. 10-6). The artificial model scattering rate is therefore at most one-tenth of the physical processes. This occurs in the source-gate region of the device which is thought to be relatively unimportant to the device behavior. In the important gate-drain region, the physical scattering rates are ten times higher and the model scattering rate is proportionally less important.

A mesh of (256×16) square cells was used to describe the potential over the $5.12 \times 0.32\,\mu\text{m}^2$ calculation region of Fig. 10-2 (*right*), using a square mesh cell of width $H = 0.02\,\mu\text{m}$. The calculation was performed with the program POT4 using the FACR(1) algorithm (see Sec. 6-5-3). The electrodes were held at the prescribed potentials by the capacity matrix technique (see Sec. 6-5-6). The electrostatic potential is given by Poisson's equation which, in a dielectric medium, is

$$\nabla^2 \phi = -\frac{\rho}{\epsilon\epsilon_0} \tag{10-30a}$$

where ε is the dielectric constant. The effect of the dielectric constant can be included most easily by appropriately reducing the charge per particle in the simulation. For the purposes of the field calculation the simulated particle represents a rod of charge with an effective charge per unit length

$$q = \frac{-enL_xL_y}{\epsilon N} \tag{10-30b}$$

where n = doping density ($10^{23}\,\text{m}^{-3}$)
L_x, L_y = dimensions of the doping region ($5.12 \times 0.16\,\mu\text{m}^2$)
ϵ = dielectric constant (~ 10)
N = number of simulated particles used in the model ($\sim 8{,}000$).

At high doping densities it may be necessary to smooth the potential calculated from Eq. (10-30*a*) before the particles are moved. A convenient method is to calculate the smoothed potential ϕ^* as the average of the surrounding eight values:

$$\phi^*_{i,j} = \tfrac{1}{8}(\phi_{i-1,j-1} + \phi_{i-1,j} + \phi_{i-1,j+1} + \phi_{i,j+1} + \phi_{i,j-1} + \phi_{i+1,j-1} + \phi_{i+1,j} + \phi_{i+1,j+1}) \tag{10-30c}$$

This model is known as the hollow NGP (HNGP) model.

We have already mentioned that hollow-particle nearest-grid-point (HNGP) charge-sharing and force interpolation are used in the FET model without problems. This has the advantage of simplicity and is computationally fast. Clearly, smoother interpolation schemes such as cloud-in-cell (CIC) or triangular shaped cloud (TSC) could be used. All these schemes have similar collision times (W varies from H to $3H$) but differ radically in their energy-conserving properties. The choice of model will therefore depend primarily on the required level of energy conservation. The relevant formulae have already been given in connection with collisionless plasma simulation, Eq. (9-22). The lack of energy conservation is measured by the heating time τ_H, which is given by

$$\frac{\tau_H}{\tau_{CM}} = \frac{K_H}{(H/\lambda_D)^2} \tag{10-31a}$$

where $K_H = 2$ (NGP), 6 (HNGP), 40 (CIC), 150 (TSC), 3,000 (QPM).

For the HNGP model we have

$$\left(\frac{H}{\lambda_D}\right) = 1.67 \qquad \text{and} \qquad \frac{\tau_H}{\tau_{CM}} = 2.15 \tag{10-31b}$$

hence

$$\tau_H = 1{,}097 \times 10^{-14}\,\text{s} = 220DT \tag{10-31c}$$

At every physical collision with the GaAs lattice, the temperature of the electron distribution is partly relaxed to the lattice temperature, hence the quantity of importance is the relative lack of energy conservation in a collision time. If we say that the physical collision time varies from about 5×10^{-14} s in the gate-drain gap to 50×10^{-14} s in the source region then the above heating times correspond to a relative energy conservation of between 0.2 and 2 percent.† If TSC charge-sharing is used then the relative energy conservation can be improved to 0.004 and 0.04 percent, but this will increase considerably the computer time for the free-flight part of the calculation. The physical effect of allowing a relatively short heating time is equivalent to the use of a slightly higher lattice temperature. This can of course be compensated for by inserting a lower lattice temperature into the calculation of the scattering processes. Because the electron temperature is primarily determined by the imposed physical scattering rates, it is evident that we may compute with heating times much shorter than would be acceptable in the simulation of a collisionless plasma by a pure particle-mesh code.

10-3-2 Monte-Carlo Scattering Selection

The selection of free-flight times between different scattering events is considerably simplified if the total scattering rate for all processes is a constant, say $\Gamma\,\text{s}^{-1}$. In this case, the probability of a free flight lying between t and $t+dt$ is

$$P(t)dt = \Gamma e^{-\Gamma t}\,dt \tag{10-32}$$

The problem of selecting a set of numbers with a given nonuniform probability $P(t)$, when one has available only a random number generator giving a uniform distribution, is met repeatedly in Monte-Carlo simulation. The general solution is to calculate first, either analytically or numerically, the cumulative distribution function

$$c(t) = \int_0^t P(t')\,dt' \tag{10-33a}$$

† In the time τ_H the kinetic energy of the electrons increases by 50 percent of the average thermal energy of an electron (see Sec. 9-2-2).

Since $P(t)$ is a probability $c(\infty) = 1$, hence

$$0 \leqslant c < 1 \qquad \text{and also} \qquad dc = P(t)\,dt \tag{10-33b}$$

If the value of c is chosen as a random number r, from a distribution that is uniform in the range $0 \leqslant r < 1$, then the probability of selecting a value in the range dc is dc. Consequently, by Eq. (10-33*b*), the probability of selecting a value within the range dt is $P(t)\,dt$ as required. Hence we set

$$r = c(t) \tag{10-34a}$$

and

$$t = c^{-1}(r) \tag{10-34b}$$

where c^{-1} is the function that is inverse to $c(r)$, that is to say, the result of solving Eq. (10-34*a*) for the variable t.

Applying this method to Eq. (10-32) we obtain

$$r = c(t) = 1 - e^{-\Gamma t} \tag{10-35a}$$

and

$$\delta t = -\Gamma^{-1} \ln(r_1) \tag{10-35b}$$

where we have introduced the symbol δt instead of t for the free-flight time. Also, in order to simplify the formula, we have defined a new random number $r_1 = (1 - r)$ that is uniformly distributed in $0 < r_1 \leqslant 1$.

Figure 10-6 shows that the total scattering rate for real processes is *not* constant and it would appear that the above simple selection procedure could not be used. However, Rees (1968, 1969) suggested the introduction of a dummy "self-scattering" process that makes no change to the **k** vector of the particle. The self-scattering rate at any energy is chosen to be that rate necessary to bring the real physical processes up to a chosen constant Γ, and is indicated by the shaded area of Fig. 10-6. If we now include self scattering as a process, then the total scattering *is* constant and free flights can be chosen according to Eq. (10-35*b*).

Having chosen a free flight, the electron drifts in the electric field obtained from the mesh for a time δt. In the mesh cell surrounding mesh point (i,j), we have

$$E_x = (\phi_{i-1,j} - \phi_{i+1,j})/2H \tag{10-36a}$$

$$E_y = (\phi_{i,j-1} - \phi_{i,j+1})/2H \tag{10-36b}$$

The new **k** vector is obtained from

$$k_x^{(t+\delta t)} = k_x^{(t)} - eE_x \delta t/\hbar \tag{10-37}$$

and similarly for k_y. The new position is then obtained from

$$x^{(t+\delta t)} = x^{(t)} + \frac{\hbar}{2m}(k_x^{(t)} + k_x^{(t+\delta t)})\delta t \tag{10-38}$$

and similarly for $y^{(t+\delta t)}$.

After the free flight the selection of the scattering process is made with a second uniformly distributed random number r_2 in the range

$$0 \leqslant r_2 < \Gamma \tag{10-39}$$

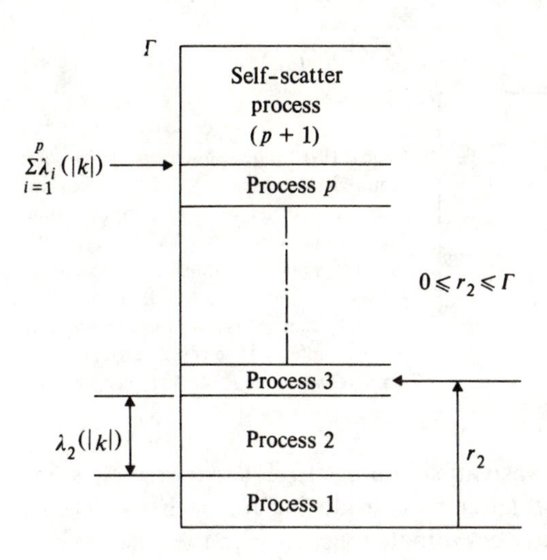

Figure 10-9 Diagram showing the selection of the scattering process. There are p real processes. The dummy self-scattering process $(p+1)$ makes the total scattering rate equal to the constant Γ. The value of r_2 is selecting process number 3. (*Courtesy of R. A. Warriner, 1976.*)

The scattering process m is selected if

$$\sum_{i=1}^{m-1} \lambda_i(k) \leqslant r_2 < \sum_{i=1}^{m} \lambda_i(k) \tag{10-40}$$

where $\lambda_i(k)$ is the scattering rate for process i. This selection method is illustrated in Fig. 10-9, where the vertical width occupied by each process is equal to the scattering rate for the process and therefore proportional to the probability of that process occurring.

The above selection method can be very wasteful of computer time if the scattering probability varies greatly with energy, as it does in a material such as GaAs. The probability of selecting the dummy self-scattering process is proportional to the shaded area of Fig. 10-6, and this is obviously large for the majority of electrons, which are in Band I with low energies. Furthermore, the formulae for $\lambda(k)$ are complex and it would save computer time if they were not recomputed for every test in Eq. (10-40) for every electron.

In order to increase the efficiency of the computer program, it is desirable to tabulate the sums in Eq. (10-40) in a two-dimensional table, as is illustrated in Fig. 10-10. The random number r_2 is then compared in turn with the tabulated sums:

$$\text{If } r_2 \geqslant \sum_{i=1}^{q} \lambda_i \text{ then select process } (q+1) \tag{10-41a}$$

Since self scattering is likely to be the most probable process for most electrons the comparison should be made starting with self scattering in the order

$$q = p, (p-1), \ldots, 1 \tag{10-41b}$$

Then, if the first test succeeds self scattering is selected, and if the last test fails process 1 is selected.

The tabulation procedure, while avoiding the repetition of arithmetic, does not

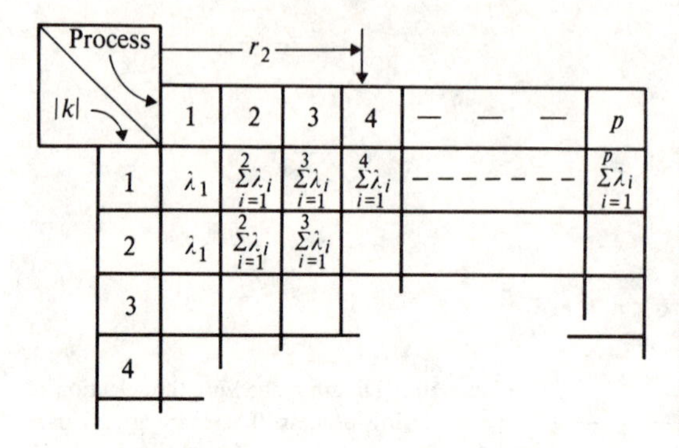

Figure 10-10 The tabulation of the sums required in the selection of the scattering process, in order to save computer time. Comparing from right to left. If r_2 is greater than the tabulated value for process q then process $(q+1)$ is selected. If the first test succeeds then self-scattering is selected. In this example r_2 is selecting process number 4. (*Courtesy of R. A. Warriner, 1976.*)

prevent the high probability of the wasteful selection of self scattering. At some cost in storage one may prevent the latter by introducing for each particle an individual Γ_j which is stored as an extra coordinate together with the position and **k** vector. The only requirement on Γ_j is that it be greater than the total scattering rate for the jth particle. Consequently, low-energy particles with low scattering rates may have a much smaller Γ_j than higher-energy particles. It may of course happen that, at some stage in the motion of the electron, the total scattering rate exceeds Γ_j. In this case Γ_j must be increased and the following heuristic is used:

guess $\Gamma_j = \sum_{i=1}^{p} \lambda_i(k^{(t)})$ real processes

select r_1 and take free-flight using Γ_j

$$\delta t = -\frac{1}{\Gamma_j} \ln(r_1)$$

compute new **k** vector

$$\mathbf{k}^{(t+\delta t)} = \mathbf{k}^{(t)} - \frac{e}{\hbar} \mathbf{E}\, \delta t \tag{10-42}$$

if $\sum_{i=1}^{p} \lambda_i(k^{(t+\delta t)}) \leqslant \Gamma_j$

Γ_j is satisfactory, proceed to next particle.

If not, increase Γ_j to $\Gamma_j + \Delta\Gamma$ and try again with same r_1

The saving in computer time in changing from constant Γ without tabulation to a variable Γ with tabulation is about a factor 10 in the particle-moving phase of the timestep (see Table 10-1).

The selection of the new **k** vector involves the calculation of its magnitude k' from Eq. (10-22) and the calculation of the angle of scattering θ according to the probability $p_i(\theta)$ for the selected scattering process. Knowing $p_i(\theta)$, we can select θ using the general method given in Eqs. (10-33) and (10-34). All scattering processes except polar optical scattering are randomizing, in which case all directions are equally probable for the final **k** vector. We do not therefore need to measure the

Table 10-1 The computer time in seconds on the IBM 360/195 for different parts of the calculational cycle.

A mesh of (256 × 32) cells is used with 12,000 particles. A region size of 700 kbytes is required.

Part of cycle	Constant Γ no tabulation	Variable Γ tabulation
QSHARE	0.07	0.07
POT4	0.35	0.35
SCATTER	9.80	1.20
Cycle	10.22	1.62

final direction with respect to the incident **k** vector, as would normally be the case in the definition of the scattering angle. Any direction may be taken as the datum and if we choose to measure the final direction in polar coordinates about the z axis then the probability $p(\theta)q(\phi)\,d\theta\,d\phi$ of the final **k** vector lying between θ and $\theta+d\theta$, ϕ and $\phi+d\phi$ is given by the fractional surface area of the unit sphere within these coordinates. That is to say

$$p(\theta)\,d\theta = \sin\theta\,d\theta/2 \qquad 0 \leqslant \theta \leqslant \pi \tag{10-43a}$$

$$q(\phi)\,d\phi = d\phi/2\pi \qquad 0 \leqslant \phi \leqslant 2\pi \tag{10-43b}$$

Calculating the cumulative distributions, one obtains

$$c_1(\theta) = \int_0^\theta p(\theta')\,d\theta' = (1-\cos\theta)/2 \tag{10-44a}$$

$$c_2(\phi) = \int_0^\phi q(\phi')\,d\phi' = \phi/2\pi \tag{10-44b}$$

and hence θ and ϕ may be selected by two random numbers r_3, r_4 uniformly distributed in $0 \leqslant r_3, r_4 < 1$

$$\theta = c_1^{-1}(r_3) = \cos^{-1}(1-2r_3) \tag{10-45a}$$

$$\phi = c_2^{-1}(r_4) = 2\pi r_4 \tag{10-45b}$$

Knowing the final direction, the components of the vector are obtained directly from r_3 and the second equality in

$$k'_z = k'\cos\theta = k'a \tag{10-46a}$$

$$k'_x = k'\sin\theta\cos\phi = k'b\cos\phi \tag{10-46b}$$

$$k'_y = k'\sin\theta\sin\phi = k'b\sin\phi \tag{10-46c}$$

where $a = (1-2r_3)$ and $b = (1-a^2)^{1/2}$.

For processes that are not randomizing, directions must be measured with respect to the incident **k** vector. The probability of scattering is independent of the

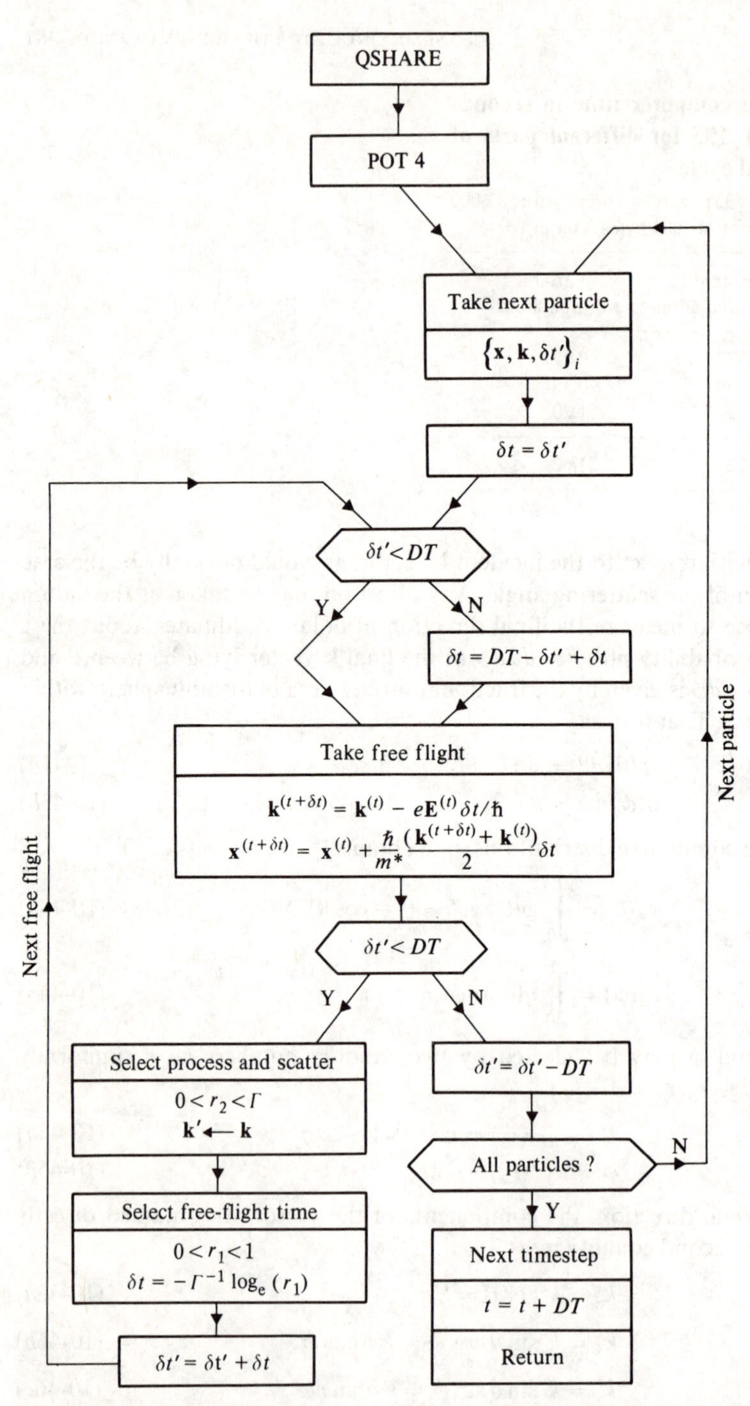

Figure 10-11 The modified particle-mesh timestep cycle for the Monte-Carlo scattering FET simulation program (i.e., the subroutine STEPON). The coordinates (x, y, k_x, k_y, k_z, $\delta t'$, Γ) are stored for each particle, where $\delta t'$ is the free-flight time left over from the last field-adjusting timestep. $\delta t'$ also accumulates the sum of the free-flight times from the beginning of the current timestep.

angle ϕ about this axis and Eq. (10-45*b*) may still be used for its selection. The reader is referred to Fawcett, Boardman, and Swain (1970, page 1971) for the method of selection of the scattering angle θ in more complicated cases such as polar-optical scattering in which there is a preference for forward scattering (θ small).

10-3-3 Modified Timestep Cycle

We are now in a position to consider the modifications to the normal particle-mesh timestep cycle that are required in order to incorporate Monte-Carlo scattering. This is illustrated in Fig. 10-11, which gives a flow diagram for the subroutine STEPON. This subroutine is called at constant intervals of DT seconds by the OLYMPUS control routine CRONUS (see Fig. 3-4). The first action of STEPON is to recalculate the electric field, hence DT is called the field-adjusting timestep. This is done by a call to QSHARE (assign charge to mesh) and to POT4 (solve for the potential) in the same way as in a normal particle-mesh algorithm. During this part of the calculation the simulation particle represents a two-dimensional rod of charge since, to a good approximation, there is no variation of charge density in the z direction and therefore no need to keep a record of the z coordinate.

After this field adjustment, an inner loop of Monte-Carlo scattering is entered during which the electric field remains unchanged. In this inner loop each particle is taken in turn and subjected to Monte-Carlo scattering until the sum of its free-flight times exceeds the time at which the next field-adjustment is to take place. The amount of the excess is stored as the coordinate $\delta t'$, so that the free flight may be correctly continued with the new field after the next field adjustment. During the inner loop each of the simulation particles is scattered in three-dimensional $\mathbf{k}$ space as though it were a single electron, with the effective mass and charge appropriate to its state. Scattering only alters the $\mathbf{k}$ vector and hence we store the three values (k_x, k_y, k_z). During scattering we regard the $\sim$8,000 simulation particles as a sample from the distribution in the real device, whereas during the field calculation the particles are regarded as superparticles [see Eq. (10-30*b*)] so that they correctly apportion the total charge to the mesh. It is important to distinguish carefully the different roles played by the simulation particles during the different parts of the cycle.

For high-energy particles in the gate-drain gap with a total scattering rate of $\sim 0.5 \times 10^{14}\,\mathrm{s}^{-1}$ there will be on average $\sim 2\frac{1}{2}$ scattering events during the field-adjusting step of $DT = 5 \times 10^{-14}$ s. This situation is illustrated in Fig. 10-12. For the low-energy Band I particles in the source region where the scattering rate is $\sim 0.02 \times 10^{14}\,\mathrm{s}^{-1}$ there will be a scattering event on average about every ten field-adjusting timesteps (or $1\frac{1}{2}$ plasma periods). In this latter case the collisional effects are rather weak and the particle executes plasma oscillations as though it were in a collisionless plasma.

Table 10-1 gives the computer characteristics of a typical Monte-Carlo scattering FET simulation program in use at Reading University. The importance

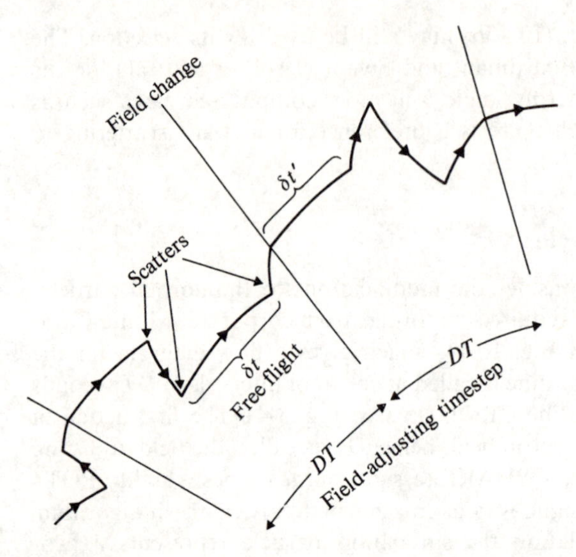

Figure 10-12 A typical electron orbit for a high scattering rate, showing several free flights per field-adjusting timestep. Scatters are indicated by a sharp change in the orbit. There is a slight change in the orbit every DT seconds due to the recalculation of the electric field.

of storing a variable Γ for each particle and of tabulating the scattering rates is obvious.

10-4 MEASUREMENTS ON FETs

Having written an FET simulation program it is necessary to measure the static and dynamic characteristics of the simulated device. These can then be compared with typical laboratory measurements in order to validate the computer program. Only after successful validation can the model be used to study FET geometries and parameter ranges that are not accessible in the laboratory. We will consider now how such measurements are made and the insights into the working of the device that can be obtained by "looking" inside the FET in a manner that is impossible in the laboratory. A method for obtaining the equivalent circuit parameters and the noise figure will also be given. Lastly, the model will be applied to the study of a potentially very fast device, operating at liquid-nitrogen temperatures, that has not yet been built—the COLDFET. In this case the computer simulation is leading the way in device development by showing the types of device it is worth building in the laboratory.

10-4-1 Static Characteristics

The static and dynamic characteristics of the simulated device are measured by applying fixed voltages to the electrodes and observing the currents that flow in response. The total current at an electrode is composed of the particle current and the displacement current. These are derived from the first and second terms, respectively, in the expression that follows, Eq. (10-47). From the start of the

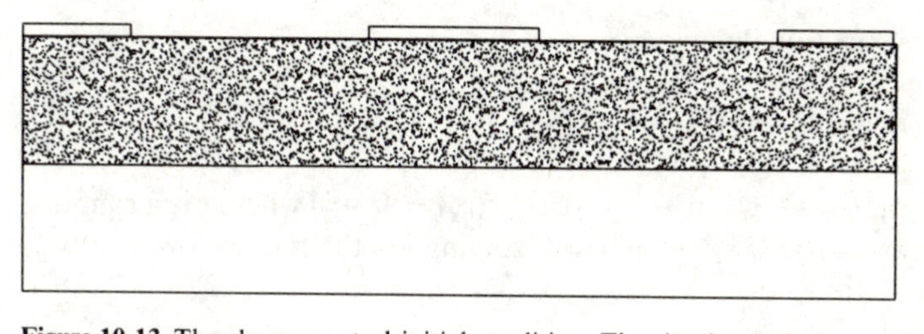

Figure 10-13 The charge-neutral initial condition. The simulated electrons that are shown are all in Band I at 300 K. The voltages on all electrodes are zero. (*Courtesy of S. J. Beard.*)

computer run, a record is kept of the net number of simulated electrons absorbed by each electrode, and at every field-adjusting timestep the integral of the normal electric field is computed along the electrode. An equivalent total charge is then calculated for each electrode from

$$Q(t) = q(N_a - N_i) + \epsilon\epsilon_0 \int E_y(x, t)\,dx \tag{10-47}$$

where the integral is taken over the surface of the electrode and

q = charge per meter of a superparticle from Eq. (10-30*b*)
N_a = total particles absorbed up to time t
N_i = total particles injected up to time t
$E_y(x, t)$ = electric field directed into the device at time t and position x

We note that during this phase of the calculation, in which we are concerned with the total flow of charge and not scattering, the particles must be treated as two-dimensional charged rods or superparticles (see Sec. 10-3-1).

The initial condition for most experiments will be the charge-neutral state shown in Fig. 10-13. Particles are distributed uniformly over the doped epitaxial region in order to neutralize the doping charge. Their (x, y) coordinates are assigned using a random-number generator with a uniform distribution. The charge-neutral condition is chosen in order to avoid the very large fields and transients that can occur if charge separations are not permitted to form naturally from the dynamics of the particles.

All electrons are initially in Band I and are assigned **k** vectors corresponding to a maxwellian distribution of velocities at the lattice temperature T (300 K). The distribution of velocities is the same in all three coordinate directions. Taking the x direction, we have that the probability of finding a velocity in the range v_x to $v_x + dv_x$ is

$$p(v_x)dv_x = (\sqrt{2\pi}v_T)^{-1} \exp(-v_x^2/2v_T^2) \tag{10-48a}$$

where $v_T = (k_B T/m_1^*)^{1/2}$ is the root-mean-square (RMS) x velocity. We now describe two methods for setting up this distribution. First we may follow the general method described in Sec. 10-3-2, and find the cumulative distribution

$$c(v_x) = \tfrac{1}{2}\operatorname{erf}(v_x/(2v_T^2)^{1/2}) \tag{10-48b}$$

where erf is the error function defined by

$$\operatorname{erf}(x) = \frac{2}{\sqrt{\pi}} \int_0^x e^{-x^2}\,dx$$

Since $c(v_x)$ is in the range $-\frac{1}{2} \leqslant c(v_x) < \frac{1}{2}$, we set $c(v_x) = (r - \frac{1}{2})$ where r is a random number in the normal range $0 \leqslant r < 1$. Then, solving for the required velocity v_x, we obtain

$$v_x = \sqrt{2}v_T \operatorname{erf}^{-1}(2r-1) \tag{10-48c}$$

This method therefore requires a subroutine for the evaluation of the inverse error function erf^{-1} or a precomputed table of this function. The value of the wave-number is then obtained from

$$k_x = \frac{m_1^*}{\hbar} v_x = \frac{\sqrt{2}v_T m_1^*}{\hbar} \operatorname{erf}^{-1}(2r-1) \tag{10-49}$$

Three random numbers are required for each particle, for the selection of each component of the **k** vector from Eq. (10-49).

An alternative method of selecting initial velocities makes use of the powerful central limit theorem of statistics (see, e.g., Ross, 1976, page 252, or most books on statistics). This theorem gives the fundamental properties of the random walk and all phenomena based upon the random walk. It states:

> "If $X_1, X_2, \ldots$ are a sequence of independent and identically distributed random variables each having mean μ and standard deviation σ, then the sum of n such random numbers, $\Sigma_{i=1}^n X_i$, tends to a normal (gaussian, or maxwellian) distribution with mean μ and standard deviation $\sigma\sqrt{n}$ as n becomes large."

Examples of phenomena controlled by this theorem are:

1. Random walk in real space (diffusion)
2. Random walk in velocity space (stochastic heating)
3. Repeated convolution of a function with itself leads to a maxwellian.

A sequence of random numbers r_i in the range 0 to 1 have $\mu = \frac{1}{2}$ and $\sigma = (12)^{-1/2}$, hence a maxwellian distribution with a mean of zero and root mean square v_T can be generated from

$$v_x = v_T\left(\sum_{i=1}^{n} r_i - n\mu\right) \bigg/ \left(\frac{n}{12}\right)^{1/2} \tag{10-50a}$$

(see Abramowitz and Stegun, 1964, page 952), or, for the case of $n = 3$, from

$$v_x = 2v_T(r_1 + r_2 + r_3 - 1.5) \tag{10-50b}$$

In selecting n as small as 3 we are deliberately minimizing the number of random numbers required and the amount of computer time. In this way we obtain an approximate maxwellian distribution as the initial condition and rely on the dynamics of the simulation to convert this to a true maxwellian distribution by the scattering processes with the lattice that are present in the simulation. This method, although more straightforward, requires the generation of nine random numbers per particle, and is probably slower than using Eq. (10-49), provided erf^{-1} is found by interpolation from a table of the function.

If typical operating voltages are now applied to the electrodes, charge flows in and out of the electrodes until a steady-flow condition is established. This is illustrated in Fig. 10-14, which shows the equivalent charge $Q(t)$ that has flowed through each electrode up to time t. Notice that the curves are proportional to the net electrons absorbed by the drain and injected by the source. Initially, electrons are repelled by the negative potential on the gate and, as the region below the gate is depleted of carriers, leave the device through both the source and the drain. During this initial transient the field is building up at the gate, which leads to the induction of a displacement charge on the gate. There is of course no particle charge or particle current at the gate, since the gate repels electrons. After about ten picoseconds a steady-flow state has been established with a steady distribution of charge and the average number of electrons leaving the source equals the

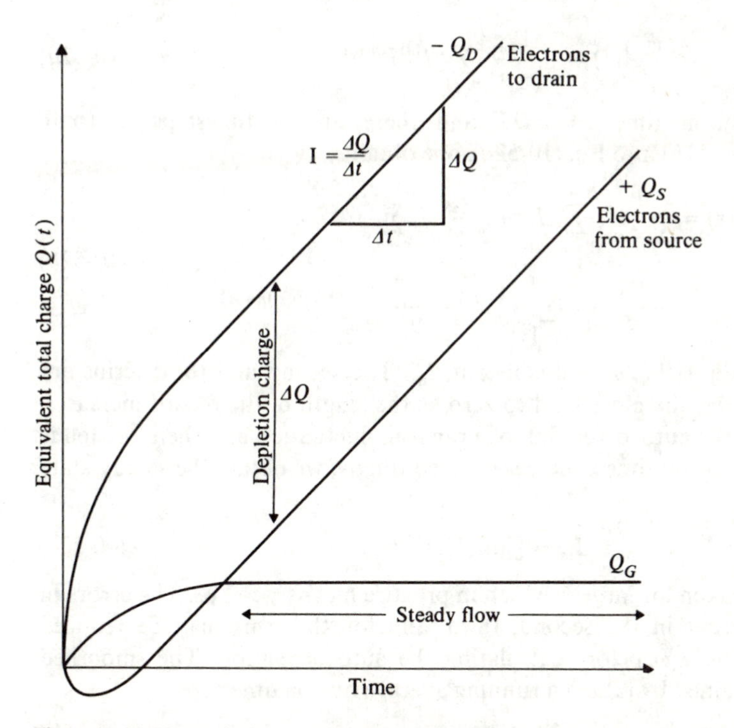

Figure 10-14 The equivalent total charge as a function of time for source, drain, and gate electrodes after a step of volts on the drain. The initial condition is the neutral device of Fig. 10-13. The difference between the electrons lost at the drain and those injected at the source is the electrons lost during the formation of the depletion under the gate.

average number entering the drain. The current through each electrode is given by

$$I(t) = \frac{dQ(t)}{dt} \tag{10-51a}$$

where $Q(t)$ is given by Eq. (10-47).

Owing to large fluctuations in $Q(t)$ it is difficult to obtain a reliable estimate for I_{ss} (the steady-state current) by fitting a straight line through a charge vs. time graph. However, I_{ss} can be estimated accurately by using the autocorrelation of the current $I(t)$. In the steady-flow condition the current can be expressed as the sum of a steady-state value, I_{ss}, plus a random noise current $i(t)$ which has a mean of zero, thus

$$I(t) = I_{ss} + i(t) \tag{10-51b}$$

$I(t)$ is obtained from $Q(t)$ by differencing. If superscripts are used to denote the time level then

$$I^{(n)} = (Q^{(n+1)} - Q^{(n)})/DT \tag{10-51c}$$

where $I^{(n)} = I(nDT)$ and $Q^{(n)} = Q(nDT)$. The autocorrelation function $A(\tau)$ is defined by

$$A(\tau) = \frac{1}{N-n'} \sum_{n=1}^{N-n'} I^{(n)} I^{(n+n')} \tag{10-52a}$$

where the correlation time $\tau = n'DT$ and there are N timesteps in total. Substituting Eq. (10-51b) into Eq. (10-52a), one obtains

$$A(\tau) = \frac{1}{N-n'} \sum_{n=1}^{N-n'} I_{ss}^2 + \frac{I_{ss}}{N-n'} \sum_{n=1}^{N-n'} i^{(n)} + \frac{I_{ss}}{N-n'} \sum_{n=1}^{N-n'} i^{(n+n')} + \frac{1}{N-n'} \sum_{n=1}^{N} i^{(n)} i^{(n+n')} \tag{10-52b}$$

The first term in Eq. (10-52b) is a constant, I_{ss}^2. The second and third terms are averages over $i(t)$ and therefore tend to zero as the length of the record increases. The fourth term is the autocorrelation of a random fluctuation and therefore tends to zero as the correlation time τ increases. Accordingly, we obtain the steady-state current from

$$I_{ss} = [\lim A(\tau)]^{1/2} \tag{10-52c}$$

where the limit is taken for large τ, which in practice means $\tau > 1$ ps. The errors in measurement inherent in the second, third, and fourth terms may be reduced further by averaging $I(t)$ before calculating the autocorrelation. The smoothed values $I_s(t)$ are obtained by taking a running average over m timesteps

$$I_s^{(n)} = \frac{1}{m} \sum_{n'=0}^{m-1} I^{(n+n')} \tag{10-52d}$$

The total capacity of the gate C_G can be found from the depletion charge ΔQ_D

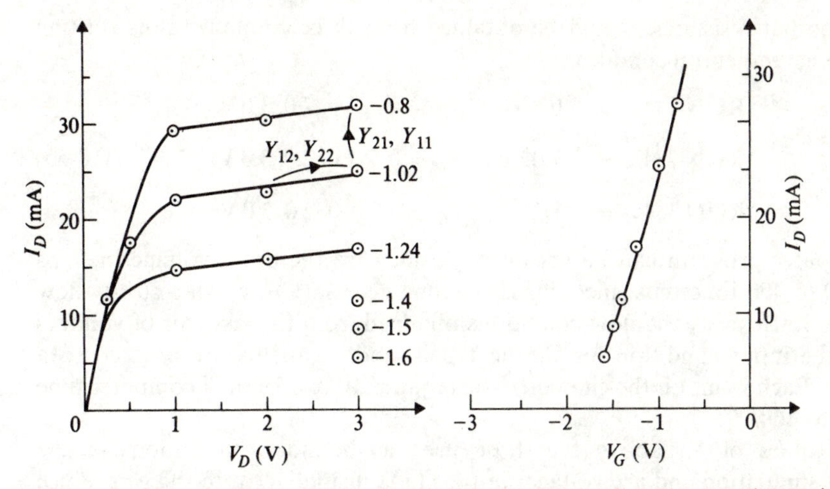

Figure 10-15 The measured static characteristics for a simulated device. The normal operating point is at $V_D = 3.0$ V and $V_G = -1.02$ V. The step changes that provide the Y parameters at the operating point are shown. The slope of the I_D/V_D characteristic, *left*, is the output resistance, the slope of the I_D/V_G characteristic, *right*, is the mutual conductance. (*From Warriner, 1977c, courtesy of Solid-State and Electron Devices, © Institution of Electrical Engineers.*)

and the voltage on the gate V_G

$$C_G = \frac{\Delta Q_D}{V_G} \tag{10-53}$$

This capacity will be the sum of the capacity with respect to the source C_{sg} and that with respect to the drain C_{gd}, which are discussed later in Sec. 10-4-4.

The result of performing a series of such steady-flow experiments is a static I_D/V_D characteristic as shown in Fig. 10-15 (*left*). Here the variation of drain current with drain voltage is obtained for a set of gate biases. Three characteristics are of primary importance: the output resistance R_{out}, the mutual conductance g_m, and the open-circuit voltage gain G_v. These are defined as follows:

$$R_{out} = \frac{\Delta V_D}{\Delta I_D} \text{ for constant } V_G \tag{10-54a}$$

$$g_m = \frac{\Delta I_D}{\Delta V_G} \text{ for constant } V_D \tag{10-54b}$$

and

$$G_v = \frac{\Delta V_D}{\Delta V_G} \text{ for constant } I_D$$

$$= g_m R_{out} \tag{10-54c}$$

The normal operating point of a FET is in the flat saturation region of the characteristic. For the device shown, it is at $V_D = 3.0$ V and $V_G = -1.02$ V. Normally the position of the start of saturation (the "knee" in the I_D/V_D curve) and the output resistance would be obtained from three computer runs starting from the charge-neutral condition:

$$\begin{aligned} &\text{RUN 1}\ V_G = -0.8\,\text{V}; \quad V_D = 0.5, 1.0, 2.0, 3.0\,\text{V} \\ &\text{RUN 2}\ V_G = -1.02\,\text{V}; \quad V_D = 0.5, 1.0, 2.0, 3.0\,\text{V} \\ &\text{RUN 3}\ V_G = -1.24\,\text{V}; \quad V_D = 0.5, 1.0, 2.0, 3.0\,\text{V} \end{aligned} \tag{10-55}$$

The computer program can be organized to increase the drain voltage in steps every 300 to 400 timesteps since this is the time necessary to reach a steady-flow condition. In this way the flow conditions obtained from the last pair of voltages become the input condition for the next pair, and transients are reduced to a minimum. Each point on the characteristic requires 10 to 15 min of computer time on the IBM 360/195.

The values of V_G above are those used as boundary conditions in the computer simulation and are voltages in the GaAs just adjacent to the gate. Since the computer simulation does not include the Schottky barrier potential of 0.8 V between the aluminum gate and the GaAs material, 0.8 V must be added to these values to obtain the equivalent laboratory voltages. The source and drain are ohmic contacts with no potential barrier and hence the values of V_D and V_S are the same in the laboratory as in the simulation.

A good FET will maintain a high g_m for low output currents, hence it is important to measure the variation of g_m as the gate voltage is made more negative. This information can be obtained by a further computer run starting from the normal operating point whose particle coordinates will have been stored:

$$\text{RUN 4}\ V_D = 3.0\,\text{V},\ V_G = -1.24, -1.40, -1.55, -1.70, -1.80\,\text{V} \tag{10-56}$$

The result of such an experiment is shown in Fig. 10-15 (*right*). The mutual conductance is the slope of the I_D/V_G curve. An objective of device design is to obtain high g_m for low I_D since this produces both high gain [see Eq. (10-54*c*)] and low noise (see Sec. 10-4-5). Most devices however show a reduction of g_m as the current is reduced and in Sec. 10-4-5 we give the results of a parameter optimization study designed to reduce this effect. This is typical of the role computer simulation can play in device design.

The computation of the characteristics from RUNS 1 to 4 for one device would take about 3 hours on the IBM 360/195. A more limited set of points, for example omitting RUNS 1 and 3, would obtain the R_{out} and g_m in about $1\frac{1}{2}$ hours. The characteristics obtained in Figs. 10-15 are typical of those measured in the laboratory.

10-4-2 Looking Inside the FET

The measurement of static characteristics is a standard laboratory procedure. Our interest in making the measurements on simulated devices is for validation and design purposes. The computer simulation, however, provides something entirely new, namely the ability to look inside the FET in a way that is quite impossible in

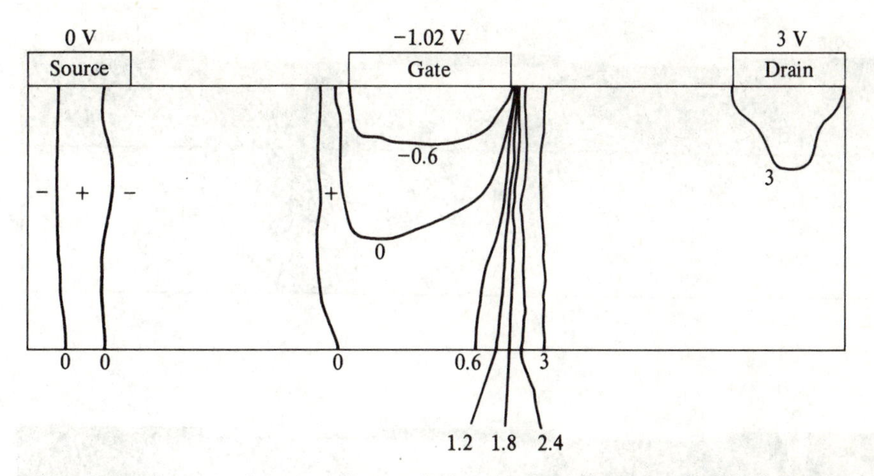

Figure 10-16 Contours of electrostatic potential in a simulated FET. Note the high-field region just to the right of the gate that drives Band I electrons to Band II to give the distributions shown in Fig. 10-17. (*After R. A. Warriner, 1977c, courtesy of Solid-State and Electron Devices, © Institution of Electrical Engineers.*)

the laboratory. At every timestep the coordinates of all particles are available together with the distribution of electrostatic potential throughout the device. This information can be analyzed to give a complete description of the distribution function and electric field. This detailed knowledge of what is happening inside the device provides the data for improved physical insight into the device behavior.

As an example we show in Fig. 10-16 contours of potential in a device at its normal operating point. One can see immediately that the fields are low except within about 0.3 μm of the drain side of the gate. In this small distance the whole gate-drain potential difference of 4 V is dropped. This is an average field of 130×10^5 V m^{-1}, which is sufficient to throw nearly all of the conduction electrons into Band II. This phenomenon is confirmed by an examination of the spatial distribution of all electrons. This is shown in Fig. 10-17. The upper picture shows the distribution of all electrons, the middle picture Band I electrons, and the bottom picture Band II electrons. The Band II electrons are seen to exist only just to the right of the high-field region, centered approximately midway between the gate and the drain. This result is in contrast to the conclusions drawn from diffusion model simulations that assume the velocity adjusts instantaneously to the field according to a $\bar{v}/E$ curve like Fig. 10-7. In these simulations the accumulation of electrons is seen in the high-field region immediately to the right of the gate. The neglect of transient relaxation effects (Sec. 10-2-5) in these models leads to an error in the position of the accumulation by about 0.5 μm.

Another conclusion to be drawn from Fig. 10-17 is the unexpectedly large current flowing through the substrate. This leads to a soft cutoff of I_D/V_G and poor values of g_m for low I_D, as previously noted (Sec. 10-4-1). The computer simulation has shown that the reason for the poor performance is conduction in the substrate. This result was not expected by traditionally trained electronics engineers because

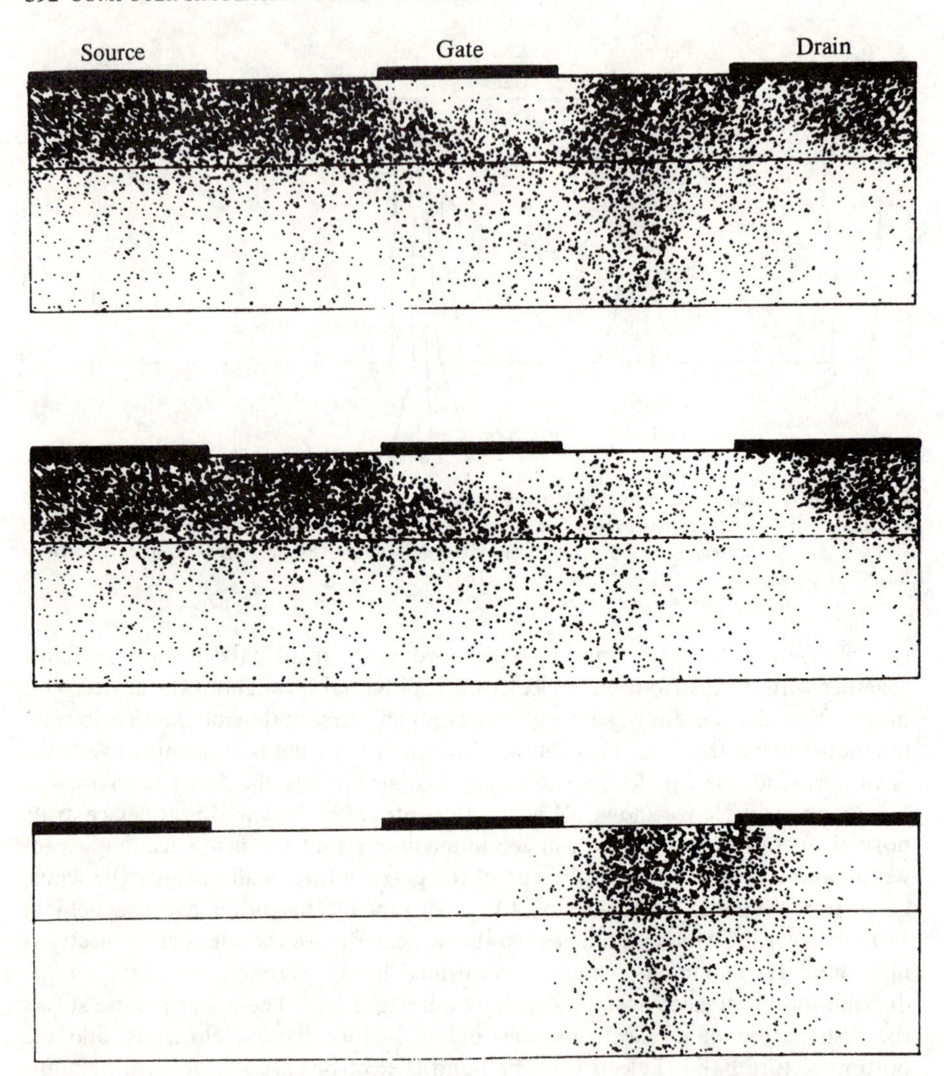

Figure 10-17 The distribution of electrons at the normal operating point ($V_D = 3.0$ V, $V_G = -1.02$ V). *Top,* all electrons; *centre,* Band I electrons; *bottom,* Band II electrons. Note the Band II electrons in the high-field region between gate and drain and the substantial current flowing in the substrate. The epitaxial/substrate boundary is shown by a horizontal line. (*From Hockney, Warriner, and Reiser, 1974, courtesy of Electronics Letters, © Institution of Electrical Engineers.*)

undoped substrates have a high resistance to current flow when measured in the laboratory. However, the computer experiment shows that, if the potential distribution is favorable, substantial currents will flow in the substrate.

The computer result has suggested the inclusion of a *p*-type buffer layer between the substrate and the epitaxial layer in order to form a potential barrier to the flow of electrons into the substrate. A series of computer experiments has been performed (see Sec. 10-4-5) to select the best thickness and doping level of the buffer layer prior to the manufacture of the device in the laboratory. Again we see the important interplay between the computer simulation and the laboratory research program.

The very rapid change of potential seen in Fig. 10-16 leads one to question whether the spatial mesh has sufficient resolution to describe the change. With a mesh spacing $H = 0.02\,\mu\text{m}$, there are 15 mesh cells in the $0.3\,\mu\text{m}$ over which the potential changes and 50 mesh cells across the gate-drain gap. This will give an adequate representation of the potential variation. However, in a model with significantly fewer mesh cells between gate and drain, say ten or less, the mesh cell would be comparable with the region over which the change takes place. Since no change of potential can take place in less than a few mesh cells, a coarse mesh of this type will incorrectly broaden the change. This will lead to an underestimate of the electric field and give an incorrect distribution of electrons in this important region of the device.

10-4-3 Dynamic Characteristics

The static characteristics discussed in Sec. 10-4-1 describe the steady currents that flow in response to constant (i.e., static) voltages applied to the electrodes. They describe the behavior of the device at zero frequency or in dc conditions. If we wish to know the behavior of the device under all conditions, it is necessary to know the current response to voltages of any frequency. One can then work out the response of the device to any voltage variation and one has a complete dynamic description of the device. The Y parameters form a convenient dynamic description and we will now show how they may be measured.

An FET is a two-port device. The input (port 1) is between the gate and source terminals, and the output (port 2) is between the drain and source terminals. The frequency-dependent Y parameters are defined by the matrix relation

$$\begin{bmatrix} \hat{i}_1(\omega) \\ \hat{i}_2(\omega) \end{bmatrix} = \begin{bmatrix} Y_{11}(\omega), Y_{12}(\omega) \\ Y_{21}(\omega), Y_{22}(\omega) \end{bmatrix} \begin{bmatrix} \hat{v}_1(\omega) \\ \hat{v}_2(\omega) \end{bmatrix} \tag{10-57}$$

where $\hat{v}_1$ and $\hat{v}_2$ are the amplitudes of small sinusoidal variations of frequency ω about the normal operating point, and $\hat{i}_1$ and $\hat{i}_2$ are the amplitudes of the small sinusoidal current responses. Some of these parameters, at zero frequency, have already been met as static characteristics

$$Y_{21}(0) = g_m \tag{10-58a}$$

$$Y_{22}(0) = (R_{\text{out}})^{-1} \tag{10-58b}$$

and

$$G_v = -\left(\frac{Y_{21}(0)}{Y_{22}(0)}\right) \tag{10-58c}$$

If we always keep the source at zero volts, then v_1 is the variation in gate voltage and v_2 the variation of drain voltage. In order to measure $Y_{p,q}(\omega)$ we should keep all voltages constant except v_q and measure the response current i_p, then

$$Y_{p,q}(\omega) = \frac{\hat{i}_p(\omega)}{\hat{v}_q(\omega)} \tag{10-59}$$

We could apply, in turn, different frequencies to port "q," but this would be very expensive in computer time. Instead we apply a step change in voltage, which contains components of all frequencies, to port "q," and obtain the Y parameter for all frequencies by Fourier transformation of the response current $\hat{i}_p$.

The Fourier expansion of a time-dependent function is (see Appendix A with $j = \sqrt{-1}$)

$$f(t) = \frac{1}{2\pi} \int_{-\infty}^{+\infty} \hat{f}(\omega) e^{j\omega t} \, d\omega \tag{10-60a}$$

where the Fourier transforms are given by

$$\hat{f}(\omega) = \int_{-\infty}^{+\infty} f(t) e^{-j\omega t} \, dt \tag{10-60b}$$

Since $v_q(t)$ is a step function

$$\begin{aligned} v_q(t) &= \Delta v_q \qquad \text{for } t \geqslant 0 \\ &= 0 \qquad \text{for } t < 0 \end{aligned} \tag{10-61}$$

the transform can be immediately found from

$$\hat{v}_q(\omega) = \int_0^{\infty} \Delta v_q e^{-j\omega t} \, dt = -\frac{\Delta v_q}{j\omega} \tag{10-62}$$

where, conventionally, ω is chosen to have a small negative imaginary part, thus ensuring that the value of the integral at the upper limit is zero. The response current at port "p" is measured and transformed numerically to give

$$\hat{i}_p(\omega) = \int_0^{+\infty} i_p(t) e^{-j\omega t} \, dt \tag{10-63}$$

The parameter is then found

$$Y_{p,q}(\omega) = \frac{\hat{i}_p(\omega)}{\hat{v}_q(\omega)} = j\omega \frac{\hat{i}_p(\omega)}{\Delta v_q} \tag{10-64}$$

The evaluation of $\hat{i}_p(\omega)$ requires some further explanation. Values of $I(t)$, the total current through the electrode, are obtained from $Q(t)$ by differencing using Eq. (10-51c). If a change is made from a steady-state current I_{ss1} to a steady-state current I_{ss2} in response to a step change in applied voltage, then the current variation about the initial state I_{ss1} is given by

$$i_p^{(n)} = \begin{cases} \frac{1}{2}(I^{(n)} - I_{ss1})\left(1 + \cos\dfrac{(n+8)\pi}{9}\right) & 1 \leqslant n \leqslant 10 \qquad (10\text{-}65a) \\ (I^{(n)} - I_{ss1}) & 10 < n < 0.8N \qquad (10\text{-}65b) \\ (I_{ss2} - I_{ss1}) + \frac{1}{2}(I^{(n)} - I_{ss2})\left(\cos\dfrac{(n-0.8N)\pi}{0.2N} + 1\right) & 0.8N \leqslant n \leqslant N \qquad (10\text{-}65c) \end{cases}$$

where $I^{(n)} = I(nDT)$ and $i_p^{(n)} = i_p(nDT)$ are the total current and the current variation at the nth timestep and there are N timesteps in total. The purpose of the cosine terms in Eq. (10-65*a*) and Eq. (10-65*c*) is to smooth the measured values into the initial and final states and is necessary to suppress aliases in the Fourier analysis of $i_p(t)$. After N timesteps (typically ~400) the final steady-state value is reached and subsequently we set

$$i_p(t) = i_p(T) = I_{ss2} - I_{ss1} \qquad t > T = NDT \tag{10-66}$$

The integral in Eq. (10-63) is then to be evaluated in two parts

$$\begin{aligned}\hat{i}_p(\omega) &= \int_0^T i_p(t)e^{-j\omega t}\,dt + \int_T^\infty i_p(T)e^{-j\omega t}\,dt \\ &= \int_0^T i_p(t)e^{-j\omega t}\,dt + i_p(T)\frac{e^{-j\omega T}}{j\omega}\end{aligned} \tag{10-67}$$

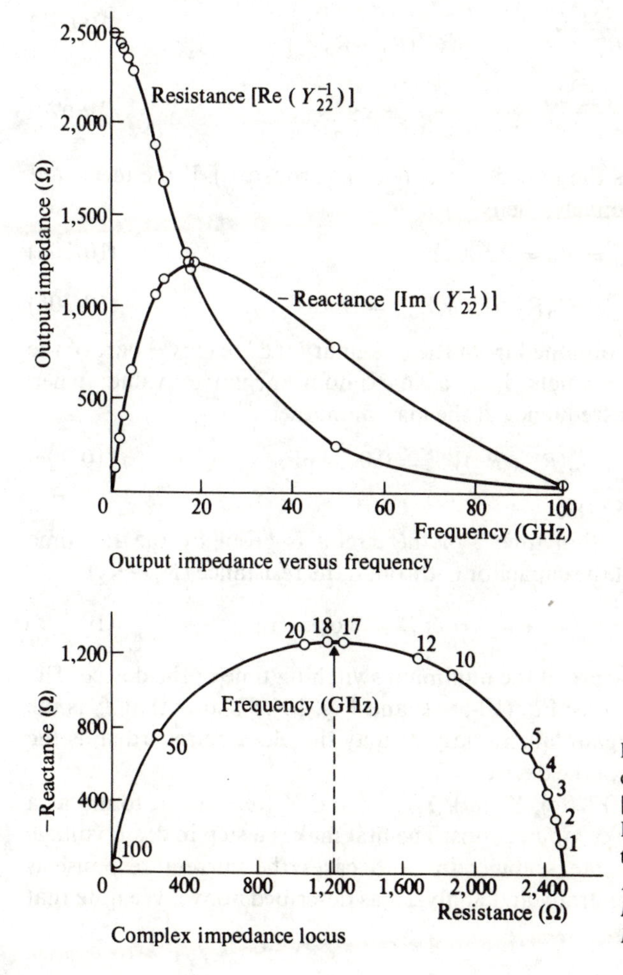

Figure 10-18 The frequency dependence of the output impedance = $[Y_{22}(\omega)]^{-1}$ obtained by the Fourier transform of the response to a step function. (*After Warriner, 1977c, courtesy of Solid-State and Electron Devices, © Institution of Electrical Engineers.*)

The first term above is evaluated by numerical quadrature as follows

$$\int_0^T i_p(t)e^{-j\omega t}\,dt = DT\sum_{n=1}^{N} i_p^{(n)}e^{-j\omega_l nDT} \tag{10-68a}$$

where
$$\omega_l = 2\pi l/T \qquad l = 0, 1, \ldots, N/2 \tag{10-68b}$$

As an example of this method we show the determination of the output impedance $[Y_{22}(\omega)]^{-1}$. A computer run of about 300 timesteps was adequate to reach the equilibrium steady-flow condition. The measured impedance, obtained by Fourier transformation of $i_2(t)$ and Eq. (10-64), is shown in Fig. 10-18. This response can be interpreted as a resistance R_1 in parallel with a series combination of a resistance R_2 and capacitor C.

The complex impedance of this circuit is

$$\begin{aligned} Z_{\text{out}} &= [Y_{22}(\omega)]^{-1} = \{R_1^{-1}+[R_2+(j\omega C)^{-1}]^{-1}\}^{-1} \\ &= \text{Re}\,(Y_{22}^{-1})+j\,\text{Im}\,(Y_{22}^{-1}) \end{aligned} \tag{10-69a}$$

where
$$\frac{\text{Re}\,(Y_{22}^{-1})}{R_1} = \frac{1+\omega^2C^2R_2(R_1+R_2)}{1+\omega^2C^2(R_1+R_2)^2} \tag{10-69b}$$

and
$$\frac{\text{Im}\,(Y_{22}^{-1})}{R_1} = \frac{-\omega CR_1}{1+\omega^2C^2(R_1+R_2)^2} \tag{10-69c}$$

At low and high frequencies the impedance is real (i.e., resistive). If the measured values are R_0 and R_∞, respectively, then

$$R_1 = R_0 = 2{,}500\,\Omega \tag{10-70a}$$

and
$$R_2 = R_0R_\infty/(R_0+R_\infty) = 25\,\Omega \tag{10-70b}$$

The capacitance can be obtained from the imaginary (i.e., reactive) part of the impedance, Eq. (10-69c), which has a maximum negative value when $\omega C(R_1+R_2) = 1$. If f_m is the frequency at the maximum then

$$C = [\omega_m(R_1+R_2)]^{-1} = 0.0034\,\text{pF} \tag{10-70c}$$

since $\omega_m = 2\pi f_m$ and $f_m = 18\,\text{GHz}$.

The characteristic relaxation time τ of the circuit is given by the RC time constant for the discharge of the capacitor C through the resistance (R_1+R_2),

$$\tau = C(R_1+R_2) = \omega_m^{-1} = 8.6\,\text{ps} \tag{10-70d}$$

The relaxation time is a measure of the minimum switching time of the device. The gain is proportional to Y_{22}^{-1} [see Eq. (10-58c)] and Fig. 10-18 shows that f_m is the frequency above which the gain falls rapidly. It may therefore be regarded as the maximum usable frequency of the device.

The four Y parameters $Y_{11}(\omega)$, $Y_{12}(\omega)$, $Y_{21}(\omega)$, and $Y_{22}(\omega)$ can be found as a function of frequency in two computer runs. The first makes a step in drain voltage and the second a step in gate voltage. In both cases the current response is measured at the gate and the drain, and analyzed as described above. We note that

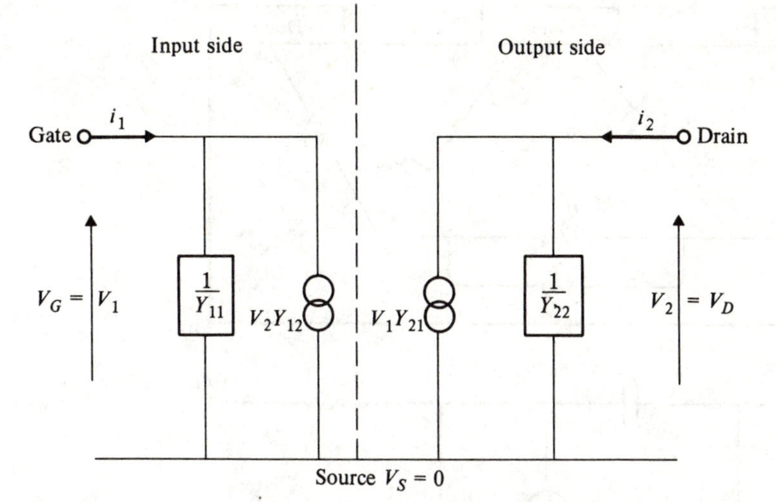

Figure 10-19 The equivalent circuit for the FET using the dynamic Y parameters. Given the four values $Y_{11}(\omega)$, $Y_{12}(\omega)$, $Y_{21}(\omega)$, and $Y_{22}(\omega)$ as a function of frequency ω, the small-signal response of the FET in any circuit can be obtained by standard circuit-analysis methods. (*Courtesy of R. A. Warriner, 1976.*)

such changes have already been computed during the measurement of the static characteristics (RUN 2 and 4) and the Y parameters can therefore be obtained by analysis of the transient responses obtained in these runs. In Fig. 10-15 we show the two step changes that provide the four Y parameters about the normal operating point. Knowing the four Y parameters, we can calculate the small-signal response of the FET in any circuit by standard circuit analysis techniques (Faulkner, 1966). The Y parameter equivalent circuit for the FET for use in this way is shown in Fig. 10-19. The response on the input side may be computed by regarding the FET as an input impedance Y_{11}^{-1} in parallel with a current source of strength $Y_{12}v_2$, and the response on the output side as an output impedance Y_{22}^{-1} in parallel with a current source of strength $Y_{21}v_1$. The Y-parameters for an FET have been obtained by Moglestue and Beard (1979) using the above techniques.

10-4-4 Lumped-Parameter Equivalent Circuit

Although the four Y parameters describe completely the small-signal response of a FET, they cannot be associated physically with particular parts of the device. For example, the imaginary part of Y_{22}^{-1} certainly represents a capacitative effect, but one has no information on where to change the geometry of the device in order to reduce it. The device designer is interested in an equivalent circuit that collects together locally the distributed resistances and capacitances into a small number of lumped parameters that are thereby associated with different regions of the device. One such equivalent circuit that is in common use is shown in Fig. 10-20. The top diagram shows the resistances and capacitors drawn on the parts of the

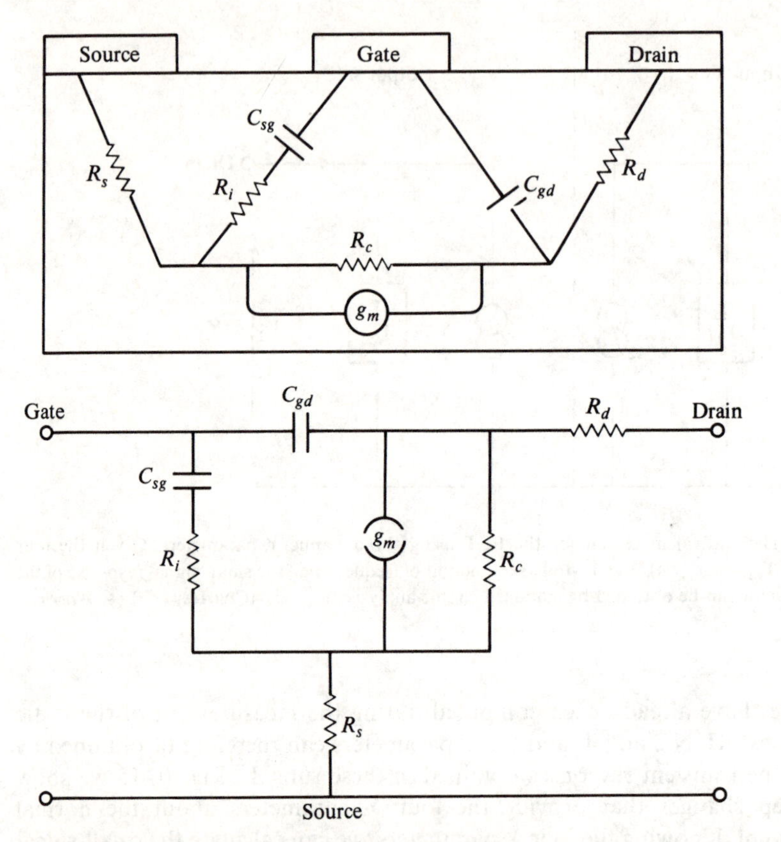

Figure 10-20 A lumped-parameter equivalent circuit for the FET. Each resistor or capacitor represents the distributed properties in a region of the device. (*After R. A. Warriner, 1977c, courtesy of Solid State and Electron Devices, © Institution of Electrical Engineers.*)

device with which they are associated. R_s and R_d are the resistances associated with the material between the source-gate and gate-drain respectively. Since these are essentially parasitic effects it is preferable to keep the conducting path in these regions wide and the associated resistances low. The region under the gate is referred to as the "channel" and here the conducting path is reduced to a small value by the negative potential on the gate. The channel resistance R_c is therefore large. The controlling action of the gate is represented by g_m, which means that the current passing in the channel is $g_m V_G$. The total capacity of the gate mentioned in connection with Fig. 10-14 is distributed between the source C_{sg} and drain C_{gd}.

The following procedure has been devised by Warriner (1977*c*) for measuring the lumped-parameter values

1. $(R_s + R_i)$ Starting from the charge-neutral condition and zero volts on all electrodes (see Fig. 10-13), a step of -0.8 V is applied to the gate. This sets up a device with two source paths to the gate. The equivalent charge flowing from

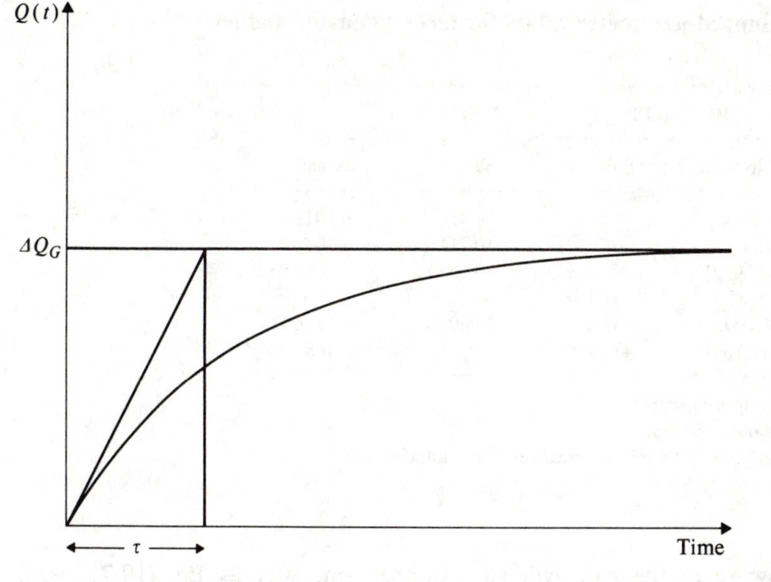

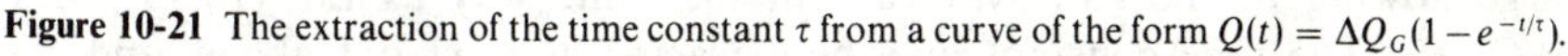
Figure 10-21 The extraction of the time constant τ from a curve of the form $Q(t) = \Delta Q_G(1-e^{-t/\tau})$.

the gate will be of the form

$$Q_G(t) = \Delta Q_G(1-e^{-t/\tau}) \tag{10-71}$$

where the time constant $\tau = \frac{1}{2}(R_s + R_i)C_{sg}$. The factor $\frac{1}{2}$ appears because of the duplicated source paths. The value of the time constant τ can be extracted by the least-squares fitting of the Eq. (10-71) or alternatively estimated very rapidly from the intercept of the initial slope of $Q(t)$ with its final value ΔQ_G, as is shown in Fig. 10-21. The capacity is obtained from its definition as the charge transferred per unit voltage, hence

$$C_{sg} = \frac{\Delta Q_G}{\Delta V_G} \tag{10-72a}$$

and

$$(R_s + R_i) = \frac{2\tau}{C_{sg}} \tag{10-72b}$$

It is not possible to separate R_s and R_i by this method, indeed when we look at the charge distribution in a device (Fig. 10-17) it is evident that the separation may have rather little meaning.

2. R_c Since $R_c \gg R_s + R_d$, the channel resistance R_c can be obtained from the slope of the static I_D/V_D (Fig. 10-15) near the operating point

$$R_c = \frac{\Delta V_D}{\Delta I_D} \tag{10-73}$$

3. R_d and C_{gd} Since $R_c \gg R_d$ the effect of a step change in drain voltage ΔV_D will be to charge the capacitor C_{gd} primarily through the resistor R_d. The equivalent

Table 10-2 Lumped-parameter values for three simulated and one real FET

Circuit parameter	Without IS		With IS	Real device
	No sub.	Sub.	Sub.	
$R_s + R_i$	2.8 Ω	2.8 Ω	9.8 Ω	~10 Ω
R_d	10 Ω	6 Ω	10.2 Ω	~10 Ω
C_{sg}	0.38 pF	0.36 pF	0.14 pF	0.1 to 1.0 pF
C_{gd}	0.02 pF	0.03 pF	—	~0.01 pF
R_c	750 Ω	857 Ω	1,000 Ω	~1,000 Ω
g_m	37.5 mS	43.2 mS	—	30 mS

IS: ionized impurity scattering.
Sub.: pure GaAs substrate.
No sub.: perfectly reflecting substrate–epitaxial interface.

charge flowing to the gate will vary in the same way as Eq. (10-71) with $\tau = C_{gd} R_d$. The time constant may be extracted as in step 1 above, and we obtain

$$C_{gd} = \frac{\Delta Q_G}{\Delta V_D} \tag{10-74a}$$

and

$$R_d = \frac{\tau}{C_{gd}} \tag{10-74b}$$

Table 10-2 compares the lumped parameters obtained from three simulated devices with values considered typical for a real device. The simulated devices are without impurity scattering and, not unnaturally, predict too small a value for the resistances R_s, R_i, R_d, and R_c, because of the omission of this important resistive effect. The comparison is made between a pure GaAs substrate and a device that reflects all electrons that are incident on the substrate boundary. The values obtained after the inclusion of ionized impurity scattering show very good agreement with values for similar laboratory devices.

10-4-5 Noise

One of the principal aims of FET design is the production of low-noise devices. The Monte-Carlo scattering simulation described in this chapter is particularly suited to the investigation of noise because all the physical processes that give rise to noise in the physical device are also present in the simulation. The noise is related to the fluctuations in the output current about its mean value that arise from the motions and scatterings of individual electrons. We note that all these effects are absent in the traditional diffusion equation simulation that treats the electron motion only through a given $\bar{v}/E$ characteristic (see Sec. 10-2-5). We give below the method of measurement suggested by Hockney.

The total current $I(t)$ at an electrode may be divided into three parts

$$I(t) = \bar{I} + \hat{i}(\omega)e^{j\omega t} + \delta i(t) \qquad (10\text{-}75)$$

where the first term represents the mean current due to the steady bias voltage, the second term a sinusoidal signal current of frequency ω and amplitude $\hat{i}(\omega)$, and the third term is a random current fluctuation due to the discrete nature of the electronic charge which we call the noise. The time averages of the signal and noise fluctuations are zero.

The total power due to this current can similarly be divided into three parts proportional to

$$\langle \bar{I}^2 \rangle \quad \text{the power dissipation due to the bias currents} \qquad (10\text{-}76a)$$

$$|\hat{i}(\omega)|^2 \quad \text{the signal power} \qquad (10\text{-}76b)$$

and

$$\langle (\delta i)^2 \rangle \quad \text{the noise power} \qquad (10\text{-}76c)$$

where the angled brackets represent a time average and the vertical bars the modulus of a possible complex value.

The quantity of prime importance is the ratio of the information-carrying signal power to the information-destroying noise power, hence we define the noise figure

$$\text{NF} = \frac{\text{(signal/noise) power at input}}{\text{(signal/noise) power at output}} \qquad (10\text{-}77)$$

as a figure-of-merit for a device, and seek to make this as small as possible.

Substituting the values for the power from Eqs. (10-76), we obtain

$$\text{NF} = \frac{|\hat{i}_G|^2}{\langle (\delta i_G)^2 \rangle} \div \frac{|\hat{i}_D|^2}{\langle (\delta i_D)^2 \rangle} \qquad (10\text{-}78a)$$

$$= \frac{\langle (\delta i_D)^2 \rangle}{\langle (\delta i_G)^2 \rangle} \times \frac{|\hat{i}_G|^2}{|\hat{i}_D|^2} \qquad (10\text{-}78b)$$

where the input, port 1, current is identified with the gate and the output, port 2, current with the drain.

The first factor in Eq. (10-78*b*) is the ratio of the mean-square current fluctuations at the drain to those at the gate. Both these quantities are directly measurable during a steady-flow computer run at the normal operating voltages, and can be obtained as a byproduct of the measurement of the device characteristics (see Secs. 10-4-1 to 10-4-3). The second factor relates the signal amplitudes at gate and drain and may be obtained from the measured Y parameters. The output signal current is maximum when the output is short-circuited to signal voltages ($v_D = v_2 = 0$). We then obtain from Eq. (10-57)

$$\hat{i}_1 = Y_{11}\hat{v}_1 \qquad (10\text{-}79a)$$

$$\hat{i}_2 = Y_{21}\hat{v}_1 \tag{10-79b}$$

Hence
$$\frac{|\hat{i}_G|}{|\hat{i}_D|} = \frac{|\hat{i}_1|}{|\hat{i}_2|} = \frac{|Y_{11}|}{|Y_{21}|} \tag{10-79c}$$

and finally, the minimum noise figure

$$\mathrm{NF}_{\min} = \left|\frac{Y_{11}}{Y_{21}}\right|^2 \frac{\langle(\delta i_D)^2\rangle}{\langle(\delta i_G)^2\rangle} \tag{10-80}$$

Equation (10-80) relates the noise figure to measured quantities in the simulated device at the normal operating point. The current fluctuations observed in the model will be magnified by many orders of magnitude due to the use of superparticles in the field calculation. However, this conversion factor between the magnitude of physical fluctuations and the magnitude of model fluctuations will cancel in Eq. (10-80). Hence the noise figure obtained from Eq. (10-80) will be the same for the model as for the physical device.

A problem with the direct measurement of noise using Eq. (10-80) is the range of frequencies over which the noise can be measured. If the timestep is DT and the noise is measured from a computer simulation of N timesteps over a time interval $T = N\,DT$, the maximum and minimum values of the frequencies that can be measured are

$$f_{\max} = \frac{1}{2DT} = 10{,}000\,\mathrm{GHz} \tag{10-81a}$$

$$f_{\min} = \frac{1}{T} = \frac{2}{N} f_{\max} = 20\,\mathrm{GHz} \tag{10-81b}$$

where we have substituted typical values of $DT = 5 \times 10^{-14}$ s and $N = 1{,}000$ timesteps. The maximum frequency is the highest frequency detectable before aliasing takes place when sampling at intervals of DT (see Sec. 7-3-2), and the minimum frequency just permits one period of the wave in the interval T and is therefore the lowest frequency measured when performing a Fourier analysis over that interval. The minimum frequency is approaching the operational frequencies of high-frequency FETs, but it is clear that excessively long computer runs would be required ($>$10,000 timesteps or 5 hours on the IBM 360/195) to measure the noise at lower frequencies.

An alternative, but less direct approach, is to relate the noise to the influence of current dipoles distributed in the channel of the device. A simple analytic theory, developed by Brewitt-Taylor, Robson, and Sitch (1978, 1980) concludes that the noise is reduced if g_m is high at low I_D. This type of relation can also be seen from Eq. (10-80) if we note that current fluctuations along the channel are like shot noise whereas current fluctuations transverse to the channel are like thermal Johnson noise. The former lead to fluctuations in drain current (hence $\langle(\delta i_D)^2\rangle \propto I_D$; Faulkner, 1966, page 108) and the latter to fluctuations in gate current (hence $\langle(\delta i_G)^2\rangle \propto k_B T$; Faulkner 1966, page 107). Remembering also that $Y_{21} = g_m$, we have

$$\mathrm{NF}_{\min} \propto I_D/g_m^2 \tag{10-82}$$

It is also found experimentally (Vokes et al., 1979) that the device noise temperature is approximately proportional to I_D/g_m.

A device optimization study has therefore been carried out by Sanghera and Chryssafis using computer simulation to optimize the doping density and geometry of an FET with the objective of producing the best g_m at low I_D, and thereby the lowest noise. The results are illustrated in Fig. 10-22. Four devices are compared. Device 0 consists only of an epitaxial layer with a reflecting lower boundary. This is an ideal device and gives the best behavior because no current is permitted to flow in the substrate of the device. However, it cannot be manufactured by the traditional processes described in Sec. 10-1-2. In order to approximate the ideal device a *p*-type buffer layer is inserted between the epitaxial layer and the substrate. This buffer layer is depleted of carriers and the negative doping charges repel incident electrons and provide a partially reflecting boundary. The geometry and doping densities for the devices are given in Fig. 10-22 (*left*) and the measured g_m/I_D curves in Fig. 10-22 (*right*). The ideal device 0 maintains a roughly constant g_m from 5 to 30 mA drain current. Device 6, while having the highest g_m at high currents, has a low g_m at low currents. This rapid change of mutual conductance as the operating conditions change is undesirable. In device 7 we have a much better behavior and a flatter g_m/I_D curve, however the g_m at low I_D (5 mA) is virtually the same as device 6. In device 8 we have quite a close approximation to the ideal behavior of device 0, and a significantly higher value of g_m at 5 mA than is found in devices 6 or 7. The above optimization study is typical of the way computer simulation can guide the choice of device geometry.

Encouraged by the above computer prediction that a device without a substrate, like device 0, would have a higher g_m and thereby lower noise, workers at the Royal Signals and Radar Establishment (Baldock) have developed a novel technology for producing such an ideal device (Vokes, Hughes, Wight, Dawsey, and Shrubb, 1979). The procedure starts with a traditional planar FET such as shown in Fig. 10-22 (*left*), comprising a doped $\sim 0.5\,\mu m$ epitaxial layer of GaAs supported on an undoped $\sim 0.7\,\mu m$ buffer layer of $Ga_{0.4}Al_{0.6}As$ and a substrate of chromiun-compensated GaAs. The upper surface of this device with its electrodes is then bonded to a glass support and inverted. Finally, both the substrate and the GaAlAs buffer layer are selectively etched away, leaving an inverted FET with no substrate bonded to a glass support. The device may be used in this inverted form (type B), or the exposed GaAs surface may be bonded to glass and the original bond removed, giving a normal noninverted device with electrodes on the upper free surface (type A). Alternatively, the inverted type B may have a second gate deposited on the exposed GaAs opposite the original gate (type C). In this case, the g_m is approximately doubled because the electron current is pinched between the two gates. The special etching and bonding techniques required in the manufacture of these novel devices are described by Griffiths, Blenkinsop, and Wight (1979).

Improved values of g_m are found for these devices, confirming the computer predictions; and the technology in principle allows any combination of source, gate and drain electrodes on both sides of a thin submicron active layer, leading to many possible new types of devices. The development of these new devices

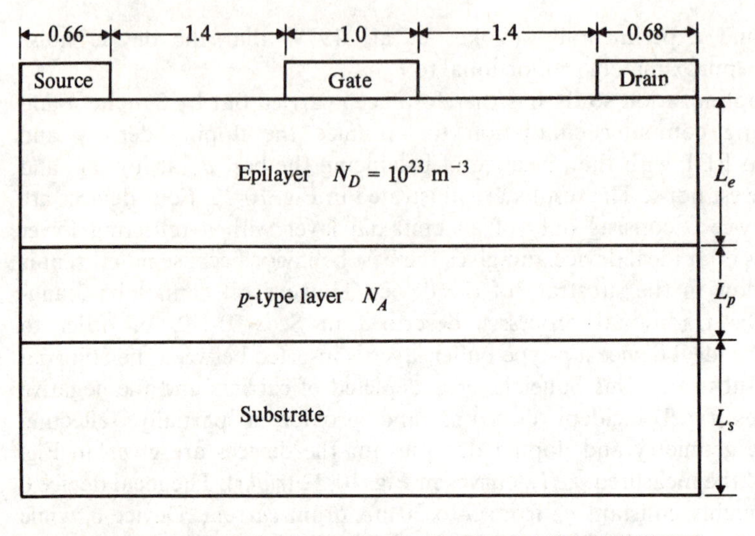

Device type	0	6	7	8
L_e	0.16	0.16	0.22	0.18
L_p	0.00	0.20	0.20	0.20
L_s	0.16	0.28	0.22	0.26
N_A (m^{-3})	–	10^{22}	10^{22}	10^{22}

Figure 10-22 The optimization of the density and thickness of a *p*-type buffer layer in order to produce high g_m at low I_D and thereby low noise. *Left*, device geometry, *right*, g_m/I_D curves. Dimensions in micrometers. (*Courtesy of G. S. Sanghera and A. Chryssafis.*)

confirms nicely the importance of a continuing program of computer simulation to any laboratory that is developing devices; for in this case the simulation has not only shown why the g_m of traditional devices falls off at low I_D because of current flow in the insulating substrate—a fact at first not believed by many device designers because substrates are deliberately made with a high resistance—but also suggested the direction in which it would be worthwhile to develop the manufacturing technology—that is to say, towards the manufacture of a device without a substrate.

The above experience bears out the statement made at the beginning of this chapter that the device manufacturer ignores computer simulation at his peril. Such a program of simulation requires the employment of a small team of computational scientists qualified to develop, modify, and use large simulation programs; and the provision of access to the fastest available computers on which to perform the simulations. Regrettably, circa 1979, few manufacturers provide this scale of computational backup to their product development, although the

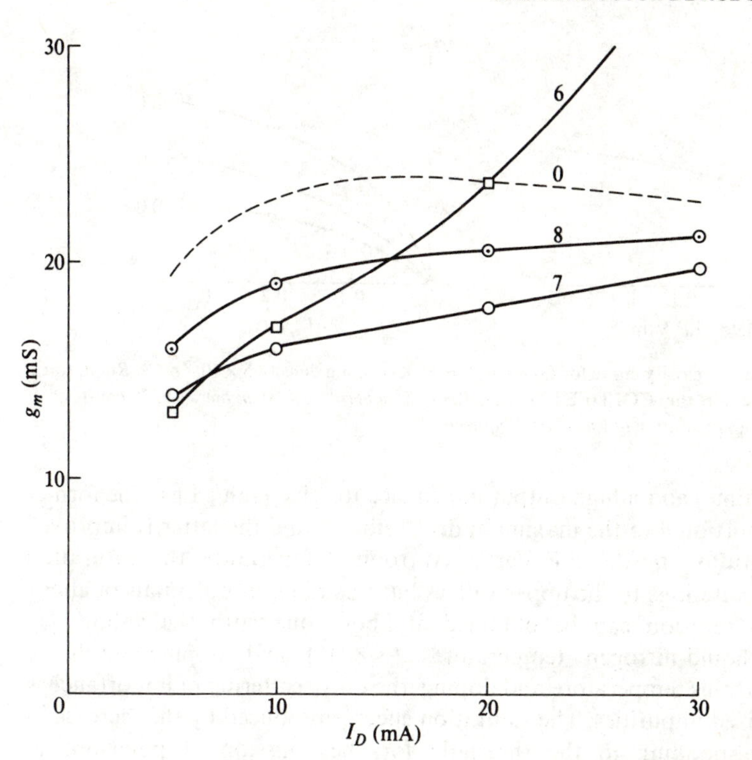

necessary computers, programs, expertise, and training have been available at several universities for many years.

10-4-6 The COLDFET

The extraction of the heat generated by the switching elements in a large-scale integrated (LSI) circuit is a major problem, particularly when the packing density of the logic is high. It is therefore a major aim of device design to build an FET switching element with substantially less power dissipation than the devices so far described. The device described in Sec. 10-4-1 has an output power dissipation of 60 mW (the product of the drain voltage and the current). We will now describe a device with approximately one-thousandth of this dissipation, the potential of which was first recognized by computer simulation. This particular device, the COLDFET, operates at liquid-nitrogen temperatures and therefore presents special problems in the laboratory; however, its investigation by computer simulation is the routine application of the methods of FET analysis described in Secs. 10-4-1 to 10-4-5. The investigation required about 3 hours of IBM 360/195 time and 3–4 man-months to complete. We report here the results published by Rees, Sanghera, and Warriner (1977).

For good performance an FET should have a high cutoff frequency (i.e., a

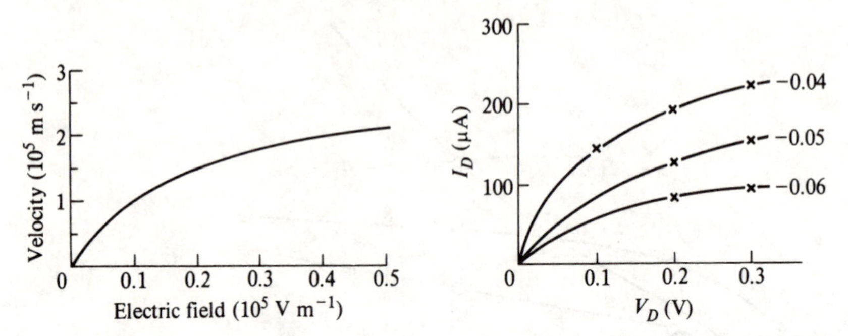

Figure 10-23 *Left*, drift velocity curve for GaAs at $T = 80$ K, doping density 5×10^{20} m^{-3}. *Right*, static I_D/V_D characteristic for the COLDFET. (*After Rees, Sanghera, and Warriner, 1977, courtesy of Electronics Letters, © Institution of Electrical Engineers.*)

short response time) and a high output impedance (to give high gain). The former is roughly proportional to the maximum drift velocity, and the latter is improved by better saturation in the $\bar{v}/E$ curve. At room temperature the saturation occurs due to scatterings to the upper valleys but this requires potentials of about 3 V. A weak saturation can be obtained at about one-tenth the voltage by operating at liquid-nitrogen temperatures (~80 K) and doping levels of 5×10^{20} m^{-3}. At this temperature and doping, the only scattering of importance is that due to ionized impurities. The saturation effect is produced by the increase in scattering corresponding to the threshold for the emission of polar-optical phonons. This shows-up as the bump in the scattering for Band I particles at $k = 0.022 \times 10^{10}$ m^{-1} in Fig. 10-6. All electron energies are low and are insufficient to scatter electrons to Band II. Thus at these temperatures GaAs acts like a single-valley material such as silicon.

The drift-velocity curve for GaAs at 80 K is shown in Fig. 10-23 (*left*). The saturation velocities obtained are similar to the room temperature curve of Fig. 10-7, hence the device will have a similar cutoff frequency. However, the low-field mobility of 10 m^2 V^{-1} s^{-1} is approximately 25 times that at room temperature, because of the much lower amplitude of lattice vibrations and consequential scattering. Of particular importance is the low level of the electric field required for normal operation, namely 0.6×10^5 V m^{-1} at 80 K compared with 30×10^5 V m^{-1} at 300 K.

The device simulated has the same geometry as the FET of Fig. 10-2 except that the epitaxial and substrate layers are both 0.32 μm deep. The static I_D/V_D characteristic for this device is shown in Fig. 10-23 (*right*). Reasonable saturation is obtained on the curve for $V_G = -0.06$ V at $V_D = 0.3$ V and output current $I_D = 100$ μA. The output power dissipation at this operating point is therefore 30 μW, which is two thousand times smaller than typical room temperature values. The computer simulation has thus been able to confirm the initial ideas of the device designer. The Y parameters have also been measured, and have permitted the prediction of the performance of the COLDFET in a logic circuit (Rees et al., 1977). A long program of laboratory work and further computer simulation will be

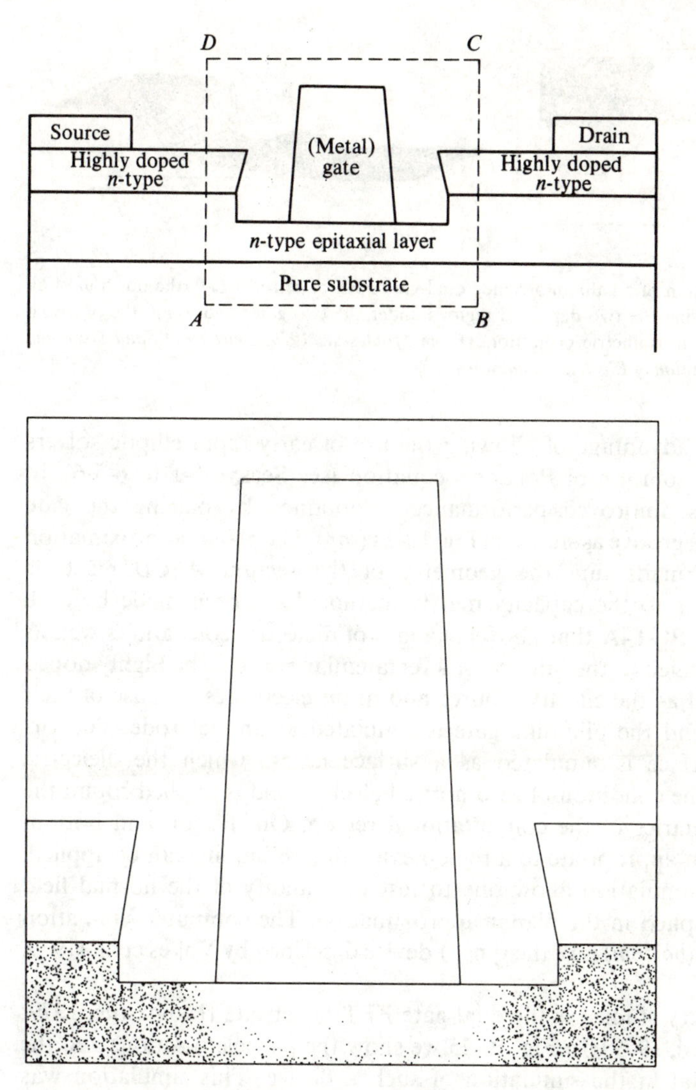

Figure 10-24 *Top*, the geometry of an etched-gate FET showing the region ABCD simulated on the computer and, *bottom*, the computer simulation. (*Courtesy of C. T. Moglestue and S. J. Beard.*)

required before this device becomes a marketable product. If it succeeds it will become a classic example of the value and role of computer simulation to the design and improvement of electronic devices.

10-4-7 Complex Geometries

In the FET simulations described up to now, the geometry has been simplified to a rectangle as shown in Fig. 10-2. This is a reasonable approximation for many

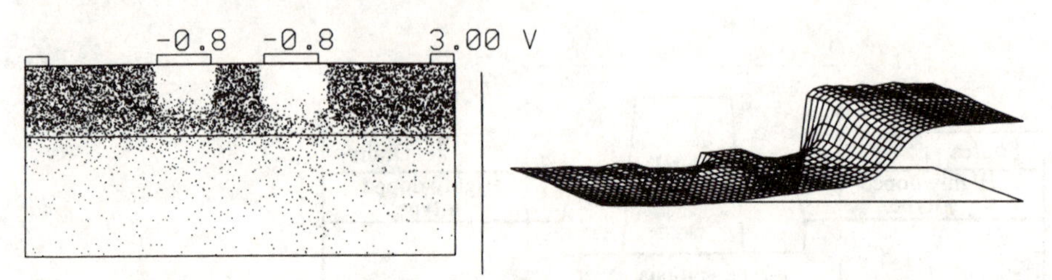

Figure 10-25 The simulation of a gallium arsenide dual-gate FET or tetrode. *Left*, the position of all simulated electrons, showing the two depletion regions under the two gates and, *right*, the potential distribution in the device in isometric projection. (*From Moglestue, 1979, courtesy of Solid-State and Electron Devices, © Institution of Electrical Engineers.*)

devices and has the advantage of allowing the use of early rapid elliptic solvers (e.g., POT4) for the solution of Poisson's equation (see Secs. 6-5-1 to 6-5-6). In some actual devices, improved performance is obtained by placing the gate electrode in an etched groove as shown in Fig. 10-24 (*top*). The planar approximation may no longer be made and the geometry of the section *ABCD* must be simulated. Extensions to the capacity matrix method have been made by S. J. Beard in a program POT4A that allows changes of dielectric constant as well as electrodes to be specified in the interior of a rectangular region. The highly doped *n*-regions are treated as the effective source and drain electrodes because of their high conductivity, and the pillarlike gate is simulated as an electrode. The air-semiconductor interface is simulated as a surface across which the dielectric constant changes. The condition of zero normal electric field is applied round the left and right boundaries of the computational region. On the top and bottom surface the condition appropriate to a region extending to infinity can be applied. The results of this simulation allow one to test the validity of the normal field condition that is applied in the planar approximation. The computer simulation shown in Fig. 10-24 (*bottom*) is of the type *A* device described by Vokes et al. (1979, Fig. 2*a*).

Another geometry studied is the dual-gate FET or tetrode (Moglestue, 1979; Moglestue and Beard, 1979). In Fig. 10-25 we show the particle positions and the electrostatic potential in the simulation of such a device. This simulation was performed by means of the potential solver POT4 used for the single-gate FET. The principal problems are those arising from the larger size of the device and consequently the larger number of simulated particles required in the simulation.

CHAPTER

ELEVEN

ASTROPHYSICS

11-1 INTRODUCTION

Astrophysics is a particularly fruitful field for the application of computer simulation because the systems under study, stars and galaxies, are not amenable to controlled laboratory experiment. With typical time and distance scales of 10^8 years and 10^{23} m, astrophysicists and astronomers can only act as passive observers of the systems they study. The only way experiments can be performed is by simulating the system on the computer and performing experiments with that simulation.

11-1-1 Stellar Evolution

Perhaps the most important contribution of computing to astrophysics has been to the study of stellar evolution (see, e.g., Hoyle, 1962, Chapters 8 and 9, for a readable introduction). This involves solving the partial-differential equations for heat transfer within a star simultaneously with equations determining the burning of the nuclear fuel. This study has been so successful that the life histories of most types of star, from birth, through main sequence and giant phases, to death as a white dwarf or neutron star, are thought to be quite well understood. As yet particle models have played no part in these developments, except in so far as they are used to determine the equation of state for the one-component plasma (see Chapter 12) which is a good model for certain classes of white dwarf stars (Horn, 1971).

11-1-2 The Gravitational *N*-Body Problem

The role of the particle model appears in the simulation of systems in which a star may be considered as a mass point with no other properties than gravitational attraction and mass—that is to say, nuclear burning, radiation, and convection, which are important in the study of stellar evolution, are not considered. The problem of determining the behavior of a collection of N mass points when they are moving under their mutual gravitational forces according to Newton's laws of motion is called the classical gravitational N-body problem. The evolution of clusters of stars or galaxies (each considered as a point mass) and the development of spiral and barred structures in single galaxies are all problems that can be approximated in this way.

11-1-3 Collisional and Collisionless Systems

N-body problems and the techniques used in their solution divide into three classes, depending on the importance of binary collisions in the evolution of the system. The binary collision rate measures the extent to which the orbit of a star in the local mean gravitational field is disturbed by the presence of individual neighboring stars.

For systems with a fixed total mass the binary collision rate ν_D is inversely proportional to the number of stars N into which that mass is divided (see Sec. 11-3-4). If N is doubled, although the number of scattering centers is also doubled, the force between any interacting pair of stars is divided by four (force is proportional to the square of the mass). The net result is that the total collisional effect is halved.

Chandrasekhar (1942, page 73) gives the formula for the binary collision time $T_D = \nu_D^{-1}$ in a three-dimensional system of point masses as:

$$T_D = \frac{v^3}{8\pi n G^2 m^2 H \ln\left(\dfrac{Dv^2}{2Gm}\right)} \tag{11-1}$$

where n = volume density of stars
m = mass of a star
v = relative velocity of stars
D = distance between stars
G = gravitational constant
H = 0.4 approximately

Since $m \propto N^{-1}$ and $n \propto N$ we have, ignoring the slight variation of the logarithm, $\nu_D = T_D^{-1} \propto N^{-1}$ as previously asserted.

If the system is composed of few stars, the orbit of any particular star depends primarily on the precise position and mass of its local neighbors and the system is said to be collision-dominated by these binary interactions. The acme of a collisional orbit is, of course, the binary star system in which two stars orbit around each other and only their mean motion responds to the mean field of the system. The evolution of a collision-dominated system is highly dependent on the masses of the individual stars, which must be properly accounted for.

If the system is composed of a large number of stars, as in a galaxy where

$N \approx 10^{11}$, Chandrasekhar (1942) estimates that the time for a stellar orbit to be deflected by a tenth of a right angle (or 9°) is of the order of 100 rotations of the galaxy. This represents about the limit of the time during which collisional effects can be regarded as negligible. It also corresponds to the estimated age of the universe, and hence also to the maximum age of a galaxy. For this reason galaxies are described as collisionless systems.

The quantities of interest in a galaxy simulation are the mean values of the mass density, potential, and field, averaged over hundreds or thousands of star separations. Hence the particles used in the simulation of a galaxy carry the mass of $\sim 10^6$ stars and their motion represents the transport of average mass density in their neighborhood rather than the orbit of any individual star. Binary star systems certainly form a significant proportion of stars in our galaxy but their detailed orbital motion is not represented in a galaxy simulation. Such stars are represented only in so far as their motion contributes to the local mean quantities. The problems associated with the time integration of tightly bound binaries, which are a major concern in small collisional systems, are therefore not important in collisionless systems.

The motion of a particle in a collisionless system is predominantly controlled by the local mean gravitational field and scarcely influenced at all by the particular distribution of other stars in its neighborhood. In fact, if one observes the motion of particles in a collisionless system, they may be seen to drift through each other like ghosts as they respond to the mean field. The particles appear to ignore each other and there are neither sharp changes of direction nor the formation of binary systems.

Thus the evolution of a collisionless system is determined by the mean mass density and not the masses of individual stars. Simulations are therefore usually carried out, for convenience, with a large number of model stars of equal mass, even though in a real galaxy stars have a distribution of masses. In a collisionless system, however, this distribution has no influence on the evolution of the local mean quantities.

11-1-4 Clustering of Stars and Galaxies

Actual stellar systems span the range from collision-dominated to collisionless. The open or galactic star clusters are groups of up to a few hundred stars (for example, the Pleiades and Hyades clusters, both containing about 100 stars) that lie close to the plane of the galactic disk. Such systems are obviously collision-dominated and simulations must integrate accurately the orbit of each star, taking into account its individual mass. Open clusters are composed of young stars with typical ages of $\sim 2 \times 10^8$ years, and are only about one-tenth the age of the galaxy. Collisional effects lead to significant loss of stars from these clusters.

At the other extreme, we have a complete galaxy of 10^{10} to 10^{12} stars which can be treated as a collisionless system and, in this case, the particles in the simulation are "superstars" with the mass of approximately 10^6 actual stars. This is a valid approximation for a collisionless system because, as we have seen above, the masses of particles in a collisionless system do not affect the evolution of the mean quantities. Intermediate between these two extremes lie the globular clusters. These are clusters of 10^4 to 10^6 stars, globular in shape, that lie out of the galactic plane. They form the most difficult group to simulate because N is not large enough for collisions to be ignored but is too large for the orbit of each star to be integrated in the same way as it is for the small galactic star cluster. Globular clusters are about as old as the galaxy.

If we look at the universe on the cosmological scale, we again have an hierarchy of cluster sizes (see Groth et al., 1977, and Hoyle, 1962, Chapter 11). Treating a particle now as representing a whole galaxy, our own Milky Way galaxy is part of a local group of several dozen galaxies that are gravitationally bound as a small cluster and lie within about 16 galactic diameters.† Our own local group of galaxies is near the edge of a cluster of small clusters known as the local supercluster. This is roughly 800 galactic diameters in extent and contains many thousands of galaxies. Above about 1,000 galactic diameters the probability of clustering fades. The total number of galaxies in the observable part of the universe is variously estimated but must be of the order 10^9.

11-1-5 The Big Bang

In order to understand the origin of the hierarchy of clustering one must consider cosmological theories of the origin of the universe. In the big-bang scenario everything originated in a colossal explosion about 15–20 billion‡ years ago when all the matter of the universe, mostly in form of radiation, was expelled at high speed in all directions from a small volume of space. As the gravitational attraction slowed the expansion, local variations of density were accentuated by local gravitational attraction and the matter of the superclusters was separated out. In turn, local variations of density in each supercluster led to the separation of clusters, which themselves separated into the matter of individual galaxies.

If the average mass density of the universe is less than a certain critical value, gravitational attraction is unable to stop the expansion and the universe is said to be open and will expand for ever. If, on the other hand, the density exceeds this critical value, gravitational attraction will eventually reverse the expansion and the universe will contract back to an enormously hot and dense condition, from which the process begins again with another big bang. The universe is then said to be closed. We mention this point now because both the examples of simulation given later in this chapter produce results that affect estimates of the mass density of the universe and hence have a bearing on the important cosmological question: Is our universe a closed or open system? (Groth, Peebles, Seldner, and Soneira, 1977, page 96.)

Returning now to the material that has separated out as an embryonic galaxy

† One galactic diameter in this discussion is 40 kiloparsecs (kpc) or 130,000 light years, which is roughly the diameter of our galaxy.

‡ One billion = 10^9.

or protogalaxy, one should imagine this as a rotating ball of gas evolving now as an independent system. Computer simulation will undoubtedly eventually show how the radius, mass, and angular momentum determine the subsequent evolution of the protogalaxy and lead to the various types of elliptical, spiral, and barred-spiral galaxies that are observed [see "The Hubble Atlas of Galaxies" (Sandage, 1961) or Hoyle 1962, Chapter 11]. It is plausible that large massive systems with little rotation develop into structureless elliptical galaxies and that systems with large amounts of rotation evolve into thin-disk galaxies, but none of this has yet been checked by simulation. The field is wide open for the computational astronomer. A case study later in this chapter is concerned with the evolution of spiral structure in a thin-disk galaxy, so we now give a possible scenario for the evolution of such a galaxy.

The first stars to condense from the rotating protogalaxy were the so-called "old" or "Population II" stars. These condensed while the gas was still roughly spherical and many were formed as large globular clusters of 10^4 to 10^6 stars. These clusters are distributed more or less uniformly throughout the sphere and form part of the halo of the galaxy. The Population II halo stars have very little rotational velocity and oscillate randomly from side to side of the spherical halo. In our simulations they are thought of as providing a fixed, imposed gravitational field through which the rest of the galaxy moves.

The remaining mass of the galaxy is still gaseous and subject to viscous damping of any relative motions. As the gas contracts under gravitational attraction, the conservation of angular momentum prevents contraction perpendicular to the axis of rotation to less than a certain radius. Contraction parallel to the axis of rotation, towards and past the equatorial plane, can however take place. Viscous effects eventually dampen out oscillations about the equatorial plane of the protogalaxy until one is left with a thin rotating disk of gas, with a thickness about one-twentieth of its radius. Stars born from this rotating disk are relatively young and are called the "Population I" stars. It is in these that gravitational instabilities produce the beautiful spiral and barred forms characteristic of the thin-disk galaxy. Most computer simulation has been devoted to the study of these instabilities. Some of the Population I stars form in small, open clusters that lie close to the plane of the rotating galactic disk. For this reason they are known as the "open" or "galactic" clusters and usually comprise a few hundred stars.

The above structure of a thin-disk galaxy can be seen in Fig. 11-1 (*top*), which shows the galaxy M104 taken by the 200-inch Hale reflecting telescope at the Mount Palomar Observatory. This galaxy is seen nearly edge-on and the disk is visible principally as a thin dark line arising from the absorption of light by dust particles that are an important constituent of the disk. Faint clumps of stars can also be seen in the disk, forming very tight spirals or ringlike structures. The halo, which is large and diffuse in this example, is seen as a spherical glow. The halo is not uniform and in the outer regions numerous blurred concentrations of stars can just be distinguished. These are the globular clusters. They form, like the rest of the halo, a spherical distribution about the center. Stars in our own galaxy which also

Figure 11-1 Two disk galaxies showing, *top*, a diffuse halo (M104, NGC 4594) and, *bottom*, a centrally condensed halo (NGC 4565). (*Photographs courtesy of the Hale Observatories, 200-inch Hale Telescope.*)

appear on the plate are distinguished by the horizontal and vertical lines radiating from their image. (Figure 11-1 (*bottom*) shows a disk galaxy with a small centrally condensed halo.)

11-1-6 Computer Simulation of Stellar Systems

If we consider computers with a floating-point arithmetic speed comparable with either the CDC 7600 or IBM 360/195 (for which a figure of half a microsecond per floating-point operation may be used to estimate computer time) then the methods most suitable for the simulation of small clusters of up to 1,000 stars, globular clusters of 10^5 stars, and galaxies of 10^{11} stars are respectively the PP, P^3M, and PM methods (see Chapter 1). A similar division based on the number of interacting bodies can be made for the simulation of clusters of galaxies of different sizes. The simulation of stellar systems, therefore, rather conveniently, makes use of all the techniques described in this volume. An excellent review of computer simulations of stellar systems is given by Aarseth and Lecar (1975).

The PP technique for the simulation of small clusters has been extensively developed by S. J. Aarseth and others since the early 1960s (von Hoerner, 1960; Aarseth, 1963; Henon, 1964). The force on a star is, of course, calculated by summing directly the interaction from all other stars and approximately $5N^2$ operations are required to obtain the force on all N stars of the cluster. Since we are concerned with less than 1,000 stars, the PP method can be used and has the important advantage of providing forces that are as accurate as the arithmetic precision of the computer. Some of the special time-integration techniques developed for the accurate integration of the stellar orbits in small clusters are described in Sec. 11-2.

The purpose of simulation is to obtain quantitative results on the evolution of star clusters, including the rate of escape of stars and the rate of binary formation, that can be used for age determination. The mechanisms for binary formation and disruption are studied by Aarseth (1971*b*) with systems of 250 and 500 simulated stars. If the initial positions are distributed randomly and uniformly within a sphere the evolution is dominated by differential mass segregation, which results in the formation of a high-density core of heavier stars surrounded by a diffuse halo of lighter stars. The effect of the equipartition of energy due to collisions is to impart high velocities to the lighter stars, some of which may reach the escape velocity and leave the cluster altogether. Others execute elongated orbits and form a halo. The central core suffers loss of particles by this mechanism and shrinks. It is in this higher-density core that the probability of binary formation becomes high. Binaries form either by multiple encounters (e.g., triple encounters) or by evaporation of stars from the core until only two stars remain. There is a long-term tendency for most of the energy to become concentrated in one heavy, tightly-bound binary. The timescale for this development is 10 to 40 crossing times, depending on the steepness of the mass spectrum. In this energy sink mechanism more than 50 percent of the total energy is absorbed by one heavy binary after 6 to 18 crossing times.

The techniques used for the simulation of small clusters of stars may obviously be applied equally well to the evolution of small clusters of galaxies. A series of computer experiments has been reported by Aarseth (1963, 1966, 1969) for a system of 100 galaxies both with and without initial rotation. The inclusion of rotation shortens the mean relaxation time and gives rise to a significant flattening. A large proportion of the angular momentum is transferred to the halo and the flattening remains frozen in. It is found that there is a significant correlation between the amount of flattening and the angular momentum. Aarseth's model has also been used by Efstathiou and Jones (1979) to study the efficiency of the tidal torque mechanism for the generation of the above galactic rotation as the universe expands.

The PP method may be used for the simulation of systems of up to about 4,000 bodies, be they stars or galaxies. Groth et al. (1977) and Aarseth, Gott, and Turner (1979) report models with 2,000 and 4,000 galaxies respectively which are used to study cluster formation in an expanding universe. Similar experiments on much larger systems of 10,000 to 20,000 galaxies have been made possible by the P^3M algorithm (Hockney et al., 1973; Eastwood et al., 1980) and undertaken by Efstathiou and Eastwood (1981) (see Sec. 11-4). A review of N-body calculations on galaxy clustering and their impact on cosmology has been given by Fall (1979).

Alternative approximate methods have been developed for large systems such as globular clusters. These are based on taking the first four moments of the velocity distribution and obtaining a set of fluidlike Fokker–Planck equations (Larson 1970*a, b*). The basic assumption is that the one-particle distribution function adequately describes the system (i.e., the pair correlation representing the formation of a binary system is ignored) and that two-body encounters dominate the relaxation. The Fokker–Planck equations may also be modeled by statistical methods and two forms of such Monte-Carlo techniques have been developed by Henon (1971*a, b*) and Spitzer and Hart (1971*a, b*), Spitzer and Shapiro (1972), and Spitzer and Thuan (1972). Excellent agreement has been obtained between N-body calculations, the fluid equations, and the Monte-Carlo techniques (Aarseth and Lecar 1975). These approximate simulation techniques are outside the scope of this book and the reader is referred to the above references for further details.

The simulation of collisionless galaxies has received a lot of attention because the PM technique is capable of simulating up to a few hundred thousand particles. Even so, the earliest work by Lindblad (1960*a, b*) with 50 particles and more recently by Ostriker and Peebles (1973) with 200 particles have used the PP method. The PM method was introduced by Hockney (1967), using a modification of a program written for the simulation of hot gas plasmas (Hockney 1966*a, b*, and Chapter 9), and enabled systems of 2,000 particles to be simulated on the IBM 7090 computer. This model was unphysical because each particle represented an infinitely long rod of mass, but it did serve to demonstrate the power of the PM technique. The extension of the PM technique to the simulation of point masses confined to move in a plane was made by Hohl and Hockney (1969), who simulated 50,000 to 200,000 particles on the CDC 6600. The extension to three dimensions was made by Hockney and Brownrigg (1974). The latter two developments were made possible by the Fast Fourier Transform algorithm (Cooley and Tukey, 1965) which enabled Fourier transform techniques to be used for the solution of the gravitational potential (Hockney, 1970). Further improvements have been made to methods for the solution of the potential by James (1977), and are described in Sec. 6-5-5. Other three-dimensional models have been reported by James and Sellwood (1978), Hohl (1978), and Miller (1978). The use of PM models in the study of spiral structure is taken as a case study in Sec. 11-3.

An entirely different approach to the simulation of a collisionless system has been taken by Miller and Prendergast (1968). These authors discretize the four-dimensional phase space (x, y, v_x, v_y) into a (256, 256, 63, 63) raster of cells. Each cell is allocated one bit of storage and if this is set to one a star is present at this position in phase space. Similarly, a zero represents the absence of a star. Thus in this method no two stars may occupy the same position in phase space. The newtonian equations determine the motion through phase space of the bits that represent the stars. The method is therefore sometimes described as bit-pushing in phase space and has obvious attractions on computers such as the ICL Distributed Array Processor (Flanders et al., 1977; Hockney, 1979) which can be programed efficiently at the bit level. The claimed advantage is that the orbits of particles in discretized phase space can be made to close exactly and in this sense can be described as exactly collisionless. However, the coarseness of the discretization is an obvious limitation. The reported simulations have also been conducted with periodic boundary conditions (Miller, Prendergast, and Quirk, 1970) which clearly impose an unrealistic square symmetry on the solution. However this is a restriction that could easily be removed by using one of the isolated-system potential solvers mentioned above and in Chapter 6. Miller (1978) has also reported a galaxy model on the ILLIAC IV parallel computer (Hockney, 1977) that has simulated 112,000 stars on a $(64 \times 64 \times 64)$ spatial mesh.

11-2 SMALL CLUSTERS

We describe now the techniques suitable for the simulation of small clusters of less than about 1,000 stars, based primarily on the work of S. J. Aarseth.

11-2-1 The Force Law

The pair potential and pair force between two particles of mass m separated by a distance r is given by:

$$\phi(r) = -\frac{Gm^2}{(r^2+\varepsilon^2)^{1/2}} \tag{11-2a}$$

$$F(r) = -\frac{\partial \phi}{\partial r} = -\frac{Gm^2 r}{(r^2+\varepsilon^2)^{3/2}} \tag{11-2b}$$

The parameter ε is introduced to soften the interaction at short range. In these expressions neither the potential nor the force goes to infinity for zero separation, as is the case for the interaction of point masses. The potential goes to the finite value $-Gm^2/\varepsilon$ and the force approaches zero like $-Gm^2r/\varepsilon^3$ as r goes to zero. Such an interaction is characteristic of two clouds of mass of radius ε. At first ε was introduced into the simulation of small clusters of galaxies (Aarseth, 1963) and represented the finite size of the galaxies. It was soon found however that it was necessary to introduce a nonzero ε in the simulation of star systems, even though a point-mass simulation ($\varepsilon = 0$) would be a better physical model. This is because it is unsound computational practice to allow any variables to reach exceptionally large values as, in these circumstances, the time integration becomes inaccurate and arithmetic overflow may occur.

In order to estimate the maximum value of ε that can be taken without disturbing the physics, consider a system with total energy TE that has reached a final state in which all but two particles have escaped to infinity with no excess energy and that these two form a tight binary system with semimajor axis a. Then, since the particles at infinity have zero energy, we have all the energy of the system in the single binary, hence

$$\frac{Gm^2}{2a} = -TE \tag{11-2c}$$

If ε is selected less than a, then clearly the tightest binary will not reach separations for which the force law Eq. (11-2b) deviates significantly from that of the point mass interaction.

11-2-2 Time Integration

The major problem in the small cluster calculation is the accurate integration of the stellar orbits, particularly when tight binary systems are formed. Since relatively few particles are to be followed, it is possible to use higher-order schemes than that of the leapfrog (see Chapter 4). Many different schemes have been used, but higher-order polynomial methods appear to be the most efficient (Lecar, 1968). We summarize here that used by Aarseth (1971a). Since stars may have widely different velocities and accelerations, each star is assigned its own variable timestep, which is chosen to achieve a prescribed relative accuracy in its orbit. Pair forces are only evaluated if the stars involved have moved since the last evaluation, and the saving in force evaluations is considerable.

The integration of the stellar orbits uses a fourth-order polynomial for the force per unit mass on each particle at time t, of the form

$$\mathbf{F} = \mathbf{F}_0 + \mathbf{B}t_r + \mathbf{C}t_r^2 + \mathbf{D}t_r^3 + \mathbf{E}t_r^4 \tag{11-3a}$$

where t_r is the time since the last evaluation of the force on the particle in question. If this was at time t_0, then

$$t_r = t - t_0 \tag{11-3b}$$

$\mathbf{F}_0$ is the value of the force calculated for the particle at $t = t_0$ and the vector quantities $\mathbf{B}$, $\mathbf{C}$, and $\mathbf{D}$ are calculated to make $\mathbf{F}$ agree with the force obtained at the previous three evaluations. The coefficient $\mathbf{E}$ is not determined until the next evaluation of the force in the loop below (i.e., it acts as a correction term).

Each particle i carries a record of the time t_i of the last evaluation of the force upon it, and the current value of its individual timestep Δt_i. The particles are moved one at a time, as follows:

1. Find next particle, say α, that needs a force evaluation. Search for

$$\min_i (t_i + \Delta t_i) \tag{11-4a}$$

where $\alpha = i$ at the minimum.

2. Advance time to

$$t^* = t_\alpha + \Delta t_\alpha \tag{11-4b}$$

3. Change the position of all particles except α to the new time t^*, using only the first two terms of the force polynomial (11-3a)

$$\mathbf{F} = \mathbf{F}_0 + \mathbf{B}t_r$$

Two integrations with respect to time give the change in position as

$$\Delta\mathbf{r} = \dot{\mathbf{r}}_0 t_r + \tfrac{1}{2}\mathbf{F}_0 t_r^2 + \tfrac{1}{6}\mathbf{B}t_r^3 \tag{11-5}$$

4. Predict the new position of the particle α by two time integrations of the cubic expression for the force upon it:

$$\mathbf{F} = \mathbf{F}_0 + \mathbf{B}t_r + \mathbf{C}t_r^2 + \mathbf{D}t_r^3 \tag{11-6}$$

5. Evaluate the force per unit mass on particle α by summing contributions from all other particles, using their positions from step 3:

$$\mathbf{F}_\alpha = -\sum_{\substack{j=1 \\ j \neq \alpha}}^{N} \frac{Gm_j(\mathbf{r}_\alpha - \mathbf{r}_j)}{(|\mathbf{r}_\alpha - \mathbf{r}_j|^2 + \varepsilon^2)^{3/2}} \tag{11-7}$$

6. Use the new force on particle α to evaluate the coefficient **E** in the quartic expression for the force (11-3a). Correct the new position for particle α by adding in the movement due to the coefficient **E**.
7. Calculate the new velocity on the particle α by one time integration of the quartic expression for the force. If $\Delta\dot{\mathbf{r}}_0$ is the change in velocity then

$$\Delta\dot{\mathbf{r}}_0 = \mathbf{F}_0 t_r + \tfrac{1}{2}\mathbf{B}t_r^2 + \tfrac{1}{3}\mathbf{C}t_r^3 + \tfrac{1}{4}\mathbf{D}t_r^4 + \tfrac{1}{5}\mathbf{E}t_r^5 \tag{11-8}$$

8. Calculate the time to the next force evaluation for particle α from

$$\Delta t_\alpha = \left[\eta\left(\frac{\mathbf{F}_0 + \dot{\mathbf{F}}_0 \Delta t}{\frac{1}{6}\dddot{\mathbf{F}}_0 + \frac{1}{24}\mathbf{F}_0^{(\mathrm{IV})}\Delta t}\right)\right]^{1/3} = \left[\eta\left(\frac{\mathbf{F}_0 + \mathbf{B}\Delta t}{\mathbf{D} + \mathbf{E}\Delta t}\right)\right]^{1/3} \tag{11-9}$$

where Δt is the value of the last timestep taken by the particle. As a protection against instability, the new timestep is not allowed to exceed $\sqrt{2}$ times the old timestep. Equation (11-9) is evaluated separately for each component of the vectors, and the minimum taken. The dots (or alternatively a superscript) give the number of time derivatives.

The above loop has moved one particle (number α) accurately and has advanced time from t to t^*. The loop is now repeated at step 1 for the selection of the next particle for advancement, and this could again be α. The storage requirement is $28N$ and 10 decimal places are recommended (this means double precision on the IBM 370) for the coordinates, velocities, and times. The force is evaluated in single precision on the IBM 370. The number of operations required to execute the loop once, that is to say, for moving one particle for one timestep is approximately $250N$. The total time required will depend on the number of star movements required per second which, in turn, will depend on the nature of the stellar orbits. The total computer time for a given physical time t_{ph} is proportional to $N^2 t_{ph}$. The timestep selection parameter η is determined empirically and a value near 2×10^{-4} leads to satisfactory integration in the absence of extremely close encounters. Relative energy errors for a binary with eccentricity 0.92 are about 10^{-5}.

Improvements on the above basic N-body integration scheme have been made by Ahmad and Cohen (1973), who include the effects of near and far objects separately, and on different timescales. In this way the integration is speeded up and the method is more efficient than the basic method for $N > 50$. Using an improved version of this method, Aarseth has a model capable of moving a system of 4,000 galaxies.

If bound pairs of stars, called binaries, form during the evolution of the stellar system, as is usually the case, the above method of integration becomes dominated by the calculation of the very short timesteps that are necessary for the accurate determination of their orbits. An efficient and comprehensive simulation program must therefore make special provision for the recognition of such binaries and for the calculation of their orbits. A suitable technique for the special treatment of binary orbits is that of two-body Levi–Civita regularization generalized to three dimensions by Kustaanheimo and Stiefel (1965). This has been used for three-body problems by Peters (1968) and modified for the N-body problem by Aarseth (1971*a*).

In the regularization procedure the equations of motion are written separately for the center-of-mass motion and the relative motion of the pair. The very large internal force between the components of the pair cancels analytically in the expression for the center of mass motion. The equations for relative motion are dominated by the internal pair force, and the external forces on a well-separated binary act as a relatively small perturbation. The introduction of a fictitious time variable τ, obeying the differential relation $d\tau = dt/r$, where r is the separation of the pair, removes the infinity in the newtonian equations of motion due to the infinite force at zero separation. In terms of the fictitious time variable, the equations of relative motion are nonsingular and their time integration may be performed with equal timesteps $\Delta\tau$. As the separation gets smaller these timesteps, of course, correspond to shorter and shorter intervals Δt in physical time. About fifty timesteps are taken per revolution in the relative motion, the period of which is twice that of the original unregularized newtonian equations of motion. The fourth-order time integration procedure, described earlier, is used for the integration of the regularized equations of motion.

It is necessary to establish criteria for accepting a pair of particles for treatment by two-body regularization and for terminating such treatment. The particle is recognized for regularization if its timestep is decreasing and less than Δt_{min}. A search is then made of other particles in the neighborhood and the other companion of the pair is accepted if its separation is less than R_{min} and its force contribution dominates all others. The latter condition is necessary to ensure that one is not dealing with, for example, a triplet. The regularization treatment of a particular pair is stopped whenever the internal force of the pair ceases to be the major force on either partner, and the members of a pair may change their partners as a result of this test. This will happen if a binary is broken up by a passing third particle.

11-3 SPIRAL GALAXIES

11-3-1 Theories of Spiral Structure

The computer simulation of spiral structure in thin-disk galaxies provides a good example of the constructive interplay between simulation, theory, and observation that can take place during the investigation of a complex problem. In the morphological classification of observed galaxies given by Hubble (1926, 1936) and shown in Fig. 11-2, the elliptical, spiral, and barred-spiral shapes observed in the sky are arranged in a sequence based upon a gradual transition of form. Although Hubble used the terms "early" and "late" in describing forms to the left and right of his classification, it still remains an open question whether the classification implies a time evolution or whether the shape of a galaxy is primarily determined by the initial conditions (e.g., angular momentum, energy, and mass) present when the protogalaxy condensed from the expanding universe (see Sec. 11-1-5). Spiral galaxies form a high proportion of those observed and are therefore thought to be long-lived structures with lifetimes comparable to that of the galaxy and therefore to the universe itself. A typical estimate of spiral lifetime would therefore be many tens of galactic rotations.

The central question facing any theory of galactic evolution is how long-lived spiral features can persist when the stars at different radii rotate with significantly different periods of rotation. It is obvious that if spiral arms always contain the same stars, this differential rotation will convert any feature into an increasingly tight spiral in a few rotations. Such "matter spirals" cannot be long-lived and a more satisfactory point of view is that of Lin and his coworkers who regard the spiral features as waves in the mass density of the galaxy arising entirely from

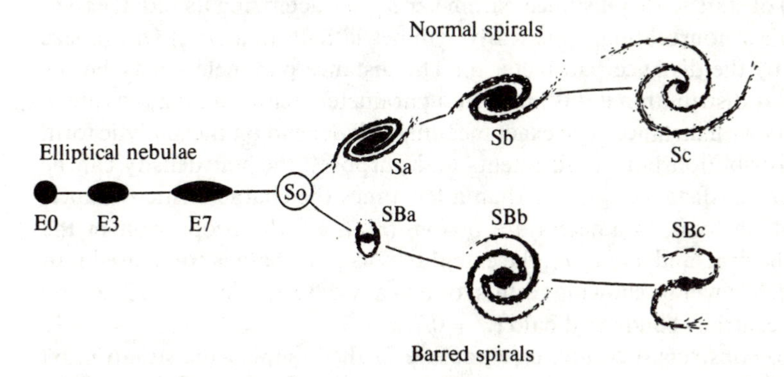

Figure 11-2 The Hubble morphological classification of galaxies. (*From Hubble, 1936, courtesy of Yale University Press.*)

gravitational effects. These density waves may travel at speeds different from that of the differentially rotating stars, in much the same way that sound waves move through the molecules of air. The winding-up problem is therefore avoided because the stars making up a given spiral pattern are continually changing.

The density wave theory of spiral structure has been given in a series of papers (see, for example, Lin and Shu, 1964; Lin, Yuan, and Shu, 1969; Lau, Lin, and Mark, 1976; Mark, 1974, 1977) which have been comprehensively reviewed by Toomre (1977). The expository paper by Lin and Lau (1979) is probably the best initial reference. The Lin–Shu theory of spiral structure is based on an asymptotic calculation that shows how spiral modes with certain characteristics (e.g., pattern speed, pitch angle) are self-consistent solutions in differentially rotating star and gas systems. The theory applies to tightly wound spirals with pitch angles less than thirty degrees, and gives the dispersion relation and growth rate of the waves.

A possible explanation of the generating mechanism for density waves, given by Lynden-Bell and Kalnajs (1972), is based on the fact that a trailing spiral arm produces a torque that transfers angular momentum outwards. For most galaxies, in which the angular velocity decreases with radius, such a transfer implies a reduction in the total rotational energy and therefore, by energy conservation, an increase in the energy of random motion. This latter represents an increase in entropy of the system and the theory appeals to the proposition, as in thermodynamics, that an isolated dynamical system tends to evolve in the direction of increasing entropy. Another mechanism based on the resonant interaction of particle orbits with a bar-like distortion has been given by Lynden-Bell (1979). As dynamical experiments cannot be performed on real galaxies, the role of the computer simulation is to provide experiments with computer models with which the above theories can be compared. A series of such experiments will now be described.

11-3-2 The Model

The idealized physical model of a thin-disk galaxy is shown in Fig. 11-3. A thin disk (mass m_d) of stars with a distance parameter a_d characterizing its radial extent rotates through a nonrotating, spherically symmetric halo (mass m_s) with a size characterized by the distance parameter a_s. The distance parameters may be the radius of the star distribution if it is finite, or a parameter characterizing the rate of decay of density with distance. The exact meaning will depend on the analytic form of the mass distribution but, to all intents and purposes, the star density can be regarded as zero at distances greater than a few times the characteristic distance. The important physical parameters are $\rho = m_d/(m_d+m_s)$, the proportion of the total mass in the disk, and $\alpha = a_d/a_s$, the extent to which the halo is condensed into a central bulge. We will be showing results for a heavy diffuse halo ($\rho = 0.2, \alpha = 1$) and for a light centrally condensed halo ($\rho = 0.7, \alpha = 3$).

In order to construct a computer model we further simplify the situation by regarding the halo stars as fixed, and represent their influence only as a fixed external gravitational field through which the disk stars rotate. The halo stars have

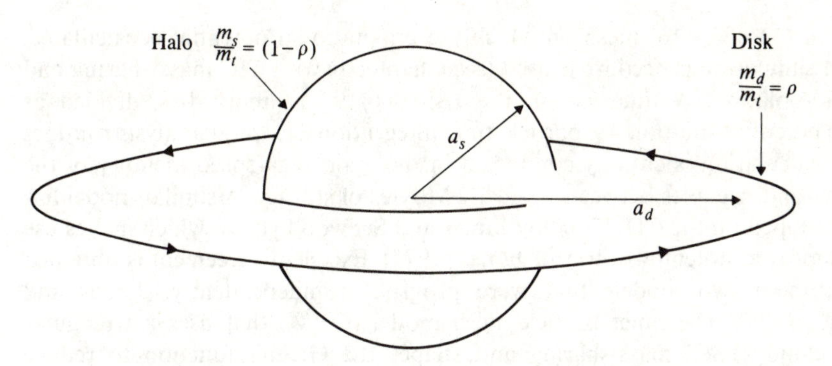

Figure 11-3 Idealized version of the structure of a disk galaxy, showing the disk and halo components and the variables $\rho = m_d/m_t$ and $\alpha = a_d/a_s$. $m_t = m_d + m_s$ is the total mass. (*After Hockney, 1979, courtesy of Contemporary Physics,* © *Taylor and Francis Ltd.*)

little rotational motion and show no discernible structure, and this approximation is good provided there is little energy exchange between the halo and the disk. In order to confirm this assertion some calculations have been performed by Hohl (1978) in which the halo and the disk are each represented by 50,000 moving particles. He observes that the halo remains essentially stationary, confirming the validity of the above approximation.

The computer model used by Hockney and Brownrigg (1974) is illustrated in Fig. 11-4 and consists of, typically, 20,000 to 25,000 simulated superstars moving

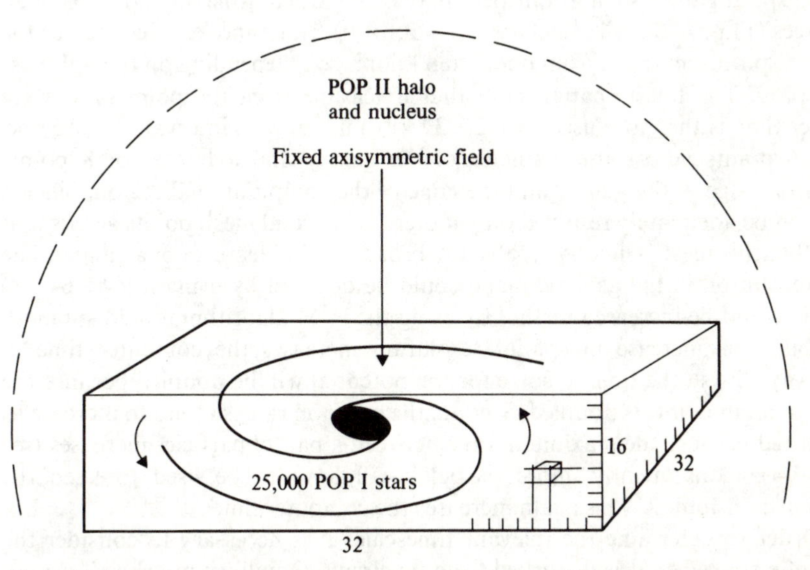

Figure 11-4 The three-dimensional computer model of a thin-disk galaxy developed by Hockney and Brownrigg (1974), showing the typical mesh used.

through a $(32 \times 32 \times 16)$ mesh on which the gravitational potential is calculated. The PM simulation procedure is used (see Chapter 1) with CIC mass-sharing and field interpolation. A timestep on the IBM 360/195 is about 10 s, divided as follows: potential solution 4 s, particle time integration 2 s, spiral analysis routines 4 s. The calculation is entirely conducted in the main high-speed memory of the computer and occupies between 1 and 1.2 Mbytes of storage. A similar model has been developed on the CDC 7600 by James and Sellwood (1978) which makes use of the improved potential-solver of James (1977). Excellent agreement is obtained between these two models that were programed independently (James and Sellwood, 1978). The quiet particle-mesh model (QPM) that uses a triangular shaped cloud (TSC) mass-sharing and shapes the Green's function to reduce further the mesh fluctuations (see Chapter 5) clearly has application to PM galaxy models. Any new models should employ these techniques.

The three modeling parameters to be chosen are the spatial mesh increment H, the timestep DT, and the total number of particles N. The space mesh must be related to the required spatial resolution and the timestep to the characteristic physical frequencies; the number of particles must be sufficient to make the system collisionless. Clearly one wishes to make H, DT as small as possible and N as large as possible, consistent with the requirements of a particular problem and the amount of computer time available. As always, a compromise must be made between the accuracy of the simulation and the cost, in computer time, of the calculation.

11-3-3 Choice of Timestep and Mesh Size

A typical spiral galaxy such as our own may be regarded as having a radius of 20 kiloparsecs (1 kpc $= 3.3 \times 10^3$ light years $= 3.1 \times 10^{16}$ km) and features such as the width of a spiral arm range from one to ten kiloparsecs depending on the tightness of the spiral. The finest spatial mesh that is feasible from the point of view of computer time is therefore used. A $(32 \times 32 \times 16)$ mesh, with interval $H = 2.5$ kpc, allows 16 points across the diameter of the galaxy and a border of 8 points between the edge of the galaxy and the edge of the computational region. Since a feature can be adequately resolved only if there are several mesh points across it, it is clear that this mesh will only resolve the broader spiral features of a galaxy. The spatial resolution in the galactic plane could be doubled by using a $(64 \times 64 \times 8)$ mesh and could be increased further by using the P^3M algorithm to add submesh resolution. This increase in spatial resolution increases the computer time in several ways. First, the time to solve for the potential will be doubled because the number of mesh points is doubled. Second, the collision rate is found to increase as H is reduced because the maximum force between a pair of particles increases (see Sec. 11-3-4). Consequently, more particles will have to be used to keep the simulation collisionless. This again increases the computer time.

In order to determine the relevant timescales it is necessary to consider the motion of a star when it is disturbed from its circular equilibrium orbit. If we use cylindrical coordinates (r, θ) lying in the plane of the galactic disk, the equation of motion for a star at radius r is

$$\ddot{r} - r\dot{\theta}^2 = F(r) \tag{11-10}$$

in which the radial and centripetal accelerations on the left-hand side are equated to the force per unit mass $F(r)$ due to gravitational attraction. If we consider an axisymmetric distribution of stars the force is independent of the axial coordinate θ, and the angular momentum per unit mass h is conserved

$$r^2\dot{\theta} = h \text{ (constant)} \tag{11-11}$$

Eliminating $\dot{\theta}$ from Eq. (11-10) and Eq. (11-11), one obtains

$$\ddot{r} - h^2/r^3 = F(r) \tag{11-12}$$

The steady-state solution to Eq. (11-12) is a circular orbit with $r = r_0$ and $\ddot{r}_0 = 0$, for which

$$-h^2 = r_0^3 F(r_0) \tag{11-13}$$

whence, eliminating h between Eq. (11-12) and Eq. (11-13), one obtains

$$\ddot{r} + F(r_0)\left(\frac{r_0}{r}\right)^3 = F(r) \tag{11-14}$$

If one now allows a perturbation r' about r_0 such that $r = r_0 + r'$ and performs a Taylor expansion to two terms about $r = r_0$, one obtains

$$\ddot{r}_0 + \ddot{r}' + F(r_0)r_0^3\left(\frac{1}{r_0^3} - \frac{3}{r_0^4}r'\right) = F(r_0) + r'\left.\frac{dF}{dr}\right|_{r=r_0} \tag{11-15}$$

Remembering that $\ddot{r}_0 = 0$, and canceling the terms $F(r_0)$, one obtains

$$\ddot{r}' = -\omega_e^2 r' \tag{11-16}$$

where $$\omega_e = \left(-\frac{3F(r_0)}{r_0} - \left.\frac{dF}{dr}\right|_{r=r_0}\right)^{1/2} \tag{11-17}$$

Equation (11-16) is the equation for simple harmonic oscillation and the frequency ω_e is called the epicyclic frequency. In Eq. (11-17) both $F(r_0)$ and its derivative are negative so that the epicyclic frequency is real. The motion of the perturbed star can be visualized as a circular rotation with frequency ω_e about a guiding center that moves along the unperturbed circular orbit. For stable integration with the leapfrog scheme we must insist that

$$\begin{aligned} &\omega_e DT < 2 && \text{for stability} \\ \text{and} \quad &\omega_e DT < 0.25 && \text{for reasonable accuracy} \end{aligned} \tag{11-18}$$

If $\Omega(r)$ is the angular velocity of the stars at radius r, then the balance of centrifugal and gravitational forces gives

$$F(r) = -r\Omega^2 \tag{11-19}$$

and $$\omega_e = \left(4\Omega^2 + 2\Omega r\frac{d\Omega}{dr}\right)^{1/2} \tag{11-20}$$

If the galaxy rotates as solid body then $\Omega(r) = \Omega_0$ is constant and

$$\omega_e = 2\Omega_0 \tag{11-21}$$

If, on the other hand, the stars move with constant azimuthal velocity v_0

$$\Omega = v_0/r \qquad d\Omega/dr = -v_0/r^2$$

and

$$\omega_e = \sqrt{2}\Omega = \sqrt{2}v_0/r \tag{11-22}$$

The rotation curve $\Omega(r)$ of most galaxies lies between these two extremes and we can conclude that a timestep that satisfactorily integrates the circular orbital motion will also satisfactorily integrate the epicyclic motion. The most troublesome situation may arise in the case of near-constant azimuthal velocity for small radii, when Eq. (11-22) predicts a very high epicyclic frequency. For this reason it may sometimes be necessary to abandon the integration of stellar orbits within a certain distance of the galactic center—stars within this distance being treated as part of the fixed halo. Since no structure is discernible in the central regions of disk galaxies, this approximation does not distort the physics. In a typical simulation one would choose between 100 and 200 timesteps per average rotation period, hence $\omega_e DT \cong \Omega DT \cong 0.05$.

A more severe constraint on the timestep is likely to arise from the requirement that a particle should not move more than the required spatial resolution in a timestep. If it does, then of course spatial features are blurred. If we are trying to achieve a resolution of about a mesh separation H, this means that a particle should not move more than about a mesh cell per timestep. Since the PM method cannot resolve changes over less than a cell, it is also true that there is nothing to be gained in resolution if the particle moves much less than a mesh cell per timestep. Hence a natural choice of timestep is likely to be $DT = H/v_0$ where v_0 is the average particle velocity. We note that this condition implies that the timestep be halved if the mesh cell is halved, which further increases the computational cost of increased spatial resolution discussed earlier. If we apply this criterion to a galaxy with a radius of 8 cells, we see that 100 steps per rotation correspond to two steps per cell and therefore satisfies the condition.

11-3-4 Collision Time and Particle Number

The number of particles used in a simulation must be sufficient to keep binary collisional effects unimportant for the duration of the computer experiment. No satisfactory theory exists for predicting the binary collision rate as a function of the parameters H, DT, and N of the computer model. We will, however, make a simple estimate based on the impulse approximation. If two particles of mass m pass each other with an impact parameter p and relative velocity v_{rel}, as shown in Fig. 11-5, the amount of perpendicular velocity $v_\perp$, acquired during the binary encounter may be obtained by equating the perpendicular change in momentum with the perpendicular impulse (force times time applied). The time of the encounter is taken as the time to travel a distance $2p$, hence

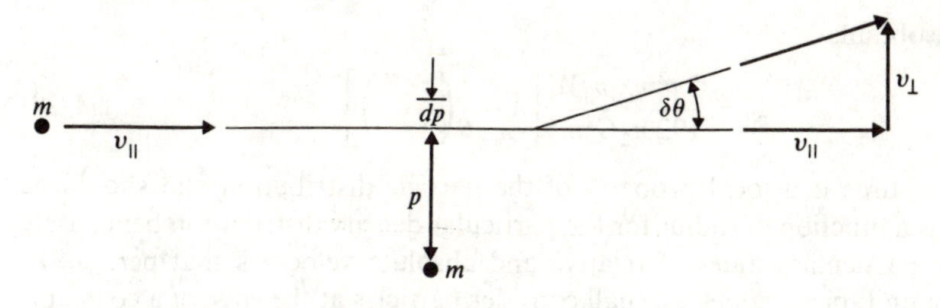

Figure 11-5 The variables used in the impulse approximation.

$$mv_\perp = Ft = F(p)\frac{2p}{v_{\text{rel}}} \tag{11-23}$$

and the deflection $\delta\theta$ caused by the encounter is

$$\delta\theta = \frac{v_\perp}{v_\parallel} = F(p)\frac{2p}{mv_{\text{rel}}v_\parallel} \tag{11-24}$$

where $v_\parallel$ is the absolute velocity of the particle.

If we consider the disk to be very thin and to have surface density n_2, the flux of particles with impact parameters between p and $p+dp$ is $2n_2v_{\text{rel}}\,dp$. This is the number of binary encounters occurring per second and, if these are assumed to be statistically independent, we may calculate the average square total deflection $\langle\Delta\theta^2\rangle$ in a time Δt by summing the squares of the individual deflections and integrating over all impact parameters

$$\begin{aligned}\langle\Delta\theta^2\rangle &= \int (2n_2v_{\text{rel}}\,dp)(2Fp/mv_{\text{rel}}v_\parallel)^2\,\Delta t \\ &= \frac{8n_2}{m^2v_{\text{rel}}}\frac{\Delta t}{v_\parallel^2}\int F^2p^2\,dp\end{aligned} \tag{11-25}$$

The collision time τ_{coll} is defined as the time for the average deflection $\langle\Delta\theta^2\rangle^{1/2}$ to reach 90 degrees of $\pi/2$ radians. Hence

$$\tau_{\text{coll}} = \frac{\pi^2m^2v_{\text{rel}}v_\parallel^2}{32n_2}\left(\int F^2p^2\,dp\right)^{-1} \tag{11-26}$$

In evaluating Eq. (11-26) one must take into account that the mesh smooths the interaction between particles at short range, and that the simulated superstars interact like clouds of mass of width W. The interaction between two such clouds is approximately

$$F(p) = \begin{cases} -\dfrac{Gm^2}{W^3}p & p < W \\ -\dfrac{Gm^2}{p^2} & p \geqslant W \end{cases} \tag{11-27}$$

inserting this interaction into Eq. (11-26) and performing the integration from zero to p_{max}, one obtains

$$\tau_{coll} = \frac{5\pi^2}{192} \frac{v_{rel} v_{\|}^2 W}{n_2 G^2 m^2} \left[1 - \frac{5}{6} \left(\frac{W}{p_{max}} \right) \right]^{-1} \tag{11-28}$$

The collision time is a local property of the particle distribution and should be calculated as a function of radius for the particular density distribution being used, and for the particular values of relative and absolute velocities that pertain. In order to obtain typical values we shall consider particles at the edge of a constant-density disk of radius R with azimuthal velocity v_{rot}, then approximately

$$\frac{GmN}{\rho R^2} = \frac{v_{rot}^2}{R} \tag{11-29a}$$

where ρ is the fraction of mass in the disk, and the rotation period is

$$\tau_{rot} = \frac{2\pi R}{v_{rot}} \tag{11-29b}$$

We also take

$$p_{max} = R \qquad v_{rel} = \alpha v_{rot} \qquad v_{\|} = v_{rot} \tag{11-29c}$$

and note that $N = n_2 \pi R^2$. One then obtains

$$\frac{\tau_{coll}}{\tau_{rot}} = \frac{5\pi^2}{384} \frac{\alpha N}{\rho^2} \left(\frac{W}{R} \right) \left[1 - \frac{5}{6} \left(\frac{W}{R} \right) \right]^{-1} \tag{11-30}$$

The width of the particle, as in the case of a plasma simulation (see Chapter 9), may be taken as the pair separation at the maximum of the force of interaction. For both NGP and CIC we have $W = H$; on the other hand, if TSC charge-sharing or QPM methods are used, W will be larger ($2H$ to $3H$). If we assume that CIC is used, and take typical values for the other variables $\rho = 0.5$, $R = 8H$, and $\alpha = 0.1$, then

$$\frac{\tau_{coll}}{\tau_{rot}} = \frac{5\pi^2}{384} \frac{4}{80} \frac{N}{0.9} = \frac{N}{140} \tag{11-31}$$

We conclude, therefore, that in a typical case ~ 140 particles are required for each collisionless rotation. If we choose $N = 25{,}000$, as was done by Hockney and Brownrigg (1974) and James and Sellwood (1978), we have a collision time of ~ 200 rotations. This means that after 200 rotations the particle motions are completely determined by the unphysical collisions of the superparticles because the average deflection due to this cause will be 90 degrees. It is more realistic to regard the system as collisionless for only one-tenth of this time, i.e., ~ 20 rotations when—because it is the average squared deflection that is proportional to time—the average deflection will be ~ 30 degrees. Fortunately this corresponds to the end of most simulations. If simulations are required over longer times, significantly more particles must be used. Equation (11-30) also shows that the collision time increases as the fraction ρ of mass in the disk decreases. Hence the problem of computational collisions is less severe for problems with heavy halos.

Although the above estimate of the collision time is only approximate it does show the importance of keeping the total number of particles N as high as possible. We note that since $W = H$ for both NGP and CIC, their collision time from Eq. (11-30) is the same. Consequently, NGP would be the favored technique since it enables a larger number of particles to be simulated in the same computer time. Simulations with several million particles will be possible with NGP on parallel computers. Most authors, however, use CIC interpolation or variations of it, because of its superior energy conservation.

There are few measurements of the collision time in galaxy simulations and there has been no comprehensive study comparable to that of Hockney (1971) for electrostatic plasma models (see Chapter 9). Hohl (1973) has measured the energy relaxation time in a simulation of a disk galaxy and finds this to be between 560 and 1,700 rotations for a system of 100,000 superstars, divided into two mass groups. The energy relaxation time for a system of particles of different masses is the time taken for the average kinetic energy per particle in each mass group to become equal to the average kinetic energy per particle of the whole system. The equipartition of energy between the different mass groups is possible because of the energy exchange during a binary collision, and the energy relaxation time is therefore expected to be of the same order of magnitude as the collision time we have defined above. If the conditions for Hohl's experiment are inserted into Eq. (11-30) we obtain $\sim$400 rotations ($\rho = 1.0$, $N = 10^5$, $\alpha = 1$, $W/R = \frac{1}{32}$), which is in reasonable agreement with the measured values.

11-3-5 The Ubiquitous Bar Instability

The computer simulation of thin-disk galaxies has been in progress since about 1967–8 and an excellent review paper has been written by Hohl (1975) describing research up to 1974. The earliest work was concerned with the evolution of a "cold," infinitely thin, disk of stars with just sufficient rotation to balance centrifugal force against gravitational attraction. The disk is described as "cold" because the motion of all stars is initially purely azimuthal and there is no random velocity dispersion about this circular motion. If the surface mass density is

$$n_2(r) = n_2(0)\sqrt{1-r^2/R^2} \qquad r \leqslant R \tag{11-32a}$$

where R is the radius of the disk, then the disk is in balance when it is rotating like a rigid body with angular velocity

$$\omega_0 = \sqrt{Gn_2(0)/2R} \tag{11-32b}$$

Although the disk is initially in balance, it is an unstable equilibrium, and the disk breaks up into several pieces in only a few rotations. This is shown in Fig. 11-6 and is caused by the Jean's instability. This is the phenomenon whereby gravitational attraction enhances any local maximum in the star density. Toomre (1964) has shown that axisymmetric instabilities of this kind can be suppressed if there exists a minimum radial velocity dispersion given by

$$\sigma_{r,\min} = 3.36Gn_2/\omega_e \tag{11-33}$$

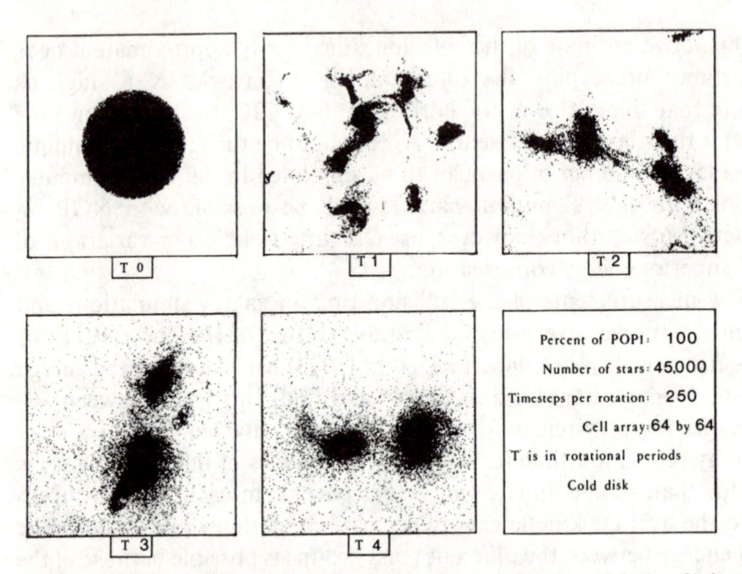

Figure 11-6 The rapid breakup of a cold disk without a halo ($\rho = 1.0$, $Q = 0$) due to Jean's instability. Time T is in units of the rotation period of the initial disk. (*From Hockney and Brownrigg, 1974, courtesy of Royal Astronomical Society Monthly Notices,* © *Royal Astronomical Society.*)

where (since $\langle v_r \rangle = 0$) $\sigma_r = \langle (v_r - \langle v_r \rangle)^2 \rangle^{1/2}$ is the root-mean-square radial velocity of stars at the radius r. If velocity dispersion is introduced, the disk is described as "warm" and the extent to which the Toomre criterion of Eq. (11-33) is satisfied is expressed by the local value of Q, the ratio of the actual radial dispersion to the minimum value required for stability

$$Q = \sigma_r / \sigma_{r,\min} \tag{11-34}$$

Many computer simulations have been performed with $Q = 1$ for a variety of initial density distributions and rotation curves, and the disk is found to be stabilized against the rapid breakup seen in Fig. 11-6 (see Hockney and Hohl, 1969; Hohl, 1971, 1972, 1975; Miller, 1971, 1976; Berman and Mark, 1979). However, in all cases a slowly growing, bar-shaped mode finally dominates the system as is seen in Fig. 11-7. Hohl (1975) has examined the evolution of mass density and angular momentum with radius as the bar mode grows. He finds that the bar transfers angular momentum outwards and that the final density distribution is well represented by a sum of two exponentials of the form

$$n_2(r) = c_s \exp(-r/a_s) + c_d \exp(-r/a_d) \tag{11-35}$$

where $\alpha = a_d/a_s = 6$. Although this experiment was conducted with a two-dimensional simulation, we have identified the terms as a nucleus with scale length a_s and a disk with scale length a_d, and take this as an example of the tendency of gravitational systems to seek centrally condensed (α large) mass distributions. This result is in line with the observations of Vacouleurs (1959) and Freeman (1970). The bar instability changes the distribution of Q, which was initially one

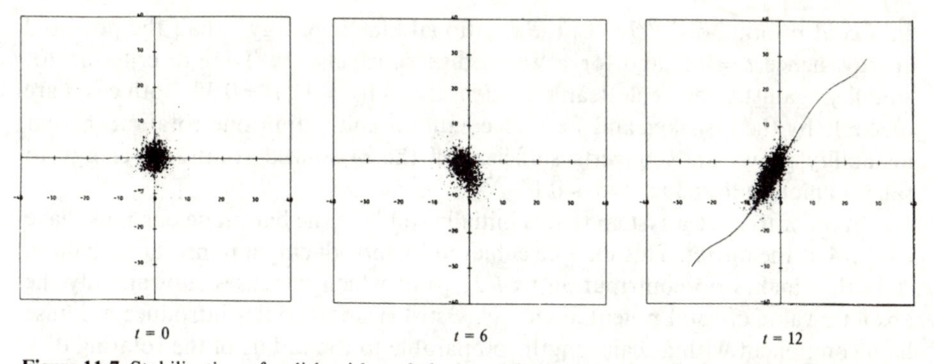

Figure 11-7 Stabilization of a disk without halo ($\rho = 1.0$) by random star motions ($Q = 1$), but note the final fate of a bar mode and the absence of spirals. Time in units 10^8 years. (*From Berman and Mark, 1979, courtesy Astronomy & Astrophysics, © Springer-Verlag.*)

everywhere, to a distribution varying from about one at the center to about seven in the outer regions. The period of rotation of the bar is approximately twice that of the initial disk.

It seems that the inevitable fate of a disk galaxy without a halo ($\rho = 1$) is to become a hot rotating bar, and it is not difficult to understand that barred galaxies form an important group of those seen in the sky (see Hubble classification; Fig. 11-2). In none of these experiments was there any evidence of long-lived spiral structure, and the final velocity dispersions in the computer systems exceed those actually observed in spiral galaxies.

In order to understand spiral structure it is obviously necessary to find some influence other than velocity dispersion to stabilize the Jean's instability and which will in addition stabilize the disk against bar formation. If the resulting disk is relatively cool, $Q \approx 1$, there will be some chance that spiral waves can grow. This new influence is that of the halo stars which we have ignored up to now. We will see below that spirals can be obtained with both heavy diffuse halos (Fig. 11-1 *top*) and light, centrally condensed halos (Fig. 11-1 *bottom*).

11-3-6 Conditions for Spiral Structure

The condition that a rotating star system be stable against the growth of the bar mode has been postulated by Ostriker and Peebles (1973) to depend on the ratio t of rotational kinetic energy to the absolute value of the potential energy

$$t = KE_{\text{rot}}/|PE| \tag{11-36}$$

The Maclaurin sequence of uniform-density, uniformly rotating fluid bodies (Lamb, 1932, Chapter XII) and many N-body computer simulations are unstable if $t > 0.14$. Ostriker and Peebles state the condition as follows:

> In the absence of counter-examples, it appears that $t = 0.14$ represents approximately the maximum rotational energy an axisymmetric stellar system can contain and remain stable to the formation of a bar.

In a cold rotating disk ($Q = 0$) the rotational kinetic energy is half the potential energy, hence $t = 0.5$, and for a warm disk satisfying the Toomre criterion for stability against small-scale Jean's condensations ($Q = 1$), $t = 0.35$. Both cases are unstable by the Ostriker and Peebles condition and within one rotation the bar instability grows and converts sufficient of the organized rotational energy to random motion to reduce t to ~ 0.14.

In order to have a system that is initially stable to the bar mode one must have $t < 0.14$ at the outset. This may be achieved by introducing a nonrotating halo of stars that makes no contribution to KE_{rot} but which increases substantially the absolute value of total potential energy. Ostriker and Peebles introduce a diffuse halo component with a scale length comparable to the radius of the rotating disk of stars, i.e., $\alpha \approx 1$, and find that if the halo mass is greater than about 67 percent of the total mass ($\rho < 0.33$) then initially $t < 0.14$ and no bar mode is observed. A halo of this mass is described as heavy or massive. Ostriker and Peebles' experiments were in three dimensions with 150 to 300 simulated stars and the halo component is represented by a fixed imposed external force. The stability condition also fits the simulation results of Hohl (1971) and Miller (1971) in which about 100,000 superstars were used.

The evolution of an initially cold disk with a heavy diffuse halo ($Q = 0$, $\rho = 0.2$, $\alpha = 1$) is shown in Fig. 11-8 (Hockney and Brownrigg, 1974; Berman, Brownrigg, and Hockney, 1978). First one notices that there is no evidence of bar formation even after 13 rotations. The cold starting condition ($Q = 0$) makes the system unstable to small-scale Jean's condensations and these take the form of dense tight spirals that persist throughout the experiment. These spiral features are not spiral waves but are material condensations that form, break, and reform in a complex way as they are twisted by the differential rotation of the galaxy. Although the qualitative overall spiral pattern has a long life, detailed features within it are rapidly changing. Many Sa and Sb galaxies (see Fig. 11-2) have filamentary features similar to those in Fig. 11-8 and may therefore be caused in a cold star distribution by a combination of Jean's condensation and differential rotation.

In order to identify a spiral wave the density distribution is Fourier-analyzed in azimuth θ around a number of annuli at different radii r, thus

$$n_2(r, \theta) = \sum_{m=0}^{\infty} A_m(r) \cos(\Phi_m - m\theta) \tag{11-37}$$

where Φ_m is the phase of the spiral mode with m arms and A_m/A_0 its fractional amplitude. The pattern speed of the spiral is

$$\Omega_p = \frac{1}{m}\frac{\partial \Phi_m}{\partial t} \tag{11-38a}$$

Figure 11-8 Evolution of a "cold" disk galaxy stabilized with a heavy diffuse halo ($Q = 0$, $\rho = 0.2$, $\alpha = 1$), showing the development of filamentary ring and spiral structures. (*From Berman, Brownrigg, and Hockney, 1978, courtesy of Royal Astronomical Society Monthly Notices, © Royal Astronomical Society.*)

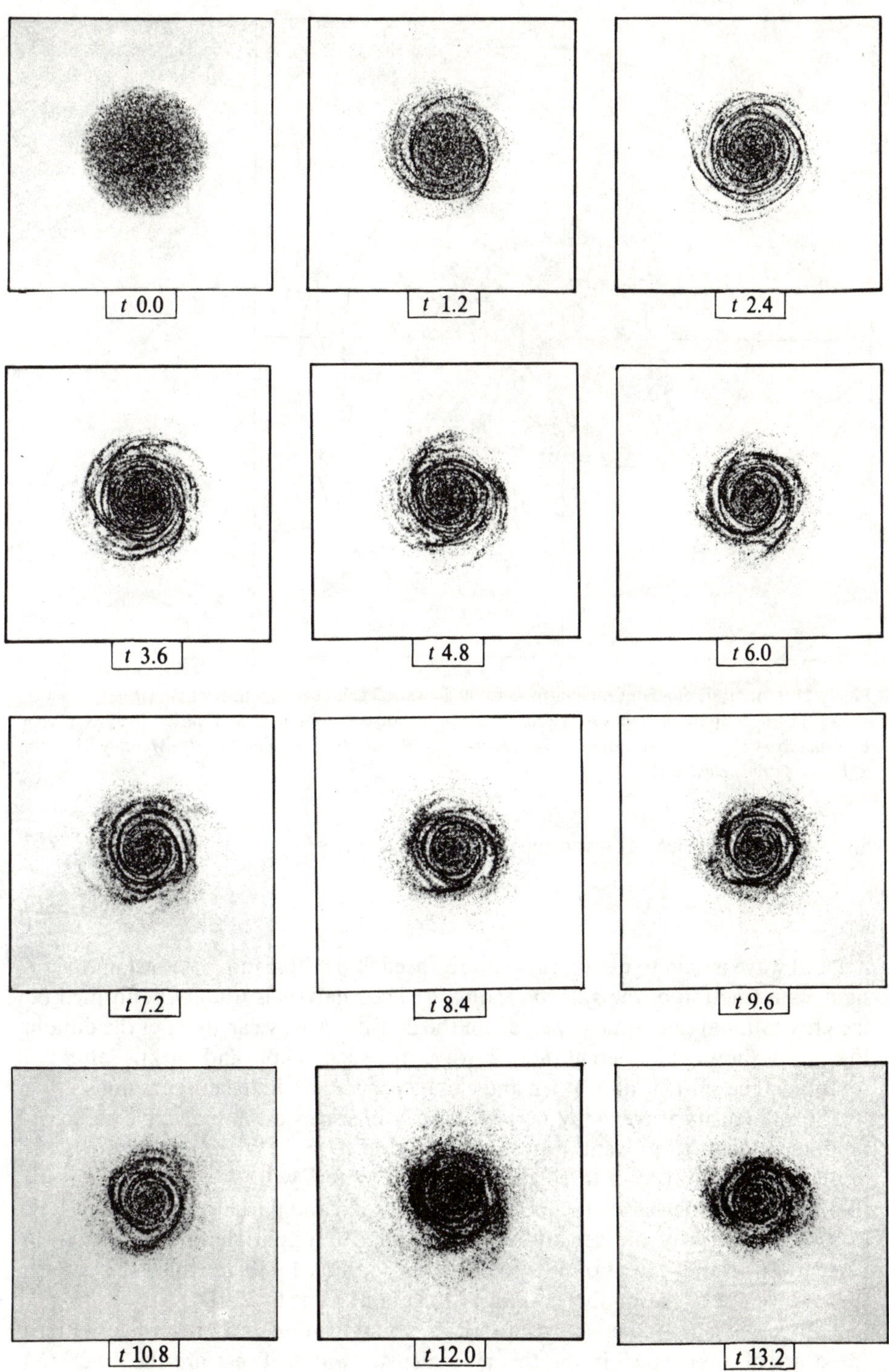
t 0.0
t 1.2
t 2.4
t 3.6
t 4.8
t 6.0
t 7.2
t 8.4
t 9.6
t 10.8
t 12.0
t 13.2

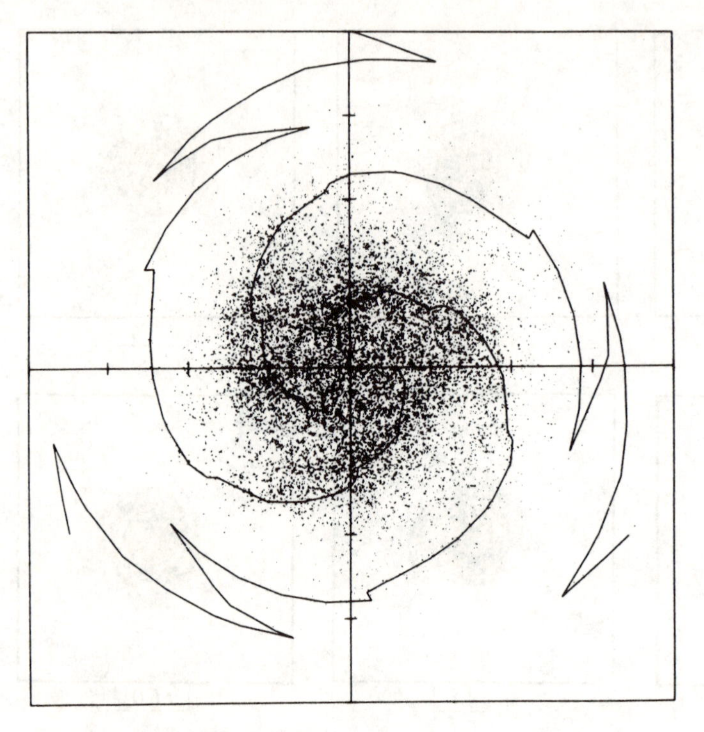

Figure 11-9 Analysis of a faint three-arm spiral in a "warm" galaxy with a heavy halo ($Q = 1$, $\rho = 0.2$, $\alpha = 4$). The lines show the locus of the maximum of the density, obtained by Fourier analysis. (*From Berman, Brownrigg, and Hockney, 1978, courtesy of Royal Astronomical Society Monthly Notices, © Royal Astronomical Society.*)

and the wavenumber k is given by

$$k = \frac{\partial \Phi_m}{\partial r} \tag{11-38b}$$

A spiral wave is said to exist if the pattern speed Ω_p is substantially constant over a significant fraction of the galactic radius R. The analysis is usually performed on the gravitational potential as well as on the density. A wave analysis of the data in Fig. 11-8 shows a coherent $m = 2$ wave between 4 kpc and 36 kpc after 13 rotations. The spiral is quite open and can be seen faintly in the outer regions.

Spiral density waves may be seen clearly in warm disks that have sufficient random velocity to prevent Jean's condensation ($Q > 1$). Figure 11-9 shows the result of the analysis of a three-arm spiral in a system with $Q = 1$, $\rho = 0.2$, $\alpha = 4$ (Berman, Brownrigg, and Hockney, 1978). The solid and partially zigzag line is the locus of the peak of the density wave. The relative amplitude of the wave is 36 percent. The radial extent of the wave is shown in Fig. 11-10, in which the pattern speed is seen to be reasonably constant from about 7 kpc to 22 kpc.

Three resonances between the stellar orbital motions and the wave pattern speed are of importance in the theory of density waves. These are the inner and

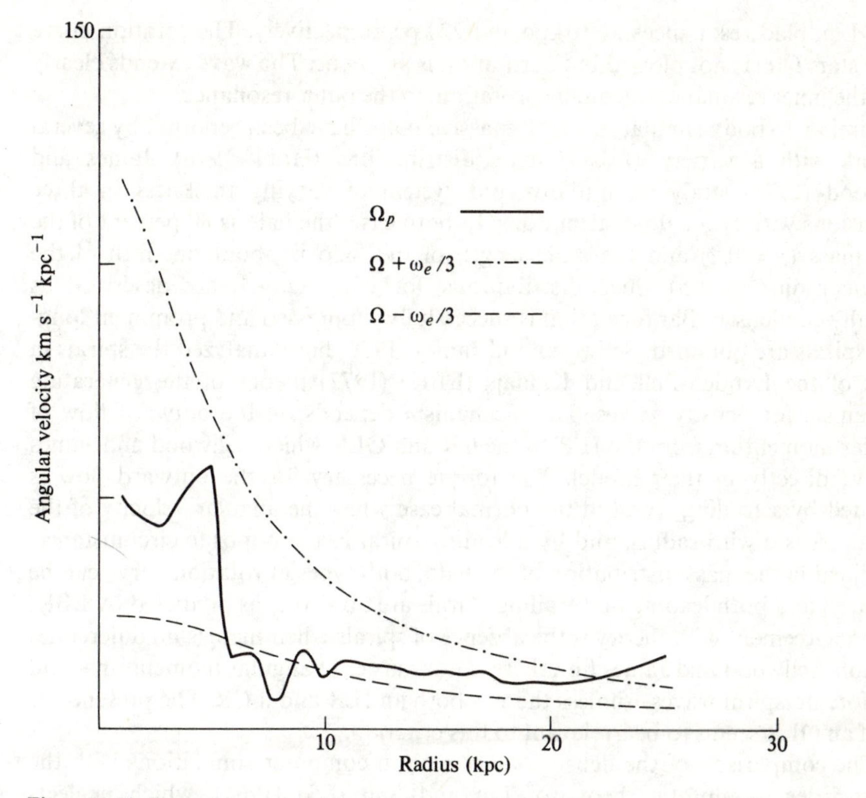

Figure 11-10 The pattern speed of the three-arm spiral of Fig. 11-9 as a function of radius showing the inner and outer Lindblad resonances at 10 kpc and 22 kpc, respectively. The corotation resonance is at 17 kpc. ω_e = epicyclic frequency, Ω = rotation frequency of the stars, Ω_p = rotation frequency of the wave. (*After Berman, Brownrigg, and Hockney, 1978, courtesy of Royal Astronomical Society Monthly Notices, © Royal Astronomical Society.*)

outer Lindblad resonances (ILR and OLR) and the corotation resonance (CR). At corotation the stars and the wave rotate with the same angular velocity. At the inner (outer) Lindblad resonance the angular velocity of the stars is greater (less) than that of the wave; however, stars moving in their epicyclic motion always pass the peak of the wave in the same phase and have a coherent effect upon it. The condition for the resonances is (Lindblad, 1961, 1962):

$$\Omega_p = \Omega + s\omega_e/m \tag{11-39}$$

where $s = 0$ at corotation

$s = -1$ at the inner Lindblad resonance

$s = +1$ at the outer Lindblad resonance

and Ω_p and Ω are, respectively, the angular velocity of the wave pattern and the stars.

Figure 11-10 shows the curves $\Omega \pm \omega_e/m$ from which we find the inner and outer Lindblad resonances at 10 kpc and 22 kpc, respectively. The rotation curve of the stars $\Omega(r)$ is not plotted but corotation is at 17 kpc. The wave extends clearly from the inner resonance, through corotation, to the outer resonance.

Similar N-body simulations with massive halos have been reported by several authors with a variety of halo mass distributions (Hohl, 1976). James and Sellwood (1978) study an initially cold system of varying thickness in three dimensions with two halo distributions. In both cases the halo is 80 percent of the total mass ($\rho = 0.2$) and the scale length of the halo is about one-fifth of the galactic radius ($\alpha = 5$). Such distributions, for which $\alpha \gg 1$, are described as centrally condensed. Bar formation is successfully suppressed and prominent long-lived spirals are obtained. Sellwood and James (1979) have analyzed the spirals in terms of the Lynden-Bell and Kalnajs (LBK) (1972) theory of the generating mechanism for density waves. The mechanism depends on the outward flow of angular momentum from the ILR to the CR and OLR which Sellwood and James observe directly in their model. The torque necessary for the outward flow is provided by a trailing spiral in the normal case when the angular velocity of the stars decreases with radius, and by a leading spiral in the opposite circumstance. By adjusting the mass distribution of the halo, both types of rotation curve can be produced and both leading and trailing spirals are observed, as predicted by LBK. Also in agreement with theory is the absence of spirals when there is no differential rotation. Sellwood and James find there is no transfer of angular momentum—and therefore no spiral waves—unless there is both an ILR and a CR. The presence or not of an OLR seems to be irrelevant to this criterion.

The comparison of the density waves seen in computer simulations with the lowest-order asymptotic theory of Lin and Shu (LS) (1964), which neglects azimuthal effects, has been made by Berman, Brownrigg, and Hockney (1978), who have measured the dispersion relation for the three-arm wave shown in Fig. 11-9. They find that the observed wave dispersion exists primarily in regions of the dispersion diagram that are forbidden to Lin–Shu waves. Sellwood and James also find that the lowest-order Lin–Shu theory is unable to explain their experiments. Not only are the observed spirals much more open than this Lin–Shu theory predicts, but they also pass smoothly through the corotation region which is a singular point in the theory and should contain spirals of infinite tightness. It is clear that computer simulation has confirmed the theoretical view that large-scale coherent spiral structure is a wave phenomenon in the density distribution. It is equally clear, however, that the spiral structure seen in computer simulations is more complex than and quite different from that assumed in the asymptotic theories of spiral waves. The detailed explanation of the waves seen in computer experiments still remains a challenge to theory.

The spiral phenomena described above are all obtained with massive halos containing approximately 80 percent of the total mass of the galaxy. The halo mass distribution ranged from diffuse ($\alpha = 1$) to centrally condensed ($\alpha = 4$). This led investigators to conclude that massive halos were a necessary condition for long-lived spiral structure and hence to the belief that the total mass of a typical galaxy was four or five times that previously assumed on the basis that most of the mass lay in the disk. This in turn implied that the average mass density of the universe had been similarly underestimated. This is an important point to the cosmologist because the mass density of the universe determines whether or not the universe is a "closed" or "open" system (see Sec. 11-1-5). Estimates of the mass density of the universe, based on visible stars, are 10 to 100 times less than the critical density required to close the universe (Gott, Gunn, and Tinsley, 1974). If the widespread existence of spiral galaxies does imply much more massive halos than were assumed in the above estimates, then cosmologists interested in a closed universe can account for more of the missing mass needed to marginally close the universe.

The conclusion that the widespread existence of spiral galaxies implies massive halos has been challenged by Berman and Mark (1979), on the basis of further computer simulation. Berman and Mark observe that, if the parameter t is calculated for a disk with surface density varying like $(r^2 + a_d^2)^{-3/2}$ and a spherical halo with volume density varying like $(r^2 + a_s^2)^{-5/2}$ then $t \rightarrow 0$ not only for the case of massive halos $(\rho \rightarrow 0)$, but also for the case of centrally condensed halos $\alpha = a_d/a_s \rightarrow \infty$. In fact, $t = 0.14$ is obtained not only for $(\rho, \alpha) = (0.3, 0.7)$, a massive diffuse halo, but also for $(\rho, \alpha) = (0.5, 10)$, a relatively light, centrally condensed halo. The second alternative follows from a suggestion of Lin that stability to bar formation can be found with a relatively light nuclear bulge as seen in galaxy NGC 4565 (Fig. 11-1 *bottom*).

The computer experiment of Berman and Mark was even more successful because a stable two-arm spiral was maintained for 33×10^8 years (≈ 15 rotations) for a halo with only 30 percent of the total mass ($\rho = 0.7$, $\alpha = 3$, $Q = 1$). This is shown in Fig. 11-11. In contrast, the evolution of the system without a halo is shown in Fig. 11-7 when a pronounced bar develops in only a few rotations. In the experiment with the nuclear bulge, the value of the t parameter is initially ~ 0.4 and drops within a rotation to 0.26, after which it falls very slowly to about 0.25 at the end of the experiment. It would appear that Berman and Mark have found a counterexample to the Ostriker and Peebles' condition for stability, in that they have a system stable to bar formation with $t > 0.14$. They have also found conditions when a relatively light nuclear bulge supports stable spiral structure, which is a counterexample to the belief that massive halos are a necessary condition for stable spiral structure. This computer experiment will undoubtedly have an important influence on the development of spiral-structure theory and indirectly on cosmology.

In the simulations of this section, the halo stars have been represented as a fixed external force through which the disk stars move. Although this is a natural approximation one should be concerned to check that the results are not significantly changed if the halo itself is simulated correctly as a collection of superstars. The counterstreaming of particles leads to instabilities in other fields (e.g., the plasma two-stream instability) and it may be that the fixed-halo approximation is suppressing important effects. This possibility has been investigated by Hohl (1978) in a three-dimensional simulation with 50,000 stars representing the disk and a further 50,000 stars representing the halo. The

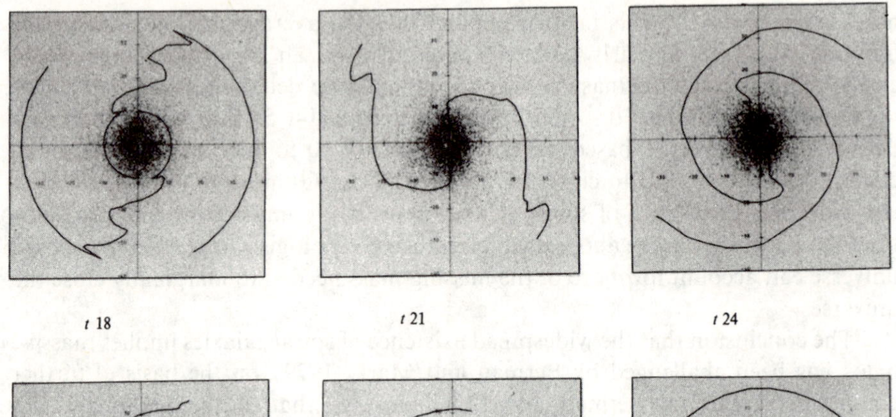

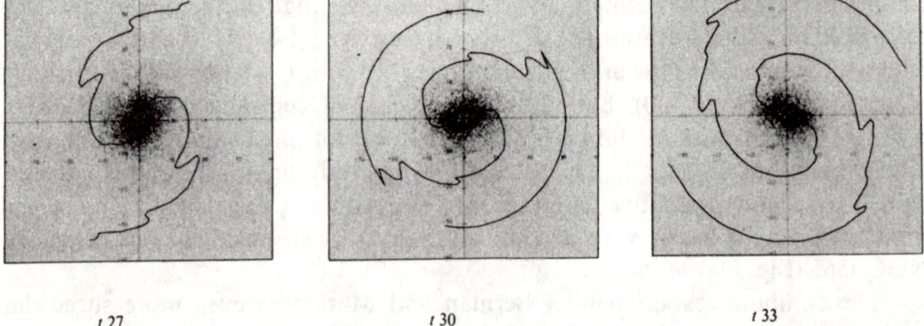

Figure 11-11 Evolution of a galaxy with a light centrally condensed halo ($\rho = 0.7$, $\alpha = 3$, $Q = 1$). Time in units of 10^8 years. (*From Berman and Mark, 1979, courtesy of Astron. & Astrophys.,* © *Springer-Verlag.*)

potential mesh comprises an array of $(64 \times 64 \times 16)$ cells. The halo and disk masses were equal ($\rho = 0.5$) and the halo was bulge-like, as in NGC 4565 ($\alpha = 4$). There were no noticeable effects in the spherical halo, which remained spherically symmetric. The disk remained flattened and supported a two-arm spiral which was still dominant at the end of the experiment (3 rotations). There was no bar formation. This experiment has confirmed that the use of a fixed halo is a good approximation.

11-3-7 The Protogalaxy

The simulations described in the last section assumed that the galaxy had already evolved from an initial rotating ball of gas into a hot, spherical steady-state halo of Population II stars through which a thin disk of Population I stars rotates. The fully three-dimensional simulation of 100,000 stars presented at the end of the section was used to simulate the whole system. It can therefore be used to simulate the evolution of the protogalaxy itself, and this has been done by Hohl and Zang (1979). They use a $(33 \times 33 \times 33)$ mesh, 100,000 superstars, and follow the evolution for three to five rotations. Other simulations of a protogalaxy have been conducted by Gott (1977), assuming axisymmetry, and Miller (1978), using the ILLIAC IV parallel computer.

If one starts with a rotating spherical ball of gas and assumes that most of the gas has condensed into stars within the first collapse, then the subsequent evolution will be well represented by a collisionless particle model of the type used above. Hohl and Zang take a uniform density sphere and vary the amount of rotation from 0 to 1.15 times that which balances gravitational attraction in the equatorial plane. This gives a range of five models with the Ostriker and Peebles' t-variable varying from 0 to 0.44. The amount of random kinetic energy ranged from 0.25 to 0.05 of the potential energy. The duration of each experiment was the time for three rotations of the system in equatorial balance. At the end of the experiment the case with no rotation is still spherically symmetric but the initial constant density has relaxed to an exponential decay with a single scale length ($\propto e^{-r/a}$). It resembles an E0 galaxy such as NGC 3379. The cases with 0.5 and 0.7 of the rotation for balance show a lenticular final shape with a shortening of about 25 percent along the axis of rotation. The final distribution is still axisymmetric and resembles an E2 galaxy such as M32. These three cases have t values of 0, 0.08, and 0.16, respectively, and are stable to bar formation or, in the last case, marginally stable. In any event, there is no sign of bar formation in three rotations.

The last two cases have 0.87 and 1.15 times the rotation for equatorial balance and t values of 0.25 and 0.44. These cases are expected to be unstable to bar formation by the Ostriker and Peebles' criterion. Both systems form bars after three and two rotations, respectively. The bars rotate roughly in the equatorial plane and in the latter case the bar generates a short two-arm spiral structure similar to the SBa barred galaxy NGC 175.

The experiments with a uniform rotating collisionless protogalaxy have shown the evolution of the common elliptical and barred elliptical forms. A complete model must necessarily separately include gaseous and stellar components and an appropriate mechanism for the birth of stars from gas, as has been done much earlier for disk galaxies by Miller, Prendergast, and Quirk (1970). When such effects are included there will be some hope of modeling the different conditions in a protogalaxy that lead to the formation of thin disks, halos, and normal spiral structure.

11-4 CLUSTERING OF GALAXIES

We return now to the phenomenon of galaxy clustering mentioned in Secs. 11-1-4 and 11-1-6. It is expected that the present distribution of galaxies contains clues to the condition of the universe when the galaxies condensed and to the nature of the subsequent expansion of the universe. In particular, it is hoped that a detailed study of clustering will provide evidence favoring either a closed or an open universe. N-body computer simulations have played a large role in investigating the consequences of different cosmological theories for both the initial conditions and the equations of the expanding universe. We present below the equations

(Secs. 11-4-1 and 11-4-2), the numerical model (Sec. 11-4-3), and some of the results (Sec. 11-4-4) obtained by Efstathiou (1979*b*) and Efstathiou and Eastwood (1981) in a study of the Friedman models of the universe using the P^3M code with 20,000 simulated galaxies. Previous work using the *PP* method was limited to 1,000, or at most 4,000, simulated galaxies.

11-4-1 Equations of the Expanding Universe

In the Friedman (1922) cosmology, the universe is assumed to be homogeneous and isotropic when averaging takes place over sufficiently large distances to smear-out the obvious inhomogeneity arising from galaxy clustering, individual galaxies, and stars. In considering the overall expansion, we speak therefore of an average density of matter $\bar{\rho}(t)$ that is a function of time but not of space. Considering the universe as a collection of mass points, any one of these may be taken as origin and the initial state of the universe is described by giving the positions $r_i(0)$ and velocities $v_i(0)$ at the datum time $t = 0$ with respect to the chosen origin. The expansion, being uniform, leads only to radial motions and velocities when viewed from any center. Hence r_i and v_i are both scalars, being the radial distance and radial velocity, respectively. We also limit discussion to Friedman models with a zero value for the Einstein cosmological constant Λ (see, e.g., Tinsley, 1977).

At a subsequent time the expansion since $t = 0$ is described by an expansion factor

$$a(t) = r_i(t)/r_i(0) \qquad a(0) = 1 \tag{11-40}$$

Because of the assumption of homogeneity, the factor $a(t)$ is the same for all particles and separations. As the expansion proceeds, all volumes expand by a factor $a^3(t)$, hence the average density of matter decreases according to

$$\bar{\rho}(t) = \bar{\rho}(0)/a^3(t) \tag{11-41}$$

At any time the Hubble velocity–distance law (Hubble, 1929; 1936, Chapter VII) applies, namely that the radial velocity of expansion is proportional to the distance of the particle

$$v_i(t) = H(t)r_i(t) \tag{11-42}$$

We note that the Hubble "constant" $H(t)$, although the same for all particles, is usually a function of time. Its inverse is a natural unit of time for cosmological problems, which we use later. It is the time that would have elapsed since the big bang if the expansion rate were constant. Substituting Eq. (11-40) into Eq. (11-42) one obtains

$$H(t) = \frac{v_i(t)}{r_i(t)} = \frac{\dot{a}(t)}{a(t)} \tag{11-43}$$

where the dot is used to denote the time derivative.

We can conclude from Gauss' theorem (Harnwell, 1949, page 19) that the force at a radius r in a spherically symmetric system comes only from the mass interior to it, acting as though it were all at the center. Hence the Newton equation of motion for the ith particle is obtained by equating the acceleration to the gravitational attraction of all the mass within the radius r_i

$$\begin{aligned} \ddot{r}_i &= -\tfrac{4}{3}\pi r_i^3 \bar{\rho} \frac{G}{r_i^2} \\ &= -\tfrac{4}{3}\pi G\bar{\rho}(t) r_i(t) \end{aligned} \tag{11-44}$$

Dividing Eq. (11-44) by $r_i(0)$, we obtain the Friedman equation for the expansion factor

$$\ddot{a}(t) = -\tfrac{4}{3}\pi G\bar{\rho}(t) a(t) \tag{11-45a}$$

or using Eq. (11-41)

$$\ddot{a}(t) = -\tfrac{4}{3}\pi \frac{G\bar{\rho}(0)}{a^2(t)} \tag{11-45b}$$

Multiplying Eq. (11-45b) by $2\dot{a}$ and integrating, one obtains

$$\frac{d}{dt}\left[\dot{a}^2 - \tfrac{8}{3}\pi \frac{G\bar{\rho}(0)}{a(t)}\right] = 0 \tag{11-46}$$

therefore

$$\dot{a}^2 - \tfrac{8}{3}\pi \frac{G\bar{\rho}(0)}{a(t)} = -k \tag{11-47}$$

where k is a constant which, in the terminology of general relativity, is the curvature of space. Equation (11-47) is an energy equation, the first term being proportional to the kinetic energy of a particle and the second to its potential energy. If the kinetic energy exceeds the potential energy ($k < 0$) the expansion will continue indefinitely, and the universe is open. Conversely, if the potential energy exceeds the kinetic energy ($k > 0$) then the expansion will eventually stop and the universe is closed. The marginal case of balance between the kinetic and potential energies ($k = 0$) implies a continued expansion, although it is usually referred to as a marginally closed universe.

Equation (11-47) with $k = 0$ allows us to compute the critical density of matter at $t = 0$ that marginally closes the universe, as

$$\bar{\rho}_c(0) = \frac{3}{8\pi G} a(t)\dot{a}^2(t) \tag{11-48}$$

or, using Eq. (11-41), the critical density at time t is

$$\begin{aligned} \bar{\rho}_c(t) &= \frac{3}{8\pi G}\left[\frac{\dot{a}(t)}{a(t)}\right]^2 \\ &= \frac{3H^2(t)}{8\pi G} \end{aligned} \tag{11-49}$$

It is usual to describe the universe in terms of the dimensionless density parameter

Ω which is the ratio of average mass density to that required to marginally close the universe

$$\Omega(t) = \frac{\bar{\rho}(t)}{\bar{\rho}_c(t)} = \frac{8\pi G\bar{\rho}(t)}{3H^2(t)} \tag{11-50a}$$

$$= \frac{8\pi G\bar{\rho}(0)}{3a(t)\dot{a}^2(t)} \tag{11-50b}$$

Combining Eq. (11-50*b*) with Eq. (11-47), we see that if $\Omega(0) = 1$ then $k = 0$ and $\Omega(t)$ remains unity throughout the expansion. This case is called the Einstein–de Sitter Universe. If, however, $\Omega < 1$, corresponding to an open universe, then Ω will decrease as the expansion proceeds. Ω is also the ratio of the potential energy to the kinetic energy and plays a similar role to that of the parameter Γ_0 of Chapter 12.

If time is measured in units of the initial Hubble time $H^{-1}(0)$ then the Friedman equation (11-47) can be expressed in terms of the single parameter $\Omega(0)$

$$\frac{da(t')}{dt'} = \left\{\frac{\Omega(0)}{a(t')} + [1 - \Omega(0)]\right\}^{1/2} \tag{11-51}$$

where $t' = H(0)t$. Given the initial dimensionless density $\Omega(0)$ and the initial condition $a(0) = 1$, this equation determines uniquely the subsequent expansion of the universe. The expansion is therefore determined by the single parameter $\Omega(0)$.

11-4-2 Comoving Coordinates

In studying the clustering of galaxies it is helpful to divide out the general expansion of the universe by introducing comoving coordinates defined by the primed variables

$$\mathbf{r}'(t) = \mathbf{r}_i(t)/a(t) \tag{11-52a}$$

$$\mathbf{r}'(0) = \mathbf{r}_i(0) \tag{11-52b}$$

Then, if there were no gravitational interaction between the particles of the universe, they would remain stationary in comoving coordinates. Any motion of particles in comoving coordinates is therefore clustering arising from gravitational forces. Of additional use is the fact that the average density of matter is a constant in the comoving coordinate system and equal to its initial value of $\bar{\rho}(0)$.

The Newton equations for the particle velocity and position are in the original or "proper" coordinates

$$\dot{\mathbf{v}} = -\nabla\phi \tag{11-53a}$$

$$\dot{\mathbf{r}} = \mathbf{v} \tag{11-53b}$$

$$\nabla^2\phi = 4\pi G\rho(\mathbf{r}, t) \tag{11-53c}$$

where ϕ is the gravitational potential and we have, for simplicity, not included the particle number subscript on $\mathbf{v}$ and $\mathbf{r}$. The mass density $\rho(\mathbf{r}, t)$, in contrast to $\bar{\rho}$, represents the effects of clustering and therefore includes variations on the scale of an individual galaxy.

Substituting Eq. (11-52*a*) into Eq. (11-53*a*) and remembering that $H = \dot{a}/a$, one obtains

$$\dot{\mathbf{v}}' + 2H(t)\mathbf{v}' = -\frac{1}{a}\nabla\phi - \frac{\ddot{a}}{a}\mathbf{r}' \tag{11-54a}$$

where by definition

$$\dot{\mathbf{r}}' = \mathbf{v}' \tag{11-54b}$$

The right-hand side of Eq. (11-54*a*) may be written as a gradient by noting $\nabla\phi = a^{-1}\nabla'\phi$ and $\nabla'\frac{1}{2}\mathbf{r}^2 = \mathbf{r}'$, where the primes show that spatial derivatives are taken with respect to the comoving coordinates. We then have

$$\dot{\mathbf{v}}' + 2H(t)\mathbf{v}' = -\frac{1}{a^3}\nabla'\phi' \tag{11-55a}$$

where

$$\phi' = a\phi + \frac{a^2\ddot{a}}{2}\mathbf{r}'^2 \tag{11-55b}$$

Substituting Eq. (11-55*b*) into (11-53*c*), and using $\nabla^2 = a^{-2}\nabla'^2$ and $\nabla'^2\frac{1}{2}\mathbf{r}'^2 = 3$, we obtain the comoving form of the Poisson equation

$$\nabla'^2\phi' = 4\pi G\rho a^3 + 3\ddot{a}a^2 \tag{11-56a}$$

$$= 4\pi G(\rho'(r', t) - \bar{\rho}(0)) \tag{11-56b}$$

where the density in the comoving frame $\rho' = a^3\rho$, and we have used Eq. (11-45*b*) for the initial average mass density $\bar{\rho}(0)$.

The Newton equations in the comoving frame, corresponding to Eqs. (11-53), are therefore

$$\dot{\mathbf{v}}' + 2H(t)\mathbf{v}' = -\frac{1}{a^3}\nabla'\phi = \mathbf{F}' \tag{11-57a}$$

$$\dot{\mathbf{r}}' = \mathbf{v}' \tag{11-57b}$$

$$\nabla'^2\phi' = 4\pi G(\rho'(r', t) - \bar{\rho}(0)) \tag{11-57c}$$

We can see that the change of coordinates has had two effects. First, a viscous-drag term proportional to the Hubble constant and the velocity has been introduced into the acceleration equation (11-57*a*), and second a constant negative mass density equal and opposite to the average mass density $\bar{\rho}(0)$ has appeared on the right-hand side of Poisson's equation (11-57*c*). This latter term makes the total mass in any system containing many galaxies equal to zero. In fact, we will simulate a triply-periodic cubic region of the universe containing 20,000 simulated galaxies. We note that there is, mathematically, no solution to the Poisson equation with triply-periodic boundary conditions unless the total mass in the system is zero. The negative mass term that has arisen during the change of coordinates is therefore a necessity and should not be regarded in any way as artificial or unphysical.

The physical equations are completed by the energy equation that has been derived in comoving coordinates by Layzer (1963) and Irvine (1961)

$$\frac{d}{dt}(a^4 T) + a\frac{dW}{dt} = 0 \tag{11-58a}$$

where the kinetic energy

$$T = \sum_i \tfrac{1}{2} m_i a v_i'^2 \tag{11-58b}$$

and the potential energy

$$W = \sum_i m_i \phi'(r_i') \tag{11-58c}$$

These are related to the corresponding energies in proper coordinates by $\hat{T} = a^2 T$, $\hat{W} = W/a$.

11-4-3 The Simulation Model

In the model we simulate a cubic sample of the expanding universe and impose triply-periodic boundary conditions to represent approximately the rest of the universe. Twenty thousand galaxies move in the cube, each being represented as a finite-sized mass cloud of diameter equal to about one-fifth of the average interparticle spacing. The periodicity ensures that all galaxies which leave the cube reenter, and the average mass density in the cube therefore remains constant and is set equal to $\bar{\rho}(0)$. The potential and forces are calculated using the P^3M algorithm (see Chapter 8) and the constant negative mass density $-\bar{\rho}(0)$ is added to each mesh point conveniently during the mass assignment phase of the algorithm. For the calculations reported here a cubical mesh of 32^3 cells was used.

The program used was that written by Eastwood, Hockney, and Lawrence (1980) for the simulation of ionic microcrystals (see Chapter 12). The program was modified as described below by Efstathiou and Eastwood (1981) in order to adapt it to the galaxy-clustering problem. This involved the introduction of the viscous term into the equations of motion, the modification of the form for the short-range force, and the addition of a routine for the integration of the Friedman equations. Also, new routines were required for measuring the energy, correlation functions, fluctuation spectrum, and velocity averages, and for setting up the initial conditions. The relative ease with which a computer simulation program could be adapted for use in an entirely different field of science emphasizes the great flexibility of computer simulation as a tool in scientific investigation.

The equations of motion (11-57a, b) are represented by central differences

$$\mathbf{v}^{n+1/2} = \mathbf{v}^{n-1/2}\frac{(1-H(t)DT)}{(1+H(t)DT)} + \frac{\mathbf{F}'^n DT}{(1+H(t)DT)} \tag{11-59a}$$

$$\mathbf{r}^{n+1} = \mathbf{r}^n + \mathbf{v}^{n+1/2} DT \tag{11-59b}$$

where the superscript designates the timelevel, e.g., $v^{n+1/2} = v((n+\frac{1}{2})DT)$. The Lazer–Irvine energy equation (11-58*a*) is represented in difference form by

$$\frac{(a^{n+1})^4 T^{n+1} - (a^n)^4 T^n}{DT} + \frac{a^{n+1} W^{n+1} - a^n W^n}{DT} - W^{n+1} \dot{a}^{n+1} = 0 \quad (11\text{-}60a)$$

and summed to obtain

$$(a^4 T)^m + (aW)^m - \sum_{n=1}^{m} DT\, W^n \dot{a}^n = C \quad (11\text{-}60b)$$

where the constant C depends on the initial conditions

$$C = (a^4 T)^0 + (aW)^0 \quad (11\text{-}60c)$$

The left-hand side of Eq. (11-60*b*) is computed and its constancy during a run is used as an energy check. The timestep chosen was 0.022 of the initial Hubble time $H^{-1}(0)$ and in a typical run of about 800 timesteps $\Delta C/\Delta(aW) \leqslant 2$ percent.

The Friedman equations for the expansion factor, Eqs. (11-45) and (11-47), are integrated by the Taylor series method. In general, the value of a at the next timestep is

$$a^{n+1} = a^n + \dot{a}^n DT + \ddot{a}^n \frac{DT^2}{2} + \dddot{a}^n \frac{DT^3}{6} + \frac{\ddddot{a}^n}{24} DT^4 + \cdots \quad (11\text{-}61a)$$

and similarly the value of $\dot{a}$ is

$$\dot{a}^{n+1} = \dot{a}^n + \ddot{a}^n DT + \dddot{a}^n \frac{DT^2}{2} + \ddddot{a}^n \frac{DT^3}{6} + \cdots \quad (11\text{-}61b)$$

Given the values of a^n and $\dot{a}^n$ from the last timestep, the higher derivatives can be found by differentiating Eq. (11-51). Using $H^{-1}(0)$ as the unit of time, one obtains

$$\ddot{a} = -\Omega(0)/2a^2 \quad (11\text{-}61c)$$

$$\dddot{a} = +\Omega(0)\dot{a}/a^3 \quad (11\text{-}61d)$$

$$\ddddot{a} = \Omega(0)(\ddot{a}/a^3 - 3\dot{a}/a^4) \quad (11\text{-}61e)$$

The accuracy of the integration can be judged by checking how well the solution satisfies Eq. (11-51). If the series for a is taken to include terms up to DT^4 and that for $\dot{a}$ to include terms up to DT^3, then the integration is found to be accurate to about six significant figures. The same timestep is used in the integration of the Friedman equations as is used in the time integration of the particles [Eqs. (11-59)].

The principal diagnostics used in the description of clustering are the two- and multi-point correlation functions. Although the three- and four-point correlations have been measured, we will present only the results for the two-point correlation. An estimator for the two-point correlation is defined by

$$\xi(r) = \frac{N_p}{\bar{n} N_C V} - 1 \quad (11\text{-}62)$$

where N_p is the number of pairs of particles with separations between r and $r+\Delta r$, V is the volume of this shell, N_C is the number of particles taken as centers, and $\bar{n}$ is the mean particle density. With this definition, the most unclustered distribution—a uniform random distribution—has a correlation $\xi(r) \equiv 0$. The two-point correlation function is therefore a measure of clustering. In a typical situation $\xi(r)$ is measured out to separations of one-seventh the cubic box side and the number of pairs of galaxies within equal intervals of $\log r$ is accumulated. The two-point correlation function defined by Eq. (11-62) is the radial (or pair) distribution function $g(r)$ of molecular dynamics minus one (see Chapter 12).

Using this program, it takes about 820 timesteps for an $\Omega = 1$ model to evolve by an expansion factor of ten. The timestep depends weakly on the extent of clustering but is typically (for $a \sim 5$) about 18 s on the IBM 360/195. This divides as follows between the different parts of the P^3M algorithm

Mass assignment	4.8 s
Potential (32^3 mesh)	3.5 s
Mesh force	1.4 s
Short-range force	6.8 s
Advance positions	1.3 s
Total cycle	17.8 s

The calculation of the two-point correlation function from the particle positions is performed in a separate analysis calculation and takes 2–5 minutes, depending on the extent of clustering. All 20,000 particles are taken as centers and the length of the calculation limits its use to four values of the expansion factor ($a = 4, 6, 9, 19.3$). It is interesting to note that the calculation of the correlations is ten times more time-consuming than the basic timestep cycle. This is because the sophisticated techniques employed to avoid an N^2 calculation in the timestep cycle—namely the P^3M algorithm—have not been employed in the analysis of the results!

11-4-4 Results and Conclusions

The basic observational result is that of Totsuji and Kihara (1969), who found the empirical relation $\xi(r) \propto r^{-1.8}$ for data obtained from the Lick observatory. Later Peebles and his coworkers at Princeton University analyzed a more extensive set of data from three observational groups: the Zwicky sample of $\sim$5,000 galaxies collected at the California Institute of Technology, the Lick observatory sample of $\sim 10^6$ galaxies assembled by Shane and Wirtanen, and the Jaggellonian University of Cracow sample of $\sim$10,000 galaxies compiled by Rudnicki. Peebles (1974) found, in agreement with Totsuji and Kihara, that the two-point correlation function has an approximate power law form over a wide range of scales

$$\xi(r) = \left(\frac{r_0}{r}\right)^{\gamma} \qquad \gamma = 1.77 \pm 0.06 \tag{11-63a}$$

where $r_0 = 5.3 \times (1.5)^{\pm 1} h^{-1}$ Mpc† over the range

$$0.1 h^{-1}\,\mathrm{Mpc} \leqslant r \leqslant 9 h^{-1}\,\mathrm{Mpc}$$

Later, a revised analysis of the Lick data by Groth and Peebles (1977) showed a sharp change of slope, or break, in the function $\xi(r)$ at

$$\xi_{\mathrm{break}} \sim 0.3 \qquad r_{\mathrm{break}} \sim 9 h^{-1}\,\mathrm{Mpc} \tag{11-63b}$$

The task of analytic theory in conjunction with computer simulation is to establish the conditions that could give rise to the observational results summarized in the simple equations (11-63).

The problem has been tackled by Davis, Groth, and Peebles (1977) and Davis and Peebles (1977) who performed a numerical integration of the BBGKY kinetic equations, based on plausible assumptions for the closure of the hierarchy of equations and for the initial conditions. These involved the use of the observed form of the three-point correlation function. They obtained good agreement with the value of γ in Eq. (11-63) and observed the break in $\xi(r)$ at about the right place. However, doubt has now been cast on this result by the N-body computer simulation results described below. Neglecting numerical errors that may be shown to be small, N-body experiments integrate the motion of 20,000 galaxies exactly and obtain a quite different result.

Another important set of theoretical results was obtained by Peebles (1974) and relates the power spectrum of the initial density fluctuations to the value of γ. Treating the galaxies as point masses, the number density is

$$n(\mathbf{r}) = \sum_{\text{galaxies}} \delta(\mathbf{r} - \mathbf{r}_i) \tag{11-64a}$$

where δ is the three-dimensional Dirac delta function and the relative density fluctuation is

$$\Delta(\mathbf{r}) = n(\mathbf{r})/\bar{n} - 1 \tag{11-64b}$$

where the mean density

$$\bar{n} = \langle n(\mathbf{r}) \rangle \tag{11-64c}$$

and the correlation function

$$\xi(\mathbf{r}') = \langle \Delta(\mathbf{r}) \Delta(\mathbf{r} + \mathbf{r}') \rangle \tag{11-64d}$$

In the above, the angular brackets mean an average taken over a large assembly. The power spectrum or structure factor $S(k)$ is obtained by Fourier-transforming the fluctuation $\Delta(\mathbf{r})$ into harmonics $\mathbf{k}$ with amplitude Δ_k, then

$$S(\mathbf{k}) = |\Delta_k|^2 \tag{11-65}$$

---

† h^{-1} Mpc is an astronomical unit of distance. h^{-1} is the constant $H/(100\,\mathrm{km\,s^{-1}\,Mpc^{-1}})$ and Mpc is megaparsecs.

is proportional to the energy or power in the kth mode. Because of the isotropy of space, $S(\mathbf{k})$ is treated as a function only of the modulus of $\mathbf{k}$.

If the initial power spectrum of fluctuations is of the power law form

$$|\Delta_k|^2 \propto |\mathbf{k}|^n \tag{11-66a}$$

then Davis and Peebles (1977) find

$$\gamma = \frac{3n+9}{n+5} \qquad \text{if } r \ll \lambda \qquad \xi \gg 1 \tag{11-66b}$$

and

$$= n+3 \qquad \text{if } r \gg \lambda \qquad \xi \ll 1 \tag{11-66c}$$

where $\lambda = (\bar{n})^{-1/3}$ is the mean intergalactic separation. Equation (11-66c) is derived from linear theory and is valid when galaxies are well separated. Equation (11-66b) is based on nonlinear theory for the case $\Omega = 1$ and is valid if the galaxies are close.

Clearly a value of $n = 0$, which gives the observed value of $\gamma = 1.8$ for small separations and steepens to $\gamma = 3$ for large separations, is attractive because with the correct amplitudes it could give very good agreement with both the observational results (11-63a) and (11-63b). The case $n = 0$ means that fluctuations have equal amplitude for all wavenumbers (i.e., it is "white noise"), and corresponds to a uniform random distribution of galaxies for which the clustering and $\xi(r) \equiv 0$. This is also known as a "Poisson distribution."

The 20,000-body computer simulations start from an $n = 0$ distribution shown in Fig. 11-12 which is the projection of the three-dimensional distribution onto a plane. Each dot represents a simulated galaxy and has an initial velocity equal to pure Hubble expansion [Eq. (11-42)]. The distribution of galaxies after evolution by an expansion factor $a = 9.9$ is shown in Fig. 11-13. At the bottom is the case of an

Figure 11-12 The initial condition for the 20,000-particle galaxy clustering experiments. Each dot represents a galaxy and these are initially positioned in a uniform random distribution in the square (Poisson distribution). (*From Efstathiou and Eastwood, 1981, courtesy of Royal Astronomical Society Monthly Notices, © Royal Astronomical Society.*)

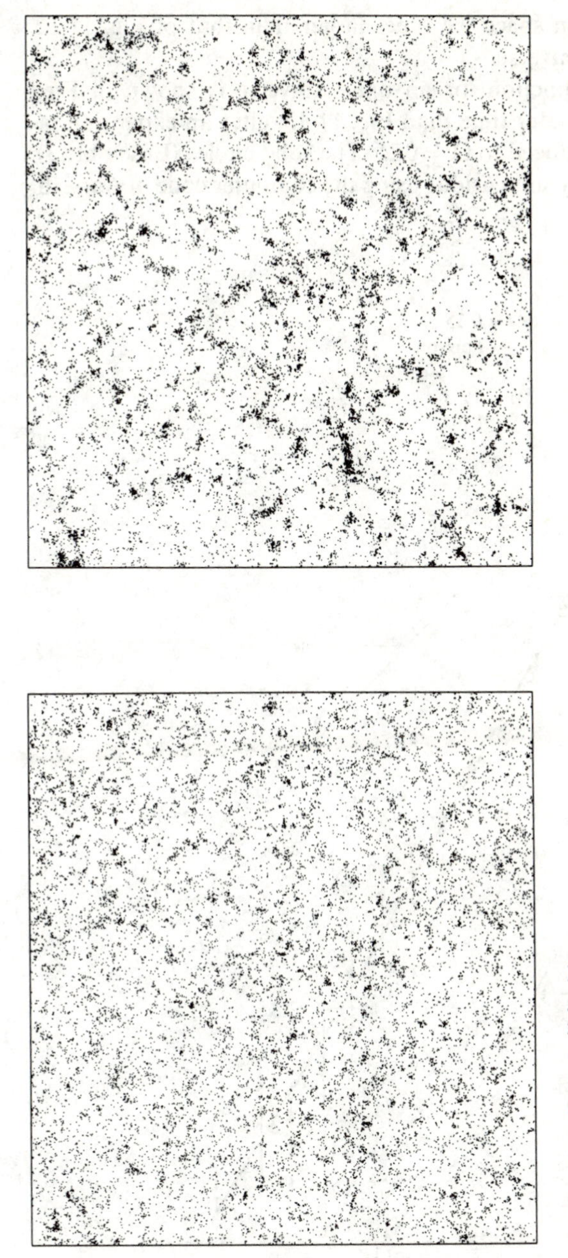

Figure 11-13 The distribution of galaxies after an expansion by a factor $a = 9.9$ from the distribution of Fig. 11-12. *Bottom*, an open universe when the density parameter $\Omega = 0.26$ and, *top*, a marginally closed universe with $\Omega = 1.0$. (*From Efstathiou and Eastwood, 1981, courtesy of Royal Astronomical Society Monthly Notices, © Royal Astronomical Society.*)

open universe which started with a density $\Omega = 0.775$ and has evolved to a density $\Omega = 0.26$. The more lumpy distribution at the top is that of a marginally closed universe, starting from the same initial positions, for which $\Omega = 1$ at all times. The plots are on the same scale in comoving coordinates but, of course, the boxes in

Fig. 11-13 represent a distance in space 9.9 times larger than that of Fig. 11-12 because of the expansion of the universe.

The two-point correlation function for an open universe is shown in Fig. 11-14. It is taken at a somewhat later time than Fig. 11-13, after an expansion of $a = 19.3$ when the density had reduced to $\Omega = 0.15$. The solid circles (Ensemble 3) are an average of three 1,000-body simulations using the computer code of Aarseth

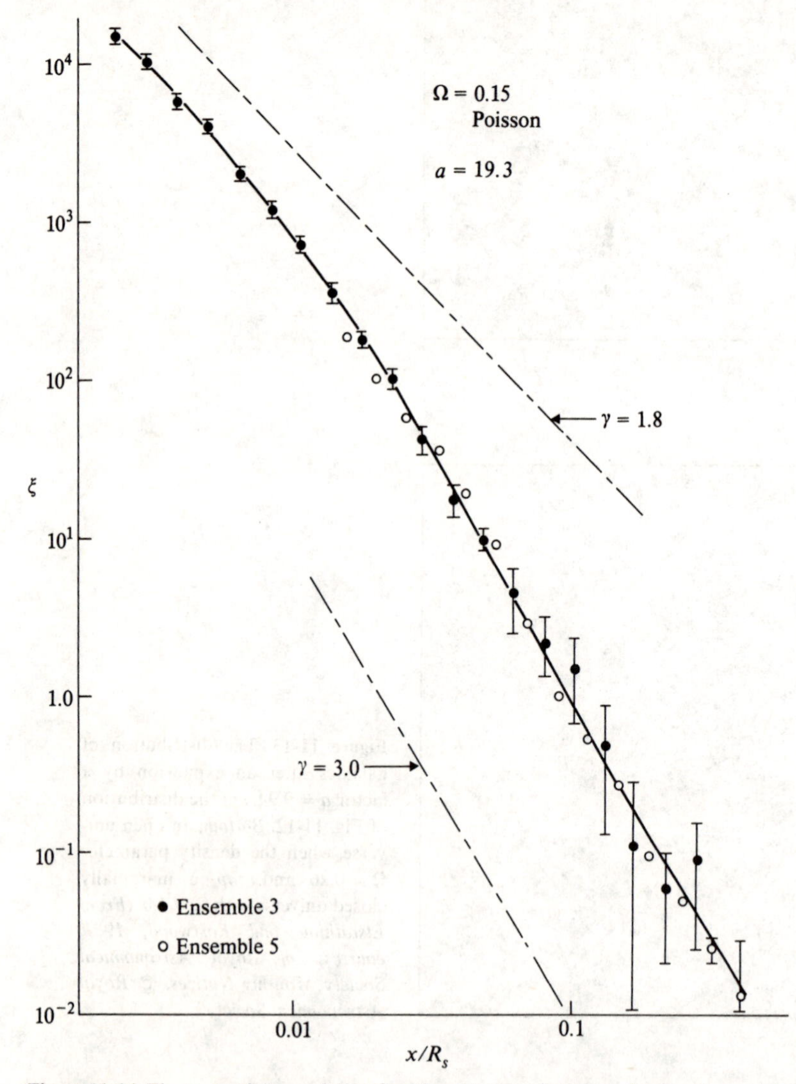

Figure 11-14 The two-point correlation function $\xi(x)$ as a function of separation x, for an ensemble of open universes after an expansion $a = 19.3$ when $\Omega = 0.15$. The solid circles are the results of 1,000-body experiments and the open circles of 20,000-body experiments. The dotted lines show variations of the form $\xi(x) \propto x^{-\gamma}$. The solid line is a cubic fit to the simulation results. (*From Efstathiou and Eastwood, 1981, courtesy of Royal Astronomical Society Monthly Notices, © Royal Astronomical Society.*)

(Secs. 11-2-1 and 11-2-2; Aarseth, Gott, and Turner, 1979), modified by Efstathiou. The open circles (Ensemble 5) are the average of two 20,000-body simulations using the P^3M code of Eastwood, Hockney, Lawrence, and Efstathiou described here. The 1,000-body data gives good statistics from separations of about 0.002 to about $0.05R_s$ ($R_s = \frac{1}{4.4}$ of the box side). The 20,000-body simulation, on the other hand, extends the data to 0.5 of the box side. At smaller separations than $0.025R_s$, the 20,000-body data are affected by the finite diameter $d_m \approx 0.007R_s$ given to each galaxy and are therefore ignored. In fact, 20,000-body calculations can be carried out with $d_m \approx 0.0024R_s$, in Fig. 11-14. The 1,000-body and 20,000-body data join

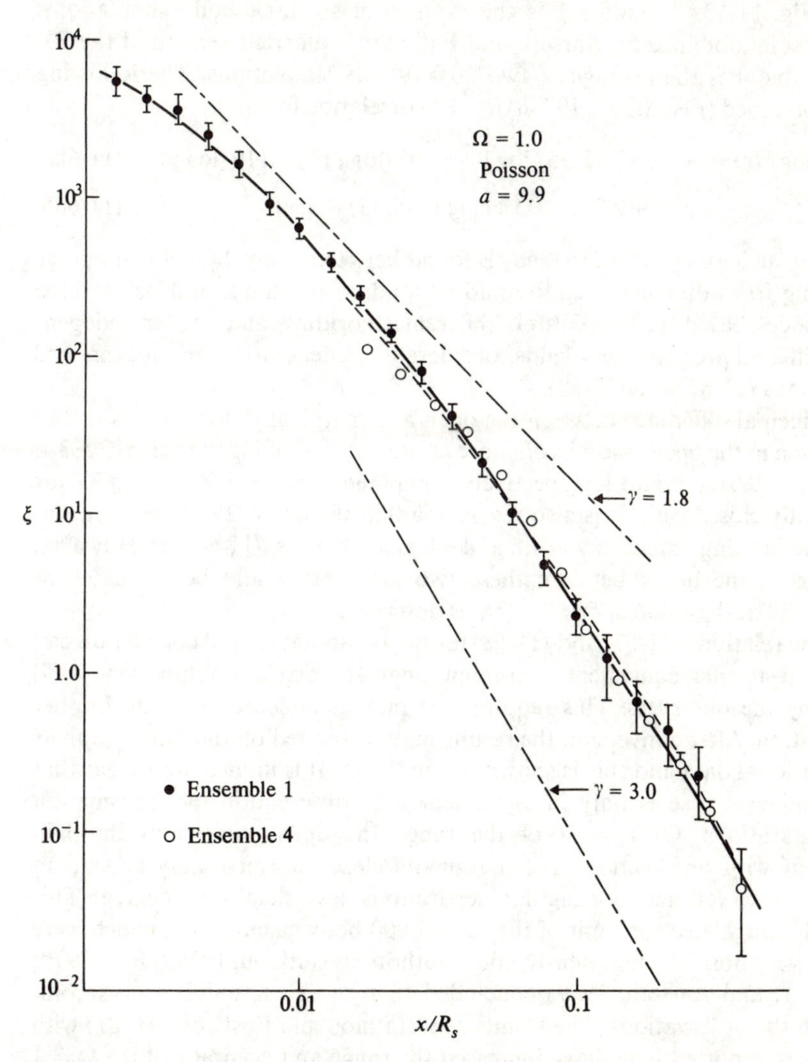

Figure 11-15 The two-point correlation function for an ensemble of marginally closed universes after an expansion $a = 9.9$. Otherwise as Fig. 11-14. (*From Efstathiou and Eastwood, 1981, courtesy of Royal Astronomical Society Monthly Notices, © Royal Astronomical Society.*)

well together, and Efstathiou and Eastwood (1981) obtain the following least-squares fit to a cubic, which is drawn as the solid curve in Fig. 11-14:

$$\log \xi(r) = -3.000 - 2.550 \log y + 0.577 (\log y)^2 + 0.200 (\log y)^3 \quad (11\text{-}67a)$$

where $0.002 \leqslant y = r/R_s \leqslant 0.5$

whence

$$\gamma = -d(\log \xi)/d(\log y)$$
$$= +2.550 - 1.154 \log y - 0.600 (\log y)^2 \quad (11\text{-}67b)$$

Similar results for the marginally closed Einstein–de Sitter universe ($\Omega = 1$) are shown in Fig. 11-15. Ensemble 1 is the average of six 1,000-body simulations, including results obtained by Aarseth and Fall with a different version of the PP code. Ensemble 4 is the average of two 20,000-body simulations. The following result was obtained (Efstathiou, 1979*b*) for the correlation function

$$\log \xi(r) = -2.537 - 2.957 \log y + 0.092 (\log y)^2 + 0.119 (\log y)^3 \quad (11\text{-}68a)$$

and

$$\gamma = +2.957 - 0.184 \log y - 0.357 (\log y)^2 \quad (11\text{-}68b)$$

The fact that such excellent consistency is found between many different computer runs, starting from different pseudo-uniform random positions, and using three computer codes based on two entirely different algorithms and written independently by different programmers, lends considerable credence to the result expressed by the relations (11-67) and (11-68).

The principal difference between the open and marginally closed cases is that the correlation in the open case falls off more rapidly, with γ taking the values 2.5, 3.1, and 2.5 for $y = 0.01$, 0.1, and 1, respectively, compared with $\gamma = 1.9$, 2.8, and 3.0 for the marginally closed case. It is also reassuring that the latter $\Omega = 1$ case satisfies the theoretic limiting values of $\gamma = 1.8$ and 3.0 given by Eqs. (11-66*b*, *c*). However, the position of the break between these two gradients would be estimated at $\xi \approx 50$ and not at the value of $\xi = 0.3$ that is observed.

Both the relations (11-67) and (11-68) for the two-point spatial correlation can be converted to the equivalent two-point angular correlation function $W(\theta)$ measured by the observers. This requires the numerical solution of the Limber (1953) equation. After conversion the results may be plotted on the same graph as the observational data and this is shown in Fig. 11-16. It is immediately clear that the open universe case is only in agreement with observation for the smallest angular separations. Over most of the range the open universe is in plain disagreement with observation. The marginally closed universe may be said to agree with observations for angular separations less than one degree. This corresponds roughly to the limit of the early 1,000-body calculations, which were in any case very noisy. Consequently, some authors (Efstathiou, 1979*a*; Fall, 1979; Gott, Turner, and Aarseth, 1979) concluded that an $\Omega = 1$ universe was compatible with the observations. The results of Efstathiou and Eastwood (1981) with 20,000 bodies, reported here, have increased the range and accuracy of the $\Omega = 1$ results to the point where there is a clear disagreement between the $\Omega = 1$ case and observation. Furthermore, an examination of the velocity distributions out to galaxy

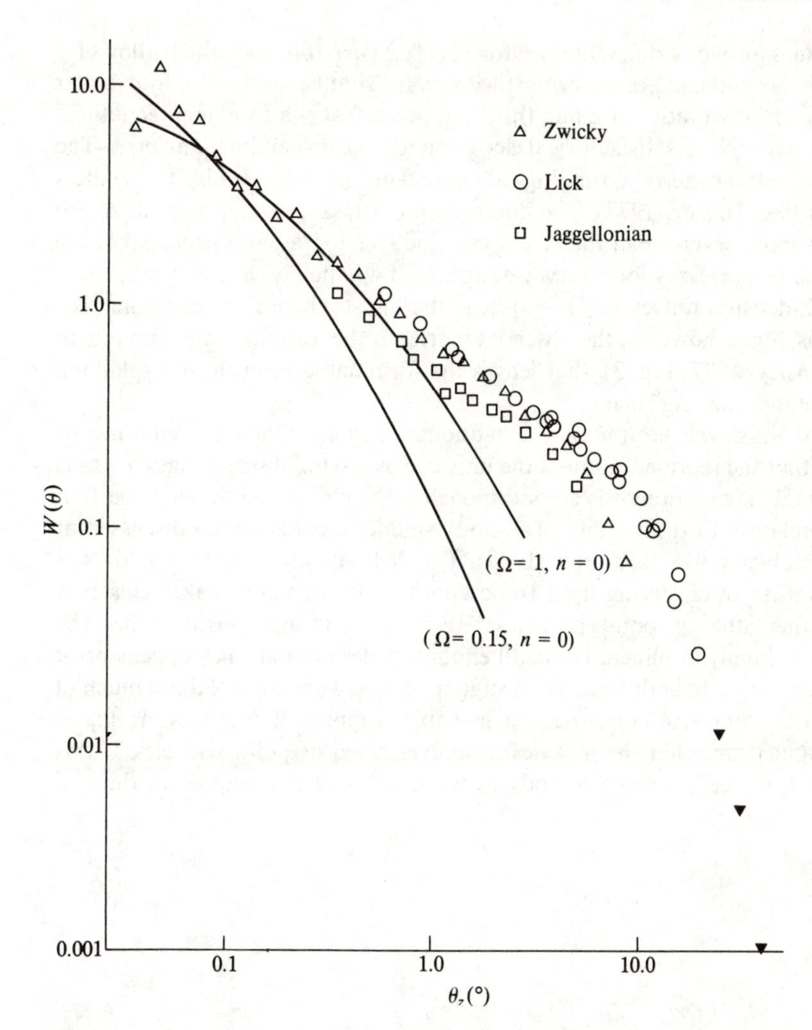

Figure 11-16 Comparison of the computer simulation measurements of the two-point correlation function (solid lines), expressed in terms of equivalent angular separation in the sky, with observations. (*From Efstathiou and Eastwood, 1981, Groth and Peebles, 1977, The Astrophysical Journal, published by the University of Chicago Press; © 1977 The American Astronomical Society; and Royal Astronomical Society Monthly Notices, © Royal Astronomical Society.*)

separations of half the box side shows excellent agreement with linear theory. One can say quite confidently that the computer simulations show no break in slope near 10 degrees, which is such a pronounced feature of the observations. On the other hand, the three-point correlation function ζ is found to be related to the two-point correlation by $\zeta = Q\xi^2$ with $Q \sim 1$. This is in good agreement with the observational value of $Q = 1.3 \pm 0.2$ (Groth and Peebles, 1977), and strengthens the argument that the galaxies are clustered hierarchially.

It is possible that a different spectrum for the initial density fluctuations, e.g., negative values for n or a different form for $S(\mathbf{k})$ [e.g., $\exp(k/k_m)\,k^n$ which allows for an adiabatic component], may bring the N-body simulation results into better agreement with observation. Failing this, it appears that the Friedman equations with $\Lambda = 0$ are not a satisfactory description of the expanding universe. The introduction of nonzero cosmological constant ($\Lambda \neq 0$) leads to endless possibilities (see Tinsley, 1977). The computational task of simulating an $\Lambda \neq 0$ model is no more severe than the $\Lambda = 0$ case, but the extra parameter makes the number of cases necessary for a survey daunting. If we stick with $\Lambda = 0$, it appears that a high-density universe ($\Omega \simeq 1$) has the most chance of satisfying the observations. Since, however, the observed matter in the universe only amounts to $\Omega \approx 0.1$ (Tinsley, 1977, Fig. 3), that leaves the formidable problem of explaining the nature of the "missing" mass.

We must finish this section with a cautionary remark that it is by no means established that the representation of the universe as a set of discrete masses—as is done in N-body simulations—is a good model of the real universe. This question and others relating to the validity of N-body simulations have been discussed at length by Peebles (1978) and Fall (1978). The N-body simulations would be a good description of clustering if, at some epoch, galaxies were weakly clustered and acted thereafter as point particles. However, it is also possible that the clustering was highly nonlinear on small enough scales even at early epochs prior to galaxy formation. In addition if (as is suggested by a wide body of data) much of the mass in the universe comprises an invisible component (the missing mass) there is no guarantee that the galaxies have ever acted as point particles. If this were the case, the results from N-body experiments would not apply to the real universe.

CHAPTER

TWELVE

SOLIDS, LIQUIDS, AND PHASE CHANGES

12-1 INTRODUCTION

12-1-1 Molecular Dynamics

One of the principal contributions made by computer simulation has been to the study of condensed matter, particularly liquids and phase changes. This has been achieved by integrating the motion of a number of representative particles in the material and is known as "molecular dynamics" (MD), even though most work has been performed with systems that cannot be said to form molecules. The motion of the particles is normally assumed to be classical and governed by Newton's laws, and the forces are normally treated as acting centrally along lines drawn between pairs of particles. Neither of these restrictions is essential to the idea of molecular dynamics and one expects to see the technique developed to nonclassical motion and noncentral, many-body forces. The discussion in this chapter will, however, be limited to classical central-force simulations.

The materials that received initially the most attention were the simplest atomic systems (the noble gases argon, neon, krypton, etc.), the simplest charged systems (the alkali-halide salts KCl, NaCl, KI, etc.) and the simplest high-density plasmas composed solely of a single charged species, e.g., electrons. The latter two cases will be taken as examples for detailed study in this chapter. From the point of view of the computational scientist the simulation is specified completely once the law of force between interacting particles is given. Like the applied mathematician or engineer, he need not concern himself with understanding in detail the physical reasoning behind the force law—that is the job of the physicist. Given the law of interaction, it is the role of the computational scientist to devise a robust model on the computer of known accuracy. As the nature of the modeling problem is determined primarily by the characteristics of the force of interaction, we will now survey briefly the force laws that have been used in molecular dynamics simulations. The computer times quoted and the capabilities of the models will refer to computers with a performance similar to an IBM 360/195 or a CDC 7600. A bibliography of 392 references to work in molecular dynamics and related fields up to 1973 has been compiled by Goel and Hockney (1974).

12-1-2 The Force Law

The earliest simulations were conducted by Alder and Wainwright (1959) with infinitely hard spheres acting like a system of billiard balls. In this case there is no interaction unless the balls touch and a sharp, impulsive change of motion if they do. Special techniques are used for hard-sphere simulations involving the calculation of the time to the next "atomic" collision, and the reader is referred to the above article for details. Particle-mesh methods, which are designed for simulating smoothly varying long-range forces, are not applicable to the simulation of hard-sphere systems.

A more realistic potential of pair interaction V is required for the simulation of real materials and the Lennard-Jones inverse power potential is found by Rahman (1964) and McDonald and Singer (1967) to predict well the properties of condensed noble gases. Although the original publication (Jones, 1924) favored different values, later work has standardized on the inverse 6–12 power potential

$$V(r_{ij}) = \varepsilon\left[\left(\frac{r_0}{r_{ij}}\right)^{12} - 2\left(\frac{r_0}{r_{ij}}\right)^{6}\right] \tag{12-1a}$$

$$= 4\varepsilon\left[\left(\frac{\sigma}{r_{ij}}\right)^{12} - \left(\frac{\sigma}{r_{ij}}\right)^{6}\right] \tag{12-1b}$$

where r_{ij} is the distance between the ith and jth atom and ε and r_0 are, respectively, the depth of the potential well and the equilibrium separation of a pair of atoms. The two forms (12-1a) and (12-1b) are equivalent, with $r_0 = 2^{1/6}\sigma = 1.12\sigma$. The force is zero at $r = r_0$ and the potential is zero at $r = \sigma$. This potential of interaction has an r^{-6} attraction at long range and a r^{-12} repulsion at short range. In most simulations the interaction is ignored if the separation of atoms is greater than a truncation radius $r_1 \simeq 3r_0$. If the force on a particle is found by summing contributions from all particles acting upon it, as in the traditional PP method, then this truncation limits the computational effort to an amount proportional to the total number of particles N. With this approximation, systems of several thousand atoms may be simulated (e.g., Schofield, 1973).

If the system contains electrically charged atoms (i.e., ions) the long-range interaction potential is coulombic and proportional to r^{-1}. In the case of such a slowly varying interaction, truncation is not permissible and the interaction between all N^2 pairs of particles must be included in the force calculation. Furthermore, if periodic boundary conditions are used, it is necessary to include the effects of all image charges. This is usually done by the efficient summation technique of Ewald (1921); see Sangster and Dixon (1976, pages 298–305). Nevertheless, the inclusion of all interactions and their images by the PP method is a major computational problem, and the system size is limited to a few hundred particles (e.g., Rahman et al., 1972, and Lewis et al., 1975). The presence of both long- and short-range forces makes the ionic problem ideally suited for solution by the P^3M technique (Hockney et al., 1973, and Chapter 8) and this makes possible the simulation of systems containing 10,648 ions with a timestep of 26 s. The P^3M method is used in the examples examined later in this chapter. In addition to the static properties studied in the above papers, the collective dynamic properties of liquid RbBr have been simulated by Copley and Rahman (1976) and those of liquid NaCl by Adams, McDonald, and Singer (1977). Further simulation measurements of the transport properties of the molten alkali halides are reported by Ciccotti, Jacucci, and McDonald (1976). Work on the structure and dynamics of simple ionic liquids has been reviewed by Parrinello and Tosi (1979).

The simplest form of interaction potential that has been employed in the study of ionic systems uses inverse powers (Hansen and McDonald, 1975)

$$V(r_{ij}) = \frac{q_i q_j}{4\pi\epsilon_0 r_{ij}}\left[1 + \frac{\operatorname{sgn}(q_i q_j)}{(p+1)}\left(\frac{s_i + s_j}{r_{ij}}\right)^p\right] \tag{12-2}$$

where q_i and s_i are, respectively, the charge and radius of the ith ion and the index p determines the hardness of the repulsive term. This form of potential, like that of Lennard-Jones, has the advantage of being scalable (see Sec. 12-3-1) and simple to analyze theoretically, while retaining the essential features of a condensed ionic system. If very close agreement is required between the simulated system and the physical then a more complex potential of the Born–Huggins–Mayer form is to be preferred (Born and Mayer, 1932; Huggins and Mayer, 1933; Born and Huang, 1954)

$$V(r_{ij}) = \frac{q_i q_j}{4\pi\epsilon_0 r_{ij}} + B_{ij}\exp(-\alpha_{ij} r_{ij}) - \frac{C_{ij}}{r_{ij}^6} - \frac{D_{ij}}{r_{ij}^8} \tag{12-3}$$

Values of the constants B, C, D, and α are given by Fumi and Tosi (1964) and Tosi and Fumi (1964) for the alkali halides. These are fitted to give the best agreement with the elastic constants of the materials.

Both the above forms of interaction assume that the electronic shells of the ion are rigidly attached to the positive nucleus and that there is no separation of the center of the electronic shells from the center of the nucleus. In real ions, particularly large ions like fluorine and iodine, the presence of the electric field from the other ions causes the center of negative charge to move with respect to the center of positive charge and the ion is said to be polarizable. In order to take this effect into account the motion of the light electronic shells must be separately computed from that of the heavy positive nuclei. Sangster and Dixon (1976) have developed a shell model in which a massless electronic shell is attached to the nucleus by a "spring." At each timestep the nuclei move according to Newton's laws of motion under the prevailing forces as in the rigid-ion model. An iterative adjustment technique is then used to move the electronic shells to a position that reduces the forces on them to zero. The introduction of polarization in this way typically lengthens the timestep several fold. In this shell model the nuclei are treated as point charges with no short-range repulsion, while the shells interact according to a Fumi–Tosi type potential, Eq. (12-3). Values for the parameters in these expressions and for the spring constants are obtained by fitting to the vibrational properties of the crystals. Values for the alkali halides are given by Sangster and Dixon (1976).

A one-component plasma (OCP) is a system in which a single species of charged particle moves in a uniform background of the opposite charge. Systems of this nature exist in the dense stellar matter of white-dwarf stars (Greenstein, 1959). In this case the motion of the electrons is constrained by the Pauli exclusion principle to be that of a uniform background charge. For certain conditions of density and temperature the motion of the heavy positive ions (e.g., ^{56}Fe) can be treated as a classical OCP (Brush, Sahlin, and Teller, 1966). At sufficiently high densities and pressures this fluid is expected to crystallize and the conditions at which this transition takes place is of considerable importance to the theory of such stars. It is obvious that laboratory experiments are impossible and the only opportunity of experimental investigation is via simulation on the computer. The potential of interaction is the simplest considered so far, namely the repulsive potential of like ions of charge q:

$$V(r) = \frac{q^2}{4\pi\epsilon_0 r} \tag{12-4}$$

This system has been simulated by Pollack and Hansen (1973) using the Monte-Carlo method of Metropolis et al. (1953). In this method many thousands of possible configurations for the system are chosen by statistical sampling, and the equilibrium properties are determined by averaging over the configurations. Unlike molecular dynamics, the dynamics of the system is not simulated and time-dependent properties cannot be measured.

Molecular dynamics simulation of the OCP have been reported by Hockney and Brown (1975) for a two-dimensional system in which point charges are confined to move in a plane. Such a system can be realized in the laboratory in two ways—either on the surface of liquid helium (predicted by Cole and Cohen, 1969, and confirmed in the laboratory by Brown and Grimes, 1972) or as an inversion layer in a silicon MOS field effect transistor (Stern, 1972). The history of the study of the two-dimensional OCP affords an interesting example of the interplay of computational science with theory and laboratory experiment. In 1974, Platzman and Fukuyama predicted, on very plausible grounds, that a phase transition between solid and liquid would occur when the ratio of potential energy to kinetic energy Γ_0 was about three. The first experimental results with which to compare the theory did not come from the laboratory, although strenuous efforts were made. They came from the computer experiments of Hockney and Brown (1975), performed with the original two-dimensional P^{3}M code (Hockney et al., 1973) modified for point charges in the plane. This molecular-dynamics experiment found the transition at $\Gamma_0 = 95 \pm 2$, a value some 34 times that of the theoretical prediction, and stimulated the theorists to think again. Subsequently, in 1978, a new theory was presented by Thouless, based on a different physical mechanism for melting, that predicted the transition at $\Gamma_0 = 78$. Considering the uncertainties in the values for some of the physical constants, this is in reasonable agreement with the molecular-dynamics experiment. The story was not complete until laboratory techniques had improved so as to permit a physical measurement. However, the experiment that was simple to do on the computer has turned out to be very difficult in the laboratory. It was not until 1979, some four years after the first computer observation of melting, that the phase transition was first observed in the laboratory by Grimes and Adams (1979) at $\Gamma_0 = 137 \pm 15$. At about the same time, Gann, Chakravarty, and Chester (1979) observed a first-order melting transition in a Monte-Carlo simulation at $\Gamma_0 = 125 \pm 15$. These results are not inconsistent but differ significantly from the earlier theoretical and molecular-dynamics results. Further laboratory and simulation work is clearly necessary to resolve these differences. The computer simulation of the two-dimensional, one-component plasma is taken as a case study in Sec. 12-2.

Molecular dynamics has also been extensively used to study phase changes, particularly melting and glass formation, in neutral and ionic systems. The melting of a two-dimensional Lennard-Jones crystal was simulated by Cotterill and Pedersen (1972) and followed by the study of the role of dislocations in phase transitions in a three-dimensional Lennard-Jones system of 396 atoms (Jensen et al., 1973; Damgaard Kristensen et al., 1974; Cotterill et al., 1974). This molecular dynamics study was an essential accompaniment to the development of a unified theory of melting and glass formation (Cotterill et al., 1975), because the particle simulation was able to test microscopically many of the basic ideas of the theory in a way that is not possible in the laboratory. In Sec. 12-3 we take as a case study the investigation by molecular dynamics of the influence of radius ratio and hardness on the melting of ionic microcrystals (Amini and Hockney, 1979; Amini, Fincham, and Hockney, 1979). In this study, computer simulation allows one to vary the hardness of the ions independently of their radii, again in a way not possible in the laboratory. Furthermore, since different theories predict different dependencies of the melting point on hardness, the computer simulations provide a new and independent test for theories of melting (see Sec. 12-3-7).

12-1-3 Time Integration

Having discussed the determination of the forces, we now consider the time integration. Newton's laws of motion are usually integrated by the leapfrog central-difference scheme (Hockney, 1970; Buneman, 1967; and Chapter 4), sometimes known as the Verlet (1967) algorithm, or by a variation due to Beeman (1976). The latter produces the same orbits as the leapfrog but has a different definition of the velocity; however, it is substantially more costly to compute than the simple leapfrog scheme. (The Beeman algorithm is analyzed in Sec. 4-5.) Some workers have used higher-order predictor-corrector methods (Rahman, 1964) at substantial cost in computer time and storage. For this reason they are only practical for rather small systems of about 100 particles. The only practicable scheme for large systems of many thousands of particles is the leapfrog scheme, as this minimizes both computer time and storage.

12-2 TWO-DIMENSIONAL ELECTRON FILM

The one-component plasma with motion confined to a plane is variously known as a two-dimensional electron liquid, solid or film. The modeling problem can be stated in one sentence: make a model of a system of identical point charges moving two-dimensionally through a uniform neutralizing background charge, and repelling each other according to Coulomb's inverse square law. This is expressed by the following equation

$$m\frac{d^2\mathbf{r}_i}{dt^2} = -\sum_j{}' \frac{q^2}{4\pi\epsilon_0|\mathbf{r}_{ij}|^2}\hat{\mathbf{r}}_{ij} \tag{12-5}$$

where m and q are, respectively, the mass and charge of the particles and $\mathbf{r}_{ij}$ is the vector from the ith to the jth particle. The summation is over all particles j not equal to i and the $\hat{\mathbf{r}}_{ij}$ denotes the *unit* vector in the direction from i to j. In Eq. (12-5) we need not include the uniform background charge because, by symmetry, this can exert no force on a particle.

12-2-1 Dimensionless Equations

If n is the two-dimensional (or areal) density then there is a natural unit of distance

$$r^* = (\pi n)^{-1/2} \tag{12-6}$$

This distance is the radius of a disk that, when associated with each particle, gives the total area of the system. It is therefore roughly half the mean distance between particles. In order to obtain a natural time unit, we consider the average velocity of the particles v^* when in thermal equilibrium at the temperature T. Equating the thermal energy of $\frac{1}{2}k_BT$ per particle with the kinetic energy in one dimension, we have

$$\tfrac{1}{2}mv^{*2} = \tfrac{1}{2}k_BT \tag{12-7a}$$

and

$$v^* = (k_BT/m)^{1/2} \tag{12-7b}$$

Because the motion is actually two-dimensional, the RMS velocity of the particles is $\sqrt{2}v^*$. However, v^* is taken as defining a unit of velocity natural to the problem. A natural unit of time t^* is then given by

$$t^* = r^*/v^* \tag{12-8}$$

and is the average time for a thermal particle to traverse roughly half the interparticle distance.

If the distances and times, r and t, are now measured in units of r^* and t^*, a dimensionless form of Eq. (12-5) is obtained

$$\frac{d^2\mathbf{r}_i}{dt^2} = -\Gamma_0 \sum_j{}' \frac{1}{|\mathbf{r}_{ij}|^2} \hat{\mathbf{r}}_{ij} \tag{12-9a}$$

where

$$\Gamma_0 = \left(\frac{q^2}{4\pi\epsilon_0 r^*}\right) \Big/ k_B T = \frac{\pi^{1/2} q^2}{4\pi\epsilon_0 k_B} \times \frac{n^{1/2}}{T} \tag{12-9b}$$

The parameter Γ_0 is a measure of the ratio of the potential energy to the kinetic energy of the system. The importance of expressing the original Eq. (12-5) in the dimensionless form of Eq. (12-9) is now apparent—a system that might have been thought to require several parameters for its description (m, q, n, T), is found in fact to require only one, the value of Γ_0. A computer experiment performed for a particular value of q, n, T can be reinterpreted as a result for any other q, n, T that has the same Γ_0. Of particular importance is the interchangeability of density and temperature. Equation (12-9b) shows that, for example, decreasing the density by a factor of one-hundredth is the same as increasing the temperature by a factor of 10. Most laboratory experiments have to be conducted by varying the density at constant temperature (liquid helium is at ~ 1 K), however in the computer experiment it is more convenient to vary the temperature at constant density (we keep a constant number of particles in a fixed computational box). Equation (12-9b) shows that it does not matter which is done and that the final result should be quoted in terms of the value of Γ_0. The recognition that the system is characterized by only one dimensionless constant clearly simplifies dramatically the computational task, since we need only complete sufficient computer runs to measure properties as a function of one parameter rather than several. Interactions that may be interpreted in this way are said to be scalable.

12-2-2 Choosing the Timestep

In order to choose the timestep of the simulation it is necessary to spot the highest oscillating frequency ω_1 in the physical system and satisfy the condition of stability [see Sec. 4-4-1, Eq. (4-42)].

$$\omega_1 DT < 2 \tag{12-10}$$

A complete analysis of the modes of oscillation of the electron fluid is very complex (see, e.g., Onuki, 1977; Baus, 1978) and in fact unnecessary for our present purpose. We only need estimates that are sufficient for the choice of a reasonable timestep and show its dependence on the principal parameters. In the final analysis the orbits of the particles and the extent of energy conservation determine the timestep used.

The following simple one-dimensional argument can be used to estimate the highest frequency, which will occur when the maximum gradient of the force comes into play. This will take place when neighboring particles move towards each other to a position of closest approach. The equations of motion of such a pair of particles are, taking $x_2 > x_1$

$$m_1 \frac{d^2x_1}{dt^2} = -F(x_2 - x_1) \tag{12-11a}$$

$$m_2 \frac{d^2x_2}{dt^2} = F(x_2 - x_1) \tag{12-11b}$$

where F is the magnitude of the repulsive force.

Multiplying the first equation by m_2, the second by m_1, and subtracting yields the equation of relative motion

$$m^* \frac{d^2r}{dt^2} = F(r) \tag{12-12}$$

where $r = x_2 - x_1$ is the separation of the particles and $m^* = m_1 m_2/(m_1 + m_2)$ is the reduced mass. In general, $F(r)$ will be nonlinear and in order to obtain an effective frequency we must consider the equation for a small perturbation to Eq. (12-12). Let $r = r + r'$ where r' is small; then

$$m^* \frac{d^2(r+r')}{dt^2} = F(r+r') \simeq F(r) + r' \frac{dF}{dr} + \cdots \tag{12-13}$$

where we have expanded the right-hand side by Taylor series, keeping only the first two terms. Subtracting Eq. (12-12) from Eq. (12-13) one obtains for the perturbation r' the linearized equation

$$\frac{d^2r'}{dt^2} = -\omega^2 r' \qquad \omega = \left(-\frac{1}{m^*}\frac{dF}{dr}\right)^{1/2} \tag{12-14}$$

If dF/dr is negative, as it is for the short-range repulsion, this is the equation for simple harmonic oscillation with a frequency ω. The maximum frequency will occur for small masses and steeper (that is to say harder) laws of repulsion between the particles.

For the case of the electron film we have $m^* = m/2$ and $F(r) = q^2/4\pi\epsilon_0 r^2$), hence

$$\omega = \left(\frac{q^2}{\pi\epsilon_0 m r^3}\right)^{1/2} \tag{12-15}$$

This expression is revealing because it shows that as the minimum separation r goes to zero the frequency of oscillation goes to infinity. Since stability requires that $\omega DT < 2$ this means that all timesteps are, in principle, unstable. However, the probability of small separations occurring depends on the temperature and is very small. In a one-dimensional maxwellian distribution 98 percent of particles have an energy less than $3k_B T$ and have a closest approach greater than $r = b$, which is given approximately by equating the potential energy of the pair to the kinetic energy

$$\frac{q^2}{4\pi\epsilon_0 b} = \beta k_B T \qquad \beta = 3 \tag{12-16a}$$

Substituting into Eq. (12-15), we obtain

$$\omega_1 = \left(\frac{q^2}{\pi\epsilon_0 m b^3}\right)^{1/2} = \frac{8\pi\epsilon_0}{m^{1/2}q^2}(\beta k_B T)^{3/2} \tag{12-16b}$$

and the timestep can be chosen from

$$\omega_1\, DT = \alpha_1 \qquad \alpha_1 < 2 \tag{12-17a}$$

$$DT = \alpha_1/\omega_1 \tag{12-17b}$$

In order to prevent the small number of particles with energies greater than $\beta k_B T$ from becoming unstable, it is necessary to limit the force of repulsion to its value at $r = b$. Hence the force law used is

$$F(r) = q^2/(4\pi\epsilon_0 b^2) \qquad r < b \tag{12-18a}$$

$$= q^2/(4\pi\epsilon_0 r^2) \qquad r \geqslant b \tag{12-18b}$$

The second condition to be satisfied is that the integration of the orbits for most particles is sufficiently accurate. Most particles will have an oscillation frequency ω_0 associated with the average separation $2r^*$, where

$$\omega_0 = \left(\frac{q^2}{8\pi\epsilon_0 m r^{*3}}\right)^{1/2} \tag{12-19}$$

If the leapfrog scheme is used to integrate Newton's laws, Eq. (12-14) is approximated as follows:

$$\frac{r'(t+DT)-2r'(t)+r'(t-DT)}{DT^2} = -\omega^2 r'(t) + \frac{DT^2}{12}\frac{d^4 r'}{dt^4} \tag{12-20}$$

For an oscillation at frequency ω

$$\frac{d^4 r'}{dt^4} = \frac{d^2}{dt^2}(-\omega^2 r') = \omega^4 r' \tag{12-21}$$

and the ratio of the truncation error [last term in Eq. (12-20)] to the real force (first term on the right-hand side) is

$$\frac{DT^2}{12}\frac{d^4 r'}{dt^4}\bigg/(-\omega^2 r') = -\tfrac{1}{12}\omega^2\, DT^2 \tag{12-22}$$

If we take $\omega_0 DT = \alpha_0 = 0.2$ as the condition for satisfactory integration of the average orbit, the relative truncation error from Eq. (12-22) is $\alpha_0^2/12 = \frac{1}{3}$ percent. This value of α_0 corresponds to $10\pi \simeq 31$ integration steps per period of oscillation and is found to be a satisfactory compromise between accuracy and economy of computer resources.

The conditions on the timestep may be expressed in terms of the dimensionless variables and summarized as follows:

1. Accuracy at the average separation:

$$\omega_0^2 DT^2 = \tfrac{1}{2}\Gamma_0\left(\frac{DT}{t^*}\right)^2 = \alpha_0^2 \qquad \alpha_0 = 0.2 \tag{12-23a}$$

2. Stability at the highest frequency:

$$\omega_1^2 DT^2 = \omega_0^2 DT^2 \left(\frac{2r^*}{b}\right)^3 = \alpha_1^2 \qquad \alpha_1 = 1 \tag{12-23b}$$

3. Very few particles with $r < b$:

$$\left(\frac{b}{r^*}\right) = \frac{\Gamma_0}{\beta} \qquad \beta = 3 \tag{12-23c}$$

The following relationships can be seen to hold between the quantities:

$$\left(\frac{\omega_1}{\omega_0}\right)^2 = \left(\frac{\alpha_1}{\alpha_0}\right)^2 = \left(\frac{2r^*}{b}\right)^3 = \left(\frac{2\beta}{\Gamma_0}\right)^3 \tag{12-24}$$

The manner of satisfying these conditions depends on the value of Γ_0 and the following procedure can be used:

4. Select b:

$$\left(\frac{b}{r^*}\right) = \left(\frac{\Gamma_0}{\beta}\right) \qquad \beta = 3 \tag{12-25a}$$

5. Select α_0:

$$\alpha_0 = \min\left[0.2, \alpha_1\left(\frac{\Gamma_0}{2\beta}\right)^{3/2}\right] \qquad \alpha_1 = 1 \tag{12-25b}$$

6. Select DT:

$$DT = \frac{\alpha_0 t^*}{(\Gamma_0/2)^{1/2}} \tag{12-25c}$$

For cold high-density cases (large Γ_0) the timestep is determined by condition 1, whereas for hot low-density cases (low Γ_0) the timestep is determined by conditions 2 and 3. Expressed in terms of the basic physical variables, we have in the first case

$$DT = \alpha_0 8^{1/2} \epsilon_0^{1/2} m^{1/2} / (\pi^{1/4} q n^{3/4}) \tag{12-26a}$$

which depends only on the density, and in the latter case

$$DT = \alpha_1 q^2 m^{1/2} / (8\pi\epsilon_0 \beta^{3/2} k_B^{3/2} T^{3/2}) \tag{12-26b}$$

which depends only on the temperature.

The frequencies considered so far are based on the interaction of a pair of charges. In a system with long-range forces there is also the possibility of frequencies arising from the collective behavior of the whole system. The plasma oscillation described in Chapter 2 is an example of such a collective mode. If the charges move in three dimensions, as in the hot-gas plasma of Chapter 9 and the alkali halide melt, the frequency of the plasma oscillation is given approximately by

$$\omega = \left(\frac{nq^2}{\epsilon_0 m}\right)^{1/2} \tag{12-27}$$

where n is the number of charges per unit volume. This frequency is independent of the wavenumber k. If, however, the motion is confined to a plane, as in the case of the two-dimensional electron film, the frequency is strongly dependent on k (Onuki, 1977) and

$$\omega = \left(\frac{nq^2}{2\epsilon_0 m}\right)^{1/2} k^{1/2} \tag{12-28}$$

where now n is the number of charges per unit area. In the analysis of collective modes the variation is assumed to be sinusoidal and proportional to $\exp[i(kx - \omega t)]$. The spatial wavelength is therefore $\lambda = 2\pi/k$. The highest plasma mode will be that with a wavelength of twice the average separation between ions ($\lambda = 4r^*$), and the lowest frequency will occur for the wave that fits just one cycle into the computational box of width L. From these two limiting cases we obtain the lowest frequency

$$\omega_2 = \left(\frac{nq^2}{2\epsilon_0 m}\right)^{1/2} \left(\frac{2\pi}{L}\right)^{1/2} \tag{12-29a}$$

and the highest frequency, with $k = 2\pi/\lambda = \pi/2r^*$

$$\omega_3 = \left(\frac{nq^2}{2\epsilon_0 m}\right)^{1/2} \left(\frac{\pi}{2r^*}\right)^{1/2} \tag{12-29b}$$

Expressed in terms of the dimensionless time unit one has

$$\omega_3 t^* = \pi^{1/2} \Gamma_0^{1/2} \tag{12-30a}$$

$$\omega_2 t^* = \omega_3 t^* \left(\frac{4r^*}{L}\right)^{1/2} \tag{12-30b}$$

It will be seen from Eq. (12-30*a*) and Eq. (12-23*a*) that the highest-frequency plasma mode will be a more severe restriction on the timestep than ω_0, the vibrational frequency at average separation.

12-2-3 Scaling the Problem

The experiment on the two-dimensional electron film was conducted at constant volume and scaled for convenience to a density $n = 10^{14}$ electrons m^{-2}. The phase change is studied by varying the temperature and the result finally expressed in terms of the dimensionless ratio Γ_0. The values of the other model parameters were as follows:

1. Number of particles $N = 10^4$.
2. Computational region $(L \times L)$; $L = 10^{-5}$ m.
3. Spatial mesh (256×256) with $H = 3.9 \times 10^{-8}$ m.

4. Timestep $DT = 0.5 \times 10^{-12}$ s.
5. Short-range correction force:
   inner limit $b = 1.5H = 5.86 \times 10^{-8}$ m,
   outer limit $r_c = 4H = 15.62 \times 10^{-8}$ m.
6. Chain cell width $= 4H$.
7. Neighbors out to radius $r_c \approx 6$.
8. Triangular lattice separation $a = 2.74H = 10.75 \times 10^{-8}$ m.
9. Distance unit $r^* = 1.44H = 5.64 \times 10^{-8}$ m.

In both the liquid and solid phases the electrons are locally in approximately a triangular close-packed lattice and their typical separation is about 3 mesh cells. It is therefore necessary to calculate the force with a spatial resolution much smaller than a mesh cell and the simple particle-mesh calculation with a resolution of approximately a mesh cell would be hopelessly inaccurate. The P^3M technique is therefore used which adds to the mesh force a short-range correction. This force extends correctly from the inner limit, $b = 1.5H$, to the outer limit, $r_c = 4H$, beyond which the mesh force alone is sufficiently accurate. The pair interaction is now represented correctly for all separations greater than b, and for $r < r_c$ the resolution is independent of the mesh separation.

The choice of timestep must be examined in relation to the frequencies introduced in the last section. We obtain from the above data:

Pair vibration near average separation at $r = 2r^*$,

$$\omega_0 \, DT = \frac{\pi^{1/4} q n^{3/4}}{2^{3/2} m^{1/2} \epsilon_0^{1/2}} \, DT = 0.42 \tag{12-31a}$$

Maximum pair vibration at $r = b$,

$$\omega_1 \, DT = \left(\frac{2r^*}{b} \right)^{3/2} \omega_0 \, DT = 1.12 \tag{12-31b}$$

Minimum plasma frequency at $k = 2\pi/L$,

$$\omega_2 \, DT = \left(\frac{nq^2}{2\epsilon_0 m} \frac{2\pi}{L} \right)^{1/2} DT = 0.157 \tag{12-31c}$$

Maximum plasma frequency at $k = \pi/(2r^*)$,

$$\omega_3 \, DT = \omega_2 \, DT \left(\frac{L}{4r^*} \right)^{1/2} DT = 1.05 \tag{12-31d}$$

The timestep is seen to satisfy the condition for stability ($\omega \, DT < 2$) for both the maximum vibrational and plasma frequencies. At $\omega \, DT \approx 1$ the orbits are oscillatory and qualitatively correct with about six integration steps per period, but the truncation error of 8 percent is large. However, the majority of particles are vibrating near the frequency ω_0 and are represented with about 15 integration steps per period and a truncation error of 1.5 percent. The largest plasma wave has a wavelength equal to the computational region and therefore a minimum wavenumber $k_{\min} = 2\pi/L$. This is integrated with about 40 steps per period and a truncation error of 0.2 percent. If we regard the integration as satisfactory up to $\omega DT \approx 0.5$, this includes the first ten wavenumbers up to $k = 10k_{\min}$. The higher wavenumbers, although stable and having qualitatively correct behavior, are inaccurately integrated. The above timestep is probably the largest that could reasonably be used in the circumstances. If economy in the use of computer time were not a major constraint, a timestep of half the above would be preferable. In this case no frequency would be integrated with a truncation error worse than about 2 percent.

12-2-4 Computer Time and Storage

With the scaling described above, the computer time on the IBM 360/195 divided between the different parts of the timestep cycle is as follows:

Assignment of charge to mesh	0.86 s
Solution for potential	2.00 s
Addition of short-range force and acceleration	5.94 s
Total computer time per timestep cycle	8.80 s

It can be seen that the short-range correction is the most expensive part of the calculation, even though the number of neighbors for which $r < r_c$ is only about six. The minimum operation count for the P^3M method is obtained theoretically when the time taken for the short-range calculation equals the time spent on solving for the potential. As this is not satisfied, one should examine whether the use of a finer mesh, say (512×512), would result in a faster calculation. This would quadruple the time for potential solving (to 8 s) but divide the short-range calculation by four (to 1.5 s). The total cycle of 10 s would now be dominated by the potential solution. It is seen that the alternative scaling is not only slower but also undesirable because it uses four times the storage for the potential. Because the total number of mesh points can be changed only by factors of four, it is not possible to reach the theoretical minimum. The storage used for the first scaling, including space for the program and all data, is 802 k bytes. With the inclusion of diagnostic routines, a typical computer run of 300 timesteps required 48 minutes on the IBM 360/195.

12-2-5 Melting the Electron Film

The computer experiment is performed by assigning initial positions and velocities to all particles and allowing the system to evolve. In order to ensure that a lattice was not artificially imposed on the system, the computer experiment was started from a random distribution of electrons. This system was allowed to relax while being kept at constant temperature (~ 1 K) by scaling the velocities at each timestep. The temperature constraint was then removed and thermal equilibrium established by computing for 200 timesteps without energy input. Heat was subsequently put into the system by multiplying the velocities at each timestep by a factor slightly greater than one (1.004) for 25 timesteps and then allowing the system to reach a new equilibrium by computing without heat input for a further 50 timesteps. The kinetic, potential, and total energies were calculated at every step giving, respectively, instantaneous values of the temperature, potential energy, and internal energy of the system. The instantaneous temperature was observed to fluctuate with a period of about 14 timesteps and fluctuation amplitude of about 2.5 percent. A running average of the energies over 28 steps was taken and this reduced the fluctuation in the measurement of temperature to ~0.5 percent. The time required for the system to reach a new equilibrium after such an input of heat was observed to be about 25 timesteps, so that the running average of the

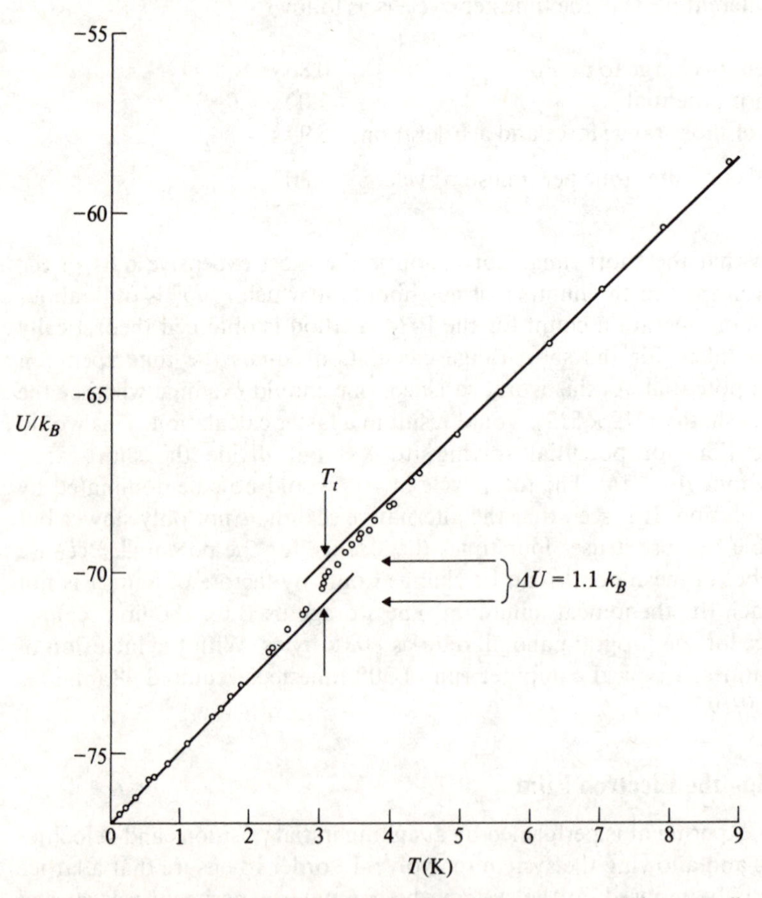

Figure 12-1 Variation of total internal energy U with temperature T in a two-dimensional electron film. Density 10^{14} electrons m^{-2}. (*After Hockney and Brown, 1975, courtesy of Journal of Physics C: Solid-State Physics, © Institute of Physics.*)

temperature T and of the internal energy U computed after 75 steps could be taken as the thermodynamic variables of the new state. The process was repeated, giving values of T and U at a succession of equilibrium states. If (T_1, U_1) and (T_2, U_2) designate the state variables at two successive states, the specific heat at $\bar{T} = (T_1 + T_2)/2$ was calculated from $C_v = (U_2 - U_1)/(T_2 - T_1)$. For measurements near the phase transition the heating factor was reduced to 1.001 and 75 timesteps without heating were allowed for equilibration.

Figure 12-1 shows the variation of internal energy with temperature. The asymptotic behavior at high and low temperatures shows clearly that a phase transition occurs around 3 K for $n = 10^{14}\,\mathrm{m}^{-2}$. Observed with temperature steps

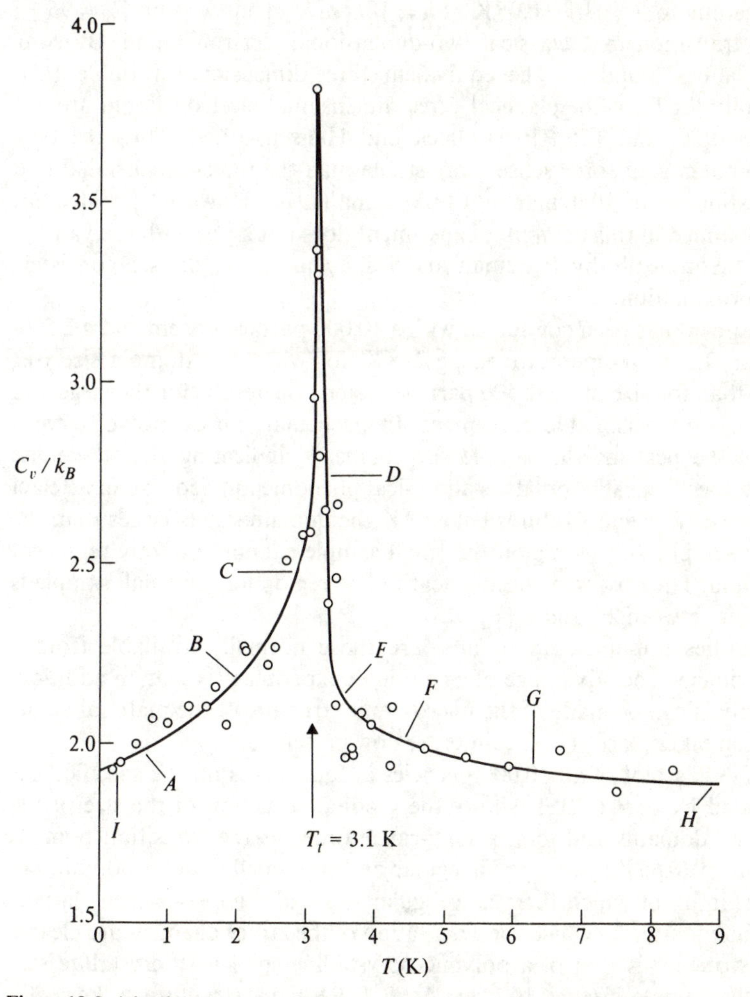

Figure 12-2 A lambda-type specific heat anomaly seen in a two-dimensional electron film. Density 10^{14} electrons m^{-2}. (*After Hocknev and Brown, 1975, courtesy of Journal of Physics C: Solid-State Physics, © Institute of Physics.*)

of ~ 0.05 K, the variation of internal energy with temperature is quite smooth through the transition. If there is an internal energy discontinuity at 3 K it is less than $0.1k_B$, compared with the total internal energy change $\Delta U = 1.1k_B$/electron associated with the transition. The nature of the transition is shown more clearly in the variation of specific heat C_v with temperature $\bar{T}$ (Fig. 12-2). We observe the gradual increase of C_v with temperature, from a value near $2k_B$ at 0 K to a maximum of about $3.8k_B$ at $T_t = 3.1$ K. There is then an almost discontinuous drop back to a value of around $2k_B$. Little further change in specific heat is observed during the rest of the experiment, in which the temperature rose to 8 K.

By integrating the excess specific heat we find a total internal energy change of $1.1 \pm 0.1k_B$/electron. The transition temperature is taken to be the peak in the C_v vs. T curve leading to $T_t = 3.1 \pm 0.05$ K at $n = 10^{14}\,\mathrm{m}^{-2}$. Thus we find $\Gamma_0 = 95 \pm 2$ at the phase transition in a classical two-dimensional electron liquid. By comparison, the values found for the equivalent three-dimensional variable $\Gamma_c = (4\pi n/3)^{1/3}q^2/(4\pi\epsilon_0 k_B T)$ in the classical three-dimensional electron liquid are 125 (Brush et al., 1966) and 155 ± 10 (Pollack and Hansen, 1973). Thus the two-dimensional system is in some sense more stable than the three-dimensional one, as has been estimated by Platzman and Fukuyama (1974). However, the absolute value of Γ_0 obtained in this computer experiment does not agree with the value of 2.8 obtained theoretically by Platzman and Fukuyama using the self-consistent harmonic approximation.

The experiment has been conducted with a 10,000-particle system and a 2,500-particle system. In the temperature range 2–8 K, for which the domain size (see below) is less than the size of the 2,500 particle system, the results for the large and small system are in reasonable agreement. In particular, we do not observe a sharpening of the peak as the sample size increases, indicating that the finite breadth of the specific heat anomaly is a physical phenomenon and not an artefact of the sample size. For temperatures below 2 K the domain size is larger than the 2,500-particle sample. In this region the small sample cannot correctly represent the phenomenon. The observed specific heat in this region for the small sample is almost constant between 2.2 and $2.3k_B$.

The quantities presented up to now are those normally available from a physical experiment. The advantage of a computer experiment is that, in addition, a detailed analysis can be made of the microscopic structure of the material as the phase transition takes place. These results are now presented.

Figure 12-3 is a plot of all 10,000 particles at four places on the specific heat curve designated by A, B, C, D. It shows the gradual reduction of the size of the microcrystalline domains and loss of long-range order as the transition point is approached. In A (0.65 K) two large irregular and one small circular domain are clearly visible, in all of which there is a regular triangular (close-packed) lattice. The domain boundaries, at which the orientation of the lattice changes, are clearly visible. This structure is that of a polymicrocrystalline solid with crystallite size larger than the sample size of 10^{-5} m. At B (1.9 K) the structure is basically unchanged, except now the crystallite size is smaller than the sample at about 0.25×10^{-5} m. At C (2.8 K), just before the transition, the crystallite structure has

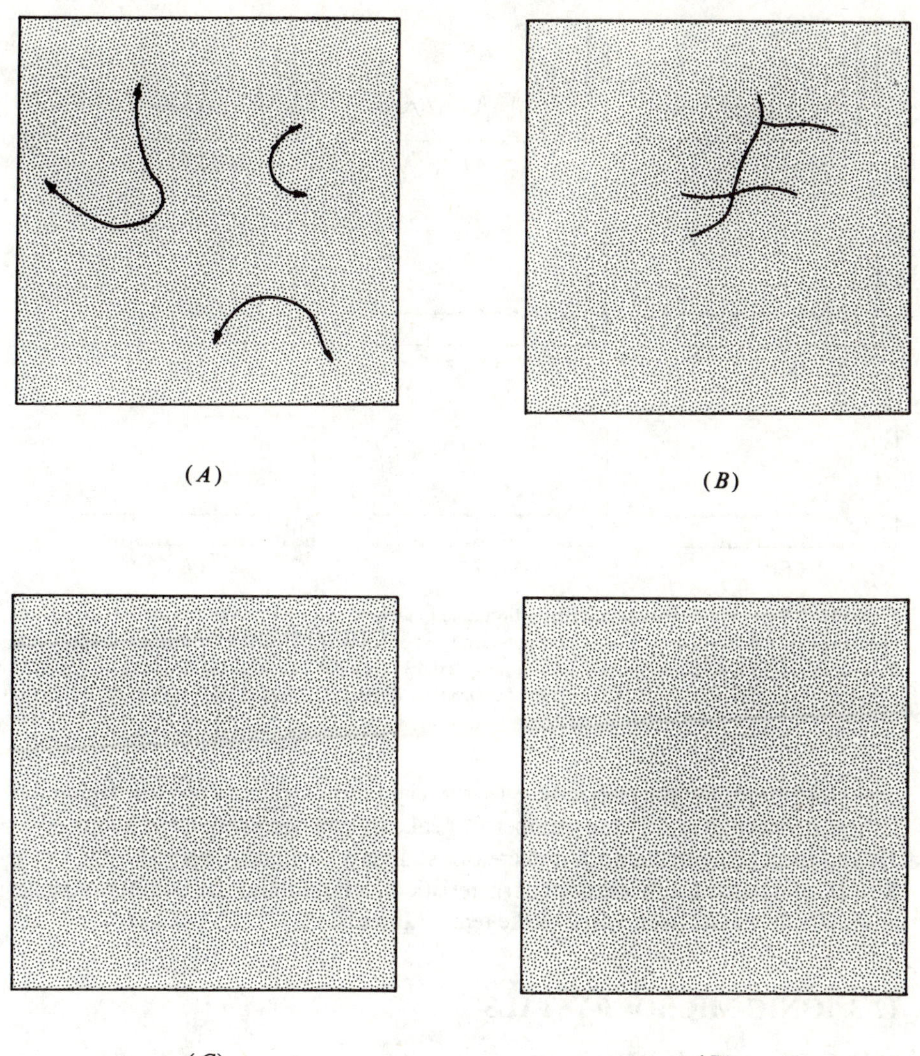

Figure 12-3 Loss of long-range structure when passing through the lambda transition. Density: 10^{14} electrons m^{-2}. The full curves trace out some of the domain boundaries. (*A*) 0.65 K; (*B*) 1.90 K; (*C*) 2.81 K; (*D*) 3.30 K. (*After Hockney and Brown, 1975, courtesy of Journal of Physics C: Solid-State Physics, © Institute of Physics.*)

almost disappeared, while at *D* (3.3 K), just past the transition, the long-range crystallite structure has gone and the structure is that which one associates with a liquid.

The short-range structure is more clearly displayed by the calculation of $g(r)$, the pair distribution function for the material. This is shown in Fig. 12-4 at nine points on the specific heat curve. At *I* and *A* (<0.65 K), very strong and persistent

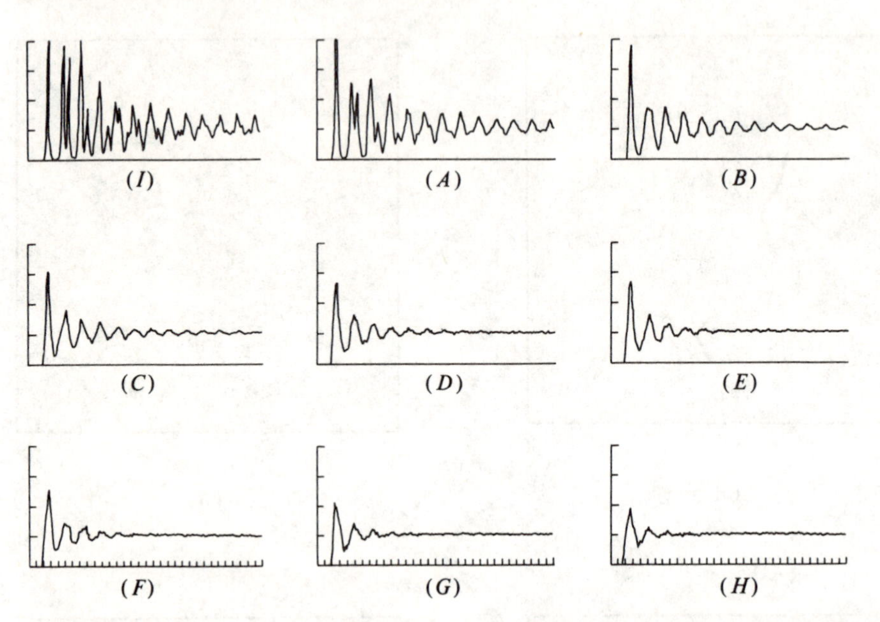

Figure 12-4 Variation of the radial distribution function when passing through the lambda transition. Density: 10^{14} electrons m^{-2}. x axis divisions are in mesh cells H. A perfect triangular lattice would have peaks at 2.74, 4.75, 5.48, and 7.3 divisions. y axis is unity per division. Letters refer to Fig. 12-2. (*After Hockney and Brown, 1975, courtesy of Journal of Physics C: Solid-State Physics,* © *Institute of Physics.*)

peaks are observed corresponding to the crystallite structure with a triangular lattice. Between B and C the number of peaks in $g(r)$ decreases from about 30 to 12, and then between D and E it decreases suddenly to 6 and 4, respectively. From F to H, $g(r)$ has the appearance characteristic of a liquid and the number of peaks gradually decreases from 4 to 2 as the temperature rises.

12-3 IONIC MICROCRYSTALS

12-3-1 Dimensionless Equations

The study of phase changes in ionic microcrystals was conducted with a three-dimensional P^3M code P3M3DP written by Eastwood, Hockney, and Lawrence (1980).

Following Pauling (1950), we have used a coulombic interaction with an inverse power to represent the repulsion of the electronic shells of the ions at short range. The force of interaction between a pair of ions, labeled i and j, separated by a distance r_{ij}, is given by

$$F(r_{ij}) = \frac{q_i q_j}{4\pi\epsilon_0 r_{ij}^2}\left[1 + \text{sgn}\,(q_i q_j)\left(\frac{s_i + s_j}{r_{ij}}\right)^p\right] \tag{12-32}$$

where q is the ionic charge and s, which measures the size of an ion, is proportional to the ionic radius. The parameter p determines the "hardness" of the repulsion between ions.

If the above interaction is combined with Newton's law of motion and the appropriate units used, the equations can be seen to depend on four parameters. For a system with a single type of positive and negative ion, designated by + and −, these are the charge ratio $\sigma(=q_+/q_-)$, the mass ratio $\mu(=m_+/m_-)$, the radius ratio $\rho(=s_+/s_-)$, and the hardness parameter p. If we limit consideration to equivalent salts (e.g., KCl, CaO) for which $\sigma = -1$, there are only three parameters μ, ρ, and p. We find the dependence on mass ratio to be negligible, hence the study is concentrated on the variation with radius ratio and hardness parameter. The above force of interaction is said to be scalable because a computer experiment performed for one set of the eight fundamental data ϵ_0, $m_\pm$, $q_\pm$, $s_\pm$, and p, can be scaled and reinterpreted as a result for any other system with the same σ, μ, ρ, and p.

The dependence on the fundamental data appears in the choice of a suitable set of units. Following the general procedure of Sec. 12-2-1, we use starred quantities to designate units natural to this problem. These are:

$r^* = s_+ + s_-$, the equilibrium separation for an isolated gaseous molecule
$m^* = m_+m_-/(m_+ + m_-)$, the reduced mass of the isolated gaseous molecule
$t^* = (4\pi\epsilon_0 r^{*3} m^*/|q_+q_-|)^{1/2}$, approximately the period of the highest optical vibration mode of the crystal
$F^* = |q_+q_-|/(4\pi\epsilon_0 r^{*2})$, the coulombic force at the separation of the isolated molecule
$E^* = |q_+q_-|/(4\pi\epsilon_0 r^*)$, the coulombic potential energy at the separation of the isolated molecule

If we express all variables in the above units we obtain the following dimensionless equations in which, for the rest of Secs. 12-3-1 to 12-3-2, the variables r, t, F, etc. now stand for the dimensionless variables r/r^*, t/t^*, F/F^* etc. We do not introduce a new symbol for the dimensionless variables in order to avoid a surfeit of superscripts. The pair interactions become

$$F_{+-}(r_{ij}) = -\frac{1}{r_{ij}^2}\left[1 - \left(\frac{1}{r_{ij}}\right)^p\right] \tag{12-33a}$$

$$F_{++}(r_{ij}) = +\frac{1}{r_{ij}^2}\left[1 + \left(\frac{2/(1+\rho^{-1})}{r_{ij}}\right)^p\right] \tag{12-33b}$$

$$F_{--}(r_{ij}) = +\frac{1}{r_{ij}^2}\left[1 + \left(\frac{2/(1+\rho)}{r_{ij}}\right)^p\right] \tag{12-33c}$$

and the equations of motion become, for positive ions,

$$(1+\mu)\frac{d^2\mathbf{r}}{dt^2} = \sum_{\text{pairs}} \mathbf{F} \tag{12-34a}$$

Table 12-1 Mass and radius ratios for the four test salts

Salt	μ	ρ	DT in units of 10^{-15} s	$\omega_0 DT$
KCl	1.10	0.73	7.5	0.42
RbCl	2.40	0.82	7.5	0.33
(LiCl)*	0.20	0.82	3.5	0.51
NaCl	0.65	0.55	6.5	0.52

Radius data from Pauling (1950). $p = 8$: DT is timestep.

and for negative ions

$$(1+\mu^{-1})\frac{d^2\mathbf{r}}{dt^2} = \sum_{\text{pairs}} \mathbf{F} \tag{12-34b}$$

For the initial survey of melting we have chosen three physical and one hypothetic substance, as specified in Table 12-1.

Most alkali halides and many other salts of interest lie within the triangle in the (μ, ρ) plane with corners at NaCl, (LiCl)*, and RbCl. (LiCl)* is a hypothetical substance with the mass ratio of LiCl and radius ratio of RbCl, and has been introduced in order to determine the effect of mass ratio. KCl is included because it is the most commonly used standard substance for which there exists both simulation and physical data. In order to compare the results more easily with laboratory data, the simulation is conducted in SI units as given in Table 12-2.

Table 12-2 Data for individual ions, as used in Eq. (12-32)

$e = +1.6 \times 10^{-19}$ C is the absolute magnitude of the electronic charge

Ion	Mass m in units of 10^{-27} kg	Charge q in units of e	Size s in units of 10^{-10} m
Cl	58.8	−1	1.55
Li	11.8	+1	0.51
Na	38.0	+1	0.77
K	64.7	+1	1.13
Rb	141.2	+1	1.27

12-3-2 Choosing the Timestep

As with the case of the electron film, the choice of timestep must be examined with respect to the pair vibration frequencies and the collective plasma frequency.

The oscillation frequency of an isolated molecule of a single positive and negative ion may be determined from Eqs. (12-33a) and (12-34a, b). Following the analysis of Eqs. (12-11) and (12-12) one has, in dimensionless form

$$\frac{d^2r}{dt^2} = -\frac{1}{r^2}\left[1-\left(\frac{1}{r}\right)^p\right] \tag{12-35}$$

Making the usual expansion about the equilibrium separation $r = 1$, by setting $r = r + r'$ [see Eqs. (12-13) and (12-14)], one has

$$\frac{d^2r'}{dt^2} = -\omega_0^2 r' \tag{12-36a}$$

where

$$\omega_0 = p^{1/2} \tag{12-36b}$$

Reverting to SI units,

$$\omega_0\, DT = p^{1/2}\, DT/t^* = \left[\frac{p|q_+q_-|}{4\pi\varepsilon_0(s_+ + s_-)^3}\left(\frac{1}{m_+}+\frac{1}{m_-}\right)\right]^{1/2} DT \tag{12-37}$$

This result shows the dependence of the oscillation frequency on the main parameters. In particular, the frequency is proportional to the square root of the hardness parameter p and inversely proportional to the $\frac{3}{2}$ power of the equilibrium separation $(s_+ + s_-)$. The dependence on mass shows that the frequency is determined by the mass of the lighter ion.

The oscillation frequency in a crystal lattice will differ from that of the molecule ω_0 for two reasons. First because the equilibrium separation, that is to say the lattice parameter, is larger than $(s_+ + s_-)$, and secondly because of the influence of the other ions in the lattice. However, these effects tend to cancel and are relatively small. It will usually be satisfactory to base the initial choice of timestep on the value of ω_0. Since most ions will be oscillating near this frequency an accurate integration will be necessary and we should select approximately

$$DT \leqslant 0.5/\omega_0 \tag{12-38}$$

The value of $\omega_0\, DT$ for $p = 8$ is given in Table 12-1 and is seen to satisfy this condition.

The equilibrium separation a between adjacent positive and negative ions in an alkali halide crystal may be determined by seeking the minimum in potential energy of the lattice. The alkali halides, with the exception of CsCl, crystallize in a cubic lattice in which the positive and negative ions occupy alternating positions at the corners of each cube, as shown in Fig. 12-5. Each ion is immediately surrounded by six neighbors of the opposite sign. This is known as the "first coordination shell" and the number of ions within it, here six, is called the "coordination number." The second-nearest neighbors form a shell of twelve ions of the same sign at a distance of $\sqrt{2}a$. Following Pauling (1950), we obtain the energy per molecule of such an infinite lattice, expressed in dimensionless units, as

$$U(r) = -\frac{A}{r} + \frac{C}{(p+1)}\frac{1}{r^{(p+1)}}G(\rho) \tag{12-39}$$

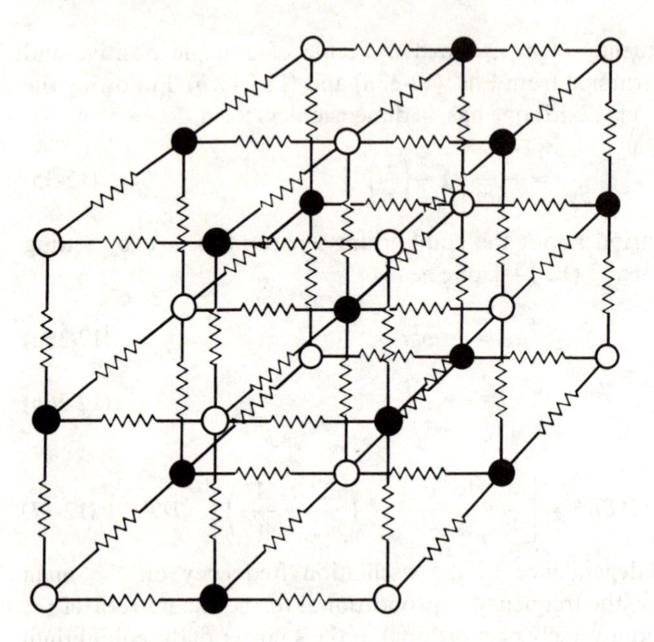

Figure 12-5 The arrangement of ions in an alkali–halide crystal. Full circles positive ions, e.g. K^+, open circles negative ions, e.g. Cl^-. The zig-zag lines indicate the imaginary springs joining neighbouring ions that are used in estimating the vibrational frequencies.

where r is the separation of neighboring positive and negative ions in the lattice, and

$$G(\rho) = 1 + \frac{1}{2^{(p+1)/2}} \left\{ \left[\frac{2}{(1+\rho^{-1})} \right]^p + \left[\frac{2}{(1+\rho)} \right]^p \right\} \tag{12-40}$$

In this expression $A = 1.747$ is the Madelung constant that takes into account the sum of the electrostatic energy for all pairs of ions in the lattice, and $C = 6$ is the coordination number. The first term in Eq. (12-39) is the electrostatic energy due to an infinite lattice of ions and the second the energy due to the repulsive interaction between ions in the first and second coordination shells. The repulsive effect of more distant ions is ignored. In the factor $G(\rho)$, the first term is due to the repulsion of unlike ions at a separation r. The second term is due to the repulsion of pairs of positive ions separated by $\sqrt{2}r$, and the third term is due to the repulsion of negative ions separated also by $\sqrt{2}r$.

The equilibrium lattice separation a at zero temperature is found by finding the separation r that minimizes $U(r)$. This occurs when

$$\left.\frac{dU}{dr}\right|_{r=a} = 0 \tag{12-41a}$$

whence

$$\frac{A}{a^2} - \frac{C}{a^{(p+2)}} G(\rho) = 0 \tag{12-41b}$$

and
$$a = \left[\frac{C}{A} G(\rho)\right]^{1/p} \tag{12-41c}$$

Since the unit of distance r^* is the equilibrium separation of a molecule, a in Eq. (12-41c) is the factor by which the lattice separation is increased over that of an isolated molecule. For $\rho = 1$ and $p = 8$, $G(\rho) = 1.088$ and $a = 1.167$.

Substituting Eq. (12-41c) into Eq. (12-39) one obtains a neater form for the lattice energy

$$U(r) = -\frac{A}{r}\left[1 - \frac{1}{(p+1)}\left(\frac{a}{r}\right)^p\right] \tag{12-42a}$$

The energy per molecule at the equilibrium lattice separation $r = a$ is therefore

$$U(a) = -\frac{p}{(p+1)}\frac{A}{a} \tag{12-42b}$$

This binding energy is the energy required to break up the crystal into its individual ions.

In order to obtain expressions for the vibration frequencies of the crystal lattice it is necessary to interpret the expression for the variation of the lattice energy with lattice separation in terms of a spring model. In this model the only interaction between ions is by way of springs with a natural length a between adjacent ions as is shown in Fig. 12-5. If the spring constant is λ and the lattice is expanded to a lattice separation of $(a+\delta r)$ then the lattice energy per pair of ions (or molecule) is

$$U(a+\delta r) = U(a) + \frac{C}{2}\lambda\,\delta r^2 \tag{12-43a}$$

where $C = 6$ is the number of springs associated with each molecule. On the other hand, the actual change of lattice energy is given by a Taylor expansion of Eq. (12-42a):

$$U(a+\delta r) = U(a) + \left.\frac{dU}{dr}\right|_a \delta r + \left.\frac{d^2U}{dr^2}\right|_a \frac{\delta r^2}{2} \tag{12-43b}$$

where
$$\left.\frac{dU}{dr}\right|_a = 0 \text{ by Eq. (12-41}a\text{)}$$

and
$$\frac{d^2U}{dr^2} = \frac{A}{r^3}\left[(p+2)\left(\frac{a}{r}\right)^p - 2\right] \tag{12-43c}$$

$$\left.\frac{d^2U}{dr^2}\right|_a = \frac{pA}{a^3} \tag{12-43d}$$

Comparing Eq. (12-43b) with Eq. (12-43a), one has

$$\lambda = \frac{1}{C}\left.\frac{d^2U}{dr^2}\right|_a = \frac{pA}{Ca^3} \tag{12-44}$$

Knowledge of the equivalent spring constant λ allows one to calculate the frequency of a number of vibrational modes. In all cases the frequency is given by

$$\omega = \left(\frac{N\lambda}{m}\right)^{1/2} = \left(\frac{N}{C}\frac{pA}{ma^3}\right)^{1/2} = \left(\frac{N}{C}\frac{A}{ma^3}\right)^{1/2}\omega_0 \tag{12-45}$$

where N is the number of springs per molecule that change their length, that is to say are "exercised" in the mode, and m is the appropriate mass in units of the reduced mass.

If the crystal breathes and all the ions move in unison, all the springs are exercised ($N = C$) and $m = 1$, hence the frequency of the breathing mode is

$$\omega_4 = \left(\left.\frac{d^2U}{dr^2}\right|_{r=a}\right)^{1/2} = \left(\frac{pA}{a^3}\right)^{1/2} = \left(\frac{A}{a^3}\right)^{1/2}\omega_0 \tag{12-46a}$$

Comparing with the vibration frequency of the isolated molecule, ω_0 in Eq. (12-36*b*), one has

$$\frac{\omega_4}{\omega_0} = \left(\frac{A}{a^3}\right)^{1/2} = 1.048 \tag{12-46b}$$

for the case $\rho = 1$, $p = 8$. The similarities of the lattice and molecule frequencies has been demonstrated.

If one ion of dimensionless mass m oscillates in its potential well, while all the others are stationary, $N = 2$ and the frequency of one-particle oscillation is

$$\omega_5 = \left(\frac{2}{C}\frac{pA}{ma^3}\right)^{1/2} \tag{12-47a}$$

If the ions are of roughly equal mass, then the reduced mass is half the mass of an ion, and therefore $m = 2$. Hence

$$\omega_5 = \left(\frac{pA}{Ca^3}\right)^{1/2} = C^{-1/2}\omega_4 \tag{12-47b}$$

If alternate planes of ions move in opposite directions, one has an "optical" mode of vibration in which $N = 2$ and $m = 1$, hence the optical frequency is

$$\omega_6 = \left(\frac{2}{C}\frac{pA}{a^3}\right)^{1/2} = \left(\frac{2}{C}\right)^{1/2}\omega_4 \tag{12-48}$$

Summarizing the above results and using the data for KCl, we have:

molecule vibration	$\omega_0 DT = 0.42$ (15 steps/period)
breathing mode	$\omega_4 DT = 0.44$ (14 steps/period)
one-particle oscillation	$\omega_5 DT = 0.19$ (33 steps/period)
optical mode	$\omega_6 DT = 0.27$ (24 steps/period)

Thus, although a choice of $\omega_0 DT \approx 0.4$ might appear rather inaccurate, we find that most particles vibrate in a crystal with slower frequencies near ω_5 and ω_6, and these are accurately integrated. The breathing mode, although present, need not be

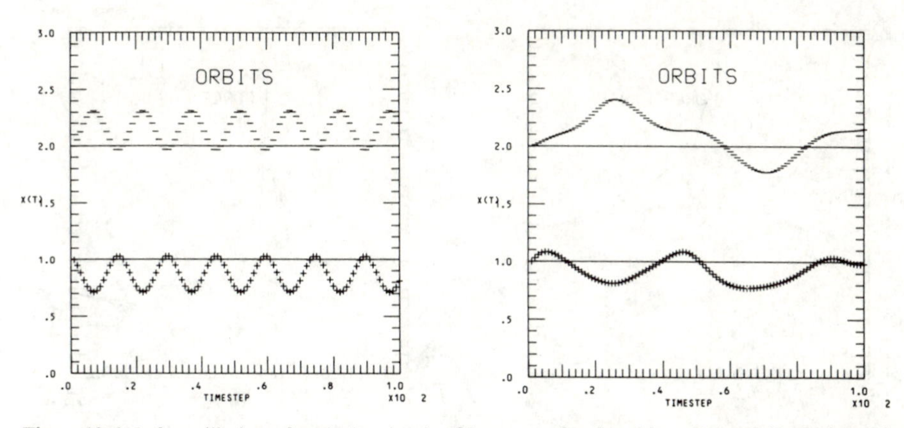

Figure 12-6 *Left*, oscillation of an isolated pair of ions, a molecule, with $\omega_0 DT = 0.42$. *Right*, typical oscillations of ions within a crystal with the same timestep. Note the slower frequency corresponding to $\omega_6 DT \simeq 0.3$. (*From Eastwood, Hockney, and Lawrence, 1979, courtesy of Computer Physics Communications,* © *North-Holland Publishing Co.*)

integrated particularly accurately because it is of small amplitude. However, it is essential to choose a timestep for which the breathing mode is stable (i.e., $\omega_4 DT < 2$) because, if this were not the case, its amplitude would grow exponentially and very quickly dominate the motion. Figure 12-6 shows orbits obtained for a KCl molecule (*left*) and a KCl crystal at 600 K (*right*) which can be seen to be in agreement with the above estimates for the vibrational modes.

In the program P3M3DP the force is held constant at $F(r) = F(b)$ for all $r < b$ where b is the radius for which

$$\left[2\left(\frac{1}{m_+} + \frac{1}{m_-}\right)\left|\frac{dF}{dr}\right|\right]^{1/2} DT = 1.8 \tag{12-49}$$

This ensures that the frequency associated with the gradient of the force never exceeds the stability limit $\omega DT = 2$. If it is found that too many ions have an energy that permits them to approach another ion closer than the distance b, the timestep must be reduced. Figure 12-7 shows the force law used in a simulation of KCl. The solid line is the total force, the dotted line that goes to the origin is the mesh force, and the dotted line that goes to zero at large separations is the short-range correction.

The collective mode of oscillation is the plasma frequency which for a particular ion is, in SI units,

$$\omega_p = \left(\frac{nq^2}{\epsilon_0 m}\right)^{1/2} \tag{12-50}$$

where n is the density of that ion and m is the mass of the ion. It is clear from Eq. (12-50) that the lightest ion will have the highest frequency and will determine the timestep. Expressed in terms of the dimensionless variables, one has

$$\omega_p = (4\pi n/m)^{1/2} \tag{12-51}$$

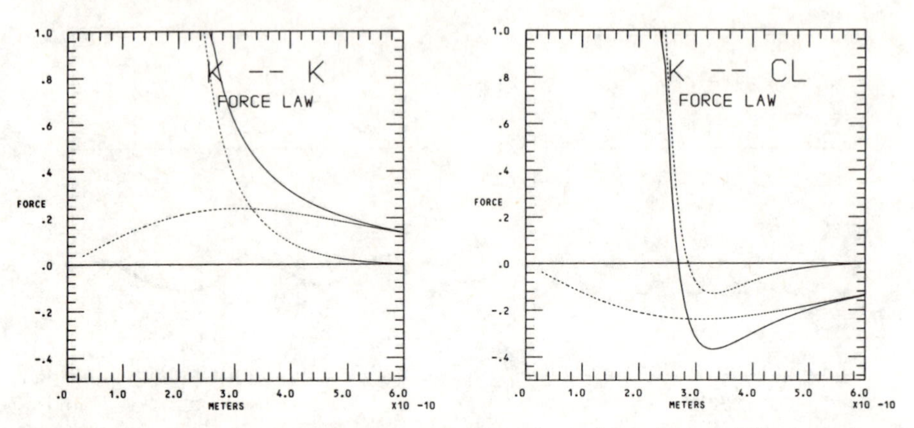

Figure 12-7 Force law between like and unlike ions in a simulation of KCl. A mesh cell $H = 1.57 \times 10^{-10}$ m and the short-range correction extends to $r_c = 6.28 \times 10^{-10}$ m. Total force (continuous line), mesh force (broken line going to origin), short-range correction (other broken line). (*From Eastwood, Hockney, and Lawrence, 1979, courtesy of Computer Physics Communications,* © *North-Holland Publishing Co.*)

and expressed as a ratio to the molecule frequency

$$\frac{\omega_p}{\omega_0} = \left(\frac{4\pi n}{mp}\right)^{1/2} \tag{12-52}$$

For the case of ions of roughly equal mass (e.g., K^+ and Cl^-), the dimensionless mass $m = 2$. The dimensionless density n is the number of positive ions in a volume $r^* \times r^* \times r^*$. For the KCl lattice $n = \frac{1}{2}$, hence

$$\frac{\omega_p}{\omega_0} = \frac{\omega_p DT}{\omega_0 DT} = \left(\frac{\pi}{p}\right)^{1/2} = 0.6 \tag{12-53}$$

for the case $p = 8$. We have therefore shown that for a typical alkali halide the plasma frequency is less than the molecule vibration frequency. A timestep selected to integrate the molecule vibration satisfactorily using Eq. (12-38) will also integrate a plasma oscillation satisfactorily. It is obvious from Eq. (12-52) that as the repulsion becomes softer ($p < 3$) the plasma frequency will exceed the molecule frequency and will determine the timestep. This will be the case for the three-dimensional one-component plasma for which $p = 0$. However, in the case of the alkali halides it is found that $p \geqslant 8$ and vibrational modes will determine the timestep.

12-3-3 Spatial Mesh and Computer Economy

The choice of spatial mesh is governed by the desire to minimize the cycle time and computer storage. The simulations reported here were performed with a $(16 \times 16 \times 16)$ spatial mesh of 4,096 mesh cells. The cell width $H = 3.36 \times 10^{-10}$ m and the outer limit to the short-range force of $r_c = 3.2H$. The chaining mesh is $(5 \times 5 \times 5)$ with mesh separation of $3.2H$. With these choices the errors in the force calculation are (see Secs. 8-3 and 8-6):

Mean-squared error in force $Q = 1.0 \times 10^{-4}$
Mean-squared force fluctuations $P = 7.7 \times 10^{-5}$
Mean-squared deviation of the force from the mean $Z = 2.3 \times 10^{-5}$

The short-range force is tabulated as a function of r^2 in order to avoid the need to take a square root. Linear interpolation is used in a table of 512 values for each pair of ions. The radial distribution function (RDF) is accumulated out to $r = r_c$ during the calculation of the short-range correction. The pressure and potential energy are calculated as integrals over the RDF and the following error arises from the truncation of the integral at $r = r_c$ [see Sec. 12-3-4(4)]:

Pressure truncation	1.6×10^4 N m^{-2}
Energy truncation	3.4×10^1 J kg^{-1}

The following program timings on the IBM 360/195 are obtained for the case of a 512 ion system:

	(1) *Filling region*	(2) *Microcrystal*
QSHARE	= 0.07 s	0.07 s
POTP3M	= 0.30 s	0.30 s
MESHFC	= 0.10 s	0.10 s
SRFORC	= 0.40 s	0.60 s
NEWENX	= 0.02 s	0.02 s
TIMESTEP	= 0.89 s	1.09 s

The separate timings are respectively for assigning charge to the mesh (QSHARE), solution for the potential (POTP3M), calculation of the mesh force (MESHFC), calculation of the short-range correction (SRFORC), and the updating of the positions (NEWENX). In both cases the short-range correction calculation exceeds the time for potential calculation and scaling of the problem is not therefore optimal. However, it is easy to see that choosing a finer mesh of $(32 \times 32 \times 32)$ would lead to a worse balance. This would increase the time for POTP3M by a factor of 8 to 2.4 s and decrease the time for SRFORC by a factor $\frac{1}{8}$ to ~ 0.05 s. The timestep would be increased to about 2.6 s. The storage required by the 16^3-mesh program and data is 476 k bytes.

12-3-4 Thermodynamic Measurements

The program P3M3DP provides the following diagnostics at each timestep:

1. *The potential energy* The total potential energy

$$V = \frac{1}{2} \sum_{i,s} \sum_{\substack{i',s' \\ (i',s') \neq (i,s)}} \phi_{i,i'} \tag{12-54}$$

is split in a similar manner to the force into a short-range and mesh part. The pair-potential energy $\phi_{i,i'}$ of particle i of species s and particle i' of species s' is written as

$$\phi_{i,i'} = \phi^{\text{sr}}_{i,i'} + \phi^{m}_{i,i'} \tag{12-55}$$

The short-range potential energy V_{sr} is calculated from the short-range pair correlation function obtained during the short-range force calculation

$$V_{\text{sr}} = \sum_{\mathbf{p}} \sum_{(i,j)\in P(\mathbf{p})} \phi^{\text{sr}}_{i,j} \tag{12-56}$$

The sum $\mathbf{p}$ is over all chaining cells and $P(\mathbf{p})$ is the set of pairs with at least one member in chaining cell $\mathbf{p}$.

The mesh-defined potential energy is obtained by approximating $\phi^{m}_{i,i'}$ by the mesh-computed value

$$\phi^{m}_{i,i'} \simeq \frac{q_s q_{s'}}{\varepsilon_0} \sum_{\mathbf{p},\mathbf{p}'} W(\mathbf{x}_i - \mathbf{x}_{\mathbf{p}}) G(\mathbf{x}_{\mathbf{p}} - \mathbf{x}_{\mathbf{p}'}) W(\mathbf{x}_{i'} - \mathbf{x}_{\mathbf{p}'}) \tag{12-57}$$

giving

$$V_m = \frac{V_c}{2} \sum_{\mathbf{p}} \rho(\mathbf{p})\phi(\mathbf{p}) - V_{\text{self}} \tag{12-58}$$

where $V_c = H^3$ is the cell volume.

Sums in Eq. (12-57) and Eq. (12-58) over $\mathbf{p}$ and $\mathbf{p}'$ are taken over all mesh points on the charge-potential mesh. The term V_{self} arises in Eq. (12-58) because the sum over mesh points includes the $(i, s) = (i', s')$ term.

The first term in Eq. (12-58) is computed from the mesh-defined charge density harmonics

$$\frac{V_c}{2} \sum_{\mathbf{p}} \rho(\mathbf{p})\phi(\mathbf{p}) \equiv \frac{1}{2V_b} \sum_{\mathbf{k}} \hat{G}(\mathbf{k}) |\hat{\rho}(\mathbf{k})|^2 \tag{12-59}$$

where $V_b = L^3$ is the box volume.

The second term is given by

$$V_{\text{self}} = \left(\frac{13}{35\pi\epsilon_0}\right) \sum_{s} N_s q_s^2 \tag{12-60}$$

where N_s is the number of particles of species s and the term in the bracket in Eq. (12-60) is the self energy of an isolated unit S2-shaped charge.

2. ***The kinetic energy*** The kinetic energy $\mathscr{T}$ is calculated from the particle momentum p_i by

$$\mathscr{T} = \sum_{i,s} \frac{p_i^2}{2m_s} \tag{12-61}$$

and is timecentered at the same timelevels n at which potential energies are known by the approximation

$$\mathscr{T}^n = \sum_{i,s} \left[\frac{(p_i^{n+1/2})^2 + (p_i^{n-1/2})^2}{4m_s} \right] + \frac{(V^{n+1} - 2V^n + V^{n-1})}{8} \tag{12-62}$$

3. ***The pressure*** The pressure is

$$P = P_K + P_\phi \tag{12-63}$$

$$= nk_B T + \frac{1}{6V_b} \sum_{(i,s)} \sum_{\substack{(i',s') \\ (i,s) \neq (i',s')}} (\mathbf{x}_i - \mathbf{x}_{i'}) \cdot \mathbf{f}_{ii'} \tag{12-64}$$

where n is the number density, k_B is the Boltzmann's constant, T is the temperature, $\mathbf{f}_{ii'}$ is the pair force and the sum (i, s) is over all particles in the computational box and the sum (i', s') is over all particles in the computation box and its periodic images.

The kinetic pressure P_K is computed from the kinetic energy

$$P_K = nk_B T = \frac{2\mathscr{T}}{3V_b} \tag{12-65}$$

The interaction pressure P_ϕ is found by splitting the sum on the right-hand side of Eq. (12-64) into the total potential energy and a correction term

$$P_\phi = \frac{V}{3V_b} + \frac{1}{3V_b} \sum_{\mathbf{p}} \sum_{(ij) \in P(\mathbf{p})} [(\mathbf{x}_i - \mathbf{x}_j) \cdot \mathbf{f}^+_{i,j} - \phi^+_{i,j}] \tag{12-66}$$

The correction term is computed using the short-range pair correlation function values obtained in the short-range force calculation. $\phi^+_{i,j}$ is the pair potential energy corresponding to the noncoulombic part $\mathbf{f}^+_{ij}$ of the short-range force.

4. ***Energy and pressure truncation errors*** In the case of coulombic and repulsive core force law the cutoff in the short-range tables at $r = r_c$ may lead to significant difference between the measured and expected values of interaction pressure and potential energy, even in cases where the dynamics are modeled accurately. The difference arises because the numerical scheme replaced the small repulsive core forces at $r \geqslant r_c$ by zero.

If the short-range corrections are truncated for pairs separated by distances greater than r_c, then the error incurred in the potential-energy calculation is

$$\Delta V = \tfrac{1}{2} \sum_{(i,s)} \sum_{\substack{(i',s') \\ r_{i,i'} > r_c}} \phi^+_{i,i'} \tag{12-67}$$

where $\phi^+_{i,i'}$ is the repulsion core contribution to the pair potential energy of particles i and i'. Rewriting Eq. (12-67) in terms of the radial distribution function $g_{s,s'}(r)$ for species s and s' gives

$$\Delta V = \tfrac{1}{2} \sum_{s,s'} N_s \frac{N_{s'}}{V_b} \int_{r_c}^{\infty} 4\pi r^2 g_{s,s'}(r) \phi^+_{s,s'}(r)\, dr \tag{12-68}$$

An estimate of the value of ΔV is obtained by replacing $g_{s,s'}$ by its asymptotic value of one and performing the integral

$$\Delta V \simeq \tfrac{1}{2} \sum_{s,s'} N_s \frac{N_{s'}}{V_b} \int_{r_c}^{\infty} 4\pi r^2 \phi^+_{s,s'}(r)\, dr \tag{12-69}$$

An estimate of the error in the interaction pressure is obtained in a similar fashion

$$\Delta P_\phi \simeq \frac{\Delta V}{3V_b} + \frac{1}{6V_b} \sum_{s,s'} N_s \frac{N_{s'}}{V_b} \int_{r_c}^{\infty} 4\pi r^2 [rf^+_{s,s'}(r) - \phi^+_{s,s'}(r)]\, dr \tag{12-70}$$

$$= \frac{1}{6V_b^2} \sum_{s,s'} N_s N_{s'} \int_{r_c}^{\infty} 4\pi r^3 f^+_{s,s'}(r)\, dr \tag{12-71}$$

The correction terms may be used in two ways: If we regard the numerical scheme as a method of accurately modeling an "approximate" force law (namely one where $f^+ \equiv 0$ for $r \geqslant r_c$) then ΔV and ΔP_ϕ should be omitted from the calculation of energy and pressure, respectively, and be used solely as a measure of the difference between the "approximate" and "exact" force laws. However, if the scheme is regarded as approximately modeling an "exact" force law (i.e., one where f^+ is very small but nonzero for $r \geqslant r_c$) then ΔV and ΔP_ϕ should be added to V and P_ϕ, respectively. In both cases, it is desirable for r_c to be chosen so that $\Delta V/V$ and $\Delta P_\phi/P_\phi$ are small. In the program, ΔV and ΔP_ϕ are computed and included in the output in the initialization stage, but not added to V and P_ϕ.

5. *The pair correlation function* At each timestep, the radial distribution function $g_{s,s'}(r_k)$ for all pairs of species (s, s') is calculated for particle separations up to the cutoff radius r_c of the short-range tables. Values of $g_{s,s'}(\mathbf{x}_i; r_k)$ are obtained by linear inverse interpolation from particles i' of species s' to a mesh of concentric spherical cells centered on particle i of species s, where cell k has a mesh point at $r_k^2 = k\Delta r^2$, cell boundaries at $r_-^2 = r_k^2 - \Delta r^2/2$ and $r_+^2 = r_k^2 + \Delta r^2/2$, and volume V_k. The radial distribution functions $g_{s,s'}(r_k)$ are obtained by averaging $g_{s,s'}(\mathbf{x}_i; r_k)$ over all particles i of species s. If $s = s'$, the self terms are excluded from the distribution function.

The radial distribution function $g_{s,s'}(r_k)$ may be written in terms of sums over pairs of particles (i, i') where i labels particles of species s and i' labels particles of species s':

$$g_{s,s'}(r_k) = \frac{2V_b}{N_s N_{s'} V_k} \sum_{\mathbf{p}} \sum_{(i,i') \in P(\mathbf{p})} U(|\mathbf{x}_i - \mathbf{x}_{i'}|^2 - r_k^2) \tag{12-72}$$

U is the linear inverse interpolation function

$$U(\xi) = \begin{cases} 1 - \dfrac{|\xi|}{\Delta r^2} & \text{if } |\xi| \leqslant \Delta r^2 \\ 0 & \text{otherwise} \end{cases} \tag{12-73}$$

where V_b = volume of the computational box
N_s = number of particles of species s
V_k = volume of the spherical cell k

The sum $\mathbf{p}$ is over chaining cells and $P(\mathbf{p})$ is the set of pairs with at least one member in chaining cell $\mathbf{p}$.

The short-range potential energy V_{sr} and the interaction pressure are computed from $g_{s,s'}(r_k)$ and tabulated values of $\phi_{s,s'}(r_k)$ and of $[r_k f^{+}_{s,s'}(r_k) - \phi^{+}_{s,s'}(r_k)]$ using the following expressions:

$$V_{sr} = \sum_{s,s'} \sum_{k} \frac{N_s N_{s'} V_k}{2V_b} g_{s,s'}(r_k)\phi_{s,s'}(r_k) \tag{12-74}$$

$$P_\phi = \frac{1}{3V_b}\left\{V + \sum_{s,s'} \sum_{k} \frac{N_s N_{s'} V_k}{2V_b} g_{s,s'}(r_k)[r_k f^{+}_{s,s'}(r_k) - \phi^{+}_{s,s'}(r_k)]\right\} \tag{12-75}$$

Sums over k in Eq. (12-74) and Eq. (12-75) are taken over all entries in the tables.

6. ***The diffusion coefficient*** Although the measurement of the diffusion coefficient is not included in the published version of the P3M3DP code, it is an important measurement that can easily be added. The equation for constant anisotropic spatial diffusion in d dimensions is

$$\sum_{i=1}^{d} D_i \frac{\partial^2 n}{\partial x_i^2} = \frac{\partial n}{\partial t} \tag{12-76}$$

where $n(x_i, t)$ is the number density of particles and we name the coordinates (x_1, x_2, x_3) instead of the usual (x, y, z). It can be shown by substitution that the solution corresponding to N particles all at $x_i = 0$ at $t = 0$ is

$$n(x_i t) = NA \exp\left[-\sum_{i=1}^{d} (x_i^2/4D_i t)\right] \tag{12-77a}$$

where
$$A = \prod_{i=1}^{d} (4\pi D_i t)^{-1/2} \tag{12-77b}$$

After a time t the displacement Δx_i of a particle at x_i from its initial position is x_i, hence the mean-squared displacement is

$$\langle(\Delta x_i)^2\rangle = N^{-1} \int_{-\infty}^{+\infty} n(x_i, t) x_i^2 \, dx_i \tag{12-78}$$

$$= 2D_i t \tag{12-79}$$

Hence for one-dimensional diffusion the diffusion coefficient is given by

$$D_i = \tfrac{1}{2}\langle(\Delta x_i)^2\rangle / t \tag{12-80}$$

If the diffusion is isotropic such that $D_1 = D_2 = D_3 = D$, the diffusion can also be measured from the squared as-the-crow-flies displacement

$$|\Delta \mathbf{r}|^2 = \sum_{i=1}^{d} (\Delta x_i)^2$$

the mean of which is

$$\langle|\Delta \mathbf{r}|^2\rangle = \sum_{i=1}^{d} \langle(\Delta x_i)^2\rangle \tag{12-81a}$$

$$= 2d\,Dt \tag{12-81b}$$

hence
$$D = \frac{1}{2d}\langle|\Delta\mathbf{r}|^2\rangle/t \tag{12-82}$$

In an ensemble of particles we measure the displacement of each particle from its initial position

$$\Delta x_{i,j} = x^n_{i,j} - x^0_{i,j} \tag{12-83}$$

where $x^n_{i,j}$ is the x_i-coordinate of the jth particle at $t = nDT$. Then the mean-squared displacement is

$$\langle(\Delta x_i)^2\rangle = N^{-1}\sum_{j=1}^{N}(\Delta x_{i,j})^2 \tag{12-84}$$

In a computer program this may be computed by storing $x^0_{i,j}$ and evaluating Eqs. (12-83), (12-84), (12-81), and (12-82). Alternatively, the displacement may be found by accumulating the velocity at every timestep separately for each particle, thus

$$\Delta x_{i,j} = \sum_{k=1}^{N} v^{k-1/2}_{i,j}\,DT \tag{12-85}$$

where $v^{k-1/2}_{i,j}$ is the velocity of the jth particle in the ith-coordinate direction at the kth timestep.

In either case one extra store is required per particle per coordinate direction. If this is more than can be afforded, a subset only of the particles can be used for the measurement. Alternatively, if the diffusion is known to be isotropic, it is only necessary to compute it in one coordinate direction and use Eq. (12-80). One then requires only one extra store per particle. If the diffusion from a set of calculated initial coordinates is required, then no extra storage need be assigned because $x^0_{i,j}$ can be recalculated whenever the diffusion is to be measured.

12-3-5 Measurements in Different Regions

The measurements of Sec. 12-3-4 are averages over all particles in the system and are most suitable if the material completely fills the computational region. In the case, however, of a microcrystal that only occupies part of the system, an alternative measurement of the radial distribution and pressure is required. This is achieved by defining two regions, examples of which are shown in Fig. 12-8. Region 1 is a rectangular parallellpiped that specifies the region under study and region 2 completely surrounds it and contains all particles within a specified range. For the calculation of the RDF, $g_{ss'}(r)$, all particles of type s in region 1, called datum particles, are paired with all particles of type s' in region 2. If N_i is the number of ss' pairs per datum particle of type s in the ith histogram box of volume V_i then

$$g_{ss'}(r_i) = N_i/(n_{s'}V_i) \tag{12-86}$$

where the RDF is normalized by the average number of density $n_{s'}$ of type s' in region 1. The RDF calculated in this way is shown for like and unlike ions in Fig.

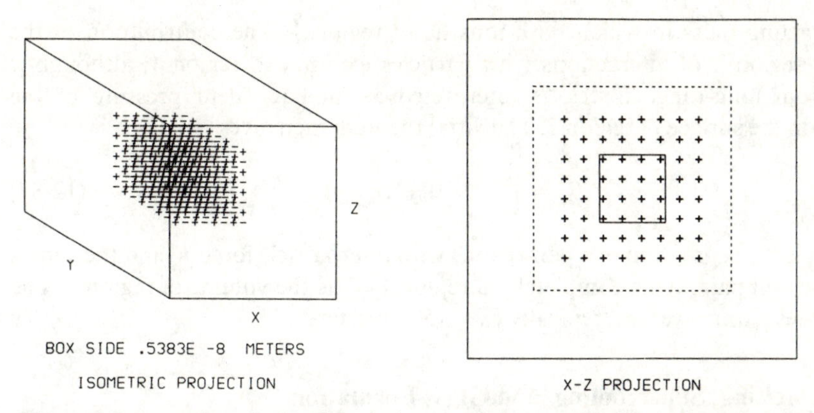

Figure 12-8 Isometric projection, *left*, of microcrystal that partially fills the computational region and, *right*, the X-Z projection showing the location of region 1 (continuous lines) and region 2 (broken lines) used in the measurement of RDF and pressure at the center of a microcrystal. (*From Eastwood, Hockney, and Lawrence, 1979, courtesy of Computer Physics Communications,* © *North-Holland Publishing Co.*)

12-9. In the A-B correlation the number on the vertical coordination bars gives the number of B ions surrounding an A datum ion out to the position of the bar. In Fig. 12-9 (*left*), which corresponds to a crystal of KCl, one sees from the coordination number bars that the first shell of chlorine ions around a potassium ion has six ions and that the second shell of chlorines has eight ions, as is expected from the crystal structure of Fig. 12-5. On the right is shown the correlation between like ions in very hot liquid KCl at 4,000 K. This temperature is so high that no peaks appear in the RDF.

The pressure on region 1 is obtained from a measurement of the virial due to

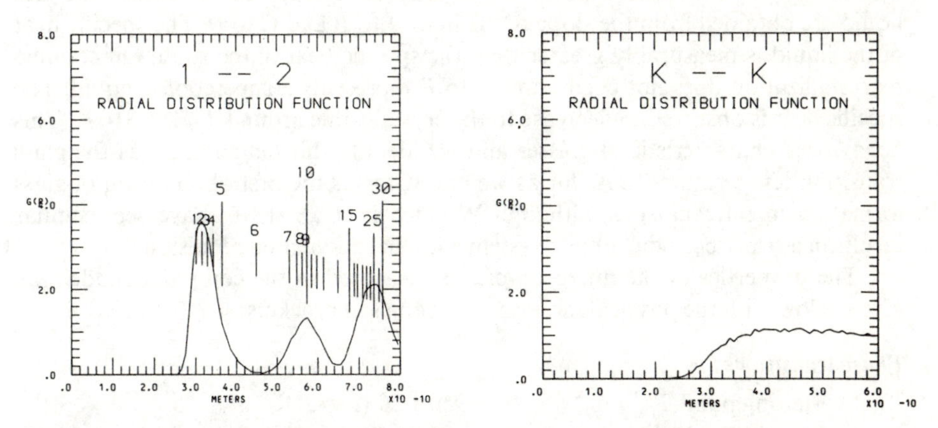

Figure 12-9 Radial distribution functions of, *left*, a KCl microcrystal near 600 K and, *right*, a hot KCl liquid at 4,000 K. Type 1 ions are K^+, type 2 ions are Cl^-. (*From Eastwood, Hockney, and Lawrence, 1979, courtesy of Computer Physics Communications,* © *North-Holland Publishing Co.*)

all interacting pairs in which both ions lie in region 1. The contribution to the virial in region 1 of interactions with particles external to region 1, although it arises from long-range forces, is interpreted as an equivalent pressure of the outside on the surface of region 1. The virial theorem then gives

$$P_1 = \frac{1}{V_1}\left(\tfrac{2}{3}\bar{\mathcal{T}} + \tfrac{1}{3}\overline{\Sigma r_{ij} f_{ij}}\right) \tag{12-87}$$

where r_{ij} is the separation of a pair of ions with interparticle force f_{ij} and the sum is taken over all pairs lying completely in region 1. V_1 is the volume of region 1. The bar denotes a time average, typically over 150 timesteps.

12-3-6 Melting, Supercooling, and Glass Formation

Amini and Hockney (1979) have studied in detail the behavior of one salt, simulated KCl, through all its phase transitions. In their computer experiment a cubic microcrystal of 512 ions was established in equilibrium at about 40 K in the center of the computational region and subjected to cyclic heating with a period of 200 timesteps. For each of the first 25 timesteps, heating is applied by multiplying all velocities by a factor HTFAC = 1.001. For the remaining 175 steps the system is allowed to equilibrate without heating. The internal energy U and temperature T are obtained by averaging the total energy of the system and the kinetic energy over the last 32 steps of the cycle.

Figure 12-10 shows the U/T curve of simulated KCl. The crystal is heated through points A to B, melts between B and C, and the liquid is heated from C through D to 1,600 K. These points are shown as solid circles in Fig. 12-10. The system was then cooled down to about 14 K by setting HTFAC = 0.999. These points are shown as the open circles. We note that, on heating, a first-order phase change occurs between B and C from which we determine the melting point $T_f = T(B \text{ or } C)$ and latent heat $L_f = U(C) - U(B)$. The specific heats for solid and liquid are obtained from the slope of the lines A to B and C to D. The specific heat of the liquid is measurably greater than the specific heat of the solid. On cooling, recrystallization does not occur, and C to E represents a supercooled liquid. The specific heat is observed to decrease to the crystal value around $T_f/3 \simeq 310$ K. This behavior is characteristic of glasses and we identify this temperature as the glass transition temperature T_g. As far as we know, this is the first observation of glass formation in microcrystals, although Woodcock et al. (1976) have seen similar results in a triply periodic infinite system with an exponential repulsion.

The properties of the microsample, as measured in the computer model, are given below with the physical large-sample values in brackets:

Phase transitions

Melting point T_f	$= 950 \pm 20$ (1,045) K
Latent heat L_f/k_B	$= 1240 \pm 50$ (1,580) K
Glass point T_g	$= 310 \pm 50$ K
Entropy change $S_f = L_f/k_B T_f$	$= 1.3 \pm 0.08$ (1.5)

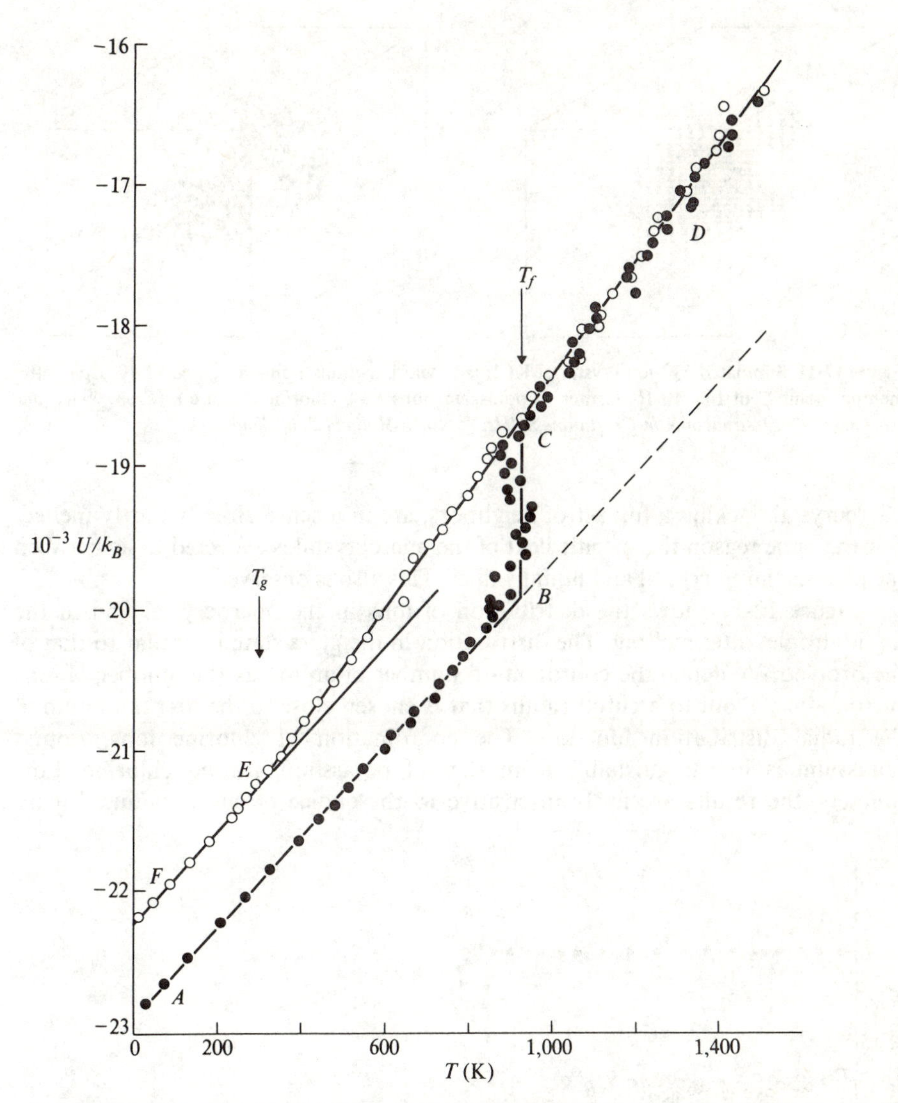

Figure 12-10 U/T curve obtained during melting and glass formation in simulated KCl. T_f = melting point, T_g = glass point. Both axes in kelvins. U has arbitrary origin. (*After Amini and Hockney, 1979, courtesy of Journal of Non-Crystalline Solids, © North-Holland Publishing Co.*)

Heat capacity per ion at zero pressure

C_p solid $= 3.24 \pm 0.05\ (3.03)\ k_B$
C_p liquid $= 4.14 \pm 0.09\ (4.03)\ k_B$
C_p glass $= 3.24 \pm 0.03\ k_B$

The observed melting point and latent heat of the microcrystal are less than the physical large-crystal values. This is to be expected, as the edge ions of the

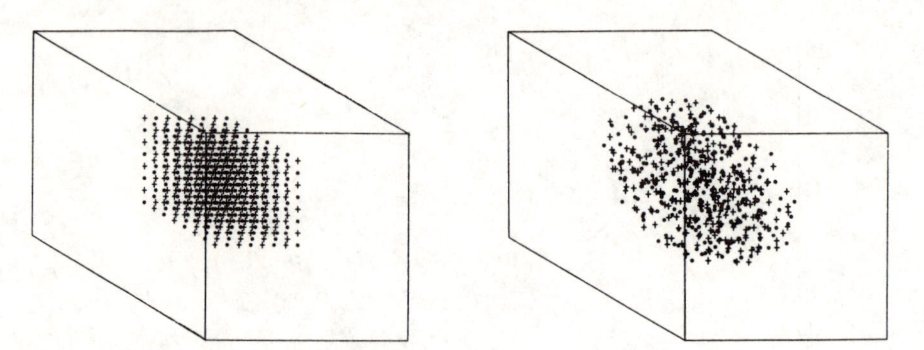

Figure 12-11 Simulated 512-ion crystal of KCl; *left*, before melting, point *A* of Fig. 12-10; *right*, after melting, point *C* of Fig. 12-10. Center of potassium ions (+), chlorine ions (●). (*From Amini and Hockney, 1979, Journal of Non-Crystalline Solids, © North-Holland Publishing Co.*)

microcrystal, lacking a full set of neighbors, are in a sense already partly melted. For the same reason the specific heat of the microcrystal is expected to lie between the physical large-crystal and liquid values. This also is observed.

Figure 12-11 shows the distribution of ions in the microcrystal and in the liquid droplet after melting. The distribution in the glass state is similar to that of the droplet. We define the coordination number of an ion as the number of ions surrounding it, out to a cutoff radius that is chosen close to the first minimum of the radial distribution function. The coordination of chlorine ions around potassium is indistinguishable from that of potassium around chlorine. Fortunately, the results are fairly insensitive to the choice of cutoff radius. Figure

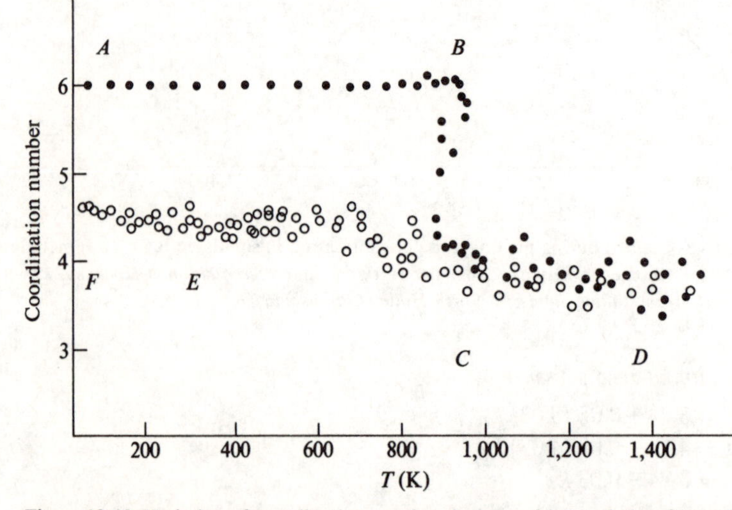

Figure 12-12 Variation of coordination number during melting and glass formation in simulated KCl. Letters indicate position in U/T curve of Fig. 12-10. (*After Amini and Hockney, 1979, courtesy of Journal of Non-Crystalline Solids, © North-Holland Publishing Co.*)

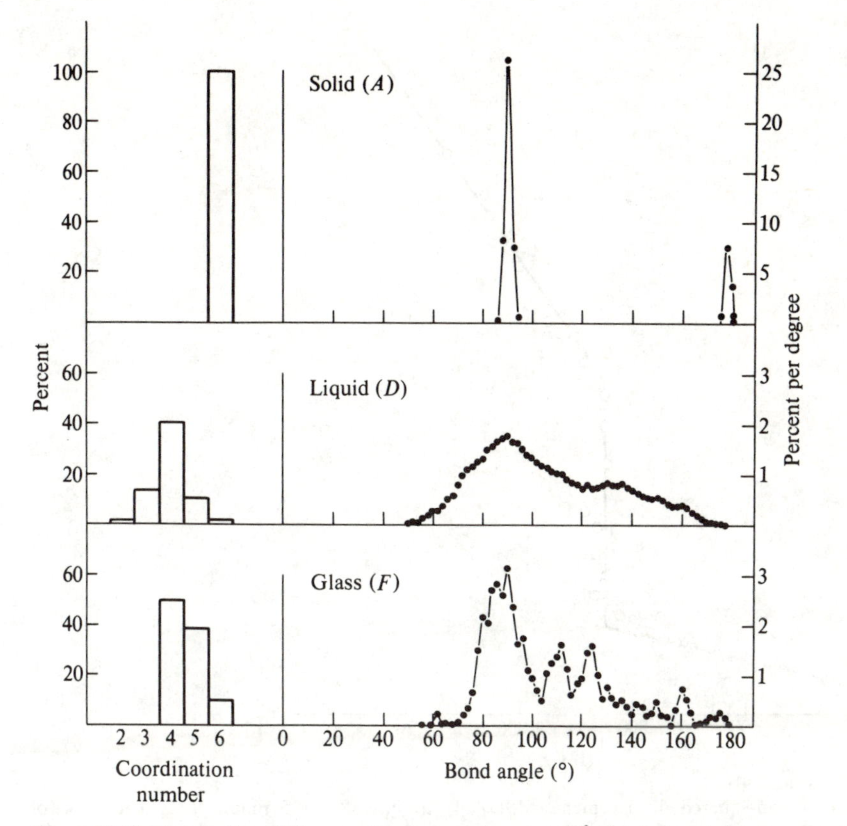

Figure 12-13 Distribution of the coordination number and ClK̂Cl bond angle in the solid, liquid, and glass states. Letters refer to position in *U/T* curve of Fig. 12-10. (*After Amini and Hockney, 1979, courtesy of Journal of Non-Crystalline Solids,* © *North-Holland Publishing Co.*)

12-12 shows the variation of the coordination number, averaged over all ions except those within $\sim 6 \times 10^{-10}$ m of the edge. It shows a sharp drop from six- to fourfold coordination on melting and a slight increase in coordination number to 4.6 on cooling to the glass state. The nearest neighbor distance decreases by 8 percent on melting.

We have also obtained (Fig. 12-13) the distribution of coordination number and the Cl—K—Cl bond angle in the solid, liquid, and glass states. In the solid, all ions are sixfold coordinated and the bond angle has peaks of 90° and 180° as expected from the crystal structure. In the liquid, there is a symmetric distribution of three-, four-, and fivefold coordinated ions with a maximum bond angle near 90° decaying monotonically to zero at 180°. The most striking characteristic of the glass is the loss of all threefold coordinated ions and the appearance of three peaks in the bond angle distribution at 90°, near the tetrahedral angle at 109°, and at 125°. All ions contribute about equally to the first peak, but the second and third peaks come primarily from fourfold coordinated ions.

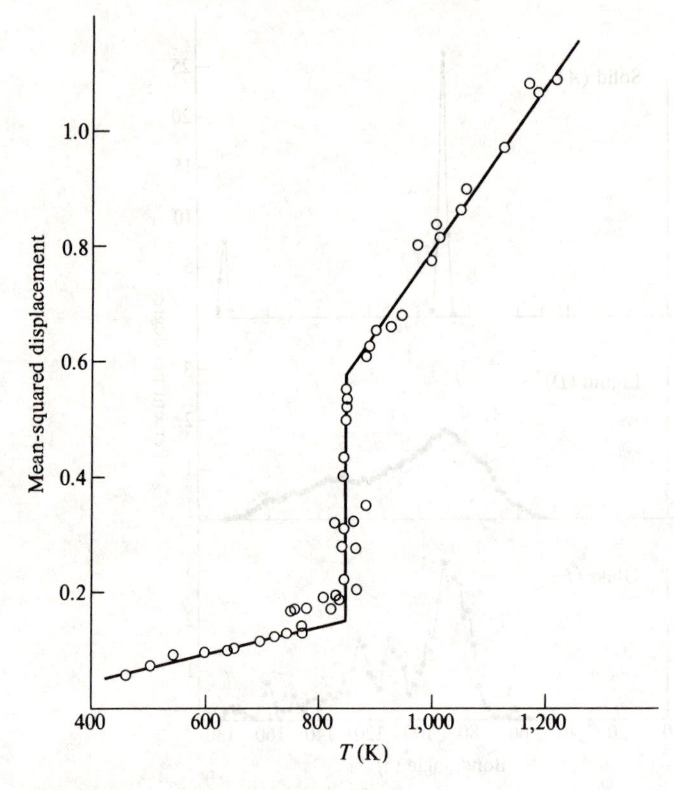

Figure 12-14 Mean-squared displacement (MSD) of the ions after 175 timesteps as a function of temperature, in a simulation of NaCl. (*After Amini, Fincham, and Hockney, 1979, courtesy of J. Phys. C: Solid State Phys.,* © *Institute of Physics.*)

The measurement of diffusion [see Sec. 12-3-4(6)] is an important diagnostic in the study of melting and glass formation. In the melting transition the diffusion coefficient suddenly increases. This is shown in Fig. 12-14, in which the mean-squared displacement after 175 timesteps is plotted as a function of temperature in a simulation of NaCl. It is interesting to note that some 100 K before the melting transition there is a noticeable increase in the diffusion. By observing the motion of all ions during this premelting phase, it can be seen that the phenomenon is due to the diffusion (or melting) of surface ions, which can occur at a lower temperature than the diffusion of ions within the interior of the microcrystal. If the diffusion coefficient is also measured during the cooling of the liquid, the glass point may be identified as the temperature at which the diffusion coefficient extrapolates to zero or becomes negligibly small compared with typical liquid values.

In this computer experiment we find it particularly interesting that the principal features of melting and glass formation are evident in a very small sample, and believe that potassium chloride will be a glass former when laboratory techniques can reach rates of cooling comparable to this computer experiment

(10^{13} K s^{-1}). It is also significant, and likely to be controversial, that we observe glass formation in a purely ionic system *without* covalent bonding.

12-3-7 Radius Ratio, Hardness, and Size Effects

The results of a series of computer experiments on the melting properties of the model salt have been compared with both laboratory measurement and theory by Amini, Fincham, and Hockney (1979). The dimensionless quantities compared were:

$T_f a$ = melting point times lattice separation
$S_f = L_f/k_B T_f$, the entropy of melting

as functions of the radius ratio ρ and the hardness parameter p. The effect of mass ratio μ was found to be negligible in the comparison between the melting point of RbCl and (LiCl)*. These "materials" have the same melting point and radius ratio, however their mass ratios differ by a factor of ten (see Table 12-1).

Figure 12-15 shows the comparison of the melting point with laboratory measurement. If the melting point depended only on the radius ratio, then all the measured points would lie on a single curve. The spread of the laboratory points shows that another parameter is important. Computer simulation results do give a single curve for a fixed value of the hardness parameter, and the curve for $p = 8$ is shown. We note that while the laboratory and simulation results both show an

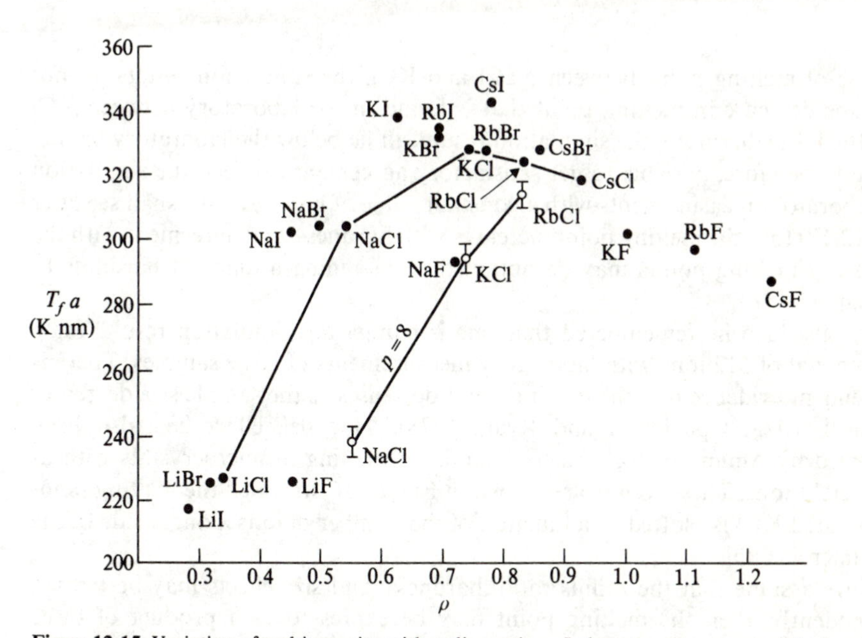

Figure 12-15 Variation of melting point with radius ratio ρ. Laboratory measurement (●), computer simulation (☿). (*After Amini, Fincham, and Hockney, 1979, courtesy of J. Phys. C: Solid State Phys., © Institute of Physics.*)

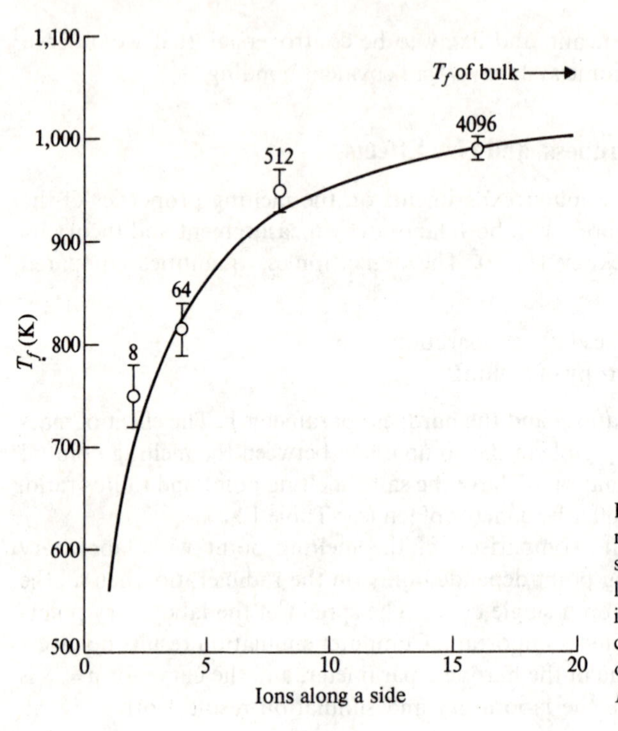

Figure 12-16 The variation of melting point with crystal size for simulated KCl with $p = 8$. Points labeled with the total number of ions in the crystal. The continuous curve is the theoretical prediction of Couchman (1979). (*Courtesy of M. Amini, 1979.*)

increase of melting point between NaCl and KCl, the simulation results do not show the decrease in melting point that is found in the laboratory between KCl and RbCl. Furthermore, the simulation results all lie below the laboratory values. It is not, therefore, possible to get satisfactory agreement between the simulation and laboratory measurements with a constant value of hardness. We shall see later (Fig. 12-18) that the melting point increases with hardness, and agreement with the laboratory melting points may be obtained by assigning a different hardness to each salt.

It must also be remembered that one is comparing simulation results for a microcrystal of 512 ions with laboratory measurements of large samples. There is independent evidence that the melting point decreases as the sample size decreases (Cotterill, 1975; Couchman and Ryan, 1978). This size effect has also been confirmed by Amini (1979), who has simulated melting in microcrystals with as few as eight ions. These results are shown in Fig. 12-16, in which the melting point of simulated KCl is plotted as a function of the number of ions along a side of the cubic microcrystal.

If we assume that the radius ratio, hardness, and size effects may be treated independently, then the melting point may be expressed as a product of three factors, each depending on only one of the variables

$$T_f a = R(\rho)H(p)S(N) \tag{12-88}$$

where N is the number of ions in the microcrystal. The curve in Fig. 12-16 is then proportional to $S(N)$. Theoretical formulae for the size effect have been obtained for a slab crystal by Couchman and Ryan (1978) and for a cubic microcrystal by Couchman (1979). The latter formula is plotted as the continuous line in Fig. 12-16 and shows good agreement with the simulation results which are shown as the open circles. Confirmation of the size effect in real metals has been reported by Couchman and Jesser (1977).

If the simulation results for 512 ions are first adjusted to values for an infinite crystal using the factor $S(N)$, agreement with the laboratory measurements can be obtained with the following values for the hardness parameter

$$\begin{array}{ll} \text{NaCl} & p = 8.9 \\ \text{KCl} & p = 7.9 \\ \text{RbCl} & p = 7.4 \end{array} \tag{12-89}$$

The need to use a varying hardness in the interaction potential of the alkali halides has also been noted by Amini, Fincham, and Hockney (1979) in the case of the Fumi–Tosi potential, using the values given by Sangster and Dixon (1976). In this case the repulsion parameter α_{ij} of Eq. (12-3) is given a different value for each salt.

The results for the entropy of melting are shown in Fig. 12-17, where again it is obvious that a constant hardness does not satisfactorily explain the laboratory results. Although the simulation results show the same trends, they are at lower values and the peak of the simulation results lies at a higher value of p. If the simulation results are adjusted to infinite crystal values and the hardness

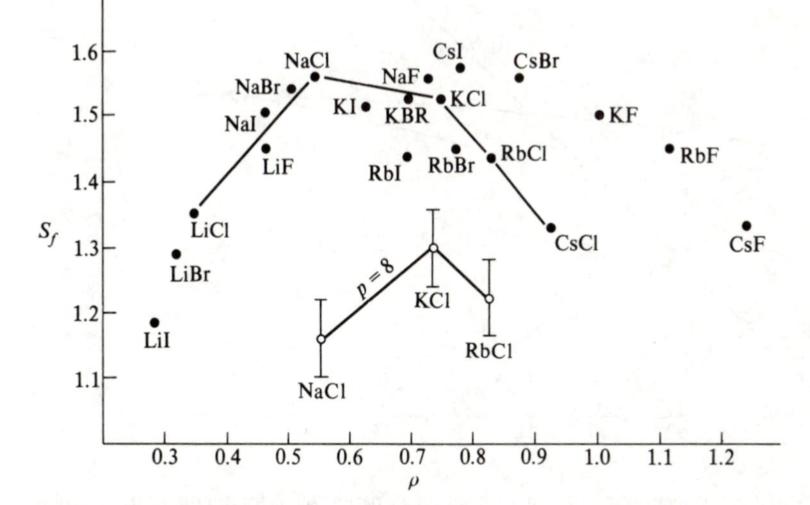

Figure 12-17 Variation of entropy of melting, $S_f = L_f/k_B T_f$, with radius ratio ρ. Laboratory measurement (●), computer simulation ($\bar{\underline{\circ}}$). (*After Amini, Fincham, and Hockney, 1979, courtesy of J. Phys. C: Solid State Phys.,* © *Institute of Physics.*)

parameters (12-89) used, we obtain the following comparison between laboratory and simulation results for the entropy of melting

$$\begin{aligned} &\text{NaCl} \quad S_f = 1.35\,(1.55) \\ &\text{KCl} \quad S_f = 1.40\,(1.51) \\ &\text{RbCl} \quad S_f = 1.27\,(1.44) \end{aligned} \tag{12-90}$$

where the physical values are given in brackets.

We see from the above results that the simple rigid-ion force law (12-32), with hardness parameter adjusted to give agreement with laboratory melting points, shows better agreement with the laboratory values for the entropy of melting. However, the simulation values are still consistently below the laboratory values by between 10 and 15 percent. This suggests that it will be necessary to introduce an additional physical effect before good agreement with both melting point and entropy can be obtained. One possibility is to allow the polarization of the ions by the use of a shell model (see Sec. 12-1-2 and Sangster and Dixon, 1976, page 306).

12-3-8 Testing Theories of Melting

The simulation results also provide a good testbed for theories of melting. These theories predict different dependencies on the hardness parameter This dependence cannot be tested satisfactorily in the laboratory because the experimenter

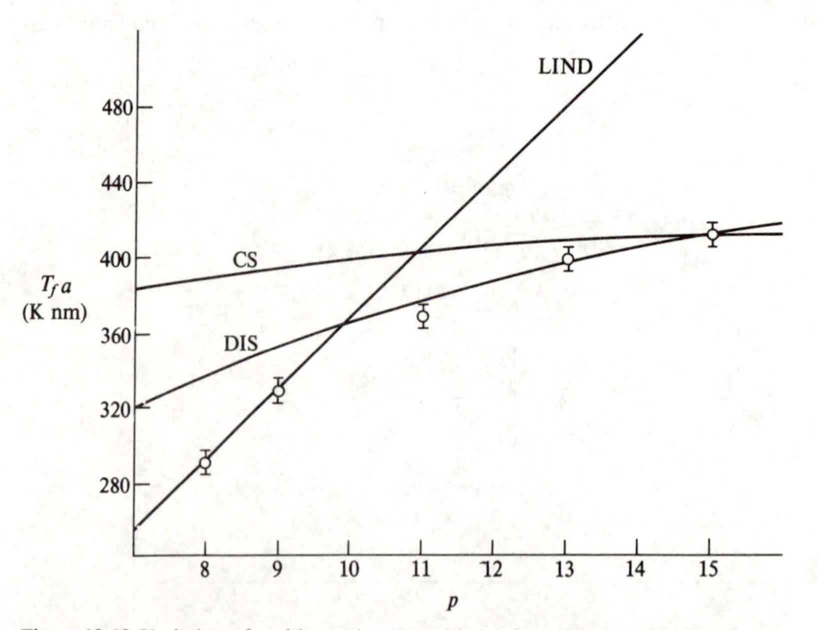

Figure 12-18 Variation of melting point $T_f a$ with hardness parameter p for simulated KCl. Comparison of dislocation theory (DIS), corresponding states (CS), and Lindemann criterion (LIND) with molecular dynamics (ǫ). (*After Amini, Fincham, and Hockney, 1979, courtesy of J. Phys. C: Solid State Phys.,* © *Institute of Physics.*)

does not have control of the hardness of the ions. On the other hand, the computational scientist does have very precise control of this parameter and the results of computer simulation may be used to test theory in a way that is not possible in the laboratory.

The three theories and their predictions are:

1. *Corresponding states theory*, in which both $T_f a$ and L_f are proportional to $U(a)$. From Eq. (12-42*b*) we obtain

$$T_f a \propto p/(p+1) \qquad S_f \text{ constant} \tag{12-91}$$

2. *Lindemann criterion* (*1910*), according to which a crystal melts when the amplitude of thermal vibrations reaches a fixed proportion θ of the lattice separation. Taking the spring constant λ of oscillation proportional to $d^2U/dr^2 = pA/a^3$ [see Eq. (12-44)] and equating the energy of vibration to $k_B T_f$ one obtains

$$k_B T_f = \tfrac{1}{2}\lambda(\theta a)^2 = \frac{1}{2}\frac{pA}{a^3}\theta^2 a^2 \tag{12-92a}$$

hence

$$T_f a \propto p \tag{12-92b}$$

The theory gives no prediction for the latent heat or entropy of melting.

3. *Dislocation theory*, in which a crystal melts when the free energy $U_d - TS_d$ for the thermal generation of dislocations becomes zero. Using a simplified approach to the dislocation theory of Kuhlman-Wilsdorf (1965), Amini, Fincham, and Hockney (1979) conclude that

$$T_f a \propto p/(p+5) \qquad \text{and} \qquad S_f \propto (p+5) \tag{12-93}$$

The above predictions are quite different, and hence the dependence of melting properties on hardness parameter should provide a good method of distinguishing

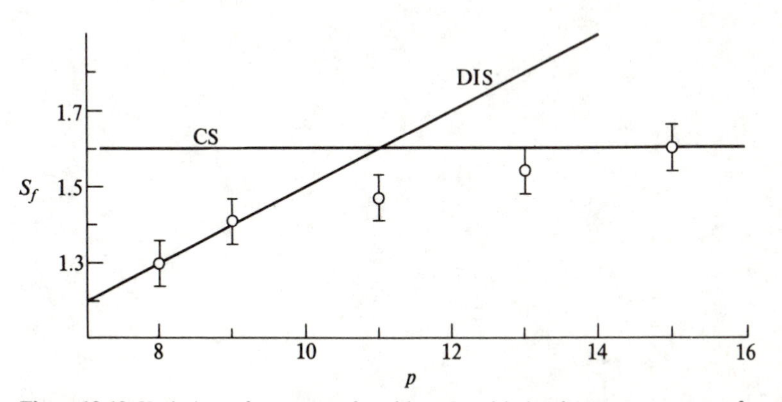

Figure 12-19 Variation of entropy of melting S_f with hardness parameter p for simulated KCl. Comparison of dislocation theory (DIS) and corresponding states (CS) with molecular dynamics (error bar symbol). (*After Amini, Fincham, and Hockney, 1979, courtesy of J. Phys. C: Solid State Phys.,* © *Institute of Physics.*)

between the theories. Figure 12-18 shows the variation of $T_f a$ against p for the three theories and the computer experiment. The simulation results are seen to be in good agreement with the Lindemann criterion for small p and with dislocation theory for large p. The predictions of the theory of corresponding states do not agree with the computer experiments. In the case of the entropy of melting shown in Fig. 12-19, the simulation results agree with dislocation theory for small p and corresponding states for large p. It is clear that none of the theories satisfactorily explain all the results.

APPENDIX

A

FOURIER TRANSFORMS, FOURIER SERIES, AND FINITE FOURIER TRANSFORMS

Fourier transforms and series are used extensively in the design and analysis of particle models. The combination of continuum and discrete quantities in the numerical models necessitates the use of either generalized functions, such as the sampling function Ш (cf. Sec. A-4-5) or the combined use of transforms and series. The summary of Fourier transforms (FT), Fourier series (FS), and finite Fourier transforms (FFT) presented below has a twofold purpose: (1) to show the interrelationship of the transforms and (2) to introduce the notations and normalizations assumed. Readers requiring a more extensive treatment of transforms should refer to Bracewell (1965), or Brigham (1974).

Throughout, we use a circumflex to indicate transformed quantities and the symbol "$\supset$" to denote a transform pair (for FT, FS, and FFT). Thus, for example, $A \supset \hat{A}$ is a shorthand for A Fourier transforms to give $\hat{A}$, etc.

A-1 TRANSFORM DEFINITIONS

A-1-1 The Fourier Transform (FT)

Let A be a continuous function of x and let $\hat{A}$ be a continuous function of k, then the FT pair are given by

$$A(x) = \int_{-\infty}^{\infty} \frac{dk}{2\pi} \hat{A}(k) e^{ikx} \tag{A-1}$$

Table A-1 The transform pairs

Transform	x space	k space	Transform pairs	
FT	x continuous	k continuous	$A(x) = \int_{-\infty}^{\infty} \frac{dk}{2\pi} \hat{A}(k) e^{ikx}$	$\hat{A}(k) = \int_{-\infty}^{\infty} dx\, A(x) e^{-ikx}$
FS(i)	x continuous. Periodic period L	k discrete spacing $= k_0 = 2\pi/L$	$B(x) = \frac{1}{L} \sum_{l=-\infty}^{\infty} \hat{B}(k) e^{ikx}$	$\hat{B}(k) = \int_{L} dx\, B(x) e^{-ikx}$
FS(ii)	x discrete spacing H	k continuous. Periodic period $= k_g = 2\pi/H$	$C(x_p) = \int_{k_g} \frac{dk}{2\pi} \hat{C}(k) e^{ikx_p}$	$\hat{C}(k) = H \sum_{p=-\infty}^{\infty} C(x_p) e^{-ikx_p}$
FFT	x discrete. Period L spacing H	k discrete. Period k_g spacing $= k_0 = 2\pi/L$	$D(x_p) = \frac{1}{L} \sum_{l=0}^{N-1} \hat{D}(k) e^{ikx_p}$	$\hat{D}(k) = H \sum_{p=0}^{N-1} D(x_p) e^{-ikx_p}$

Note: For x discrete, x takes values $x_p = pH$, p integer, and for k discrete, k takes values $k = lk_0$, l integer.

$$\hat{A}(k) = \int_{-\infty}^{\infty} dx\, A(x) e^{-ikx} \tag{A-2}$$

If x measures position, then k may be identified as wavenumber. k is related to the wavelength λ of the harmonic by

$$k = \frac{2\pi}{\lambda} \tag{A-3}$$

A-1-2 The Fourier Series (FS)

If B is a periodic function of the continuous variable x, with a period length L, then the integral transform pair is replaced by the Fourier series pair (FS(i)):

$$\hat{B}(k) = \int_{L} dx\, B(x) e^{-ikx} \tag{A-4}$$

$$B(x) = \frac{1}{L} \sum_{l=-\infty}^{\infty} \hat{B}(k) e^{ikx} \tag{A-5}$$

The wavenumber k now takes only those values permitting integral numbers of wavelengths to fit in period length L;

$$k = \frac{2\pi l}{L} \qquad l = \text{integer} \tag{A-6}$$

Similarly, if the continuous variable x is replaced by a set of discrete values $\{x_p = pH;\ p = \text{integer}\}$, k space becomes periodic with period $k_g = 2\pi/H$, and the Fourier series pair (FS(ii)) are

$$C(x_p) = \int_{k_g} \frac{dk}{2\pi} \hat{C}(k) e^{ikx_p} \tag{A-7}$$

$$\hat{C}(k) = H \sum_{p=-\infty}^{\infty} C(x_p) e^{-ikx_p} \tag{A-8}$$

A-1-3 The Finite Fourier Transform (FFT)

If D is a periodic mesh of values defined at points $x_p = pH$ and the mesh has period length L, then a finite transform pair (FFT) may be defined relating values of D to period $k_g = 2\pi/H$. If N is the number of mesh points in one periodic length, then

$$D(x_p) = \frac{1}{L} \sum_{l=0}^{N-1} \hat{D}(k) e^{ikx_p} \tag{A-9}$$

$$\hat{D}(k) = H \sum_{p=0}^{N-1} D(x_p) e^{-ikx_p} \tag{A-10}$$

The finite Fourier transform is efficiently computed by the fast Fourier transform algorithm (Brigham, 1974) which is conveniently denoted by the same abbreviation.

Table A-2 Symmetries of a function and its transform

$f(x) \supset \hat{f}(k)$ or $\hat{f}(k) \supset f(x)$	
Real + even	Real + even
Real + odd	Imaginary + odd
Imaginary + even	Imaginary + even
Imaginary + odd	Real + odd
Real	Hermitian
Imaginary	Antihermitian
Even	Even
Odd	Odd

A-2 SYMMETRIES

All four transform pairs exhibit the same symmetry properties. For example, if $f \supset \hat{f}$ and f is real and even, then $\hat{f}$ is also real and even. More generally, the symmetries of a function f may be summarized by the diagram:

$$\begin{array}{ccccccccc} f(x) & = & E_r(x) & + & iE_i(x) & + & O_r(x) & + & iO_i(x) \\ & & \updownarrow & & \updownarrow & & \updownarrow & \times & \updownarrow \\ \hat{f}(k) & = & \hat{E}_r(k) & + & i\hat{E}_i(k) & + & \hat{O}_r(k) & + & i\hat{O}_i(k) \end{array} \qquad \text{(A-11)}$$

Functions E and O represent the even and odd parts of f and subscripts r and i refer to real and imaginary parts, respectively. The arrows in Eq. (A-11) indicate the correspondence between the parts of f and $\hat{f}$. These correspondences are listed in Table A-2.

A-3 THEOREMS

The convolution theorem is central to the application of transform analysis of particle models. The convolution integral $h(x)$ of two functions $f(x)$ and $g(x)$ is defined as

$$h(x) = \int_{-\infty}^{\infty} f(x')g(x-x')\,dx' \qquad \text{(A-12)}$$

A useful shorthand notation for convolution is

$$h(x) = f * g \qquad \text{(A-13)}$$

where the asterisk means perform the integral given on the right-hand side of Eq.

Table A-3 Convolutions and their transforms

Transform	x space	k space
FT	$\int_{-\infty}^{\infty} dx' f(x')g(x-x')$	$\hat{f}(k)\hat{g}(k)$
	$f(x)g(x)$	$\int_{-\infty}^{\infty} \frac{dk'}{2\pi} \hat{f}(k')\hat{g}(k-k')$
FS(i)	$\int_L dx' f(x')g(x-x')$	$\hat{f}(k)\hat{g}(k)$
	$f(x)g(x)$	$\frac{1}{L} \sum_{l'=-\infty}^{\infty} \hat{f}(k')\hat{g}(k-k')$
FS(ii)	$H \sum_{p'=-\infty}^{\infty} f(x_{p'})g(x_p - x_{p'})$	$\hat{f}(k)\hat{g}(k)$
	$f(x_p)g(x_p)$	$\int_{k_s} \frac{dk'}{2\pi} \hat{f}(k')\hat{g}(k-k')$
FFT	$H \sum_{p'=0}^{N-1} f(x_{p'})g(x_p - x_{p'})$	$\hat{f}(k)\hat{g}(k)$
	$f(x_p)g(x_p)$	$\frac{1}{L} \sum_{l'=0}^{N-1} \hat{f}(k')\hat{g}(k-k')$

Table A-4 Theorems

Convolution	If $f \supset \hat{f}$ and $g \supset \hat{g}$ then $f * g \supset \hat{f}\hat{g}$ and $\hat{f} * \hat{g} \supset fg$
Similarity	If $f(x) \supset \hat{f}(k)$, then $f\left(\frac{x}{a}\right) \supset \lvert a\rvert \hat{f}(ka)$
Multiply by constant	If $f \supset \hat{f}$ then $bf \supset b\hat{f}$
Addition	If $f \supset \hat{f}$ and $g \supset \hat{g}$ then $f + g \supset \hat{f} + \hat{g}$
Shift	If $f(x) \supset \hat{f}(k)$ then $f(x+a) \supset \hat{f}(k)e^{ika}$
Power	If $f \supset \hat{f}$ and $g \supset \hat{g}$ then $\int_{-\infty}^{\infty} f(x)g^*(x)\,dx = \int_{-\infty}^{\infty} \hat{f}(k)\hat{g}^*(k)\frac{dk}{2\pi}$ (similar results hold for FS and FFT— cf. Table A-3)
Derivative	If $f \supset \hat{f}$ then $\frac{df}{dx} \supset ik\hat{f}$
Reciprocity	If $f \supset \hat{f}(k) = g(k)$ then $g(x) \supset \hat{g}(k) = 2\pi f(-k)$

(A-12) or the corresponding sum for discrete quantities, as appropriate (cf. Table A-3). The convolution operator "$*$" is commutative, associative, and distributive:

$$f*g = g*f \tag{A-14}$$

$$f*(g*h) = (f*g)*h \tag{A-15}$$

$$f*(g+h) = f*g+f*h \tag{A-16}$$

The convolution theorem states that if $f \supset \hat{f}$ and $g \supset \hat{g}$, then $f*g \supset \hat{f}\hat{g}$ and $\hat{f}*\hat{g} \supset fg$. Convolutions and the associated products of the transformed quantities are listed in Table A-3 for the FT, FS, and FFT.

Table A-4 contains a summary of some useful theorems which apply (apart

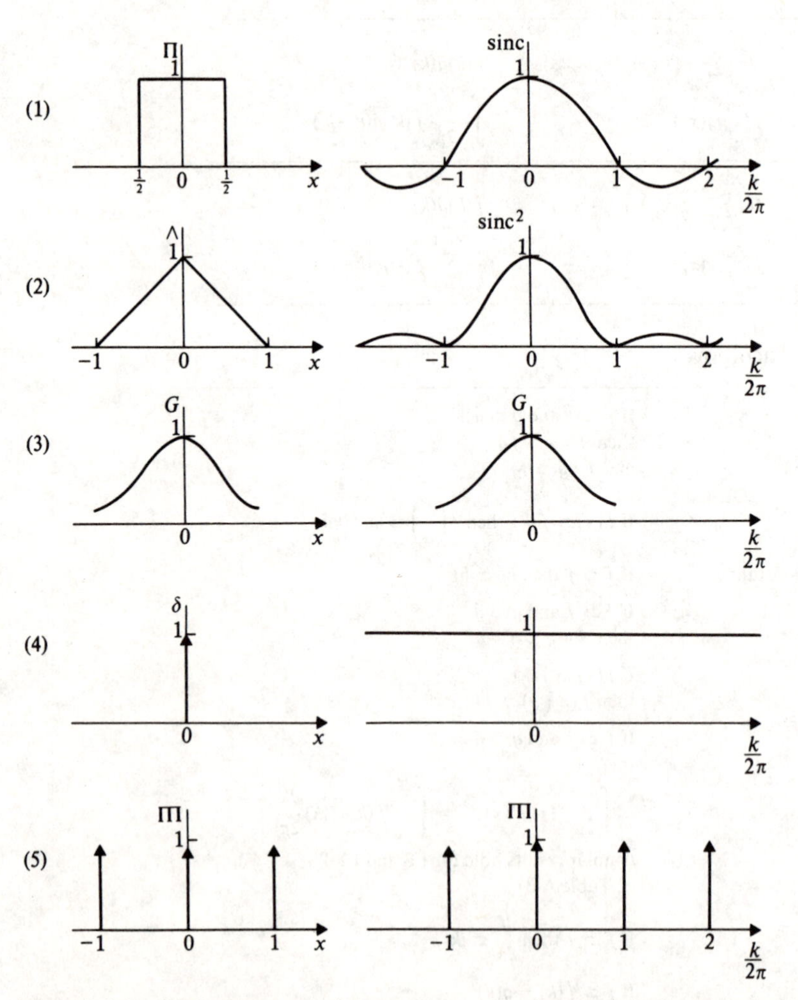

Figure A-1 Special functions and their transforms. The height of the impulses in $\delta(x)$ and $\text{Ш}(x)$ represent the areas under the impulses.

from the derivative theorem) equally well to the FT, FS, or FFT when the appropriate transform is used to relate the x space and k space quantities.

A further important theorem relating to transforms, the sampling theorem, is discussed in the context of its application to particle models in Chapter 5.

A-4 SPECIAL FUNCTIONS

Sketched in Fig. A-1 are some particular functions and their Fourier transforms:

A-4-1 The "Top-Hat" Function $\Pi(x)$

$$\Pi(x) = \begin{cases} 0 & |x| > \frac{1}{2} \\ \frac{1}{2} & |x| = \frac{1}{2} \\ 1 & |x| < \frac{1}{2} \end{cases} \tag{A-17}$$

$$\Pi(x) \supset \operatorname{sinc}\left(\frac{k}{2\pi}\right) = \frac{\sin k/2}{k/2} \tag{A-18}$$

A-4-2 The "Triangle" Function $\wedge(x)$

$$\wedge(x) = \begin{cases} 0 & |x| > 1 \\ 1 - |x| & |x| < 1 \end{cases} \tag{A-19}$$

$$\wedge(x) = \Pi * \Pi \supset \operatorname{sinc}^2\left(\frac{k}{2\pi}\right) \tag{A-20}$$

A-4-3 The Gaussian $G(x)$

$$G(x) = e^{-\pi x^2} \tag{A-21}$$

$$G(x) \supset \hat{G}(k) = G\left(\frac{k}{2\pi}\right) \tag{A-22}$$

A-4-4 The Dirac Delta Function $\delta(x)$

$$\delta(x) = 0 \qquad x \neq 0 \tag{A-23}$$

$$\int_{-\epsilon}^{\epsilon} \delta(x)\, dx = 1 \qquad \epsilon > 0 \tag{A-24}$$

$$\delta(x) = \int_{-\infty}^{\infty} \frac{dy}{2\pi} e^{\pm ixy} \tag{A-25}$$

$$\delta(x) \supset 1 \tag{A-26}$$

The δ function may be interpreted as the limit of a sequence of increasingly narrow and tall gaussians, each gaussian in the sequences having unit area beneath it.

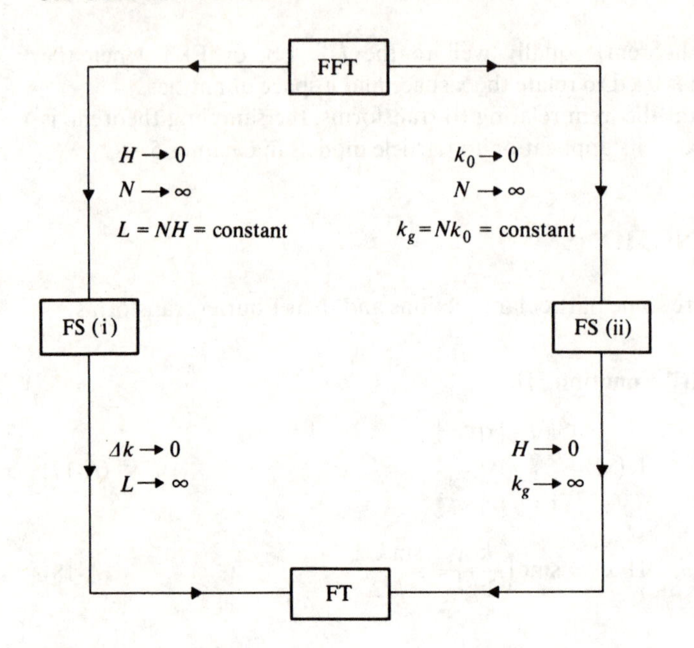

Figure A-2 The FFT, FS, and FT are related by the limiting processes indicated beside the flow lines.

$$\delta(x) = \lim_{a \to 0} \frac{1}{(2\pi a^2)^{1/2}} \exp\left(-\frac{x^2}{2a^2}\right) \tag{A-27}$$

The quantity used in the discrete transforms corresponding to the Dirac δ function is the Kronecker delta:

$$\delta_{l,0} = \begin{cases} 1 & l = 0 \\ 0 & \text{otherwise} \end{cases} \tag{A-28}$$

A-4-5 The Sampling Function Ш(x)

The sampling function is an infinite row of δ functions spaced at unit intervals

$$Ш(x) = \sum_{n=-\infty}^{\infty} \delta(x-n) \tag{A-29}$$

$$Ш(x) \supset \hat{Ш}(k) = Ш\left(\frac{k}{2\pi}\right) \tag{A-30}$$

Similar sampling functions may be defined for FS and FFT. For instance, for FS(i)

$$Ш(x) \supset Ш_s\left(\frac{k}{2\pi}\right) \tag{A-31}$$

where, for consistency, the periodic length L is an integral multiple of the spacing of the x-space δ-functions and

$$\text{III}_s\left(\frac{k}{2\pi}\right) = L \sum_{n=-\infty}^{\infty} \delta_{l,nL} \tag{A-32}$$

$$\left[\text{cf. III}\left(\frac{k}{2\pi}\right) = L \sum_{n=-\infty}^{\infty} \delta(l-nL) \text{ for the FT}\right].$$

A-5 RELATIONSHIP BETWEEN TRANSFORMS

Fourier transforms may be obtained from Fourier series, and Fourier series may be obtained from finite Fourier transforms by a limiting process, as is summarized in Fig. A-2. The reverse links are obtained by using the sampling function, as shown in Fig. A-3. The FS and FFT expressions for a single periodic length are obtained from the sampled infinite systems by multiplying by a top-hat function whose width is scaled to the periodic lengths L or k_g, as appropriate.

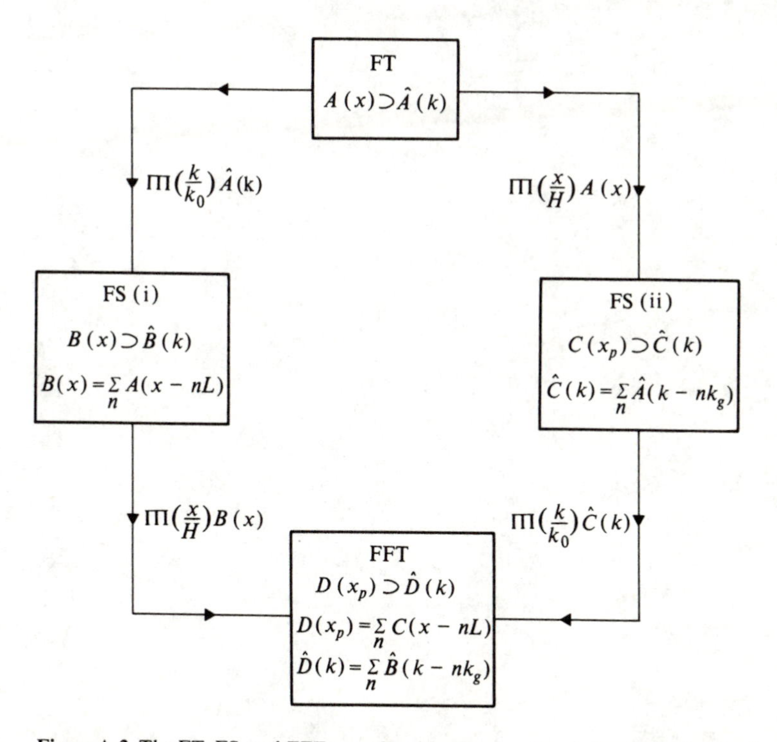

Figure A-3 The FT, FS, and FFT are related by the sampling indicated beside the flow lines. The sums over n shown in the boxes are known as "alias sums."

A-6 MULTIDIMENSIONAL TRANSFORMS

The one-dimensional transforms and their properties are generalized straightforwardly to two, three, or more dimensions. The three-dimensional results are obtained by replacing

1. x by $\mathbf{x} = (x_1, x_2, x_3)$
2. k by $\mathbf{k} = (k_1, k_2, k_3)$
3. kx by $\mathbf{k} \cdot \mathbf{x}$
4. dx by $d\mathbf{x}$, $\dfrac{dk}{2\pi}$ by $\dfrac{d\mathbf{k}}{(2\pi)^3}$
5. L by $V_b = L_1 L_2 L_3$
6. H by $V_c = H_1 H_2 H_3$

and so forth. Subscripts 1, 2, and 3 refer to the cartesian-axes directions.

BIBLIOGRAPHY

The following is an alphabetical list of papers and books referred to in the text. The abbreviations follow the British Standards Institution (1975) rules, BS 4148, which are also an American ANSI standard. They are used by INSPEC in Physics Abstracts (published by the Institution of Electrical Engineers, London).

Aarseth, S. J. (1963): Dynamical Evolution of Clusters of Galaxies—I, *Mon. Not. R. Astron. Soc.*, vol. 126, pp. 223–255.

——— (1966): Dynamical Evolution of Clusters of Galaxies—II, *Mon. Not. R. Astron. Soc.*, vol. 132, pp. 35–65.

——— (1969): Dynamical Evolution of Clusters of Galaxies—III, *Mon. Not. R. Astron. Soc.*, vol. 144, pp. 537–548.

——— (1971*a*): Direct Integration Methods of the *N*-body Problem, *Astrophys. & Space Sci.*, vol. 14, pp. 118–132.

——— (1971*b*): Binary Evolution in Stellar Systems, *Astrophys. & Space Sci.*, vol. 13, pp. 324–334.

———, J. R. Gott, and E. L. Turner (1979): *N*-Body Simulations of Galaxy Clustering I: Initial Conditions and Galaxy Collapse Times, *Astrophys. J.*, vol. 228, pp. 664–683.

——— and M. Lecar (1975): Computer Simulations of Stellar Systems, *Annu. Rev. Astron. & Astrophys.*, vol. 13, pp. 1–21.

Abe, H., J. Miyamoto, and R. Itatani (1975): Grid Effects on the Plasma Simulation by the Finite-Sized Particle, *J. Comput. Phys.*, vol. 19, pp. 134–149.

Abernathy, F. H., and R. E. Kronauer (1962): The Formation of Vortex Streets, *J. Fluid Mech.*, vol. 13, pp. 1–20.

Abramowitz, M., and I. A. Stegun (1964): "Handbook of Mathematical Functions," U.S. Department of Commerce, National Bureau of Standards, AMS 55, Washington D.C.

Adams, E. M., I. R. McDonald, and K. Singer (1977): Collective Dynamical Properties of Molten Salts: Molecular Dynamics Calculations on Sodium Chloride, *Proc. R. Soc. London*, ser. A, vol. 357, pp. 37–57.

Agajanian, A. H. (1975): A Bibliography on Semiconductor Device Modeling, *Solid-State Electron.*, vol. 18, pp. 917–929.

Ahmad, A., and L. Cohen (1973): A Numerical Integration Scheme for the *N*-Body Gravitational Problem, *J. Comput. Phys.*, vol. 12, pp. 389–402.

Alder, B. J., and T. E. Wainwright (1959): Studies in Molecular Dynamics. I. General Method, *J. Chem. Phys.*, vol. 31, pp. 459–466.

Alwin, V. C., D. H. Navon, and L. J. Turgeon (1977): Time-dependent Carrier Flow in a Transient Structure under Nonisothermal Conditions, *IEEE Trans. Electron. Devices*, vol. ED-24, pp. 1297–1304.

Amini, M. (1979): "Computer Simulation of Phase Changes," Ph.D. Thesis, Reading University.

———, D. Fincham, and R. W. Hockney (1979): A Molecular Dynamics Study of the Melting of Alkali-Halide Crystals, *J. Phys. C: Solid State Phys.*, vol. 12, pp. 4707–4720.

——— and R. W. Hockney (1979): Computer Simulation of Melting and Glass Formation in a Potassium Chloride Microcrystal, *J. Non-Cryst. Solids*, vol. 31, pp. 447–452.

Andrew, R. (1972): Improved Formulation of Gummel's Algorithm for Solving the 2-Dimensional Current Flow Equations in Semiconductor Devices, *Electron. Lett.*, vol. 8, pp. 536–538.

Balescu, R. (1960): Irreversible Processes in Ionised Gases, *Phys. Fluids*, vol. 3, pp. 52–63.

Bank, R. E. (1977): Marching Algorithms for Elliptic Boundary Value Problems, II: The Variable Coefficient Case, *SIAM J. Numer. Anal.*, vol. 14, pp. 950–970.

——— and D. J. Rose (1977): Marching Algorithms for Elliptic Boundary Value Problems, I: The Constant Coefficient Case, *SIAM J. Numer. Anal.*, vol. 14, pp. 792–829.

Barnes, C. W. (1972): Virtual Memory Utilisation in Large Plasma Simulation Codes, *SUIPR report* No. 497, Stanford University.

Barnes, J. J. and R. J. Lomax (1974): Two-Dimensional Finite Element Simulation of Semiconductor Devices, *Electron. Lett.*, vol. 10, pp. 341–343.

———, ———, and G. I. Haddad (1976): Finite-Element Simulation of GaAs MESFET's with Lateral Doping Profiles and Submicron Gates, *IEEE Trans. Electron Devices*, vol. ED-23, pp. 1042–1048.

Baus, M. (1978): Collective Modes and Dynamic Structure Factor of a Two-Dimensional Electron Fluid, *J. Stat. Phys.*, vol. 19, pp. 163–176.

Beeman, D. (1976): Some Multistep Methods for use in Molecular Dynamics Calculations, *J. Comput. Phys.*, vol. 20, pp. 130–139.

Berk, H. L., and K. V. Roberts (1970): The Water-Bag Model, *Methods Comput. Phys.*, vol. 9, pp. 88–134.

Berman, R. H., D. R. K. Brownrigg, and R. W. Hockney (1978): Numerical Models of Galaxies I. The Variability of Spiral Structure, *Mon. Not. R. Astron. Soc.*, vol., 185, pp. 861–875.

——— and J. W.-K. Mark (1979): Are Massive Haloes Necessary to Prevent Rapid, Global Bar Formation?, *Astron. & Astrophys.*, vol. 77, pp. 31–36.

Bernstein, I. B., and F. Engelmann (1966): Quasi-Linear Theory of Plasma Waves, *Phys. Fluids*, vol. 9, pp. 937–952.

Birdsall, C. K., and W. B. Bridges (1961): Space-Charge Instabilities in Electron Diodes and Plasma Converters, *J. Appl. Phys.*, vol. 32, pp. 2611–2618.

——— and ——— (1966): "Electron Dynamics of Diode Regions," Academic Press, New York.

——— and J. M. Dawson (1970): Plasma Physics, in "Computers and their Role in the Physical Sciences" (ed. S. Fernbach and A. Taub), University of California, Gordon and Breach, Science Publishers Inc. New York, pp. 247–310.

——— and D. Fuss (1969): Clouds-in-Clouds, Clouds-in-Cells Physics for Many-Body Plasma Simulation, *J. Comput. Phys.*, vol. 3, pp. 494–511.

——— and A. B. Langdon (1981): "Plasma Physics via Computer Simulation," McGraw-Hill, New York.

——— and N. Maron (1980): Plasma Self-Heating and Saturation due to Numerical Instability, *J. Comput. Phys.*, vol. 36, pp. 1–19.

———, A. B. Langdon, and H. Okuda (1970): Finite-Size Particle Physics Applied to Plasma Simulation, *Methods Comput. Phys.*, vol. 9, pp. 241–258.

Blakemore, J. S. (1969): "Solid State Physics," W. B. Saunders Co., London.

Blum, E. K. (1962): A Modification of the Runge–Kutta Fourth Order Method, *Math. Comput.*, vol. 16, pp. 176–187.

Boris, J. (1970): Relativistic Plasma Simulation—Optimisation of a Hybrid Code, *Proc. 4th Conf. on Numer. Simulation of Plasmas*, Office of Naval Research, Arlington, Va., pp. 3–67.

——— and K. V. Roberts (1969): The Optimisation of Particle Calculations in 2 and 3 Dimensions, *J. Comput. Phys.*, vol. 4, pp. 552–571.

Born, M., and K. Huang (1954): "Dynamical Theory of Crystal Lattices," Oxford University Press, London.

——— and J. E. Mayer (1932): Zur Gittertheorie der Ionen Kristalle, *Z. Phys.*, vol. 75, pp. 1–18.

Bott, I. B., and W. Fawcett (1968): The Gunn Effect in Gallium Arsenide, *Adv. Microwaves*, vol. 3, pp. 223–300.

Bracewell, R. (1965): "The Fourier Transform and its Application," McGraw-Hill, New York.

Brackbill, J. U. (1976): Numerical Magnetohydrodynamics for High-Beta Plasmas, *Methods Comput. Phys.*, vol. 16, pp. 1–41.

Brewitt-Taylor, C. R. (1979): Velocity Overshoot Effects in 2-dimensional Device Simulation, in "Numerical Analysis of Semi-conductor Devices" (eds. B. T. Browne and J. J. H. Miller), Boole Press, Dublin, pp. 191–193.

Brewitt-Taylor, C. R., P. N. Robson, and J. E. Sitch (1978): New Estimate of the Minimum Noise Figure of a MESFET, *Electron. Lett.*, vol. 14, pp. 818–820.

———, ———, and ——— (1980): The Noise Figure in MESFETs, *IEE Proc., part I*, vol. 127, pp. 1–8.

Brigham, E. O. (1974): "The Fast Fourier Transform," Prentice-Hall Inc., Englewood Cliffs, N.J.

British Standards Institution (1975): BS 4148: "The Abbreviation of Titles of Periodicals. Pt 2. Word-Abbreviation List," British Standards Institution, London.

Brown, T. R., and C. C. Grimes (1972): Observation of Cyclotron Resonance in Surface-Bound Electrons on Liquid Helium, *Phys. Rev. Lett.*, vol. 29, pp. 1233–1236.

Brownrigg, D. R. K. (1975): "Computer Modelling of Spiral Structure in Galaxies," Ph.D. Thesis, University of Reading.

Brush, S. G., H. L. Sahlin, and E. Teller (1966): Monte-Carlo Study of a One-Component Plasma. I, *J. Chem. Phys.*, vol. 45, pp. 2102–2118.

Buneman, O. (1959): Dissipation of Currents in Ionised Media, *Phys. Rev.*, vol. 115, pp. 503–517.

——— (1967): Time-Reversible Difference Procedures, *J. Comput. Phys.*, vol. 1, pp. 517–537.

——— (1969): A Compact Non-Iterative Poisson-Solver, *SUIPR report* No. 294, Stanford University.

——— (1971): Analytic Inversion of the Five-Point Poisson Operator, *J. Comput. Phys.*, vol. 8, pp. 500–505.

——— (1973): Analytic Inversion of Nine-Point Poisson Operator, *J. Comput. Phys.*, vol. 11, pp. 447–448.

——— (1976): The Advance from 2D Electrostatic to 3D Electromagnetic Particle Simulation, *Comput. Phys. Commun.*, vol. 12, pp. 21–31.

——— (1979): Vectorization and Parallelism in Stanford's 3D, EM Particle Code, *MFECC Buffer*, vol. 3, No. 7, pp. 7–12. Magnetic Fusion Energy Computer Center Monthly Bulletin, Livermore, Cal.

——— (1980): Tetrahedral Finite Elements for Interpolation, *SIAM J. Sci. & Stat. Comput.*, vol. 1, pp. 223–248.

———, C. W. Barnes, J. C. Green, and D. E. Nielsen (1980): Principles and Capabilities of 3-D, E-M Particle Simulations, *J. Comput. Phys.*, vol. 38, pp. 1–44.

——— and D. A. Dunn (1966): Computer Experiments in Plasma Physics, *Sci. J. (U.K.)*, vol. 2, No. 7, pp. 34–40.

Burger, P. (1964): Nonexistence of DC States in Low-Pressure Thermionic Converters, *J. Appl. Phys.*, vol. 35, pp. 3048–3049.

——— (1965): Theory of Large-Amplitude Oscillation in the One-Dimensional Low-Pressure Cesium Thermionic Converter, *J. Appl. Phys.*, vol. 36, pp. 1938–1943.

——— (1967): Elastic Collisions in Simulating One-Dimensional Plasma Diodes on the Computer, *Phys. Fluids*, vol. 10, pp. 658–666.

———, D. A. Dunn and A. S. Halstead (1966): Computer Experiments on the Randomization of Electrons in a Collisionless Plasma, *Phys. Fluids*, vol. 8, pp. 2263–2272.

Butcher, J. C. (1964): Implicit Runge–Kutta Processes, *Math. Comput.*, vol. 18, pp. 50–64.

Buzbee, B. L., F. W. Dorr, J. A. George, and G. H. Golub (1971): The Direct Solution of the Discrete Poisson Equation on Irregular Regions, *SIAM J. Numer. Anal.*, vol. 8, pp. 722–736.

———, G. H. Golub, and C. W. Nielson (1970): On direct Methods for Solving Poisson's Equation, *SIAM J. Numer. Anal.*, vol. 7, pp. 627–656.

Byers, J. A., B. I. Cohen, W. C. Condit, and J. D. Hanson (1978): Hybrid Simulations of Quasineutral Phenomena in Magnetised Plasma, *J. Comput. Phys.*, vol. 27, pp. 363–396.

Chandrasekhar, S. (1942): "Principles of Stellar Dynamics," Dover Publications Inc., New York.

Chen, L., A. B. Langdon, and C. K. Birdsall (1974): Reduction of Grid Effects in Simulation Plasmas, *J. Comput. Phys.*, vol. 14, pp. 200–222.

——— and H. Okuda (1975): Theory of Plasma Simulation Using Multipole-Expansion Scheme, *J. Comput. Phys.*, vol. 19, pp. 339–352.

Christiansen, J. P. (1973*a*): "The Non-Linear Dynamics of Vortex Flows by Numerical Methods," Ph.D. Thesis, Warwick University.

——— (1973*b*): Numerical Simulation of Hydrodynamics by the Method of Point Vortices, *J. Comput. Phys.*, vol. 13, pp. 363–379.

——— and R. W. Hockney (1971): DELSQPHI, a 2D Poisson-Solver Program, *Comput. Phys. Commun.*, vol. 2, pp. 139–155.

——— and K. V. Roberts (1974): OLYMPUS—A Standard Control and Utility Package for Initial Value FORTRAN Programs, *Comput. Phys. Commun.*, vol. 7, pp. 245–270.

——— and N. J. Zabusky (1973): Instability, Coalescence and Fission of Finite-Area Vortex Structures, *J. Fluid Mech.*, vol. 61, pp. 219–243.

Chryssafis, A., and W. Love (1979): A Computer-Aided Analysis of One-Dimensional Thermal Transients in *n-p-n* Power Transistors, *Solid-State Electron.*, vol. 22, pp. 249–256.

Ciccotti, G., J. Jacucci, and I. R. McDonald (1976): Transport Properties of Molten Alkali Halides, *Phys. Rev.*, ser. A., vol. 13, pp. 426–436.

Clemmow, P. C., and J. P. Dougherty (1969): "Electrodynamics of Particles and Plasmas," Addison-Wesley, Reading, Mass.

Cole, M. W., and M. H. Cohen (1969): Image-Potential-Induced Surface Bands in Insulators, *Phys. Rev. Lett.*, vol. 23, pp. 1238–1241.

Concus, P., and G. H. Golub (1973): Use of Fast Direct Methods for the Efficient Numerical Solution of Non-Separable Elliptic Equations, *SIAM J. Numer. Anal.*, vol. 10, pp. 1103–1120.

———, G. H. Golub, and D. P. O'Leary (1976): A Generalized Conjugate Gradient Method for the Numerical Solution of Elliptic Partial Differential Equations, in "Sparse Matrix Computations" (eds. J. R. Bunch and D. J. Rose), Academic Press, New York, pp. 309–332.

Cooley, J. W., and J. W. Tukey (1965): An Algorithm for the Machine Calculation of the Complex Fourier Series, *Math. Comput.*, vol. 19, pp. 297–301.

Copley, J. R. D., and A. Rahman (1976): Density Fluctuations in Molten Salts: Molecular Dynamics Study of Liquid RbBr, *Phys. Rev.*, ser. A, vol. 13, pp. 2276–2286.

Cotterill, R. M. J. (1975): Melting Point Depression in Very Thin Lennard-Jones Crystals. A Molecular Dynamics Study, *Philos. Mag.*, vol. 32, pp. 1283–1288.

———, W. Damgaard Kristensen, and E. J. Jensen (1974): Molecular Dynamics Studies of Melting. III. Spontaneous Dislocation Generation and the Dynamics of Melting, *Philos. Mag.*, vol. 30, pp. 245–263.

———, E. J. Jensen, W. Damgaard Kristensen, R. Paetsch, and P. O. Esbjorn (1975): A Unified Theory of Melting, Crystallization and Glass Formation, *J. Phys.* (*Fr.*), Coll. C2, Suppl. 4, vol. 36, pp. 35–48.

——— and L. B. Pedersen (1972): A Molecular Dynamics Study of the Melting of a Two-Dimensional Crystal, *Solid State Commun.*, vol. 10, pp. 439–441.

Couchman, P. R. (1979): The Lindemann Hypothesis and the Size-Dependence of Melting Temperature II, *Philos. Mag.*, ser. A, vol. 40, pp. 637–643.

——— and W. A. Jesser (1977): Thermodynamic Theory of Size Dependence of Melting Temperature in Metals, *Nature*, vol. 269, pp. 481–483.

——— and C. L. Ryan (1978): The Lindemann Hypothesis and the Size-Dependence of Melting Temperature, *Philos. Mag.*, vol. 37, pp. 369–373.

Crank, J., and P. Nicolson (1947): A Practical Method for Numerical Evaluation of Solutions of Partial Differential Equations of the Heat Conduction Type, *Proc. Cambridge Philos. Soc.*, vol. 43, pp. 50–67.

Curtis, A. R., and J. K. Reid (1971): FORTRAN Subroutines for the Solution of Sparse Sets of Linear Equations, *UKAEA report* R6844, H.M. Stationery Office, London.

Damgaard Kristensen, W., E. J. Jensen, and R. M. J. Cotterill (1974): Molecular Dynamics Studies of Melting: II. Dislocation Density and Thermodynamic Functions, *Philos. Mag.*, vol. 30, pp. 229–243.

Daniel, J. W., and R. E. Moore (1970): "Computation and Theory in Ordinary Differential Equations," W. H. Freeman & Co., San Francisco.

Davis, M., E. J. Groth, and P. J. E. Peebles (1977): Study of Galaxy Correlations: Evidence for the Gravitational Instability Picture in a Dense Universe, *Astrophys. J.*, vol. 212, pp. L107–L111.

——— and P. J. E. Peebles (1977): On the Integration of the BBGKY Equations for the Development of Strongly Nonlinear Clustering in an Expanding Universe, *Astrophys. J. Suppl.*, vol. 34, pp. 425–450.

Dawson, J. M. (1962): One-Dimensional Plasma Model, *Phys. Fluids*, vol. 5, pp. 445–459.

——— (1970): The Electrostatic Sheet Model for a Plasma and its Modification to Finite-Size Particles, *Methods Comput. Phys.*, vol. 9, pp. 1–28.

———, H. Okuda, and B. Rosen (1976): Collective Transport in Plasmas, *Methods Comput. Phys.*, vol. 16, pp. 281–325.

Dekker, A. J. (1958): "Solid State Physics," Macmillan & Co. Ltd, London.

Detyna, E. (1979): Point Cyclic Reduction for Elliptic Boundary Value Problems I: The Constant Coefficient Case, *J. Comput. Phys.*, vol. 33, pp. 204–216.

Dorr, F. W. (1970): The Direct Solution of the Discretised Poisson Equation on a Rectangle, *SIAM Rev.*, vol. 12, pp. 248–263.

Drangeid, K. E., and R. Sommerhalder (1970): Dynamic Performance of Schottky-Barrier Field-Effect Transistors, *IBM J. Res. & Dev.*, vol. 14, pp. 82–94.

Drummond, W. E., and D. Pines (1964): Nonlinear Plasma Oscillations, *Ann. Phys. (U.S.A.)*, vol. 28, pp. 478–499.

Duff, I. S., A. M. Erisman, and J. K. Reid (1976): On George's Nested Dissection Method, *SIAM J. Numer. Anal.*, vol. 13, pp. 686–695.

Dungey, J. W. (1961): Interplanetary Magnetic Field and the Auroral Zones, *Phys. Rev. Lett.*, vol. 6, pp. 47–48.

Eastwood, J. W. (1972): Consistency of Fields and Particle Motion in the Speiser Model of the Current Sheet, *Planet. & Space Sci.*, vol. 20, pp. 1555–1568.

——— (1974): The Warm Current Sheet Model, and its Implications on the Temporal Behaviour of the Geomagnetic Tail, *Planet. Space Sci.*, vol. 22, pp. 1641–1668.

——— (1975): Optimal Particle Mesh Algorithms, *J. Comput. Phys.*, vol. 18, pp. 1–20.

——— (1976*a*): Value for Money in Particle-Mesh Plasma Simulations, in "Computational Methods in Classical and Quantum Physics" (ed. M. B. Hooper), Advance Publications Ltd, London, pp. 196–205.

——— (1976*b*): Optimal P^3M Algorithms for Molecular Dynamics Simulations, in "Computational Methods in Classical and Quantum Physics" (ed. M. B. Hooper), Advance Publications Ltd, London, pp. 206–228.

——— and D. R. K. Brownrigg (1979): Remarks on the Solution of Poisson's Equation for Isolated systems, *J. Comput. Phys.*, vol. 32, pp. 24–38.

——— and R. W. Hockney (1974): Shaping the Force Law in Two-Dimensional Particle-Mesh Models, *J. Comput. Phys.*, vol. 16, pp. 342–359.

———, R. W. Hockney, and D. N. Lawrence (1980): P3M3DP—the Three-Dimensional Periodic Particle-Particle/Particle-Mesh Program, *Comput. Phys. Commun.*, vol. 19, pp. 215–261.

——— and K. V. Roberts (1979): The Authentication of Computer Programs, *Computer. Phys. Commun.*, to be submitted.

Efstathiou, G. (1979*a*): The Clustering of Galaxies and its Dependence upon Ω, *Mon. Not. R. Astron. Soc.*, vol. 187, pp. 117–127.

——— (1979*b*): "On the Rotation and Clustering of Galaxies," Ph.D. Thesis, University of Durham.

———, and J. W. Eastwood (1981): On the Clustering of Particles in an Expanding Universe, *Mon. Not. R. Astron. Soc.*, vol. 194, pp. 503–526.

——— and B. J. T. Jones (1979): The Rotation of Galaxies: Numerical Investigations of the Tidal Torque Theory, *Mon. Not. R. Astron. Soc.*, vol. 186, pp. 133–144.

Ehrlich, L. W. (1979): A Marching Technique for Non-Separable Equations, *Math. Comput.*, vol. 33, pp. 881–890.

Eldridge, O. C., and M. Feix (1962): One-Dimensional Plasma Model at Thermodynamic Equilibrium, *Phys. Fluids*, vol. 5, pp. 1076–1080.

Ewald, P. P. (1921): Die Berechnung Optischer und Elektrostatische Gitterpotentiale, *Ann. Phys. (Ger.)*, vol. 64, pp. 253–287.

Faddeev, D. K., and V. N. Faddeeva (1963): "Computational Methods in Linear Algebra," translated by R. C. Williams, W. H. Freeman & Co., San Francisco.

Fall, S. M. (1978): On the Evolution of Galaxy Clustering and Cosmological *N*-Body Simulations, *Mon. Not. R. Astron. Soc.*, vol. 185, pp. 165–177.

——— (1979): Galaxy Correlations and Cosmology, *Rev. Mod. Phys.*, vol. 51, pp. 21–42.

Faulkner, E. A. (1966): "Principles of Linear Circuits," Chapman & Hall Ltd, London.

Fawcett, W., A. D. Boardman, and S. Swain (1970): Monte-Carlo Determination of Electron Transport Properties in Gallium Arsenide, *J. Phys. Chem. Solids*, vol. 31, pp. 1963–1990.

Flanders, P. M., D. J. Hunt, S. F. Reddaway, and D. Parkinson (1977): Efficient High Speed Computing with the Distributed Array Processor, in "High Speed Computer and Algorithm Design," Academic Press Ltd, London, pp. 113–128.

Forsythe, G. E., and W. R. Wasow (1960): "Finite-Difference Methods for Partial Differential Equations," John Wiley & Sons, Inc., New York.

Freeman, K. C. (1970): On the Disks of Spiral and SO Galaxies, *Astrophys. J.*, vol. 160, pp. 811–830.

Friedman, A. (1922): Uber die Krümmung des Raumes, *Z. Phys.*, vol. 10, pp. 377–386.

Fumi, F. G. and M. P. Tosi (1964): Ionic Sizes and Born Repulsion Parameters in the NaCl-Type Alkali Halides—I, *J. Phys. Chem. Solids*, vol. 25, pp. 31–43.

Fyfe, D. J. (1966): Economic Evaluation of Runge–Kutta Formulae, *Math. Comput.*, vol. 20, pp. 392–398.

Gann, R. C., S. Chakravarty, and G. V. Chester (1979): Monte-Carlo Simulation of the Two-Dimensional One Component Plasma, *Phys. Rev.*, ser. B, vol. 20, pp. 362–344.

Gaur, S. P., and D. H. Navon (1976): Two-Dimensional Carrier Flow in a Transistor Structure under Nonisothermal Conditions, *IEEE Trans. Electron Devices*, vol. ED-23, pp. 50–57.

George, J. A. (1973): Nested Dissection of a Regular Finite-Element Mesh, *SIAM J. Numer. Anal.*, vol. 10, pp. 345–363.

Gill, S. (1951): A Process for the Step-by-Step Integration of Differential Equations in an Automatic Digital Computing Machine, *Proc. Cambridge Philos. Soc.*, vol. 47, pp. 96–108.

Goel, S. P., and R. W. Hockney (1974): A Resource Letter CSSMD-I, Computer Simulation Studies by the Method of Molecular Dynamics, *Rev. Bras. Fis.*, vol. 4, pp. 121–157.

Goldstein, H. (1959): "Classical Mechanics," Addison-Wesley, Inc., Reading, Mass.

Gott, J. R. (1977): Recent Theories of Galaxy Formation, *Annu. Rev. Astron. & Astrophys.*, vol. 15, pp. 235–266.

———, J. E. Gunn, and B. M. Tinsley (1974): An Unbound Universe?, *Astrophys. J.*, vol. 194, pp. 543–553.

———, E. L. Turner, and S. J. Aarseth (1979): *N*-Body Simulations of Galaxy Clustering III: The Covariance Function, *Astrophys. J.*, vol. 234, pp. 13–26.

Greenstein, J. L. (1959): Dying Stars, *Sci. Am.*, vol. 200, No. 1, 46–53.

Griffiths, R. J. M., I. D. Blenkinsop, and D. R. Wight (1979): Preparation and Properties of GaAs Layers for Novel F.E.T. Structures, *Electron. Lett.*, vol. 15, pp. 629–630.

Grimes, C. C., and G. Adams (1979): Evidence for a Liquid-to-Crystal Phase Transition in a Classical Sheet of Electrons, *Phys. Rev. Lett.*, vol. 42, pp. 795–798.

Groth, E. J., and P. J. E. Peebles (1977): Statistical Analysis of Catalogs of Extra-Galactic Objects. VII. Two- and Three-Point Correlation Functions for the High-Resolution Shane-Wirtanen Catalog of Galaxies, *Astrophys. J.*, vol. 217, pp. 385–405.

———, ———, M. Seldner, and R. M. Soneira (1977): The Clustering of Galaxies, *Sci. Am.*, vol. 237, No. 5, pp. 76–98.

Guernsey, R. L. (1962): Kinetic Equation for a Completely Ionised Gas, *Phys. Fluids*, vol. 5, pp. 322–328.

Gula, W. P., and C. K. Chu (1973): Effect of Krook Model Collisions on Two-Stream Instability, *Phys. Fluids*, vol. 16, pp. 1135–1141.

Gummel, H. K. (1964): A Self-Consistent Iterative Scheme for One Dimensional Steady-State Transistor Calculations, *IEEE Trans. Electron Devices*, vol. ED-11, pp. 455–465.

Gunn, J. B. (1963): Microwave Oscillation of Current in III–V Semiconductors, *Solid State Commun.*, vol. 1, pp. 88–91.

Hamilton, D. J., F. A. Lindholm, and A. H. Marshak (1971): "Principles and Applications of Semiconductor Device Modeling," Rinehart and Winston, New York.

Hamilton, J. M. (1976): The Numerical Solution of Ampere's Equation on a 2-Dimensional Mesh. Part I—Theoretical Discussion, *Reading University Computer Science Report*, RCS 49.

——— (1977): COIRCE: The Numerical Solution of Ampere's Equation on a 2-Dimensional Mesh. Part II—Further Theoretical Results, Timing and Convergence Properties, *Reading University Computer Science Report*, RCS 50.

——— (1980): "Computer Simulation of the Geomagnetic Current Sheet," Ph.D. Thesis, University of Reading.

Hansen, J. P., and I. R. McDonald (1975): Statistical Mechanics of Dense Ionized Matter. IV. Density and Charge Fluctuations in a Simple Molten Salt, *Phys. Rev.*, ser. A, vol. 11, pp. 2111–2123.

——— and ——— (1976): "Theory of Simple Liquids," Academic Press Ltd, London.

Harlow, F. H. (1964): The Particle-in-Cell Computing Method in Fluid Dynamics, *Methods Comput. Phys.*, vol. 3, pp. 319–343.

Harnwell, G. P. (1949): "Principles of Electricity and Magnetism," McGraw-Hill, New York.

Hartree, D. R. (1950): Some Calculations of Transients in an Electronic Valve, *Appl. Sci. Res.*, col. B1, pp. 379–390.

——— and P. Nicolson (1941–1944): *CVD Reports Mag.* 3, 12, 18, 23, 36, British Admiralty, London.

Henon, M. (1964): L'Evolution Initiale d'un Amas Sphérique, *Ann. Astrophys.* (*Fr.*), vol. 27, pp. 83–91.

——— (1971*a*): Monte-Carlo Models of Star Clusters, *Astrophys. & Space Sci.*, vol. 13, pp. 284–299.

——— (1971*b*): The Monte-Carlo Method, *Astrophys. & Space Sci.*, vol. 14, pp. 151–167.

Henrici, P. (1962): "Discrete Variable Methods in Ordinary Differential Equations," John Wiley & Sons, Inc., New York.

Hewett, D. W., and C. W. Nielson (1978): A Multidimensional Quasineutral Plasma Simulation Model, *J. Comput. Phys.*, vol. 29, pp. 219–236.

Hill, G., P. N. Robson, A. Majerfeld, and W. Fawcett (1977): Effect of Ionised Impurity Scattering on the Electron Transit Time in GaAs and InP FETs, *Electron. Lett.*, vol. 13, pp. 235–236.

Himsworth, B. (1972): A Two-Dimensional Analysis of Gallium Arsenide Junction Field Effect Transistors with Long and Short Channels, *Solid-State Electron.*, vol. 15, pp. 1353–1361.

Hockney, R. W. (1965): A Fast Direct Solution of Poisson's Equation using Fourier Analysis, *J. Assoc. Comput. Mach.*, vol. 12, pp. 95–113.

——— (1966*a*): "The Computer Simulation of Anomalous Plasma Diffusion and the Numerical Solution of Poisson's Equation," Ph.D. Thesis, Stanford University.

——— (1966*b*): Computer Experiment of Anomalous Diffusion, *Phys. Fluids*, vol. 9, pp. 1826–1835.

——— (1967): Gravitational Experiments with a Cylindrical Galaxy, *Astrophys. J.*, vol. 150, pp. 797–806.

——— (1968): Formation and Stability of Virtual Electrodes in a Cylinder, *J. Appl. Phys.*, vol. 39, pp. 4166–4170.

——— (1970): The Potential Calculation and Some Applications, *Methods Comput. Phys.*, vol. 9, pp. 135–211.

——— (1971): Measurements of Collision and Heating Times in a Two-Dimensional Thermal Computer Plasma, *J. Comput. Phys.*, vol. 8, pp. 19–44.

——— (1977): Super-Computer Architecture, in "Infotech State of the Art Report: Future Systems I," Infotech International, Maidenhead, Berks, pp. 277–305.

——— (1978*a*): Computers, Compilers and Poisson-Solvers, in "Computers, Fast Elliptic Solvers and Applications" (ed. U. Schumann), Advance Publications Ltd, London, pp. 75–97.

——— (1978*b*): POT4—A FACR(1) Algorithm for Arbitrary Regions, in "Computers, Fast Elliptic Solvers and Applications" (ed. U. Schumann), Advance Publications Ltd, London, pp. 141–169.

——— (1979): The Large Parallel Computer and University Research, *Contemp. Phys.*, vol. 20, pp. 149–185.

——— (1980): Rapid Elliptic Solvers, in "Numerical Methods in Applied Fluid Dynamics" (ed. B. Hunt), Academic Press Ltd, London.

——— and T. R. Brown (1975): A Lambda Transition in a Classical Electron Film, *J. Phys. C: Solid State Phys.*, vol. 8, 1813–1822.

——— and D. R. K. Brownrigg (1974): Effects of Population II Stars and Three-Dimensional Motion on Spiral Structure, *Mon. Not. R. Astron. Soc.*, vol. 167, pp. 351–357.

———, S. P. Goel, and J. W. Eastwood (1973): A 10000 Particle Molecular Dynamics Model with Long-Range Forces, *Chem. Phys. Lett.*, vol. 21, pp. 589–591.

———, S. P. Goel, and J. W. Eastwood (1974): Quiet High-Resolution Computer Models of a Plasma, *J. Comput. Phys.*, vol. 14, pp. 148–158.

——— and F. Hohl (1969): Effects of Velocity Dispersion on the Evolution of a Disk of Stars, *Astron. J.*, vol. 74, pp. 1102–1104 & 1119–1124.

——— and C. R. Jesshope (1981): "Parallel Computers—Architecture, Programming and Algorithms," Adam Hilger, Bristol.

———, R. A. Warriner, and M. Reiser (1974): Two-Dimensional Particle Models in Semiconductor Device Analysis, *Electron. Lett.*, vol. 10, pp. 484–486.

Hoerner, S. von (1960): Die Numerische Integration des *N*-Körper-Problemes für Sternhaufen I, *Z. Astrophys.*, vol. 50, pp. 184–214.

Hohl, F. (1971): Numerical Experiments with a Disk of Stars, *Astrophys. J.*, vol. 168, pp. 343–359.

——— (1972): Evolution of a Stationary Disk of Stars, *J. Comput. Phys.*, vol. 9, pp. 10–25.

——— (1973): Relaxation Time in Disk Galaxy Simulations, *Astrophys. J.*, vol. 184, pp. 353–359.

——— (1975): *N*-Body Simulations of Disks, in "Dynamics of Stellar Systems" (ed. A. Hayli), Reidel Co., Dordrecht, pp. 349–366.

——— (1976): Suppression of Bar Instability by a Massive Halo, *Astron. J.*, vol. 81, pp. 30–36.

——— (1978): Three-Dimensional Galaxy Simulations, *Astron. J.*, vol. 83, pp. 768–778.

——— and R. W. Hockney (1969): A Computer Model of Disks of Stars, *J. Comput. Phys.*, vol. 4, pp. 306–323.

——— and T. A. Zang (1979): Collapse and Relaxation of Rotating Stellar Systems, *Astron. J.*, vol. 84, pp. 585–600.

Horn, H. M. van (1971): "White Dwarfs" (ed. W. J. Luyten), Springer-Verlag, Berlin, p. 96.

Hoyle, F. (1962): "Astronomy," Doubleday and Co., Inc., New York.

Hubble, E. (1926): Extra-Galactic Nebulae, *Astrophys. J.*, vol. 64, pp. 321–369.

——— (1929): A Relation between Distance and Velocity among Extra Galactic Nebulae, *Proc. Natl. Acad. Sci. U.S.A.*, vol. 15, pp. 168–173.

——— (1936): "The Realm of the Nebulae," Yale University Press, reprinted (1958) by Dover Publications, Inc., New York.

Huggins, M. L., and J. E. Mayer (1933): Interatomic Distances in Crystals of the Alkali Halides, *J. Chem. Phys.*, vol. 1, pp. 643–646.

Hughes, M. H. (1971): Solution of Poisson's Equation In Cylindrical Coordinates, *Comput. Phys. Commun.*, vol. 2, 157–167.

——— (1980*a*): COMPOS—the OLYMPUS Fortran Compositor, *Comput. Phys. Commun.*, submitted.

——— (1980*b*): GENSIS—A Program for Automatic Construction of Standard OLYMPUS Subprograms, *Comput. Phys. Commun.*, to be submitted.

——— and A. P. V. Roberts (1974*a*): TIMER—A Software Instrumentation Routine for Making Timing Measurements, *Comput. Phys. Commun.*, vol. 8, pp. 118–122.

——— and K. V. Roberts (1974*b*): OLYMPUS Restart Facilities, *Comput. Phys. Commun.*, vol. 8, pp. 123–129.

———, ——— and P. D. Roberts (1975): OLYMPUS and Preprocessor Package for an IBM 370/165, *Comput. Phys. Commun.*, vol. 9, pp. 51–58.

Irvine, W. M. (1961): "Local Irregularities in a Universe Satisfying the Cosmological Principle," Ph.D. Thesis, Harvard University.

Isihara, A. (1971): "Statistical Physics," Academic Press, Inc., New York.

James, R. A. (1977): The Solution of the Poisson Equation for Isolated Source Distributions, *J. Comput. Phys.*, vol. 25, pp. 71–93.

——— and J. A. Sellwood (1978): Galactic Models with Variable Spiral Structure, *Mon. Not. R. Astron. Soc.*, vol. 182, pp. 331–344.

Jensen, E. J., W. Damgaard Kristensen, and R. M. J. Cotterill (1973): Molecular Dynamics Studies of Melting: I. Dislocation Density and the Pair Distribution Function, *Philos. Mag.*, vol. 27, pp. 623–632.

Jensen, K., and N. Wirth (1978): "Pascal—User Manual and Report," 2nd ed., Springer-Verlag, New York.

Jesshope, C. R. (1975): Numerical Solutions to the 2-Dimensional Time-Dependent Semiconductor Equations, *Electron. Lett.*, vol. 11, pp. 431–433.

——— (1979): SIPSOL—A suite of Subprograms for the Solution of the Linear Equations Arising from Elliptic Partial Differential Equations, *Comput. Phys. Commun.*, vol. 17, pp. 383–391.

Johnson, F. S. (1960): The Gross Character of the Geomagnetic Field in the Solar Wind, *J. Geophys. Res.*, vol. 65, pp. 3049–3051.

Jones, J. E. (1924): On the Determination of Molecular Fields—II. From the Equation of State of a Gas, *Proc. R. Soc.*, ser. A, vol. 106, pp. 463–477.

Kennedy, D. P., and R. R. O'Brien (1970): Computer-Aided Two-Dimensional Analysis of the Junction Field-Effect Transistor, *IBM J. Res. & Dev.*, vol. 14, pp. 95–116.

Kilpatrick, J. A., and W. D. Ryan (1971): Two-Dimensional Analysis of Lateral-Base Transistors, *Electron. Lett.*, vol. 7, pp. 226–227.

Kittel, C. (1971): "Introduction to Solid State Physics," 4th ed., John Wiley and Sons, Inc., New York.

Kruer, W. L., J. M. Dawson, and B. Rosen (1973): The Dipole Expansion Method in Plasma Simulation, *J. Comput. Phys.*, vol. 13, pp. 114–129.

Kuhlman-Wilsdorf, D. (1965): Theory of Melting, *Phys. Rev.*, ser. A, vol. 140, pp. 1599–1610.

Kustaanheimo, P., and E. Stiefel (1965): Perturbation Theory of Kepler Motion Based on Spinor Regularization, *J. Reine Angew. Math.*, vol. 218, pp. 204–219.

Kutta, W. (1901): Beitrag zur Näherungsweisen Integration Totaler Differentialgleichungen, *Z. Math. Phys.*, vol. 46, pp. 435–453.

Lamb, H. (1932): "Hydrodynamics," Cambridge University Press, republished (1954) by Dover Publications Inc., New York.

Langdon, A. B. (1970*a*): Non-Physical Modification to Oscillations, Fluctuations and Collisions due to Space Time Differencing, *Proc. 4th Annu. Conf. Numer. Simulation of Plasmas*, Office of Naval Research, Arlington, Va., pp. 467–495.

——— (1970*b*): Effect of the Spatial Grid in Simulation Plasmas, *J. Comput. Phys.*, vol. 6, pp. 247–267.

——— (1973): Energy Conserving Plasma Simulation Algorithms, *J. Comput. Phys.*, vol. 12, pp. 247–268.

——— (1979*a*): Kinetic Theory for Fluctuations and Noise in Computer Simulation of Plasma, *Phys. Fluids*, vol. 22, pp. 163–171.

——— (1979*b*): Analysis of the Time Integration in Plasma Simulation, *J. Comput. Phys.*, vol. 30, pp. 202–221.

——— and C. K. Birdsall (1970): Theory of Plasma Simulation Using Finite-Size Particles, *Phys. Fluids*, vol. 13, pp. 2115–2122.

——— and F. Lasinski (1976): Electromagnetic and Relativistic Plasma Simulation Models, *Methods Comput. Phys.*, vol. 16, 327–366.

Larson, R. B. (1970*a*): A Method for Computing the Evolution of Star Clusters, *Mon. Not. R. Astron. Soc.*, vol. 147, pp. 323–337.

——— (1970*b*): The Evolution of Star Clusters, *Mon. Not. R. Astron. Soc.*, vol. 150, pp. 93–110.

Lau, Y. Y., C. C. Lin, and J. W.-K. Mark (1976): Unstable Spiral Modes in Disk-Shaped Galaxies, *Proc. Natl. Acad. Sci. U.S.A.*, vol. 73, pp. 1379–1381.

Layzer, D. (1963): A Preface to Cosmology. I. The Energy Equations and the Virial Theorem for Cosmic Distributions, *Astrophys. J.*, vol. 138, pp. 174–184.

Le Bail, R. C. (1971): Use of Fast Fourier Transforms for Solving Partial Differential Equations, *J. Comput. Phys.*, vol. 9, pp. 440–465.

Leboeuf, J. N., T. Tajima, and J. M. Dawson (1979): A Magnetohydrodynamic Particle Code for Fluid Simulation of Plasmas, *J. Comput. Phys.*, vol. 31, pp. 379–408.

———, ———, C. F. Kennel, and J. M. Dawson (1978): Global Simulation of the Time-Dependent Magnetosphere, *Geophys. Res. Lett.*, vol. 5, 609–612.

———, ———, ———, and ——— (1979): Global Magnetohydrodynamic Simulation of the Two-Dimensional Magnetosphere, *Geophys. Monogr. Ser.*, vol. 21, pp. 536–556.

Lebwohl, P. A., and P. J. Price (1971): Direct Microscopic Simulation of Gunn Domain Phenomena, *Appl. Phys. Lett.*, vol. 19, pp. 530–532.

Lecar, M. (1968): A Comparison of Eleven Numerical Integrations of the same Gravitational 25-Body Problem, *Bull. Astron. (Fr.)*, 3e sér., t. 3, pp. 91–104.

Lenard, A. (1960): On Bogoliubov's Kinetic Equation for Spatially Homogeneous Plasma, *Ann. Phys. (U.S.A.)*, vol. 10, pp. 390–400.

Levy, R. H., and R. W. Hockney (1968): Computer Experiments of Low-Density Crossed-Field Electron Beams, *Phys. Fluids*, vol. 11, pp. 766–771.

Lewis, H. R. (1970*a*): Energy-Conserving Numerical Approximations for Vlasov Plasmas, *J. Comput. Phys.*, vol. 6, pp. 136–141.

——— (1970*b*): Application of Hamilton's Principle to the Numerical Analysis of Vlasov Plasmas, *Methods Comput. Phys.*, vol. 9, pp. 307–338.

———, A. Sykes, and J. A. Wesson (1972): A Comparison of Some Particle-in-Cell Plasma Simulation Methods, *J. Comput. Phys.*, vol. 10, pp. 85–106.

Lewis, J. W. E., K. Singer, and L. V. Woodcock (1975): Thermodynamic and Structural Properties of Liquid Ionic Salts Obtained by Monte-Carlo Computation, *J. Chem. Soc., Faraday II*, vol. 71, pp. 301–312.

Limber, D. N. (1953): The Analysis of Counts of the Extragalactic Nebulae in Terms of a Fluctuating Density Field, *Astrophys. J.*, vol. 117, pp. 134–144.

Lin, C. C., and Y. Y. Lau (1979): Density Wave Theory of Spiral Structure of Galaxies, *Stud. Appl. Math.*, vol. 60, pp. 97–163.

——— and F. H. Shu (1964): On the Spiral Structure of Disk Galaxies, *Astrophys. J.*, vol. 140, pp. 646–655.

———, C. Yuan, and F. H. Shu (1969): On the Spiral Structure of Disk Galaxies. III. Comparison with Observation, *Astrophys. J.*, vol. 155, pp. 721–746.

Lindblad, B. (1961): On the Formation of Dispersion Rings in the Central Layer of a Galaxy, *Stockholms Obs. Ann.*, vol. 21, No. 8.

Lindblad, P. O. (1960*a*): On Tidal Interaction between Galaxies, *Stockholms Obs. Ann.*, vol. 21, No. 3.

——— (1960*b*): The Development of Spiral Structure in a Galaxy Approached by Numerical Computations, *Stockholms Obs. Ann.*, vol. 21, No. 4.

——— (1962): Gravitational Resonance Effects in the Central Layer of a Galaxy, in "The Distribution and Motion of Interstellar Matter in Galaxies" (ed. L. Woltjer), W. A. Benjamin Inc., New York, pp. 222–233.

Lindemann, F. A. (1910): Über die Berechnung Molekularer Eigenfrequenzen, *Phys. Z.*, vol. 11, pp. 609–612.

Lindsey, C. H., and S. G. Van der Meulen (1971): "Informal Introduction to Algol 68," North-Holland Publ. Co., Amsterdam.

Loeb, H. W., R. Andrew, and W. Love (1968): Application of Two-Dimensional Solutions of the Shockley-Poisson Equation to Inversion-Layer MOST Devices, *Electron. Lett.*, vol. 4, pp. 352–354.

Lomax, R. J. (1960): Transient Space Charge Flow, *J. Electron. & Control*, vol. 9, pp. 127–140.

Longmire, C. L. (1963): "Elementary Plasma Physics," John Wiley and Sons, Inc., New York.

Lorenz, E. N. (1976): A Rapid Procedure for Inverting Del-Square with Certain Computers, *Mon. Weather Rev.*, vol. 104, pp. 961–966.

Lynden-Bell, D. (1979): On a Mechanism that Structures Galaxies, *Mon. Not. R. Astron. Soc.*, vol. 187, pp. 101–107.

——— and A. J. Kalnajs (1972): On the Generating Mechanism of Spiral Structure, *Mon. Not. R. Astron. Soc.*, vol. 157, pp. 1–30.

Marder, R. M. (1975): GAP–A PIC-Type Fluid Code, *Math. Comput.*, vol. 29, pp. 434–446.

Mark, J. W.-K. (1974): On Density Waves in Galaxies. I. Source Terms and Action Conservation, *Astrophys. J.*, vol. 193, pp. 539–559.

——— (1977): On Density Waves in Galaxies. V. Maintenance of Spiral Structure and Discrete Spiral Modes, *Astrophys. J.*, vol. 212, pp. 645–658.

Martin, E. D. (1974): A Generalised-Capacity-Matrix Technique for Computing Aerodynamic Flows, *Comput. & Fluids*, vol. 2, pp. 79–97.

Maruhn, J. A., T. A. Welton, and C. Y. Wong (1976). Remarks on the Numerical Solution of Poisson's Equation for Isolated Charge Distributions, *J. Comput. Phys.*, vol. 20, pp. 326–335.

Matsuda, Y., and H. Okuda (1975): Collisions in Multidimensional Plasma Simulations, *Phys. Fluids*, vol. 18, pp. 1740–1747.

McCrory, R. L., R. L. Morse, and K. A. Taggart (1977): Growth and Saturation of Instability of Spherical Implosion Driven by Laser or Charged Particle Beams, *Nuclear Sci. & Eng.*, vol. 64, pp. 163–176.

McDonald, I. R., and K. Singer (1967): Calculation of Thermodynamic Properties of Liquid Argon from Lennard-Jones Parameters by a Monte-Carlo Method, *Discuss. Faraday Soc.*, vol. 43, pp. 40–49.

Meijerink, J. A., and H. A. van der Vorst (1977): An Iterative Solution Method for Linear Systems of which the Coefficient Matrix is a Symmetric M-Matrix, *Math. Comput.*, vol. 31, pp. 148–162.

Metropolis, N., A. W. Rosenbluth, M. N. Rosenbluth, A. H. Teller, and E. Teller (1953): Equation of State Calculations by Fast Computing Machines, *J. Chem. Phys.*, vol. 21, pp. 1087–1092.

Miller, R. H. (1971): Numerical Experiments in Collisionless Systems, *Astrophys. & Space Sci.*, vol. 14, pp. 73–90.

——— (1976): Predominance of Two-Arm Spirals, *Astrophys. J.*, vol. 207, pp. 408–413.

——— (1978): Free Collapse of a Rotating Sphere of Stars, *Astrophys. J.*, vol. 223, pp. 122–128.

——— and K. H. Prendergast (1968): Stellar Dynamics in a Discrete Phase Space, *Astrophys. J.*, vol. 151, pp. 699–709.

———, ———, and W. J. Quirk (1970): Numerical Experiments on Spiral Structure, *Astrophys. J.*, vol. 161, pp. 903–916.

Moglestue, C. (1979): Computer Simulation of a Dual Gate GaAs Field-Effect Transistor using the Monte-Carlo Method, *Solid State & Electron Devices*, vol. 3, pp. 133–136.

——— and S. J. Beard (1979): A Particle Model Simulation of Field Effect Transistors, in "Numerical Analysis of Semiconductor Devices," (eds. B. T. Browne and J. J. H. Miller), Boole Press, Dublin, pp. 232–236.

Montgomery, D. C., and D. A. Tidman (1964): "Plasma Kinetic Theory," McGraw-Hill Book Company, New York.

Morse, R. L. (1970): Multidimensional Plasma Simulation by the Particle-in-Cell Method, *Methods Comput. Phys.*, vol. 9, pp. 213–239.

——— and C. W. Nielson (1969): Numerical Simulation of Warm Two-Beam Plasma, *Phys. Fluids*, vol. 12, pp. 2418–2425.

N.C.C. (1970): Standard Fortran Programming Manual, Computer Standards Series, National Computing Centre Ltd., Manchester, U.K.

Nielsen, D. (1978): "Three-Dimensional Electromagnetic Particle Simulation of Fusion Plasmas," Ph.D. Thesis, Stanford University.

———, J. Green, and O. Buneman (1978): Dynamic Evolution of a Z Pinch, *Phys. Rev. Lett.*, vol. 42, pp. 1274–1277.

Nielson, W. C., and H. R. Lewis (1976): Particle-Code Models in the Nonradiative Limit, *Methods Comput. Phys.*, vol. 16, pp. 367–388.

Okuda, H. (1970): Nonphysical Instabilities in Plasma Simulation Due to Small $\lambda_D/\Delta x$, *Proc. 4th Annu. Conf. on Numer. Simulation of Plasma*, Office of Naval Research, Arlington, Va., pp. 515–526.

——— (1972*a*): Nonphysical Noises and Instabilities in Plasma Simulation due to a Spatial Grid, *J. Comput. Phys.*, vol. 10, pp. 475–486.

——— (1972*b*): Verification of Theory for Plasma of Finite-Sized Particles, *Phys. Fluids*, vol. 15, pp. 1268–1274.

——— and C. K. Birdsall (1970): Collisions in a Plasma of Finite-Size Particles, *Phys. Fluids*, vol. 8, pp. 2123–2134.

O'Leary, D. P., and O. Widlund (1979): Capacity Matrix Methods for the Helmholtz Equation on General Three-Dimensional Regions, *Math. Comput.*, vol. 33, pp. 849–879.

Oliphant, T. A., and C. W. Nielson (1970): Simulation of Binary Collision Processes in Plasmas, *Phys. Fluids*, vol. 13, pp. 2103–2107.

Onuki, A. (1977): Collective Modes of a Two-Dimensional Electron Fluid, *J. Phys. Soc. Jpn*, vol. 43, pp. 396–405.

Orens, J. H., J. P. Boris, and I. Haber (1970): Optimisation Techniques for Particle Codes, in *Proc. 4th Annu. Conf. Numer. Simulation of Plasmas*, Office of Naval Research, Arlington, Va., pp. 526–558.

Ostriker, J. P., and P. J. E. Peebles (1973): A Numerical Study of the Stability of Flattened Galaxies: or, Can Cold Galaxies Survive?, *Astrophys. J.*, vol. 186, pp. 467–480.

Parrinello, M., and M. P. Tosi (1979): Structure and Dynamics of Simple Ionic Liquids, *Riv. Nuovo Cimento, Ser. 3*, vol. 2, No. 6, pp. 1–69.

Pauling, L. (1950): "The Nature of the Chemical Bond," Oxford University Press, London.

Peebles, P. J. E. (1974): The Gravitational-Instability Picture and the Nature of the Distribution of Galaxies, *Astrophys. J.*, vol. 189, pp. L51–L53.

——— (1978): Comment after paper by S.-J. Aarseth, in "The Large Scale Structure of the Universe," IAU Symposium No. 79 (eds. M. Longair and J. Einasto), Reidel Co., Dordrecht, p. 194.

Peiravi, A., and C. K. Birdsall (1978): "Self Heating in 1-D Thermal Plasma, Comparison of Weights: Optimal Parameter Choice," Memorandum No. UCB/ERL M78/32, ERL, College of Engineering, University of California, Berkeley.

Peters, C. F. (1968): Numerical Regularisation, *Bull. Astron. (Fr.), ser. 3*, vol. 3, 167–175.

Petravic, M., and G. Kuo-Petravic (1979): An ILUCG Algorithm which Minimises in the Euclidean Norm, *J. Comput. Phys.*, vol. 32, pp. 263–269.

Platzman, P. M., and H. Fukuyama (1974): Phase Diagram of the Two-Dimensional Electron Liquid, *Phys. Rev. B*, vol. 10, pp. 3150–3158.

Pollack, E. L., and J. P. Hansen (1973): Statistical Mechanics of Dense Ionised Matter. II. Equilibrium Properties and Melting Transition of the Crystallised One-Component Plasma, *Phys. Rev., ser. A*, vol. 8, pp. 3110–3122.

Potter, D. (1973): "An Introduction to Computational Physics," John Wiley and Sons, Ltd, London.

——— (1976): Waterbag Methods in Magnetohydrodynamics, *Methods Comput. Phys.*, vol. 16, pp. 43–83.

Proskurowski, W. (1977): "Numerical Solution of the Helmholtz's Equation by Implicit Capacity Matrix Methods," *Report* LBL-6402, Lawrence Berkeley Laboratory, University of California, Berkeley.

——— and O. Widlund (1976): On the Numerical Solution of Helmholtz's Equation by the Capacity Matrix Method, *Math. Comput.*, vol. 30, pp. 433–468.

Rahman, A. (1964): Correlations in the Motion of Atoms in Liquid Argon, *Phys. Rev., ser. A*, vol. 136, pp. 405–411.

———, R. H. Fowler, and A. H. Narten (1972): Structure and Motion in Liquid BeF_2, $LiBeF_3$ and LiF from Molecular Dynamics Calculations, *J. Chem. Phys.*, vol. 57, pp. 3010–3011.

Rathmann, C. E., J. L. Vomvoridis, and J. Denavit (1978): Long-Time-Scale Simulation of Resonant Particle Effects in Langmuir and Whistler Modes, *J. Comput. Phys.*, vol. 26, pp. 408–442.

Rees, H. D. (1968): Calculation of Steady-State Distribution Function by Exploiting Stability, *Phys. Lett.*, vol. 26A, pp. 416–417.

——— (1969): Calculation of Distribution Functions by Exploiting the Stability of the Steady State, *J. Phys. Chem. Solids*, vol. 30, pp. 643–655.

———, G. S. Sanghera, and R. A. Warriner (1977): Low Temperature FET for Low-Power High-Speed Logic, *Electron. Lett.*, vol. 13, pp. 156–158.

Reid, J. K. (1971): On the Method of Conjugate Gradients for the Solution of Large Sparse Systems of Equations, in "Large Sparse Sets of Linear Equations," Proc. of Conf. in Oxford, April, 1970, Academic Press Ltd, London.

——— (1976): Sparse Matrices, in Proc. Conf. on "The State-of-the-Art in Numerical Analysis," York, April, Academic Press Ltd, London.

Reiser, M. (1970): Two-Dimensional Analysis of Substrate Effects in Junction FETs, *Electron. Lett.*, vol. 6, pp. 493–494.

——— (1971): Difference Methods for the Solution of the Time-Dependent Semi-Conductor Flow Equations, *Electron. Lett.*, vol. 7, pp. 353–355.

——— (1972): Large Scale Numerical Simulation in Semi-Conductor Device Modelling, *Comput. Methods Appl. Mech. & Eng.*, vol. 1, pp. 17–37.

——— (1973): On the Stability of Finite Difference Schemes in Transient Semiconductor Problems, *Comput. Methods Appl. Mech. & Eng*, vol. 2, pp. 65–68.

Reitz, J. R., and F. J. Milford (1962): "Foundations of Electromagnetic Theory," Addison-Wesley, Reading, Mass.

Rice, S. A., and P. Gray (1965): "The Statistical Mechanics of Simple Liquids," Interscience Publishers, John Wiley and Sons, Inc., New York.

Richtmyer, R. D., and K. W. Morton (1967): "Difference Methods for Initial-Value Problems," Interscience Publishers, John Wiley and Sons, Inc., New York.

Roberts, K. V. (1974): An Introduction to the OLYMPUS System, *Comput. Phys. Commun.*, vol. 7, pp. 237–243.

——— (1975): The OLYMPUS Programming System, *ATOM*, vol. 226, pp. 137–147.

——— and D. E. Potter (1970): Magnetohydrodynamic Calculations, *Methods Comput. Phys.*, vol. 9, pp. 339–420.

Rosen, B., W. L. Kruer, and J. M. Dawson (1970): A New Version of the Dipole Expansion Scheme, *Proc. 4th Annu. Conf. Numer. Simulation of Plasmas*, Office of Naval Research, Arlington, Va., pp. 561–573.

Ross, S. (1976): "A First Course in Probability," Macmillan, New York.

Runge, C. (1895): Ueber die Numerische Auflösung von Differentialgleichungen, *Math. Ann.* (*Ger.*), vol. 46, pp. 167–178.

Sagdeev, R. Z. (1979): The 1979 Oppenheimer Lectures: Critical Problems in Plasma Astrophysics. II. Singular Layers and Reconnection, *Rev. Mod. Phys.*, vol. 51, pp. 11–20.

Sandage, A. (1961): "The Hubble Atlas of Galaxies," Publication 618, Carnegie Institute of Washington, Washington, D.C.

Sangster, M. J. L., and M. Dixon (1976): Interionic Potentials in Alkali Halides and their use in Simulation of Molten Salts, *Adv. Phys.*, vol. 25, pp. 247–342.

Scharfetter, D. L., and H. K. Gummel (1969): Large-Signal Analysis of Silicon Read Diode Oscillator, *IEEE Trans. Electron Devices*, vol. ED-16, pp. 64–77.

Schofield, P. (1973): Computer Simulation Studies in the Liquid State, *Comput. Phys. Commun.*, vol. 5, pp. 17–23.

Schumann, U. (Ed.) (1978): "Computers, Fast Elliptic Solvers and Applications," Advance Publications Ltd, London.

Sellwood, J. A., and R. A. James (1979): Angular Momentum Redistribution by Spiral Waves in Computer Models of Disc Galaxies, *Mon. Not. R. Astron. Soc.*, vol. 187, pp. 483–496.

Shampine, L. F. (1979): Storage Reduction for Runge–Kutta Codes, *ACM Trans. Math. Software*, vol. 5, pp. 245–250.

Shur, M. (1976): Influence of Non-Uniform Field Distribution on Frequency Limits in GaAs FETs, *Electron. Lett.*, vol. 12, pp. 615–616.

Slotboom, J. W. (1969): Iterative Scheme for 1- and 2-Dimensional D.C.—Transistor Simulation, *Electron. Lett.*, vol. 5, pp. 677–678.

——— (1973): Computer-Aided Two-Dimensional Analysis of Bipolar Transistors, *IEEE Trans. Electron Devices*, vol. ED-20, pp. 669–679.

Spitzer, L., and M. H. Hart (1971*a*): Random Gravitational Encounters and the Evolution of Spherical Systems. I. Method, *Astrophys. J.*, vol. 164, pp. 399–409.

——— and ——— (1971*b*): Random Gravitational Encounters and the Evolution of Spherical Systems. II. Models, *Astrophys. J.*, vol. 166, pp. 483–511.

——— and S. L. Shapiro (1972): Random Gravitational Encounters and the Evolution of Spherical Systems. III. Halo, *Astrophys. J.*, vol. 173, pp. 529–547.

——— and T. X. Thuan (1972): Random Gravitational Encounters and the Evolution of Spherical Systems. IV. Isolated Systems of Identical Stars, *Astrophys. J.*, vol. 175, pp. 31–61.

Stern, F. (1972): Self-Consistent Results for *n*-Type Si Inversion Layers, *Phys. Rev. ser. B*, vol. 5, pp. 4891–4899.

Stix, T. H. (1962): "The Theory of Plasma Waves," McGraw-Hill, New York.

Stone, H. J. (1968): Iterative Solution of Implicit Approximations of Multi-Dimensional Partial Differential Equations, *SIAM J. Numer. Anal.*, vol. 5, pp. 530–558.

Strang, G., and G. Fix (1973): "An Analysis of the Finite Element Method," Prentice-Hall, Inc., Englewood Cliffs, N.J.

Swarztrauber, P. N. (1974): A Direct Method for the Discrete Solution of Separable Elliptic Equations, *SIAM J. Numer. Anal.*, vol. 11, pp. 1136–1150.

——— (1977): The Methods of Cyclic Reduction, Fourier Analysis, and the FACR Algorithm for the Discrete Solution of Poisson's Equation on the Rectangle, *SIAM Rev.*, vol. 19, pp. 490–501.

——— and R. A. Sweet (1975): Efficient FORTRAN Subprograms for the Solution of Elliptic Partial Differential Equations, *NCAR Technical Note*, 1A-109, National Center for Atmospheric Research, Boulder, Col.

Sweet, R. A. (1974): A Generalised Cyclic-Reduction Algorithm, *SIAM J. Numer. Anal.*, vol. 11, pp. 506–520.

——— (1977): A Cyclic Reduction Algorithm for Solving Block Tridiagonal Systems of Arbitrary Dimension, *SIAM J. Numer. Anal.*, vol. 14, pp. 706–719.

Sze, S. M. (1969): "Physics of Semiconductor Devices," Wiley-Interscience, New York.

Temperton, C. (1979*a*): A Fast Poisson-Solver for Large Grids, *J. Comput. Phys.*, vol. 30, pp. 145–148.

——— (1979*b*): Direct Methods for the Solution of the Discrete Poisson Equation: Some Comparisons, *J. Comput. Phys.*, vol. 31, pp. 1–20.

——— (1980): On the FACR(*l*) Algorithm for the Discrete Poisson Equation, *J. Comput. Phys.*, vol. 34, pp. 314–329.

Thouless, D. J. (1978): Melting of the Two-Dimensional Wigner Lattice, *J. Phys. C: Solid State Phys.*, vol. 11, pp. L189–L190.

Tien, P. K., and J. Moshman (1956): Monte Carlo Calculation of Noise Near the Potential Minimum of a High-Frequency Diode, *J. Appl. Phys.*, vol. 27, pp. 1067–1078.

Tinsley, B. M. (1977): The Cosmological Constant and Cosmological Change, *Phys. Today*, vol. 30, No. 6, pp. 32–38.

Todd, A. M. M. (1975): Numerical Simulation of Shock-Heated Plasma with a Magnetic Dam, *Phys. Fluids*, vol. 18, pp. 453–457.

Toomre, A. (1964): On the Gravitational Stability of a Disk of Stars, *Astrophys. J.*, vol. 139, pp. 1217–1238.

——— (1977): Theories of Spiral Structure, *Annu. Rev. Astron. & Astrophys.*, vol. 15, pp. 437–478.

Tosi, M. P., and F. G. Fumi (1964): Ionic Sizes and Born Repulsion Parameters in the NaCl-Type Alkali Halides—II, *Phys. Chem. Solids*, vol. 25, pp. 45–52.

Totsuji, H., and T. Kihara (1969): The Correlation Function for the Distribution of Galaxies, *Publ. Astron. Soc. Jpn.*, vol. 21, pp. 221–229.

Vacouleurs, G. de (1959): General Physical Properties of External Galaxies, *Handb. Phys. (Ger.)*, vol. 53, pp. 311–372.

Varga, R. S. (1962): "Matrix Iterative Analysis," Prentice-Hall, Inc., Englewood Cliffs, N.J.

Verlet, L. (1967): Computer Experiments on Classical Fluids. I. Thermodynamic Properties of Lennard-Jones Molecules, *Phys. Rev.*, vol. 159, pp. 98–103.

Vokes, J. C., B. T. Hughes, D. R. Wight, J. R. Dawsey, and S. J. W. Shrubb (1979): Novel Microwave GaAs Field-Effect Transistors, *Electron. Lett.*, vol. 15, pp. 627–629.

Wada, T., and J. Frey (1979): Physical Basis of Short-channel MESFET Operation, *IEEE J. Solid-state Circuits*, vol. SC-14, pp. 398–412.

Wadhwa, R. P., O. Buneman, and D. F. Brauch (1965): Two-Dimensional Computer Experiments on Ion-Beam Neutralization, *Am. Inst. Aeronaut. & Astronaut.*, vol. 3, pp. 1076–1081.

Warriner, R. A. (1976): "Computer Simulation of Gallium Arsenide Semi-Conductor Devices," Ph.D. Thesis, Reading University.

——— (1977*a*): Distribution Function Relaxation Times in Gallium Arsenide, *Solid State & Electron Devices*, vol. 1, pp. 92–96.

——— (1977*b*): Computer Simulation of Negative-Resistance Oscillators Using a Monte-Carlo Model of Gallium Arsenide, *Solid State & Electron Devices*, vol. 1, pp. 97–104.

——— (1977*c*): Computer Simulation of Gallium Arsenide Field-Effect Transistors Using Monte-Carlo Methods, *Solid State & Electron Devices*, vol. 1, pp. 105–110.

Widlund, O. (1972): On the Use of Fast Methods for Separable Finite-Difference Equations for the Solution of General Elliptic Problems, in "Sparse Matrices and their Applications" (eds. D. J. Rose and R. A. Willoughby), Plenum Press, London, pp. 121–131.

Wijngaarden, A., B. J. Mailioux, J. E. L. Peck, C. H. A. Koster, M. Sintzoff, C. H. Lindsey, L. G. L. T. Meertens, and R. G. Fisker (1976): "Revised Report on Algol 68," Springer-Verlag, Berlin.

Williamson, J. H. (1980): Low Storage Runge–Kutta Schemes, *J. Comput. Phys.*, vol. 35, pp. 48–56.

Woodcock, L. V., C. A. Angell, and P. Cheeseman (1976): Molecular Dynamic Studies of the Vitreous State: Simple Ionic Systems and Silica, *J. Chem. Phys.*, vol. 65, pp. 1565–1577.

Yamaguchi, K., S. Asai, and H. Kodera (1976): Two-Dimensional Numerical Analysis of Stability Criteria of GaAs FETs, *IEEE Trans. Electron Devices*, vol. ED-23, pp. 1283–1290.

Young, D. (1962): The Numerical Solution of Elliptic and Parabolic Partial Differential Equations, in "Survey of Numerical Analysis" (ed. J. Todd), McGraw-Hill, New York, pp. 380–418.

Yu, S. P., G. P. Kooyers, and O. Buneman (1965): Time-Dependent Computer Analysis of Electron-Wave Interaction in Crossed-Fields, *J. Appl. Phys.*, vol. 36, 2550–2559.

AUTHOR INDEX

SUBJECT INDEX